George Guekos (Ed.)

Photonic Devices for Telecommunications

Springer

Berlin
Heidelberg
New York
Barcelona
Hong Kong
London
Milan
Paris
Singapore
Tokyo

George Guekos (Ed.)

Photonic Devices for Telecommunications

How to Model and Measure

With 211 Figures and 27 Tables

Springer

Prof. Dr. George Guekos
ETH Zürich
Institut für Quantenelektronik,
Mikro- und Optoelektronik
ETH-Hönggerberg, HPT
CH-8093 Zürich
e-mail: guekos@iqe.phys.ethz.ch

Library of Congress Cataloging-in-Publication Data

Guekos, George.
Photonic devices for telecommunications: how to model and measure / George Guekos.
Includes bibliographical references and index.
ISBN 3-540-64318-4 (hc.: alk. paper)
1. Telecommunication systems. 2. Photonics. I. Title.
TK5103.G84 1998 621.382-dc21 98-44450

ISBN 3-540-64318-4 Springer-Verlag Berlin Heidelberg New York

Printed in Germany

Typesetting: Camera-ready copy from editor
Cover-Design: MEDIO, Berlin
SPIN 10650857 62/3020 5 4 3 2 1 0 Printed on acid-free paper

Καὶ τίς τόδ' ἐξίκοιτ' ἂν ἀγγέλων τάχος;

Ἥφαιστος, Ἴδης λαμπρὸν ἐκπὲμπων σέλας·
φρυκτὸς δὲ φρυκτὸν δεῦρ' ἀπ' ἀγγὰρου πυρός ἔπεμπεν

And who could bring the message here with such speed ?

Hephaestos, sending from Ida a bright light,
beacon to beacon transmitting to us by courier fire

Aeschylos, *Agamemnon*, 458 B.C.

Authors

George Guekos	ETHZ - Swiss Federal Institute of Technology, Zurich, Switzerland

Part I

Jiri Ctyroký, ed.	Institute of Radio Engineering and Electronics, Prague, The Czech Republic
H.V.Baghdasaryan	State Engineering University of Armenia, Yerevan
Carlo De Bernardi	CSELT, Torino, Italy
Jan Haes	Telenet Operaties N.V., Mechelen, Belgium
Stefan Helfert	FernUniversität, Hagen, Germany
H.J.W.M.Hoekstra	University of Twente, The Netherlands
Alain Küng	EPFL - Ecole Polytechnique Fédérale, Lausanne, Switzerland
Ottokar Leminger	Deutsche Telekom, Darmstadt, Germany
Hans-Peter Nolting	Heinrich-Hertz-Institut für Nachrichtentechnik, Berlin
Reinhold Pregla	FernUniversität, Hagen, Germany
Wolf von Reden	Heinrich-Hertz-Institut für Nachrichtentechnik, Berlin
Philippe Robert	EPFL - Ecole Polytechnique Fédérale, Lausanne, Switzerland
Phillip Sewell	University of Nottingham, UK
Aasmund Sudbø	University of Oslo, Norway

Part II

Geert Morthier, ed.	University of Gent, Belgium
Roel Baets	University of Gent, Belgium
Arthur Lowery	University of Melbourne, Australia
Roberto Paoletti	CSELT, Torino, Italy
Paolo Spano	FUB - Fondazione Ugo Bordoni, Roma, Italy

Part III

François Girardin, ed.	ETHZ - Swiss Federal Institute of Technology, Zurich
Thomas Ducellier	Alcatel Alsthom Recherche, Marcoussis, France
Stefan Diez	Heinrich-Hertz-Institut für Nachrichtentechnik, Berlin
Dominique Marcenac	British Telecom Laboratories, Ipswich, United Kingdom
Antonio Mecozzi	FUB - Fondazione Ugo Bordoni, Roma, Italy
Jesper Mørk	Technical University of Denmark, Lyngby, Denmark
Kristof Obermann	Technical University of Berlin, Germany
Simona Scotti	FUB - Fondazione Ugo Bordoni, Roma, Italy

Appendix

P. Verhoeve	University of Gent, Belgium

Prolegomenon

Photonic devices use various interaction mechanisms between light and electricity and have been the basis of the tremendous developments of fibre communications in the last two decades of the 20th century. The breathtaking increase of the world-wide demand for information transmission capacity and the obvious advantages of the optical technology have been and still are today the main driving forces behind the continuing interest in photonic devices. They are used as lasers and light-emitting diodes to generate light, as detectors of optical signals, as components to impress information on a light beam, to route it to its proper destination through fibres, to switch light signals in space, time, and wavelength, and as devices for signal amplification. Photonic devices are attractive because of the many possibilities they offer to the communication engineer, as robust, low-power consumption, small, reliable elements, suitable for high-speed operation and allowing very large modulation bandwidths. Their characteristics are getting better every day and the repercussions are reflected in the continuous improvements of the performance of modern fibre systems and networks. Their "value-for-money" is high, since they do not only enable the modulation, transmission, and reception of huge amounts of data that would have been impossible with the old technology of copper-wire pairs or coaxial cables, but they do so at a low and even decreasing price.

Photonics have radically changed the world of telecommunications. They opened the door to numerous developments and helped to realise the information society. Of course, photonics is not confined to telecommunications only. Thanks to the unique potential of the devices, it has become part of everyday life at home and in business. Consumer products, office equipment, medical apparatus, and other achievements have made the photonic technology accessible for the general public. The role of photonics in telecommunications, on the other hand, is known more to specialists than to the general consumer.

Photonic devices are by their nature objects of interdisciplinarity. Their fabrication requires the team work of physicists, chemical engineers and electrical engineers, and their application challenges the creativity of network designers, system specialists, and field engineers. Understanding the physics of the material combinations involved, the ways to shape the appropriate materials into practical devices to perform certain functions, and the characteristics of the devices when put into circuits and systems, requires vast amounts of knowledge and is a necessity for an innovative process towards new devices. If successful, this process can produce properties that can be used to address specific demands, better operational parameters, or higher degrees of integration. The possibilities offered by the marriage of electrons and photons, especially in semiconductors, seem to be unlimited. Perhaps only the innovative horizon of the scientists and the technological tools at their disposal are likely to set boundaries.

This book is intended for engineers and physicists with a good general background in photonics who would like to update and enlarge their knowledge in the field of modelling and measurement of key devices, like semiconductor lasers and optical amplifiers. Since the field is broad, we had to concentrate on devices that are widely considered to be important for modern optical communications. The typical reader may be a research scientist, a PhD student, or a company engineer wishing to advance beyond the knowledge provided by a classical undergraduate textbook.

Zurich, 14th July 1998

George Guekos
ETH Zurich

Acknowledgements

I wish to thank all chapter authors for their enthusiasm and the care of their contributions. It has been my privilege to work with them in a truly international and exciting atmosphere. The framework was set by COST which stands for "European Cooperation in the Field of Scientific and Technical Research". This framework for scientific and technical research collaboration among European countries is bottom-up oriented and widely used today.
In the Action COST 240, physicists and engineers specialised in photonic devices have worked together. This book is the result of this co-operation. The names of the authors are given in the beginning of each of the three parts into which this book is subdivided. It is a particular pleasure to acknowledge Jiri Ctyroký of The Institute of Radio Engineering and Electronics, Academy of Sciences of the Czech Republic, Geert Morthier of the University of Gent, Belgium, and François Girardin from our ETH Zurich, for co-ordinating the contributions to Parts I, II, and III, respectively. Further, I wish to sincerely thank Hans Melchior, Head of our group at the Institute of Quantum Electronics at the ETH Zurich, for his continuous and warm support of research activities in the frame of COST. The Swiss Federal Office for Education and Science in Berne provided significant financial contribution to the work of Swiss scientists for COST Actions. Many former and present colleagues at the ETHZ were involved in COST and have produced fine scientific work: Roberto Dall'Ara, Jürg Eckner, Emilio Gini, Christoph Holtmann, Werner Hunziker, Jürg Leuthold, and Stephan Pajarola. Lorenzo Occhi's and Patrick Zenklusen's contribution in editing was invaluable.
It is my duty to thank the European COST authorities for providing the framework for this co-operation, and in particular the Technical Committee for Telecommunications TCT, and the COST Secretariat at the General Directorate XIII of the European Commission in Brussels.
Finally, it is to my wife Marion that I wish to address my deepest appreciation and thanks for her understanding, patience, and love.

George Guekos

Foreword

This book is subdivided into three main **Parts**. The common spirit in these parts is to provide, at the beginning of each, a comprehensive introduction into the subject treated, followed by specific aspects pertaining to the modelling and/or measuring particularities arising from the investigation of photonic devices for telecommunications. Some of the devices treated here can be considered as widely known and well established. Others are rather new and their potential for applications is not yet fully exploited. The methods to model and measure photonic devices and structures outlined in this book and the comparison of results obtained by applying such methods are likely to interest both the engineer investigating the behaviour of a device in a system and the engineer looking for new ways to explore the possibilities offered by emerging devices.

Many authors have contributed to this book. There are two main reasons for this. First, the book addresses two broad fields in photonic device research, modelling and measurements, for which a vast knowledge exists in many research groups that was not integrated in a book before. Second, a significant number of laboratories decided to closely co-operate in order to gain additional information on merits and drawbacks of their own methods for simulation and experimentation of devices as compared to the methods used by their colleagues in other laboratories. The outcome are new aspects and approaches that would not have been investigated in the absence of a framework for a co-operative programme. This framework was offered by COST which stands for European Co-operation in the Field of Scientific and Technical Research, and by the specific Action COST 240 "Modelling and Measuring Advanced Photonic Devices for Telecommunications" in which over 80 laboratories participated from 1991 to 1998.

Part I investigates methods to simulate and to measure the main characteristics of photonic waveguide structures. ***Chapter 1*** is devoted to numerical calculation methods, known as "mode solvers". Three of the most powerful are discussed: An improved method of lines, the mode matching method, and the free space radiation mode method. This chapter provides a comprehensive discussion into the potential and the relative merits and drawbacks of these methods. ***Chapter 2*** deals with beam propagation methods (BPMs) for modelling longitudinally nonuniform waveguide structures and discusses the possibilities offered by three frequently encountered BPMs: The improved finite difference BPM, the method of lines BPM, and the bi-directional mode expansion propagation method. A new bi-directional method, useful especially for structures incorporating Kerr nonlinearities, closes the chapter. ***Chapter 3*** presents results of original benchmark tests and modelling tasks which were extensively investigated in the frame of COST 240. This chapter is unique since it provides a plethora of results obtained in different laboratories that are presented in such a manner as to help the reader appreciate the potential of the various methods used. ***Chapter 4*** outlines the main methods to experimentally characterise photonic waveguide structures and should interest all those dealing with such measurements when they are faced with the problem of choosing the appropriate method to determine a certain parameter. ***Chapter 5***, finally, deals with comparison of measurement results relative to experimental investigations of loss, group index, mode profile, and cut-off wavelength performed on waveguides made from semiconductors or glass.

Part II treats the distributed feedback (DFB) laser. ***Chapter 6*** provides the knowledge necessary to understand the physics underlying the operation of the laser and the main characteristics of the device. This chapter sets also the general perspective under which the modelling and measurement of DFBs can be developed and introduces in this way the work outlined in the subsequent chapters. A particular aspect of the chapter is the presentation of the outcome of the work performed by the group of scientists participating in the Action COST 240 for the standardisation of parameters used to describe the operation of a DFB. ***Chapter 7*** treats the modelling of DFB laser diodes. The reader will find a useful description of the various numerical methods used to simulate the behaviour of laser diodes together with indications on their advantages and drawbacks. Numerical case studies were performed in order to compare various laser models. This chapter outlines and compares the results obtained by several industrial and university laboratories and provides valuable results on the accuracy of models. ***Chapter 8*** compares and discusses measurements on DFB laser diodes undertaken in several laboratories. The devices were supplied by various sources, their parameters that are most important for applications were experimentally determined, and the results obtained at the laboratories participating in the round robin were compared. The reader will find information of much practical importance to assist him when performing DFB characterisation. ***Chapter 9***, finally, treats the subject of parameter extraction from measurements. From this chapter the reader can learn how to reliably extract those parameters that he can subsequently use to model with accuracy and efficiency his DFB laser.

Part III addresses the nonlinear effect of four-wave mixing (FWM) in semiconductor optical amplifiers (SOA). ***Chapter 10*** presents an introduction to the FWM phenomenon and defines the parameters for its characterisation. ***Chapter 11*** is addressed to the reader interested in the theory of FWM. Rate equations are used to arrive at a simple description of FWM in SOAs. An overview of the main analytical models of the FWM conversion efficiency - a parameter widely used in practical investigations - based on the coupled mode approach is outlined in detail.

This chapter closes with the description of different analytical models of the amplified spontaneous emission of a SOA. ***Chapter 12*** shows experimental investigations on FWM performance of SOAs. A few samples from different sources were measured in different laboratories. The reader will find information on experimental set-ups and on the comparison of the results which are of interest to the much discussed potential of SOAs for telecommunication. ***Chapter 13***, finally, serves to trigger the interest of the reader interested in FWM for some aspects in the applicability of this effects that are usually not broadly discussed in the literature but can nevertheless become important as the potential of FWM finds a growing general acceptance.

In the ***Appendix***, an interchange format for optical data is proposed. By using this format, the difficulties often encountered when exchanging data from one laboratory to another can be alleviated.

George Guekos

Contents

Abbreviations

1D	one-dimensional
2D	two-dimensional
3D	three-dimensional
AM	amplitude modulation
AR	antireflection
ASCII	american standard code for information interchange
ASE	amplified spontaneous emission
BEP	bi-directional eigenmode propagation (method)
BPM	beam propagation method
CDP	carrier density pulsation
CH	carrier heating
COST	European cooperation in the field of scientific and technical research
CW	continuos wave
DBR	distributed Bragg reflector
DC	direct current
DFB	distributed feedback
EDFA	erbium doped fibre amplifier
EIC	efficient interface conditions
EMBH	etched mesa buried heterostructure
EO	electro-optic
FCA	free carrier absorption
FD	finite difference
FDM	finite difference method
FE	finite elements
FEM	finite elements method
FFT	fast Fourier transform
FM	frequency modulation
FMM	film mode matching (method)
FP	Fabry-Perot
FPH	facet phase
FSR	free spectral range
FSRM	free space radiation mode (method)
FT BPM	Fourier transform beam propagation method
FWHM	full width at half maximum
FWM	four-wave mixing

GEEP	guided eigenmode expansion propagation (method)
GRIN	graded index
GS	ground signal
GSG	ground signal ground
HR	high reflection
IM	intensity modulation
IOC	integrated optics circuit
MEP	mode expansion propagation method
MI	Michelson interferometer
MM	mode matching
MMM	mode matching method
MMT	mode matching technique
MoL	method of lines
MQW	multiple quantum well
NF	noise figure
ODIF	optoelectronic data interchange format
OEIC	optoelectronic integrated circuit
OLCR	optical low coherence reflectometry
PC	polarisation controllers
P-I	power versus current
PMM	power matrix method
RF	high frequency (prober)
RIN	relative intensity noise
RRM	Rayleigh-Ritz method
SA	saturable absorber
SBR	signal to background ratio
SCH	separate confinement heterostructure
SHB	spectral hole burning *or* spatial hole burning
SHG	(optical) second harmonic generation
SI	international system of units
SMSR	side mode suppression ratio
SNR	signal to noise ratio
SOA	semiconductor optical amplifier
SOP	state of polarisation
SP	surface plasmon
SVE	slowly varying envelope
SVEA	slowly varying envelope approximation
TD	time domain
TDM	time-domain multiplexing
TE	transverse electric
TLLM	transmission line laser model
TM	transverse magnetic
TMM	transfer matrix method
TPA	two photon absorption

TRM	transverse resonance method
UF	ultrafast (process)
VCSEL	vertical cavity surface emitting laser
V-I	voltage versus current
XGM	cross gain modulation
WDM	wavelength division multiplexing
WG	working group

Introduction

G. Guekos

The field of photonic devices has witnessed tremendous advances in the last two decades of the 20th century that made the enormous success of the optical fibre communications possible. Many types of photonic devices have been designed and investigated in the laboratories, a significant number found its way into the commercial market and became an appreciable success. These devices were fundamental to the rapid proliferation of fibre systems and networks over the world. Key examples are the "active" devices like the diode laser, the optical amplifier, and the optical modulator, and the "passive" devices made of glass or semiconductor material. Work on devices is becoming more focused as the demand for high-performance photonic components and sub-systems needed for concrete applications is increasing. At the same time we discern a tendency in many companies to decrease the funding for research and development in new devices and to rely increasingly on what is available both as a knowledge on components and as hardware itself. This calls for an effort by specialists to compound the main findings of research of the last years of the 20th century and to make them available to engineers and physicists who are confronted in their daily work with such devices.

Perhaps the two most commonly asked questions by scientists wishing to deepen their knowledge in the field are "how are photonic devices designed" before they enter the fabrication stage and "how are they experimentally characterised" once they leave the fabrication process. Several textbooks were written on photonic devices that offer an excellent starting basis to those entering the field. On the other hand, many scientific publications in specialised journals are available that treat specific aspects of design and measurements presented by specialists and addressed to specialists. However, a book that offers a wide overview on modelling tools and of measuring methods to treat this kind of devices, and containing at the same time practical examples from broad international interlaboratory investigations on devices did not exist. The present book tries to fill this gap and positions itself between the classical textbook and the publication for the specialist. This book is written by people working many years on modelling and measurements of photonic devices who came together to examine and treat problems that often surpassed manpower availability, know-how basis, software and hardware tool capability in their own laboratories, and that could finally be solved thanks to the mutual exchange of information and to the co-ordination of the work. The authors of this book estimated that the information gathered during the several years of their co-operation is of primary importance to those working in the field and should be compounded in a book that covers both aspects encountered in the study of photonic waveguide structures and devices: how to model and how to measure.

Photonic waveguide structures. Modelling a photonic device is a complex and challenging task. The reason lies in the diversity of devices to be modelled and of the tools available towards this end. The choice of the right tool to model a certain device presupposes an accurate knowledge of the capabilities of those tools and of their applicability for the work in question. The common situation, however, is that the scientist can have access to know-how accumulated in his company or institute, stemming from developments carried-out by applying only a limited number of modelling tools that happened to be in use or were developed in his group. An accurate description of the potential of other tools is often not at hand and, although parts of the information are scattered around in periodicals and conference proceedings, compounding such information requires costly additional work. The result is that the scientist is obliged to use what is readily available in his group without knowing how far he operates from the optimum situation. For this reason, the preparation of a book where the information is consolidated and results of comparative examinations of simulation tools are described, seemed to the authors of Part I a highly motivating endeavour.

An important information for a photonic device designer is the optical field distribution inside the structure, the relation between the optical field and the externally applied electrical field and/or charge carriers flowing through the structure, and the interfacing of the structure fields with external optical and electrical elements. An analytical approach to calculate the optical fields and their evolution in space and time is usually unpractical since it requires an unacceptably large effort, even for simple device structures. More flexible approaches, capable to calculate with sufficient accuracy the field distributions are necessary. The most well known, and some other lesser known, are brought together and discussed in detail in this book. The presentation is done in such a manner as to help the engineer - who is supposed to have a basic knowledge about simulation tools - find the way to handle his own problem by comparing capabilities, advantages and drawbacks of each method. Particular attention is given to the various beam propagation methods (BPMs) and, consequently, the benchmark tests and the modelling tasks described in this book have been used by the laboratories participating in the round robin exercises to test the performance of the various BPMs.

Semiconductor distributed feedback laser diodes. From the very early developments of fibre communications it became clear that the realisation of high transmission capacity systems would have to address the question of signal dispersion. The introduction of single mode fibres has solved only part of the problem, the intermodal dispersion, which was due to the many optical ray paths in the core of the early multimode fibres. The chromatic dispersion due to the interplay of the wavelength dependence of the fibre refractive index, the spectral width of the optical source and its dynamic behaviour, and the waveguiding of the optical mode, remained, however, and produced a wild broadening and overlapping of high bit-rate optical signal pulses in the fibre line.

Two approaches are commonly used to minimise the effects of chromatic dispersion. The first pertains to the fibre, the second to the laser. By tailoring the refractive index profile, dispersion shifted or dispersion flattened fibres are produced which can significantly reduce the pulse broadening at wavelengths around 1550 nm where modern high-capacity systems are installed. Of crucial importance for the quality of the transmission is, in addition, the dynamic spectral behaviour of the laser. A laser that emits in a single longitudinal mode and has a narrow linewidth is a necessity for high bit rates. The search for such a light source went through many stages and generated interesting proposals. Although many of the results are per se very valuable, the only laser type that left the laboratory stage with a clear commercial success is the one that incorporates diffraction gratings in the laser waveguide structure. Two types were established over the years. The Distributed Feedback (DFB) laser, now widely used in high-capacity longhaul transmission, and the rather less widespread Distributed Bragg Reflector (BDR) laser. The integrated diffraction gratings play the role of frequency/wavelength-selective reflectors that replace the end-facet mirrors in the usual Fabry-Perot type laser cavity which show practically no frequency selectivity and allow several modes to oscillate under the gain curve. The grating causes small fractions of the propagating wave to reflect at each segment of the corrugation and to built constructive interference in case the wavelength satisfies the Bragg condition, i.e. the grating period equals an integer of half-wavelengths in the semiconductor. This condition results in a single lasing wavelength that comes closest to satisfy the Bragg condition whereas the strong suppression of the gain for all other wavelengths produces only very weak side modes. In the DFB device, both the grating and the active layer extend over the whole length of the laser cavity, see Fig.1, whereas in the DBR laser the active amplifying layer and the feedback grating layer are axially separated and cover each only part of the cavity length.

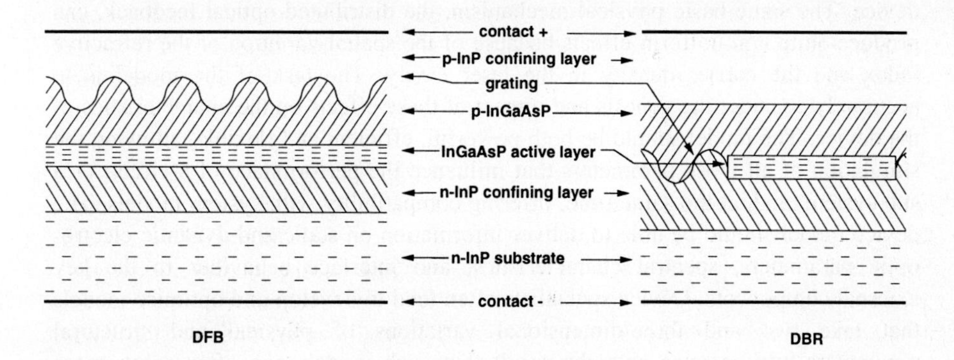

Fig.1. Schematic cross-section of a DFB and a DBR laser.

The fabrication of DFBs remains a rather sophisticated process. However, the technologist has at his disposal very powerful tools and that allow him not only to precisely control the growth and construction of the device but also to design and to try new structures and material combinations in order to further improve the characteristics required by a high-performance system. These are usually a low threshold current, a light-current dependence offering ease of modulation for a current-modulated transmitter, enough optical power offering a good system margin, linearity in case of analogue modulation systems, spectral purity and stability even at GHz modulation, and low optical noise. Modern commercially available DFBs meet the requirements imposed on them by the system and show impressive performance, either under direct current modulation up to a few GHz or as a dc operated laser followed by an optical modulator for multi-GHz bandwidths. However, the steadily growing demand for transmission capacity imposes more stringent requirements on the performance of all parts of a system, and mostly on the transmitter. The design and fabrication of DFBs turns out to be even more demanding, the experimental characterisation of the device intended as optical source in the transmitter quite complex, time consuming and cumbersome.

It becomes imperative for the specialist associated with the development of DFBs to know better the potential and limitations of the tools at his disposal for modelling new devices and for measuring their behaviour. The designer and experimentalist can draw on a vast reservoir of published and unpublished knowledge, and so, quite naturally, various methods tend to establish themselves in different laboratories, where they are further developed and optimised, thus becoming "house-made" tools. An objective comparison of advantages and drawbacks of the methods is rarely done and even more rarely published. With this in mind, many scientists from industrial and university laboratories came together, motivated by the need to compare the performance of their methods on a number of well defined, commonly accepted and relevant device problems.

Accurate numerical models are a crucial help to the designer of reliable DFB lasers who cannot simply apply analytical formulae to predict the behaviour of the device. The same basic physical mechanism, the distributed optical feedback, can produce quite non-uniform effects because of the spatial variation of the refractive index and the carrier density in the laser cavity. The task of the model is to accurately consider the genesis and impact of these effects on the characteristics of the device. The model should be both powerful, efficient and flexible, allowing the simulation of all those parameters that influence the performance of a circuit or a sub-system, and, at the same time, offering compatibility with system models. The device model should be able to deliver information on static and dynamic electro-optic parameters, spectral characteristics, and interface behaviour to the key external components. Device specialists often tend to develop and optimise models that take two and three-dimensional variations of physical and structural parameters into account with the result that such models can offer much more information, but they are more complex and time consuming to implement. On the other hand, the system engineer prefers accurate, yet simple and easy to use device models as part of his system simulation software. Twelve models are compared in this book through numerical case studies. The comparison enables to adapt models in order to make them more easy to use for applications in actual devices.

The measurements of those laser characteristics that are important when the laser is intended to be a light source in a transmitter sub-system, and the description of the associated experimental set-ups have been the subject of numerous books and scientific publications in magazines and conference proceedings. As in the case of simulation tools, various techniques for the determination of laser parameters are in use in different laboratories. The comparison of the results obtained with different set-ups and the discussion of the reasons for the divergences observed is presented in this book. This presentation enables to determine some critical points that can affect the measurements, to search for solutions, and to check their applicability in an iterative process. This was done in the laboratories that participated in the round robin exercise of Part II. Some general requirements became apparent that should be met in order to perform accurate static and dynamic laser measurements and to extract reliable laser parameters that can be used in the numerical modelling. The large number of parameters involved necessitated a standardisation of the measurements. The result is presented in this book and should be of interest for the physicist and engineer involved in photonic device experimentation.

Four-wave mixing in semiconductor optical amplifiers. As in the case of some other high-performance photonic devices, the SOA has become commercially available (around the mid-nineties) only after a relatively long period of design and laboratory tests. The reason is that the SOA could not offer in its earlier stages of development a viable solution to amplification problems of weak optical signals, and certainly not an alternative to the doped fibre amplifier. Now that the technology of the SOA is more refined and its operation better understood, this device does not only begin to offer solutions to various signal amplification cases, but it also enables the exploration and application of physical phenomena that would not have been possible to address (or, at least, very difficult) with the fibre amplifier. The small size of the chip - typically one millimetre in length -, the compactness of the packaged device, and the simplicity of use make the SOA an increasingly attractive component for the investigation and application in telecommunications of electro-optic effects that were not an issue in this domain before. One of them is the generation of phase conjugated waves which is due to the strong nonlinear behaviour of the SOA. The reason for this behaviour is the interplay between the electrical carriers injected in the SOA and the photons entering the SOA and belonging to two or more wavelengths, typically one of them carrying the signal.

In this book, the four-wave mixing (FWM) phenomenon in SOAs is presented and discussed in some detail. The motivation is that the study of FWM enables to understand a multitude of nonlinear effects and to devise ways to apply these effects in telecommunications. The nonlinear mechanism in a SOA can make the device attractive to address a number of tasks in fibre communication systems, such as chromatic dispersion compensation, wavelength conversion, optical demultiplexing, clock recovery, and others. The implementation of the FWM in SOAs is just emerging thanks to the broader availability of the device and to the better understanding of the phenomena involved and of their potential. In general, the nonlinear optical processes of interest in the SOA arise because the incident electrical field induces a polarisation in the semiconductor material due mainly to the nonlinear oscillation of the free electrons with the optical field. This oscillation produces new spectral components of the field at frequencies 2ω, 3ω, etc., where ω is the angular frequency of the incident field. If two or more fields enter the SOA at different frequencies, e.g. ω_1 and ω_2, the intermodulation products of the type $2\omega_1 \pm \omega_2$, $2\omega_2 \pm \omega_1$, $3\omega_1 \pm 2\omega_2$, $3\omega_2 \pm 2\omega_1$, etc. appear, some of which may fall under the gain curve of the SOA. Attractive for applications is the case where the four spectral components, ω_1, ω_2, $2\omega_1 - \omega_2$, and $2\omega_2 - \omega_1$ appear at the output of the SOA because their position and intensity can be easily manipulated by changing the input frequencies and powers, and the amplifier gain. Fig.2 shows the two phase conjugated spectral components, one of which - indicated with the full line towards lower frequencies - is of particular interest for the application.

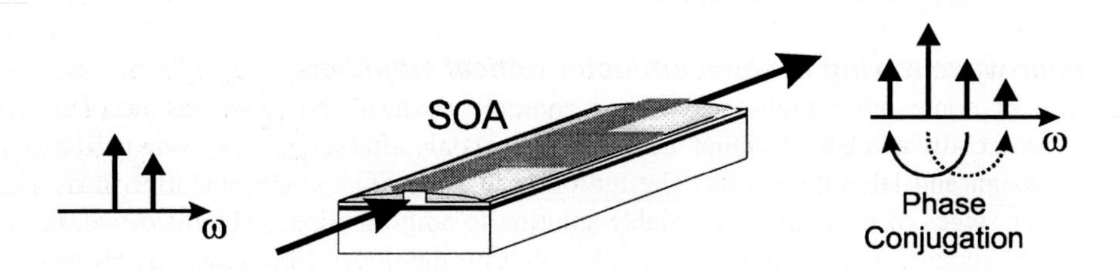

Fig.2. Four-wave mixing in a SOA.

How far the input frequencies can be from each other and still produce measurable mixing spectral products is an important issue for the application of SOAs in telecommunications. The answer depends on several factors, such as material composition, geometry of the waveguide, and physical mechanisms generating the mixing process. The main mechanisms have different efficiencies and time constants. The most efficient, the interband process of carrier recombination in the semiconductor, is also the slowest with response times around 100 picoseconds. Faster is the carrier-phonon scattering, an intraband process, due to carrier heating, with response times around 500 femtoseconds, and even faster, but less important, is the carrier-carrier scattering because of the spectral hole burning, also an intraband process, with response times around 100 femtoseconds. This means that the input frequency difference can extend well into the Terahertz range, albeit with decreasing efficiency. In practical terms, usable mixing products can be expected if the frequency difference of the two input waves does not exceed about 5 THz. The early SOA technology did not allow to have an efficient mixing process but advanced SOA devices offer quite impressive values of efficiency when converting the input signal power to the conjugate output power, often exceeding 0 dB for conversion ranges over 2 THz and with good signal-to-background ratio at the output. Performances like these are important for applications. Another issue is the sensitivity of the mixing process to the changing state of polarisation of the optical wave at the SOA input. Elegant solutions to the polarisation problem were proposed lately by a few research groups. Perhaps the most attractive characteristic of the FWM for fibre communications is the optical transparency offered by this process to the modulation format of the signals. Although not yet quite an issue, the transparency can play a significant role in the future as fibre communication networks become increasingly hungry for bandwidth and the all-optical data processing advances into the implementation stage.

The treatment of FWM in SOAs in this book starts with a short introduction to the phenomenon, followed by an extensive theoretical excursion. Care has been taken to present theoretical models that can be used to develop simulation tools and to interpret experimental results. Rigorous theoretical analysis can be superbly elegant in itself but at the same time quite complex and cumbersome to relate to the actual situation. This is why the theories discussed here have made certain simplifications that bring them closer to the real situation while keeping the base of theoretical rigorosity and presenting the outcomes in an easy-to-use form. Broadly speaking, two approaches are followed. One is based on the rate equations derived from density matrix equations which, after consideration of the propagation equation for the field envelope and its phase, allow a general description of the FWM effect. The other is based on the coupled-mode theory and discusses the various analytical models resulting from it and related to the FWM in a SOA.

Measuring FWM is simple, in principle, but it can become tricky in the real situation. This book offers an introduction to the general case where two optical inputs are used, one representing the signal carrying wave, the other being the "pump" wave. The quantities of interest are the efficiency of the signal to the conjugate conversion and the signal-to-noise ratio. They were measured in different laboratories that participated in a round robin where devices from three manufacturers were used. The results are discussed in terms of parameters of interest for the application, such as input frequency detuning, optical power and SOA driving current. The investigations of the round robin were extended to cover aspects that are related to the FWM phenomenon and can be helpful for the understanding of physical and operational characteristics of SOAs in general. Some of the topics, such as the approach to extract parameters for numerical modelling, cross-gain modulation to provide data for comparison with FWM experiments, nearly degenerate FWM experiments to analyse the SOA behaviour for small optical frequency differences at the input, and first results from the measurements of the influence of birefringence on FWM, can be of particular interest to the reader who wishes to gather more detailed information on this fascinating field.

Part I

Photonic Waveguide Structures

Edited by:

Jiří Čtyroký

Institute of Radio Engineering and Electronics
Academy of Sciences of The Czech Republic

Authors

Hovik V. Baghdasaryan	State Engineering University of Armenia, Yerevan, Armenia
Carlo De Bernardi	CSELT S.p.A., Torino, Italy
Jan Haes	Telenet Operaties N.V., Mechelen, Belgium
Stefan Helfert	FernUniversität, Hagen, Germany
Hugo J. W. M. Hoekstra	University of Twente, The Netherlands
Alain Küng	École Polytechnique Fédérale de Lausanne, Switzerland
Ottokar Leminger	Deutsche Telekom, Darmstadt, Germany
Hans-Peter Nolting	Heinrich Hertz Institut für Nachrichtentechnik, Berlin, Germany
Reinhold Pregla	FernUniversität, Hagen, Germany
Wolf von Reden	Heinrich Hertz Institut für Nachrichtentechnik, Berlin, Germany
Philippe Robert	École Polytechnique Fédérale de Lausanne, Switzerland
Phillip Sewell	Nottingham University, UK
Aasmund Sudbø	University of Oslo, Norway

Further contributors

Roel Baets	University of Gent, Belgium
Mario Bertolotti	Università di Roma "La Sapienza", Rome, Italy
Jiří Čtyroký	Institute of Radio Engineering and Electronics, Prague, The Czech Republic
Gijs J. M. Krijnen	University of Twente, The Netherlands
Paul V. Lambeck	University of Twente, The Netherlands
Martin Reed	Nottingham University, UK
Concita Sibilia	Università di Roma "La Sapienza", Rome, Italy

Fabrication of a novel optical guided-wave device like a semiconductor laser, a wavelength-division multi/demultiplexor, a modulator, a distributed Bragg grating spectral filter or a grating-assisted directional coupler, to name some of them, is a highly specialised and expensive "high-tech" process. The design of any well-working device requires an iterative closed-loop process in which the device designed and fabricated in the first iteration is to be thoroughly characterised, then re-designed and re-fabricated *etc.* Within each iteration, a new set of expensive electron-beam-written microlithographic masks must be often designed, fabricated and checked. Besides the fabrication itself, the characterisation of samples is a time-consuming procedure that is to be done by a highly qualified personnel. Consequently, the existence of reliable, accurate and flexible modelling and design tools is of key importance as it helps considerably reduce the number of iterations needed for a successful design.

A number of software packages have already been developed for this purpose in many research laboratories, at universities, and in specialised software firms. Many of them are currently commercially available. The methods used for modelling have been subject of an intense development for more than two decades, and the basics of some of them are traceable back to the early days of microwave engineering some forty years ago. A large number of methods have already been described in literature, new ones are continuously appearing in journals and at conferences. Recent monographs devoted to this subject can be also found[1]. Thus, why "just another book" on optical waveguide modelling? Let us try to explain this.

Chapters 1 to 3 of this book devoted to waveguide modelling were written in a co-operative effort of a number of specialists. Many of them made original contributions to the development of the modelling methods, all of them have written their own computer codes and have applied the methods to solve practical problems. A good deal of experience with their application has thus been collected and is presented in the next pages. As the text does not represent a completely system-

---

[1] C. Vassallo, *Optical Waveguide Concepts*, Amsterdam, Elsevier, 1991,
R. März, *Integrated optics design and modelling*, Boston, Artech House, 1994.

atic and self-contained tutorial textbook, some reader's preliminary knowledge of the methods is to advantage.

The basic theoretical problem in the analysis of optical waveguide structures consists of determining the propagation constants and optical field distributions of their guided modes. Methods for their numerical calculation — analytical solutions do not exist even for the simplest case of a planar waveguide — are generally known as mode solvers. In Ch. 1, three "mode solving" methods are presented. Section 1.1 devoted to an improved mode solver based on the well-known method of lines (MoL) is written by R. Pregla. A complementary approach of film-mode matching based on the very general mode-matching method (MMM) is described in Sec. 1.2 by A. Sudbø. Mutual relations of MMM with the MoL and a bi-directional mode expansion propagation method (BEP) that will be described later in Ch. 2 are also discussed in some detail. The principles of an approximate but efficient free space radiation mode method (FSRM) are described by P. Sewell in Sec. 1.3, together with its applications for calculating not only guided modes but also their propagation and reflection in longitudinally inhomogeneous waveguide structures.

Chapter 2 is devoted to beam propagation methods used to model light distribution in longitudinally nonuniform optical waveguide structures. The introduction explaining similarities and differences among various BPM methods is written by R. Pregla. Then, three frequently used BPM methods are discussed: section 2.1 devoted to the finite difference (FD) BPM and its improvements was written by H. J. W. M. Hoekstra. The BPM based on the method of lines (MoL BPM) is presented in Sec. 2.2 by R. Pregla. Section 2.3 on the fully bi-directional mode expansion propagation method (BEP) is written by W. von Reden based on the original work of G. Sztefka. H. V. Baghdasaryan contributed in Sec. 2.4 with another bi-directional method of backward calculation useful especially for the analysis of single-mode structures with Kerr nonlinearity.

The core of the "modelling part" of the book is Ch. 3 that contains results of original benchmark tests and modelling tasks that run within the Action COST 240. An extensive set of four different BPM benchmark tests is presented in Sec. 3.1. This section has been written by Hans-Peter Nolting, based on results of his own and other COST 240 members. Section 3.2 is a bit exceptional as it does not describe the benchmark test but explains properties of a special waveguide structure with a balance of gain and loss which forms the basis of some BPM benchmark tests. It has been written by W. v. Reden, using the original theoretical analysis by J. Čtyroký. An important problem of modelling waveguide tapers is treated in Sec. 3.3; the results of a modelling task are described by J. Haes, in co-operation with R. Baets. Section 3.4 is devoted to a special modelling task on an electrooptic polymer waveguide modulator based on the resonance excitation of surface plasmons. The necessity to model light propagation in strongly lossy and very thin metal layers supporting surface plasmons makes this task challenging for many methods. This section has been written by H. Hoekstra in collaboration with P. V. Lambeck and G. Krijnen. In the last Sec. 3.5 of this chapter, modelling of a very deeply etched waveguide Bragg grating is described by S. Helfert in co-operation with J. Čtyroký. The comparative computational results by G. R. Hadley (Sandia National Laboratories) are greatly appreciated. We are grateful also to a number of other COST 240 members that contributed to this chapter with their results and descriptions of their modelling methods.

In the fourth chapter, the most important methods for optical waveguide characterisation that were realised and used in the authors' laboratories are described. In Sec. 4.1, the widely used and powerful Fabry-Perot resonator method for the measurement of loss and group velocity in channel waveguides is presented by C. De Bernardi. The next Sec. 4.2 written by Alain Küng is devoted to the method of optical low-coherence reflectometry applied to group index, chromatic dispersion and loss measurement. Problems of mode field profile measurement using near-field imaging are described in detail by C. De Bernardi in Sec. 4.3. A greatly improved alternative method of transverse offset for near-field measurement is presented in Sec. 4.4 by O. Leminger. The last Sec. 4.5 in this chapter is devoted to the measurement of cut-off wavelengths of guided modes by the spectral transmission method.

The fifth chapter brings mutual comparison of the results of measurement methods used in different laboratories, based on experience gained from round-robin measurements of several types of photonic waveguide structures. In a single Sec. 5.1, the comparison of results of measurement of loss, group index, mode profile and cut-off wavelength is discussed by Ph. Robert.

The main task of the editor of this Part I was to moderate the very enthusiastic effort of a number of authors and contributors with the demanding aim of creating an original, interesting and reasonably balanced text that would simultaneously allow for smooth enough reading.

1 Mode solvers and related methods

R. Pregla, A. Sudbø, Ph. Sewell

To determine propagation constants and spatial optical field distributions of eigenmodes of integrated optical waveguide structures is a fundamental problem of their modelling. The knowledge of (generally complex) propagation constants, or the phase and attenuation constants, and spatial field distributions of guided modes of the waveguide structure helps make important decisions, *e.g.*, whether the waveguide can be used for a particular application. In particular, the field distribution shows whether a waveguide is prone to radiation loss due to irregularities introduced in the fabrication process. Bends of homogeneous waveguides can be analysed by eigenmode solvers; the spatial field distribution of modes furnishes a physical insight into the radiation loss.

Usually, the dimensions of the waveguide cross-sections are small compared to the length of the waveguide sections in optical circuits. For modelling such waveguide devices, special methods known as "beam propagation methods" [1] were developed; some of them will be presented in Ch. 2. However, the most accurate approach to describe optical field distribution in such devices is to use eigenmode propagation techniques together with transfer procedures of wave impedances, admittances, or reflection coefficient matrices. In these cases, the eigenmodes and particularly the propagation constants are to be determined very precisely. Nevertheless, in many cases, approximate results are also useful.

The modes propagating in the photonic waveguides are often quasi-TE or -TM polarised. In these cases, scalar or semi-vectorial solutions are often good approximations. The well known procedure based on the effective index method is frequently used for obtaining such approximate solutions; in this approach, two-dimensional cross-sections are modelled by successive solutions of one-dimensional ones. The algorithm consists of two steps. In the first step the effective lateral refractive index profile has to be determined. For the case of a rib waveguide, the three effective indices are the effective indices of guided modes of the layered regions at both sides of the rib and that of the rib region. For the calculation of these effective indices the lateral widths of the regions are assumed to be infinite. As a suitable procedure, the transfer matrix method [2], [3] can be used. In the next step, each of these regions is replaced by a homogeneous slab with the refractive index equal to the corresponding effective index obtained in the first step. The propagation constant of modes of this equivalent waveguide can be obtained in the same way as the effective indices in the first step.

In principle, the eigenmode solvers which were first developed for integrated microwave and millimetre wave techniques can also be used for photonic devices. We can thus learn a lot from the microwave literature. A rich source is, *e.g.*, the book [3]. Also some points for which a special care is necessary — *e.g.*, the relative convergence phenomenon that takes place when mode expansion is applied to some waveguide structures and which is not widely known among researchers in the area of photonics — are given a detailed description there. A comprehensive review of photonic mode solvers developed in last years is given in [4].

Generally, the most important eigenmode solvers can be classified in two main groups. The first one, in contrast to the second, contains no special analytical work: only the cross-section is discretised. In the finite element method (FEM) [2, 3] the cross-section is subdivided into polygons (*e.g.*, triangles or rectangles). Because of this, no restriction is to be imposed on the shape of the cross-section. Instead of finite differences, the corresponding functionals are used and variational expressions are applied to each polygon.

In the finite difference method (FDM) [2] the partial derivatives in the lateral directions are replaced by finite differences obtained from Taylor expansion. Homogeneous or inhomogeneous grids are used. The propagation constants are then obtained by the solution of an eigenvalue problem. To obtain correct results with good convergence behaviour, the interface conditions for permittivity steps have to be taken into account.

One of the most well-known and widely spread procedures is the mode matching method (MMM) [5] that belongs to the second group. In this method the cross-section is subdivided into suitable subregions in such a way that in each of the subregions the field can be described by a superposition of an infinite series of exact solutions of the wave equations. At the interfaces between the subregions the tangential components of field vectors have to be matched. From the matching procedure an indirect eigenvalue problem is formulated for the propagation constant. In the photonic waveguides all subregions or at least some of them are composed of layered structures. Therefore, the solutions of the wave equations can *e.g.* be constructed by a superposition of the solutions for equivalent homogeneous regions following the Rayleigh-Ritz method [5]. This algorithm was used long ago for dielectric and microstrip waveguides at microwave frequencies.

The method of lines (MoL) [2, 3] is a special kind of a finite-difference method (FDM). In contrast to the standard FDM the discretisation is not done completely but only as far as necessary. The discretisation is used to transform partial differential equations into ordinary ones that are then solved analytically. The fields in subregions are described by means of these eigensolutions. The tangential fields at the interfaces between the subregions are matched at the crossing points of the discretisation lines with these interfaces. This point matching procedure results in a smoother convergence behaviour than MMM avoiding also the well-known Gibbs phenomenon. Even if the permittivity in the subregion is a complicated function of one of the directions, the solution of the wave equations is obtained very easily, while the overlap integrals in the Rayleigh-Ritz method cannot be solved anymore analytically. Furthermore, the convergence behaviour and the accuracy of the field calculation can be improved by using a higher order approximation for the difference operators. In most of the eigenmode solvers the material parameters may exhibit also anisotropic or gyrotropic behaviour.

Three mode solvers from the wide spectrum of algorithms of this class are presented in this chapter. In the first section a new algorithm for the eigenmode analysis based on the method of lines is described. In this algorithm, two mutually perpendicular discretisation line systems are used. The cross-section is subdivided by means of horizontal as well as vertical cuts into subregions in which the fields can be described by two terms. The principle of the algorithm is applicable also to MMM and has also been used previously for the analysis of waveguides in integrated microwave and millimetre wave circuits.

In the second section, one of the mode-matching methods for layered (or film) structures is reported. The method makes an efficient use of the impedance/admittance transfer concept that was originally developed for the MoL to avoid numerical problems. Common points and similarities of this method with the FDM, MoL and the bi-directional eigenmode propagation (BEP) method are stressed there.

An approximate eigenmode solver known as the free space radiation mode method is presented in the third section. As in the above mentioned MMM the waveguide is divided into film guide slices. The fields and wave equations in each of the slices are then Fourier-transformed by the co-ordinate parallel to the cuts and matched at the interfaces. The algorithm requires minimum computational effort but is restricted to low refractive index contrasts. Propagation in longitudinally nonuniform 3D structures is treated in this contribution, too, making thus a smooth transition to the BPM methods treated in detail in chapter 2.

1.1 MoL mode solver using enhanced line algorithm

Passive components of integrated optics consist of concatenations of waveguide sections. The planar or quasi-planar waveguides are arranged in inhomogeneous layers. Two examples are shown in Fig. 1.1, on the left side a channel waveguide and on the right side a waveguide structure with two ribs, one of them with a metal sheet on the top. Usually, the dimensions of the waveguide cross-sections are small compared to the length of sections in the circuits. To describe the circuit behaviour correctly we have to accurately determine the eigenmodes and particularly their propagation constants. In this contribution a new procedure based on the method of lines (MoL) is described.

The proposed approach can also be used in the framework of the mode matching techniques (MMT) [6, 7]. In the MoL we discretise the field only in one direction. In the remaining direction the field distribution is calculated analytically. The modal fields of the waveguides are concentrated in the channel or in the rib and adjacent layers under it, respectively. This requires a realistic modelling of the fields in these parts of the waveguides. Special effort is necessary for the region containing metal.

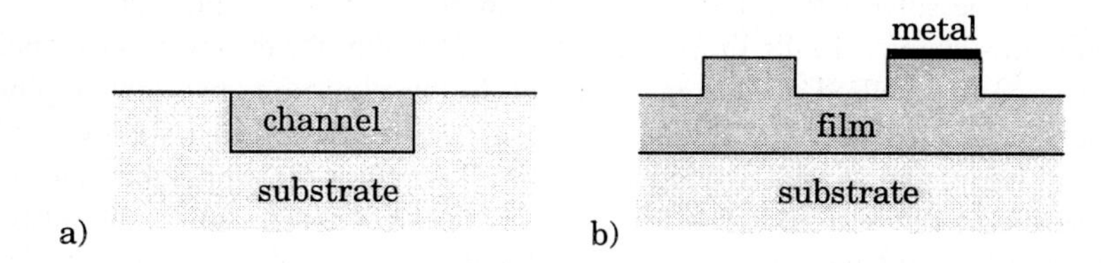

Fig. 1.1. Cross-sections of channel (a) and rib waveguides (b). The right rib has a metal sheet on the top.

The two ways of subdivision of the cross-sections for the purpose of analysis by the MMT or by the MoL are shown in the right part of Fig. 1.2. In the first (second) case this subdivision is done by vertical (horizontal) lines and the discretisation lines would have horizontal (vertical) direction. Which type of subdivision and discretisation has to be preferred cannot generally be recommended *a priori.*

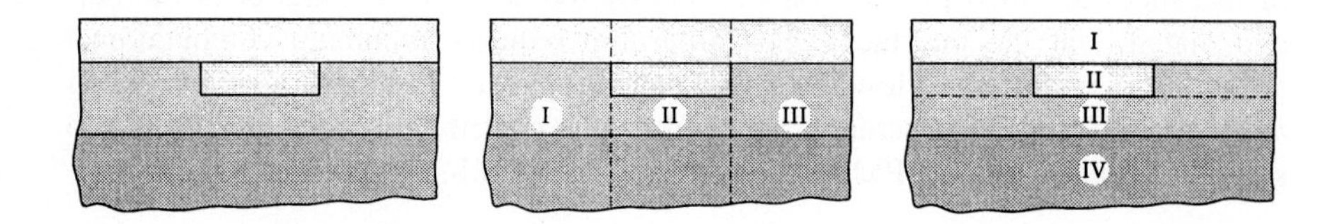

Fig. 1.2. Cross-section of a channel waveguide and alternative subdivision possibilities.

A question arises as to why not use subdivisions (discretisation lines) in both directions simultaneously, as shown in Fig. 1.3.

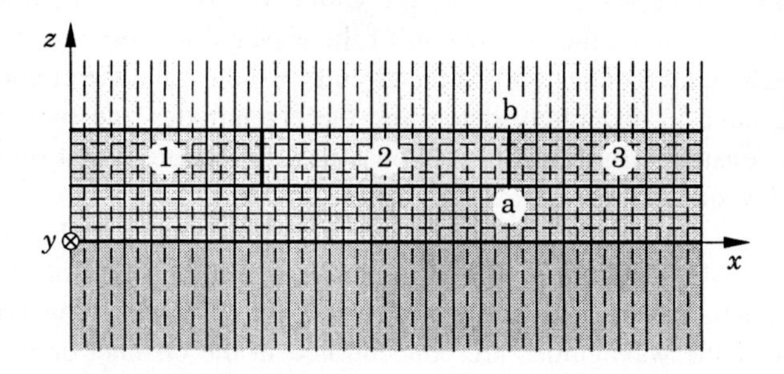

Fig. 1.3. Discretisation with both vertical and horizontal lines in some regions of the waveguide cross-section shown in Fig. 1.1.a.

In this case, the fields in the channel, rib and also metal regions can be determined more accurately. Note, however, that the application of two sets of crossed discretisation lines does not mean a 2D discretisation. The other layers of the structure considered here, especially the film under the rib, can also be discretised in this way. If two sets of crossed lines are used simultaneously, the fields are expected to be composed of two parts, each corresponding to one of the two discretisation line systems. The numbers of discretisation lines in horizontal and vertical directions can be chosen separately. Therefore, if necessary, these numbers can be chosen large enough to obtain high accuracy without increasing the numerical effort appreciably.

1.1.1 Theoretical foundation of the algorithm

First, the theoretical background for the proposed algorithm will be discussed. The field in the region with crossed lines can be described in the following way. Fig. 1.4 shows a general region R as *e.g.* region 2 in Fig. 1.3. If the tangential field components on the whole surface of this region are known, the fields outside R can be determined exactly, as stated by a uniqueness theorem. Because of the linearity of the problem, the following matrix relation holds among the tangential fields at the ports A, B, C and D of the general region R:

$$\begin{bmatrix} \mathbf{H}_A \\ -\mathbf{H}_B \\ \mathbf{H}_C \\ -\mathbf{H}_D \end{bmatrix} = \begin{bmatrix} \mathbf{y}_{AA} & \cdots & \mathbf{y}_{AD} \\ \vdots & & \vdots \\ \mathbf{y}_{DA} & \cdots & \mathbf{y}_{DD} \end{bmatrix} \begin{bmatrix} \mathbf{E}_A \\ \mathbf{E}_B \\ \mathbf{E}_C \\ \mathbf{E}_D \end{bmatrix}, \tag{1.1}$$

where $\mathbf{E}_U$ and $\mathbf{H}_U$ (U = A, B, C, D) are supervectors (see below) of the discretised tangential fields at port U.

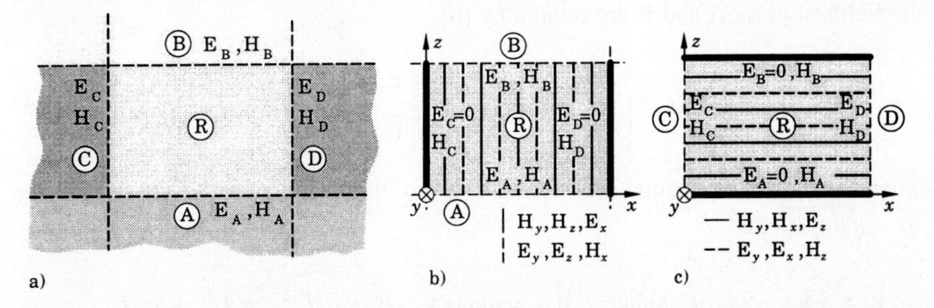

Fig. 1.4. Field calculation in a region R using crossed lines: a) general region R with introduced notations; b) ports C and D short-circuited: analysis with vertical lines; c) ports A and B short-circuited: analysis with horizontal lines.

Each supervector consists of two vectors corresponding to the two tangential components. In a more compact form we can write

$$\begin{bmatrix} \hat{\mathbf{H}}_{\mathrm{AB}} \\ \hat{\mathbf{H}}_{\mathrm{CD}} \end{bmatrix} = \begin{bmatrix} \hat{\mathbf{y}}_{\mathrm{AB}}^{\mathrm{AB}} & \hat{\mathbf{y}}_{\mathrm{AB}}^{\mathrm{CD}} \\ \hat{\mathbf{y}}_{\mathrm{CD}}^{\mathrm{AB}} & \hat{\mathbf{y}}_{\mathrm{CD}}^{\mathrm{CD}} \end{bmatrix} \begin{bmatrix} \hat{\mathbf{E}}_{\mathrm{AB}} \\ \hat{\mathbf{E}}_{\mathrm{CD}} \end{bmatrix}, \tag{1.2}$$

$$\hat{\mathbf{H}}_{\mathrm{AB,\,CD}} = \left[\mathbf{H}_{\mathrm{A,C}}^{t}, -\mathbf{H}_{\mathrm{B,D}}^{t}\right]^{t}, \quad \hat{\mathbf{E}}_{\mathrm{AB,\,CD}} = \left[\mathbf{E}_{\mathrm{A,C}}^{t},\ \mathbf{E}_{\mathrm{B,D}}^{t}\right]^{t}.$$

The admittance matrices $\hat{\mathbf{y}}$ can be calculated in the following way: By short-circuiting the ports C and D (using metallic walls) we obtain magnetic field parts at the ports A and B and at the ports C and D excited by the electric field $\hat{\mathbf{E}}_{\mathrm{AB}}$. All these field parts are determined using vertical discretisation lines parallel to the metallic side walls (see Fig. 1.4.b). From these partial fields the matrices $\hat{\mathbf{y}}_{\mathrm{AB}}^{\mathrm{AB}}$ and $\hat{\mathbf{y}}_{\mathrm{CD}}^{\mathrm{AB}}$ are obtained. Similarly, short-circuiting ports A and B and using horizontal discretisation lines (see Fig. 1.4.c) we obtain the matrices $\hat{\mathbf{y}}_{\mathrm{AB}}^{\mathrm{CD}}$ and $\hat{\mathbf{y}}_{\mathrm{CD}}^{\mathrm{CD}}$. Instead of Eq. (1.2) we can alternatively use the inverted form with impedance submatrices $\hat{\mathbf{z}}$. These open circuiting submatrices are obtained in a dual form to $\hat{\mathbf{y}}$ submatrices by open-circuiting the ports (using magnetic walls). For the regions with unidirectional discretisation lines, equations analogous to the above can be written [8].

1.1.2 Determination of matrices

In the following it will be described how the supermatrices $\hat{\overline{\mathbf{y}}}_{\mathrm{AB}}^{\mathrm{AB}}$ and $\hat{\overline{\mathbf{y}}}_{\mathrm{CD}}^{\mathrm{AB}}$ in the transform domain can be determined. For this purpose the ports C and D are short circuited. The remaining matrices can be determined in a similar way. Using the following definitions for the supervectors in the transform domain

$$\hat{\overline{\mathbf{H}}} = \left[-j\overline{\mathbf{H}}_{y}^{t}, \overline{\mathbf{H}}_{x}^{t}\right]^{t}, \qquad \hat{\overline{\mathbf{E}}} = \left[\overline{\mathbf{E}}_{x}^{t}, -j\overline{\mathbf{E}}_{y}^{t}\right]^{t}, \tag{1.3}$$

the fields at ports A and B are related by [6]

$$\hat{\overline{\mathbf{H}}}_{\mathrm{AB}} = \begin{bmatrix} \hat{\overline{\mathbf{H}}}_{\mathrm{A}} \\ -\hat{\overline{\mathbf{H}}}_{\mathrm{B}} \end{bmatrix} = \begin{bmatrix} \overline{\mathbf{y}}_1 & \overline{\mathbf{y}}_2 \\ \overline{\mathbf{y}}_2 & \overline{\mathbf{y}}_1 \end{bmatrix} \begin{bmatrix} \hat{\overline{\mathbf{E}}}_{\mathrm{A}} \\ \hat{\overline{\mathbf{E}}}_{\mathrm{B}} \end{bmatrix} = \hat{\overline{\mathbf{y}}}_{\mathrm{AB}}^{\mathrm{AB}} \hat{\overline{\mathbf{E}}}_{\mathrm{AB}}, \tag{1.4}$$

where the H-components are normalised by the free space wave impedance $\eta_0 = \sqrt{\mu_0/\varepsilon_0}$ and

$$\begin{aligned} \overline{\mathbf{y}}_1 &= \hat{\gamma}\Lambda, \quad \hat{\gamma} = \left[\hat{\Gamma}\tanh\left(\hat{\Gamma}\overline{d}\right)\right]^{-1} = \mathrm{diag}\,(\gamma_{\mathrm{h}}, \gamma_{\mathrm{e}}), \quad \Gamma_{\mathrm{e,h}}^{2} = \lambda_{\mathrm{e,h}}^{2} - \varepsilon_{\mathrm{d}}\boldsymbol{I}_{\mathrm{e,h}}, \\ \overline{\mathbf{y}}_2 &= -\hat{\alpha}\Lambda, \quad \hat{\alpha} = \left[\hat{\Gamma}\sinh\left(\hat{\Gamma}\overline{d}\right)\right]^{-1} = \mathrm{diag}\,(\alpha_{\mathrm{h}}, \alpha_{\mathrm{e}}), \quad \hat{\Gamma} = \mathrm{diag}\left(\Gamma_{\mathrm{h}}, \Gamma_{\mathrm{e}}\right), \end{aligned} \tag{1.5}$$

$$\boldsymbol{\Lambda} = \begin{bmatrix} \varepsilon_{\mathrm{d}} \boldsymbol{I}_{\mathrm{h}} & \tilde{\boldsymbol{\delta}}^{t} \\ \tilde{\boldsymbol{\delta}} & \varepsilon_{\mathrm{r}} \boldsymbol{I}_{\mathrm{e}} - \overline{\boldsymbol{\lambda}}_{\mathrm{e}}^{2} \end{bmatrix}, \quad \begin{matrix} \tilde{\boldsymbol{\delta}}^{t} = \boldsymbol{T}_{\mathrm{e}}^{t} \overline{\boldsymbol{D}}_{\mathrm{h}} \boldsymbol{T}_{\mathrm{h}}, \\ \tilde{\boldsymbol{\delta}} = \sqrt{\varepsilon_{\mathrm{re}}} \overline{\boldsymbol{\delta}}, \end{matrix} \quad \begin{matrix} \boldsymbol{T}_{\mathrm{e,h}}^{t} \overline{\boldsymbol{D}}_{\mathrm{e,h}}^{t} \overline{\boldsymbol{D}}_{\mathrm{e,h}} \boldsymbol{T}_{\mathrm{e,h}} = \overline{\boldsymbol{\lambda}}_{\mathrm{e,h}}, \\ \varepsilon_{\mathrm{d}} = \varepsilon_{\mathrm{r}} - \varepsilon_{\mathrm{re}}, \end{matrix} \tag{1.6}$$

where ε_{r} is the relative permittivity of the region R and $\varepsilon_{\mathrm{re}}$ denotes the effective permittivity of the whole waveguide. $\boldsymbol{I}_{\mathrm{e,h}}$ are identity matrices, $\overline{\boldsymbol{D}}_{\mathrm{h}}$ ($\overline{\boldsymbol{D}}_{\mathrm{e}}$) is the first-order difference operator for the field component $\mathbf{H}_y$ ($\mathbf{E}_y$) for the line in vertical direction which has to fulfil Neumann (Dirichlet) boundary conditions. From the field vectors $\hat{\overline{\mathbf{H}}}_{\mathrm{A}}$ and $\hat{\overline{\mathbf{H}}}_{\mathrm{B}}$ at the ports A and B in Eq. (1.4) we can determine the field components $\overline{\mathbf{H}}_{\mathrm{yx}}$ and $\overline{\mathbf{H}}_{\mathrm{zx}}$ at every position in the region R by

$$\overline{\mathbf{H}}_{\mathrm{yx}}(z) = \frac{\sinh\left(\boldsymbol{\Gamma}_{\mathrm{h}}(\overline{d}_{\mathrm{AB}} - \overline{x})\right)}{\sinh\left(\boldsymbol{\Gamma}_{\mathrm{h}} \overline{d}_{\mathrm{AB}}\right)} \overline{\mathbf{H}}_{\mathrm{yxA}} + \frac{\sinh\left(\boldsymbol{\Gamma}_{\mathrm{h}} \overline{x}\right)}{\sinh\left(\boldsymbol{\Gamma}_{\mathrm{h}} \overline{d}_{\mathrm{AB}}\right)} \overline{\mathbf{H}}_{\mathrm{yxB}}, \tag{1.7}$$

where $\overline{d}_{\mathrm{AB}} = k_0 d_{\mathrm{AB}}$ is the distance between the ports A and B normalised with the angular repetency k_0. A similar equation can be written for $\overline{\mathbf{H}}_{\mathrm{zx}}$ with $\boldsymbol{\Gamma}_{\mathrm{e}}$ instead of $\boldsymbol{\Gamma}_{\mathrm{h}}$. The subscript x denotes that these field parts are obtained from discretisation in x direction (in Eq. (1.4) this index is omitted). The fields at the ports C and D from these parts are obtained by transformation with suitable row vectors of $\boldsymbol{T}_{\mathrm{h}}$ which must be generated separately. With the normal row vectors the fields on full vertical discretisation lines are obtained. For the boundaries C and D, $\boldsymbol{N}_{\mathrm{R}x} = N + 1$, $i = 1/2$, and $i = \boldsymbol{N}_{\mathrm{Rx}} + 1/2$, respectively, must be introduced into Eq.(227) of [9]. The result is

$$\mathbf{T}_{\mathrm{C,D}} = \sqrt{2 / \boldsymbol{N}_{\mathrm{R}x}} \left[1/\sqrt{2}, \pm 1, 1, \pm 1, \ldots \right], \tag{1.8}$$

where $\boldsymbol{N}_{\mathrm{R}x}$ is the number of full discretisation lines in x direction. Now we obtain

$$\mathrm{H}_{\mathrm{yC}}(z) = \mathbf{T}_{\mathrm{hC}} \overline{\mathbf{H}}_{\mathrm{y}}(z), \qquad \mathrm{H}_{\mathrm{yD}}(z) = \mathbf{T}_{\mathrm{hD}} \overline{\mathbf{H}}_{\mathrm{y}}(z). \tag{1.9}$$

The vectors of the discretised fields at ports C and D are given by

$$\begin{bmatrix} -j\overline{\mathbf{H}}_{\mathrm{yC}} \\ j\overline{\mathbf{H}}_{\mathrm{yD}} \end{bmatrix} = \begin{bmatrix} \overline{\boldsymbol{V}}_{\mathrm{CA}}^{\mathrm{N}} & \overline{\boldsymbol{V}}_{\mathrm{CB}}^{\mathrm{N}} \\ \overline{\boldsymbol{V}}_{\mathrm{DA}}^{\mathrm{N}} & \overline{\boldsymbol{V}}_{\mathrm{DB}}^{\mathrm{N}} \end{bmatrix} \begin{bmatrix} -j\overline{\mathbf{H}}_{\mathrm{yA}} \\ j\overline{\mathbf{H}}_{\mathrm{yB}} \end{bmatrix}, \quad \begin{bmatrix} \overline{\mathbf{H}}_{\mathrm{zC}} \\ -\overline{\mathbf{H}}_{\mathrm{zD}} \end{bmatrix} = \begin{bmatrix} \overline{\boldsymbol{V}}_{\mathrm{CA}}^{\mathrm{D}} & \overline{\boldsymbol{V}}_{\mathrm{CB}}^{\mathrm{D}} \\ \overline{\boldsymbol{V}}_{\mathrm{DA}}^{\mathrm{D}} & \overline{\boldsymbol{V}}_{\mathrm{DB}}^{\mathrm{D}} \end{bmatrix} \begin{bmatrix} \overline{\mathbf{H}}_{\mathrm{zA}} \\ -\overline{\mathbf{H}}_{\mathrm{zB}} \end{bmatrix}, \tag{1.10}$$

with

$$\begin{matrix} \overline{\boldsymbol{V}}_{\mathrm{CA}} = \boldsymbol{T}_{\mathrm{H}} \boldsymbol{\Lambda}_{\mathrm{A}} \boldsymbol{T}_{\mathrm{C}}^{\mathrm{d}}, & \overline{\boldsymbol{V}}_{\mathrm{CB}} = -\boldsymbol{T}_{\mathrm{H}} \boldsymbol{\Lambda}_{\mathrm{B}} \boldsymbol{T}_{\mathrm{C}}^{\mathrm{d}}, \\ \overline{\boldsymbol{V}}_{\mathrm{DA}} = -\boldsymbol{T}_{\mathrm{H}} \boldsymbol{\Lambda}_{\mathrm{A}} \boldsymbol{T}_{\mathrm{D}}^{\mathrm{d}}, & \overline{\boldsymbol{V}}_{\mathrm{DB}} = \boldsymbol{T}_{\mathrm{H}} \boldsymbol{\Lambda}_{\mathrm{B}} \boldsymbol{T}_{\mathrm{D}}^{\mathrm{d}}, \end{matrix} \tag{1.11}$$

where $\boldsymbol{T}_{\mathrm{C,D}}^{\mathrm{d}}$ are diagonal matrices obtained from $\mathbf{T}_{\mathrm{C,D}}$ in Eq. (1.9). $\boldsymbol{T}_{\mathrm{H}}$ is the transformation matrix for the discretisation lines in horizontal direction. For $\overline{\boldsymbol{V}}^{\mathrm{N}}$ ($\overline{\boldsymbol{V}}^{\mathrm{D}}$) the matrix for h (e) lines must be used.

The components of matrices $\boldsymbol{\Lambda}_{\mathrm{A}}$ and $\boldsymbol{\Lambda}_{\mathrm{B}}$ are given by the following expressions:

$$\left(\boldsymbol{\Lambda}_{\mathrm{A}}\right)_{ik} = \sinh\left[\boldsymbol{\Gamma}_{\mathrm{h}k}\left(\bar{d}_{\mathrm{AB}} - \bar{z}_i\right)\right]\left[\sinh\left(\boldsymbol{\Gamma}_{\mathrm{h}k}\,\bar{d}_{\mathrm{AB}}\right)\right]^{-1}, \tag{1.12}$$

$$\left(\Lambda_{\mathrm{B}}\right)_{ik} = \sinh\left(\Gamma_{\mathrm{h}k}\,\bar{z}_i\right)\left[\sinh\left(\Gamma_{\mathrm{h}k}\,\bar{d}_{\mathrm{AB}}\right)\right]^{-1}. \tag{1.13}$$

The positions z_i have to be taken along z-axis at the intersections with the corresponding horizontal lines. They are different for the $\mathbf{H}_y$ (superscript N for Neumann BC: full lines in Fig. 1.3c) and $\mathbf{H}_z$ (superscript D for Dirichlet BC: dashed lines in Fig. 1.3c) vectors. The matrix $\mathbf{y}_{\mathrm{CD}}^{\mathrm{AB}}$ is constructed from the two systems of equations in (1.10) by replacing $\overline{\mathbf{H}}_{y\mathrm{A,B}}$ from Eq. (1.4) and $\overline{\mathbf{H}}_{z,\mathrm{AB}}$ by

$$\overline{\mathbf{H}}_{z\mathrm{A,B}} = \left[-\sqrt{\varepsilon_{\mathrm{re}}}\boldsymbol{I}_{\mathrm{h}} \quad \boldsymbol{\delta}^t\right]\hat{\overline{\mathbf{E}}}_{\mathrm{A,B}}\,. \tag{1.14}$$

Taking into account the definitions in Eqs. (1.2) and (1.3) we can write the matrix $\mathbf{y}_{\mathrm{CD}}^{\mathrm{AB}}$ in the following form

$$\hat{\overline{\mathbf{y}}}_{\mathrm{CD}}^{\mathrm{AB}} = \begin{bmatrix} \overline{V}_{\mathrm{CA}}^{\mathrm{N}} & \mathbf{0} & \overline{V}_{\mathrm{CB}}^{\mathrm{N}} & \mathbf{0} \\ \mathbf{0} & \overline{V}_{\mathrm{CA}}^{\mathrm{D}} & \mathbf{0} & \overline{V}_{\mathrm{CB}}^{\mathrm{D}} \\ \overline{V}_{\mathrm{DA}}^{\mathrm{N}} & \mathbf{0} & \overline{V}_{\mathrm{DB}}^{\mathrm{N}} & \mathbf{0} \\ \mathbf{0} & \overline{V}_{\mathrm{DA}}^{\mathrm{D}} & \mathbf{0} & \overline{V}_{\mathrm{DB}}^{\mathrm{D}} \end{bmatrix} \begin{bmatrix} \varepsilon_{\mathrm{d}}\boldsymbol{\gamma}_{\mathrm{h}} & \tilde{\boldsymbol{\delta}}^t\boldsymbol{\gamma}_{\mathrm{e}} & -\varepsilon_{\mathrm{d}}\boldsymbol{\alpha}_{\mathrm{h}} & -\tilde{\boldsymbol{\delta}}^t\boldsymbol{\alpha}_{\mathrm{e}} \\ -\sqrt{\varepsilon_{\mathrm{re}}}\boldsymbol{I}_{\mathrm{h}} & \boldsymbol{\delta}^t & \mathbf{0} & \mathbf{0} \\ -\varepsilon_{\mathrm{d}}\boldsymbol{\alpha}_{\mathrm{h}} & -\tilde{\boldsymbol{\delta}}^t\boldsymbol{\alpha}_{\mathrm{e}} & \varepsilon_{\mathrm{d}}\boldsymbol{\gamma}_{\mathrm{h}} & \tilde{\boldsymbol{\delta}}^t\boldsymbol{\gamma}_{\mathrm{e}} \\ \mathbf{0} & \mathbf{0} & \sqrt{\varepsilon_{\mathrm{re}}}\boldsymbol{I}_{\mathrm{h}} & \tilde{\boldsymbol{\delta}}^t \end{bmatrix}. \tag{1.15}$$

The whole analysis can now be performed as an impedance/admittance matching process [10]. For this purpose, first the fields of all regions of a layer at its common ports, *i.e.*, on its vertical boundaries, must be matched. Then, the fields between adjacent layers have to be matched. This should be done using the impedance/admittance transfer concept [10]. If the layers consist of subregions of different widths, the components in the supervectors must be ordered in advance to become equally positioned. Special care should be taken at the edges of the region R. The corresponding components must be calculated separately. Accordingly, the $\mathbf{H}_{\mathrm{x}}$ components at the corners a and b of the common wall between the regions 2 and 3 (cf. Fig. 1.3), given by

$$\mathrm{H}_{x\mathrm{A,B}} = \mathbf{T}_{\mathrm{A,B}}\left[\sqrt{\varepsilon_{\mathrm{re}}}\boldsymbol{I}_{\mathrm{h}}^{\mathrm{H}} \quad -\boldsymbol{\delta}_{\mathrm{hH}}^t\right]\hat{\overline{\mathbf{E}}}_{\mathrm{C,D}} \tag{1.16}$$

must be introduced into the final system of equations. $\mathbf{T}_{\mathrm{A,B}}$ is constructed similarly as in Eq. (1.8).

1.1.3 Numerical results

The proposed algorithm is verified by numerical results for the rib waveguide of the benchmark test in [4] (see Table 1 in [4]). The normalised propagation constants B differ only in the fourth digit after the decimal point. Fig. 1.5 shows the distribution of the field components of fundamental quasi TE mode along the central line of the film layer demonstrating a highly accurate solution. Since the structure is symmetric, only the right parts of all curves are shown.

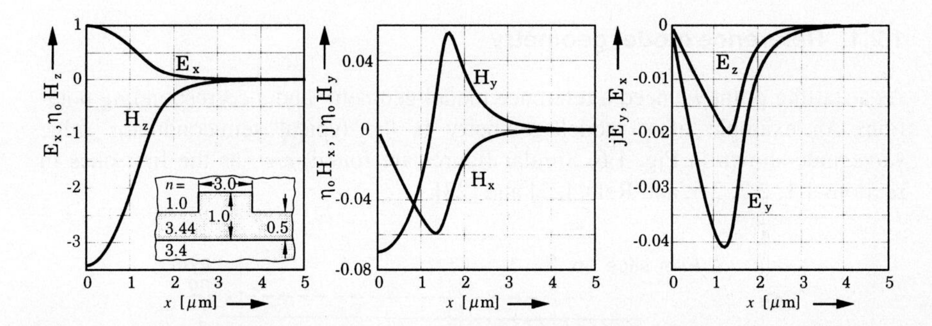

Fig. 1.5. Normalised field components of the fundamental TE mode in the middle of the film of rib waveguide (benchmark test [4]) vs. x — the distance from symmetry line of the structure; dimensions in μm.

1.2 Film-mode matching with relation to BEP and MoL

The transverse resonance method (TRM) is well established for calculating mode fields in optical waveguides [3], [11]. The basic idea of the method was used already in Unger's pioneering work on rectangular dielectric waveguides [12]. The popular effective index method [13] may be considered the lowest order approximation of the TRM. An important development was given by Peng and Oliner [14], who used the method to calculate the radiation loss of the leaky modes of deeply etched rib waveguides. In [15] and [16] the descriptive term 'film mode matching' (FMM) and a systematic mathematical notation were adopted. Based on this work, Gallagher has developed a computer program for vectorial mode field calculations and made it commercially available [17]. The close relationship between the TRM and another extensively developed method for mode field calculations, the method of lines (MoL) [18], [19] has also been noted [20], [4]. Still another close relative in this family is the spectral index (SI) method [21], an approximate but fast method restricted to rib waveguides. In his mode solver review [4] Vassallo recommends the TRM as a reference method for vectorial mode field calculations in rib waveguides, based on a benchmark test for a variety of geometries.

The bi-directional eigenmode propagation method (BEP) [22] and the MoL-BPM [23], [24] are methods for calculating optical field distribution in longitudinally nonuniform optical devices. They are both discussed in the next chapter, and may be viewed as a special case of the TRM, as we shall see below. All of these methods may be generically termed mode-matching methods (MMMs). In the following, we shall use this term to encompass the TRM, MoL, BEP, and MoL-BPM, and discuss the common merits of the MMMs, their strengths, potentials and limitations.

1.2.1 Reference model geometry

As a starting point we need a reference model geometry and a corresponding notation. An example of a model geometry is the typical semiconductor ridge waveguide shown in Fig. 1.6. Similar figures are found, *e.g.*, as the first ones in sections 1.1, 2.2, 2.3, and Refs. [22] and [23].

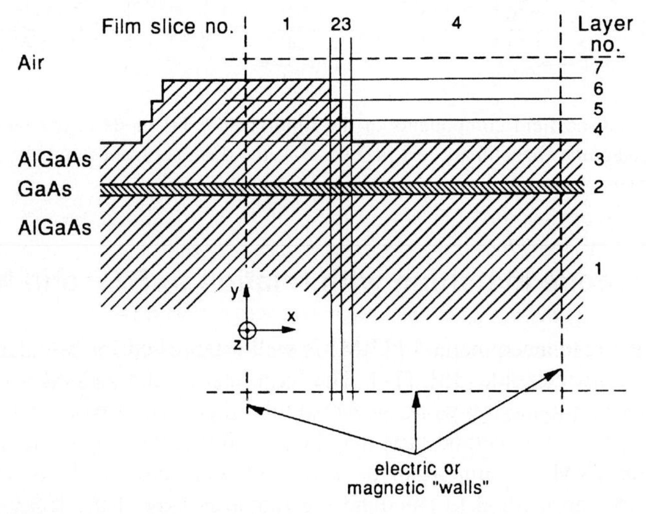

Fig. 1.6. Typical semiconductor ridge waveguide cross-section, modelled by a number of film guide 'slices'. The slopes on each side of the ridge have been approximated by staircases. (From Ref. [15].)

The structures analysed by mode matching methods (MMMs) in rectangular coordinates are always modelled by a 'sandwich' of M 'slices' each of thickness ${}^{m}d_x, m = 1, 2, \ldots, M$, like in Fig. 1.6. Let ${}^{m}x$ be the position of the interface between slice no. m and no. m+1, so that

$$ {}^{m}d_x = {}^{m}x - {}^{m-1}x . \tag{1.17} $$

Let the total thickness of the 'sandwich' be w_x, and let us consider each slice to be cut from a planar film (slab) waveguide of total thickness w_y. It is natural to attach the label m not only to slice no. m, but to the film that it is cut from, as well. Let us choose our x-axis parallel to the film layers and perpendicular to the film slice interfaces, our y-axis perpendicular to the layers and parallel to the slice interfaces, and our z-axis along the waveguide, parallel to the layers and slices, as in Fig. 1.6. We have made an effort to use the same notation here and in Sec. 2.3, and it is therefore slightly different from the one used in Refs. [15] and [16]. (Since we shall first consider propagation *perpendicular* to the waveguide, choosing the z-direction along the waveguide is in conflict with the convention used in Sec. 2.3, where the propagation direction is chosen as the z-direction.)

In the scattering problem discussed in [22] and [23], light in the fundamental mode of the rightmost (no. M) film waveguide is incident (backward travelling) from the right. Some light is reflected by the scattering structure (*e.g.*, the ridge in Fig. 1.6) back into the fundamental mode and some is transmitted past the scattering structure and continues in the fundamental mode. The rest is scattered into radiating (and, if existing, higher order bound) modes, forward travelling in the rightmost slice and backward travelling in the leftmost slice. The same type of scattering problem is analysed in detail in Sec. 2.3.

1.2.2 Theory

For the purpose of introducing a unifying notation and outlining the common principles of the BEP, the TRM, and the MoL, we start our discussion with the BEP in the scalar (weakly guiding) approximation. Let c be the speed of light, ω the angular optical frequency, λ_0 the corresponding vacuum wavelength, and k_0 the corresponding angular repetency [25], so that $k_0 = 2\pi/\lambda = \omega/c$. With reference to Fig. 1.6, the optical field within slice no. m can be expressed as a sum of products of a y-dependent function and an x-dependent function:

$$^{m}F(x,y) = \sum_{i=1}^{\infty} {}^{m}u_i(x) \cdot {}^{m}\varphi_i(y), \tag{1.18}$$

where we have omitted the harmonic time dependence $\exp(j\omega t)$ of the field, and where ${}^{m}\varphi_i$ is the field distribution of mode no. i in film no. m. In numerical calculations, the sum in (1.18) must of course be restricted to a finite range I, so that i runs from 1 through the total number I of film modes considered. For the BEP each film is usually modelled by a stack of homogeneous layers, so that analytic expressions are available for the film mode fields, whereas for the MoL-BPM a finite-difference approximation is used to obtain the film mode fields. This approximation is what distinguishes the MoL from the BEP and the TRM.

Introducing the propagation constant ${}^{m}\beta_i$ of film mode no. i allows us to write for the corresponding amplitude ${}^{m}u_i(x)$

$$ {}^{m}u_i(x) = {}^{(m,l)}u_i^- \exp\left[j\,{}^{m}\beta_i\left(x - {}^{m-1}x\right)\right] + {}^{(m,l)}u_i^+ \exp\left[-j\,{}^{m}\beta_i\left(x - {}^{m-1}x\right)\right]. \quad (1.19) $$

As in [15] one may define

$$ {}^{(m,l)}u_i^s = {}^{(m,l)}u_i^- + {}^{(m,l)}u_i^+ \quad (1.20) $$

and

$$ {}^{(m,l)}u_i^a = j\,{}^{m}\beta_i\left({}^{(m,l)}u_i^- - {}^{(m,l)}u_i^+\right), \quad (1.21) $$

to obtain the equivalent of (1.19),

$$ {}^{m}u_i(x) = {}^{(m,l)}u_i^s \cos[{}^{m}\beta_i(x - {}^{m-1}x)] + \left({}^{(m,l)}u_i^a / {}^{m}\beta_i\right)\sin[{}^{m}\beta_i(x - {}^{m-1}x)], \quad (1.22) $$

${}^{(m,l)}u_i^s$ is the amplitude of mode no. i on the left side of slice no. m at ${}^{m-1}x$, and ${}^{(m,l)}u_i^a$ is the corresponding x-derivative of the amplitude. ${}^{(m,l)}u_i^-$ is the corresponding left-moving component of the amplitude, and ${}^{(m,l)}u_i^+$ is the corresponding right-moving component. The amplitudes on the other side of slice no. m, at ${}^{m}x$ are

$$ {}^{(m,r)}u_i^- = {}^{(m,l)}u_i^- \exp(j\,{}^{m}\beta_i\,{}^{m}d_x), \quad (1.23) $$

$$ {}^{(m,r)}u_i^+ = {}^{(m,l)}u_i^+ \exp(-j\,{}^{m}\beta_i\,{}^{m}d_x), \quad (1.24) $$

$$ {}^{(m,r)}u_i^s = {}^{(m,l)}u_i^s \cos({}^{m}\beta_i\,{}^{m}d_x) + \left({}^{(m,l)}u_i^a / {}^{m}\beta_i\right)\sin({}^{m}\beta_i\,{}^{m}d_x), \quad (1.25) $$

$$ {}^{(m,r)}u_i^s = {}^{(m,l)}u_i^a \cos({}^{m}\beta_i\,{}^{m}d_x) - {}^{(m,l)}u_i^s \cdot {}^{m}\beta_i \sin({}^{m}\beta_i\,{}^{m}d_x). \quad (1.26) $$

We introduce vector notation, so that ${}^{(m,l)}\mathbf{u}^s$ is a vector with the elements ${}^{(m,l)}u_i^s$, $i = 1, 2, \ldots, I$, with corresponding definitions for the other u's. The modal amplitudes and their derivatives at one side of slice interface no. m are then related to the ones on the other side of the interface via a coupling matrix ${}^{m}\mathbf{O}$,

$$ {}^{(m,r)}\mathbf{u}^s = {}^{m}\mathbf{O} \cdot {}^{(m+1,l)}\mathbf{u}^s, \quad (1.27) $$

$$ {}^{(m,r)}\mathbf{u}^a = {}^{m}\mathbf{O} \cdot {}^{(m+1,l)}\mathbf{u}^a. \quad (1.28) $$

The connection between the above notation and notation of sections 1.1 and 2.2 is that in line with a tradition in microwave theory the quantities ${}^{m}\gamma_i = j\,{}^{m}\beta_i$ are considered in Sec. 2.2, and show up as diagonal elements of a diagonal matrix Γ. The 'slice' or section label m is omitted. Furthermore, for TE modes, *e.g.*, the vectors of mode amplitudes on each side of a section, ${}^{(m,l)}\mathbf{u}^s$ and ${}^{(m,r)}\mathbf{u}^s$, are denoted by $\overline{\mathbf{E}}_A$ and $\overline{\mathbf{E}}_B$ in Sec. 2.2, and the x-derivatives ${}^{(m,l)}\mathbf{u}^a$ and ${}^{(m,r)}\mathbf{u}^a$ are denoted by $jk_0\overline{\mathbf{H}}_A$ and $jk_0\overline{\mathbf{H}}_B$. Whereas the coupling matrix ${}^{m}\mathbf{O}$ for the BEP and the TRM is expressed as a matrix of overlap integrals between mode fields of neighbouring slices, for the MoL it is obtained via point matching of the field values obtained in the finite difference approximation used also to obtain the mode fields.

The BEP in the scalar approximation considered above is of very limited practical interest since it is valid only in structures where reflections tend to be negligible. Then unidirectional beam propagation methods discussed elsewhere in this chapter are applicable, and they are in general faster than mode propagation methods, as will be discussed below.

If we leave the scalar approximation and consider the vector case, as in Sec. 2.3, we have two possible polarisations, TE and TM. The coupling matrices ${}^{m}\mathbf{O}$ are no longer simple overlap integrals of film mode fields, and depend on the propagation constants ${}^{m}\beta_i$ of the film modes.

For scattering problems with normal incidence on the scatterer (the ridge in Fig. 1.6) TE and TM modes do not couple. With a skew incidence, however, TE-TM mode coupling does occur, and has to be introduced via cross coupling matrices that in addition to coupling the two polarisations also couple (1.27) and (1.28). With a skew incidence, the mode propagation vector of the incident film mode has a nonzero component k_z parallel to the scatterer (*e.g.*, along the ridge in Fig. 1.6). This component is conserved in transmission, scattering, and reflection, and gives rise to an $\exp(j\omega t - jk_z z)$ variation of the field instead of the $\exp(j\omega t)$ variation. Perpendicular to the scatterer, the component of the propagation vector of mode no. i in film no. m is then

$$ {}^{m}k_{xi} = \sqrt{{}^{m}\beta_i^2 - k_z^2}\,, \tag{1.29} $$

which replaces ${}^{m}\beta_i$ in (1.19) to (1.26). Furthermore, the coupling matrices ${}^{m}\mathbf{O}$ in (1.27) and (1.28) depend on both k_z and ${}^{m}k_{xi}$ (*i.e.*, on the angle of incidence).

The formalism just outlined for vectorial BEPM with skew incidence is identical to the one used for mode calculations with the TRM [15]. The scatterer (*e.g.*, the ridge of Fig. 1.6) is then considered a waveguide running in the z direction, k_z suddenly becomes the propagation constant of the waveguide mode, the effective index of the mode is k_z/k_0, the set of linear equations determining the mode amplitudes ${}^{(m,l)}u_i^s$ turns into a nonlinear eigenvalue problem for k_z, and suitable boundary conditions (*e.g.*, exponentially decaying film mode amplitudes in the leftmost ($m = 1$) and rightmost ($m = M$) slices) have to be imposed instead of the travelling wave boundary conditions of the scattering problem. The full vector field formalism for the TRM is presented in [15] and for the MoL in [18].

The most straightforward way of solving the scattering problem outlined above is with transfer matrices, as in [22], [23], and in Sec. 2.3. This approach works well for a localised scatterer like a narrow ridge, but, as discussed in [26], it is numerically unstable for extended structures with multiple scatterers separated by many wavelengths, like a wide ridge or a distributed Bragg reflector (DBR) structure. An approach that works with such structures is to define auxiliary single-sided 'scattering' matrices relating ${}^{(m,l)}\mathbf{u}^-$ and ${}^{(m,l)}\mathbf{u}^+$ (or equivalently, 'impedance' matrices relating ${}^{(m,l)}\mathbf{u}^a$ and ${}^{(m,l)}\mathbf{u}^s$) at each slice interface, as discussed in Sec. 2.3. A numerically stable nonlinear recursion relation connecting the scattering (or impedance) matrices of neighbouring interfaces can then be set up instead of the transfer matrices, as explained in detail in [16] and [19]. (The relevant equations are (2.24) and (2.25) of Sec. 2.2, or (18) of [16]). A completely different approach that works equally well is described in [27].

1.2.3 Numerical considerations

The rank (dimension) of the coupling matrices ${}^m\mathbf{O}$ in (1.27) is equal to the number of film modes I used to describe the field distribution in each slice (see Fig. 1.6). This number is roughly equal to the number p of polarisations considered (one or two) times the number of sampling points necessary to describe the field variation along a vertical line. For the strongly scattering structures typically analysed with the vectorial MMMs, two considerations indicate that a sampling density of several samples per optical wavelength (in the highest-index materials of the structure) is necessary to ensure a faithful representation of the optical field. (In a metal the relevant length scale is skin depth instead of wavelength.) Firstly, interference fringes may occur in the field distribution, and they may have a period as short as a wavelength for counter-propagating waves. Secondly, the electric field at a sharp edge, *e.g.*, as found along a rectangular waveguide or a rib, is known to diverge along the edge [28], [29]. The singularity in the field distribution may appear mainly within a fraction of a wavelength from the edge, as seen, *e.g.*, in [20] and [29]. If the Nyquist limit of 2 samples per wavelength is assumed as a minimum density, the total number of film modes necessary to describe the field faithfully is at least

$$I_{\text{min}} = 2p\, n_{\text{max}} w_y / \lambda_0 , \tag{1.30}$$

where n_{max} is the maximum of the absolute value of the refractive index in the structure, and w_y is the height of the 'computational window' considered in the calculation.

At the very heart of MMMs is the fact that the number of floating-point multiplications encountered scales as I^3, the cube of the number of modes considered, and hence roughly with the cube of the linear dimension of the region analysed. This scaling is manifest already in (1.18), if ${}^mF(x,y)$ is sampled within the whole 'computational window' of size $w_x \times w_y$, with the sampling density implied in (1.30). Furthermore, the computations that have to be performed with the MMMs are standard matrix manipulations like multiplication and inversion, solution of linear equation systems, and eigenvalue problems. In all of the matrix operations just mentioned, the number of floating-point multiplications in general scales as

I^3, the cube of the rank of the matrices [30]. Reliable matrix manipulation software is available in the public domain [31], and in the form of more user-friendly commercial packages like Matlab [32].

1.2.4 Discussion

On present-day desktop computers, matrices with a rank of a hundred can be multiplied in seconds, so that (1.30) implies that dielectric structures several microns high may be comfortably analysed on such computers using MMMs. Higher structures rapidly get impractical, because of the I^3 (rank cubed) dependence of the number of multiplications in matrix operations encountered. There are no corresponding limits to the horizontal extent of the structure that can be analysed, but of course a more complex structure will take longer to analyse, since the number of matrix multiplications encountered is proportional to the number of slices needed to model the structure.

One trivial but important consequence of the I^3 scaling is that fully vectorial calculations take about eight times as long as semi-vectorial or scalar calculations. According to (1.30), the size of the computations in the MMMs scale as the cube of the height w_y of the 'computational window.' Hence it is very important to limit this height, and bring the artificial top and bottom boundaries as close as possible to the structure of interest. The boundaries used in mode field calculations with the TRM have often been perfectly reflecting, like the artificial electric or magnetic 'walls' discussed, *e.g.*, in [15] and in section 2.3. Some distance to the 'walls' is needed, so that the evanescent mode fields extending into the waveguide cladding are negligible at the 'walls'. A significant improvement is the introduction of 'absorbing' boundary conditions [33], whereby the boundary can be moved much closer to the waveguide without disturbing the calculated mode field distributions significantly. Further improvements of this idea are reviewed and discussed in [34]. As discussed elsewhere in this chapter and in [35], in the scattering problems treated with the BEP or MoL-BPM, 'transparent' or 'absorbing' boundary conditions are even more important than in mode field calculations.

The vectorial MMMs discussed here are two-dimensional. Formally, the extension to three dimensions is straightforward, but this is in general not practical today (not even on a supercomputer) because of the large size of the matrices encountered. One exception is structures that can be reduced to two dimensions using co-ordinate systems other than the rectangular one. The MoL in cylindrical coordinates, *e.g.*, has yielded the radiation loss from curved waveguides [36], [37] and the eigenmodes of vertical-cavity surface emitting lasers (VCSELs) [38].

For purely non-absorbing structures of dielectric materials and with perfectly reflecting boundary conditions, no computations involving complex numbers are needed in the BEP and the TRM, the search for film mode indices can be performed along the real axis, and represents a computation that scales directly with the number of modes. The introduction of absorbing boundary conditions or film materials forces the search into the complex plane, and the size of the search then scales roughly as the square of the number of modes. For the MoL there is no such search, however, instead a linear eigenvalue problem for all the film modes is solved, a task that in general scales [30, p. 235] as the cube of the number of modes, or equivalently, the number of sampling points in the film mode field distributions. This scaling with the cube of the number of sampling points is a strong disincentive to trying, as proposed in [23], to improve the accuracy of the MoL by using a high sampling density for finding the film modes. Nevertheless, as already discussed above, the I^3 scaling is a general property of all MMMs.

Worth noting in this context is also the fact that the introduction of absorption into the structure by itself carries a penalty of about a factor of four in computer time, because a multiplication of two complex numbers involves four floating-point multiplications.

Finally, as discussed elsewhere in this chapter and as shown in [39], the number of multiplications in well-designed finite-difference method (FDM) calculations is directly proportional to the number of grid points, i.e., to the area of the 'computational window'. Hence one may conclude that FDMs in general will be faster than MMMs for sufficiently large structures, especially three-dimensional ones.

1.2.5 Conclusion

Numerical methods based on mode matching make accurate calculations of field distributions in optical devices accessible on desktop computers. For the calculations to be successful, a number of pitfalls have to be circumvented. Pitfalls that have often been overlooked are the possibility of analytic singularities in the vector field distributions and the inherent numerical instability of the popular transfer matrix formalism. Mode matching is only an option for problems that can be given a two-dimensional formulation; for truly three-dimensional problems finite-difference methods have greater potential.

1.3 Free space radiation mode method

In recent years the free space radiation mode (FSRM) method has been developed into a very accurate and computationally efficient analysis tool for buried optical waveguide structures. This section will briefly discuss its features, formulation and application to a variety of practical optical structures. The FSRM method is applicable to many structures that consist of a weakly guiding region surrounded by a homogeneous half space which supports a plane wave spectrum. The fields in the former region often decompose into a local guided mode and a radiation field. In principle, straightforward mode matching at the interface yields a solution to the problem if the radiation modes in the guiding region are available. Unfortunately, even if this is the case, the resultant analysis becomes numerically difficult. The philosophy of the FSRM method is to simplify the representation of the radiation field in the guiding region by assuming that it propagates in a medium of uniform refractive index and therefore can also be expressed as a superposition of plane waves. This approximation is found to yield excellent results if the index contrast of the guiding region is less than 10%, a common occurrence in many practical buried heterostructure components. Further details of the FSRM analysis and some results obtained are now presented for the modal properties of buried waveguides, the demanding problem of facet reflectivity and a more general propagation algorithm applied to a 3D waveguide taper.

1.3.1 Polarised and vectorial modal analysis of buried waveguides

Theory. Consider the problem of finding the propagation constant, β, and field profile of a single guided mode of the structure shown in Fig. 1.7 in which symmetry allows the analysis to be restricted to the half space $x > 0$.

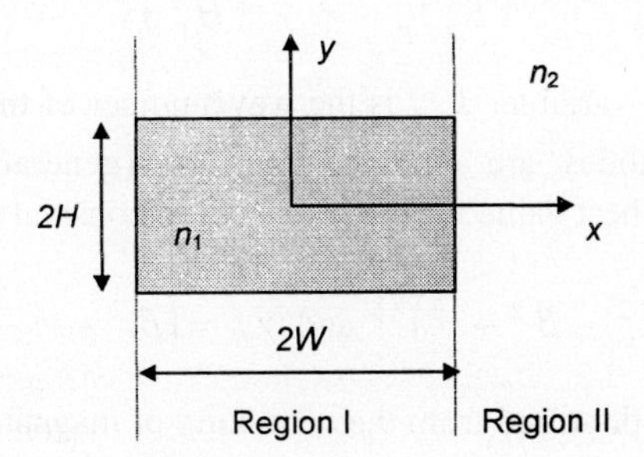

Fig. 1.7. Uniformly buried rectangular waveguide.

A detailed examination of the scalar case can be found in [40, 41] and here we derive the case for the TE dominant vector mode. The solution may be uniquely defined from the two transverse field components E_y and E_z, the remaining four field components may then be found directly from Maxwell's equations. The structure is divided into two regions; region I, which includes the waveguide core $(x \le W)$, and region II, which is the outer uniform cladding $(x \ge W)$. The field in region I consists of the guided mode of the three layer slab there, along with the FSRM radiation spectrum discussed above. The field in region II is exactly represented as a plane wave spectrum. Continuity of the tangential electric fields at $x = W$ requires that

$$\tilde{\mathbf{E}}_1^I + \tilde{\mathbf{E}}_R^I = \tilde{\mathbf{E}}^{II}, \tag{1.31}$$

where the subscripts 1 and R refer to the guided slab and radiation fields in region I and the tilde indicates the Fourier transform with respect to the y-co-ordinate.

Each term is a two-component vector $\begin{pmatrix} \tilde{E}_y \\ \tilde{E}_z \end{pmatrix}$.

It is straightforward to define from Maxwell equations 2×2 admittance matrices, $\mathbf{Y}_R^I$ and $\mathbf{Y}^{II}$

$$\mathbf{Y}_R^I = \frac{\cot(\gamma_{un} W)}{j\omega\mu\gamma_{un}} \begin{pmatrix} -\beta s & \gamma_{un}^2 + \beta^2 \\ -\gamma_{un}^2 - s^2 & \beta s \end{pmatrix}, \tag{1.32}$$

$$\mathbf{Y}^{II} = \frac{1}{j\omega\mu\gamma_{II}} \begin{pmatrix} -\beta s & -\gamma_{II}^2 + \beta^2 \\ \gamma_{II}^2 - s^2 & \beta s \end{pmatrix}, \tag{1.33}$$

such that

$$\begin{pmatrix} \tilde{H}_y^I \\ \tilde{H}_z^I \end{pmatrix}_R = \mathbf{Y}_R^I \cdot \begin{pmatrix} \tilde{E}_y^I \\ \tilde{E}_z^I \end{pmatrix}_R \quad \text{and} \quad \begin{pmatrix} \tilde{H}_y^{II} \\ \tilde{H}_z^{II} \end{pmatrix} = \mathbf{Y}^{II} \cdot \begin{pmatrix} \tilde{E}_y^{II} \\ \tilde{E}_z^{II} \end{pmatrix}, \tag{1.34}$$

where s is the Fourier variable, k_{un} is the wavenumber of the uniform medium in which the FSRM modes are assumed to exist (generally, it is found that $k_{un} = k_0 n_2$ gives the best value for the propagation constant),

$$\gamma_{un} = \left(k_{un}^2 - \beta^2 - s^2\right)^{1/2} \text{ and } \gamma_{II} = \left(\beta^2 + s^2 - k_2^2\right)^{1/2}.$$

The radiation field is identified from the continuity of magnetic fields at $x = W$,

$$\tilde{\mathbf{H}}_1^I + \mathbf{Y}_R^I \cdot \tilde{\mathbf{E}}_R^I = \mathbf{Y}^{II} \cdot (\tilde{\mathbf{E}}_1^I + \tilde{\mathbf{E}}_R^I), \tag{1.35}$$

giving

$$\tilde{\mathbf{E}}_R^I = (\mathbf{Y}_R^I - \mathbf{Y}^{II})^{-1} \cdot (\mathbf{Y}^{II} \cdot \tilde{\mathbf{E}}_1^I - \tilde{\mathbf{H}}_1^I). \tag{1.36}$$

The radiation and guided slab mode fields in region I must be orthogonal at $x = W$ and so the following transcendental equation for the propagation constant, β, is obtained:

$$\int_{-\infty}^{\infty} (\tilde{\mathbf{E}}_R^I \times \tilde{\mathbf{H}}_1^I + \tilde{\mathbf{E}}_1^I \times \tilde{\mathbf{H}}_R^I) \cdot \mathbf{x}^0 \, ds = 0. \tag{1.37}$$

For pure TE polarisation, the admittance matrices reduce to scalar quantities as there is no y component to the field. A similar analysis follows for the quasi-TM case. The general approach described here is easily extended to deal with multilayered structures including rib waveguides and results have also been determined for guides with loss or gain [42].

Example results. In Table 1.1, the normalised propagation constant, b, is tabulated against the normalised waveguide height for a rectangular waveguide with $W/H = 2$, $n_1 = 1.5$ and $n_2 = 1.45$, where b and h are defined by

$$b = \left(\beta^2 - k_2^2\right)/\left(k_1^2 - k_2^2\right), \quad h = 4H\sqrt{n_1^2 - n_2^2}/\lambda. \tag{1.38}$$

Results obtained from the FSRM method for the scalar, polarised and vectorial cases are compared with those from a finite difference (FD) method [43]. It can be seen that the results are in excellent agreement and it should be noted that the FSRM results can be obtained in less than 5 seconds on a PC whereas each of the FD results takes several hours. Field profiles are readily available from the FSRM method and Fig. 1.8 shows the electric field components of a typical quasi-TE vectorial mode.

	h	0.4	0.5	0.6	0.7	0.8	0.9	1.0
Scalar	FSRM	0.0360	0.1116	0.2046	0.2961	0.3787	0.4507	0.5124
	FD	0.0361	0.1118	0.2048	0.2962	0.3789	0.4508	0.5125
x-polarised (quasi-TE)	Polarised FSRM	0.0337	0.1071	0.1991	0.2907	0.3739	0.4465	0.5089
	Polarised FD	0.0334	0.1069	0.1991	0.2908	0.3740	0.4466	0.5090
	Vectorial FSRM	0.0332	0.1065	0.1987	0.2905	0.3738	0.4464	0.5088
	Vectorial FD	0.0334	0.1068	0.1990	0.2907	0.3739	0.4465	0.5089
y-polarised (quasi-TM)	Polarised FSRM	0.0303	0.1001	0.1898	0.2803	0.3634	0.4364	0.4995
	Polarised FD	0.0305	0.1004	0.1900	0.2806	0.3635	0.4365	0.4996
	Vectorial FSRM	0.0303	0.1001	0.1878	0.2804	0.3634	0.4364	0.4995
	Vectorial FD	0.0306	0.1005	0.1901	0.2806	0.3636	0.4366	0.4997

Table 1.1. Normalised propagation constant, b, of rectangular waveguide with $W/H = 2$, $n_1 = 1.5$, $n_2 = 1.45$ and $\lambda = 1.15\mu$m.

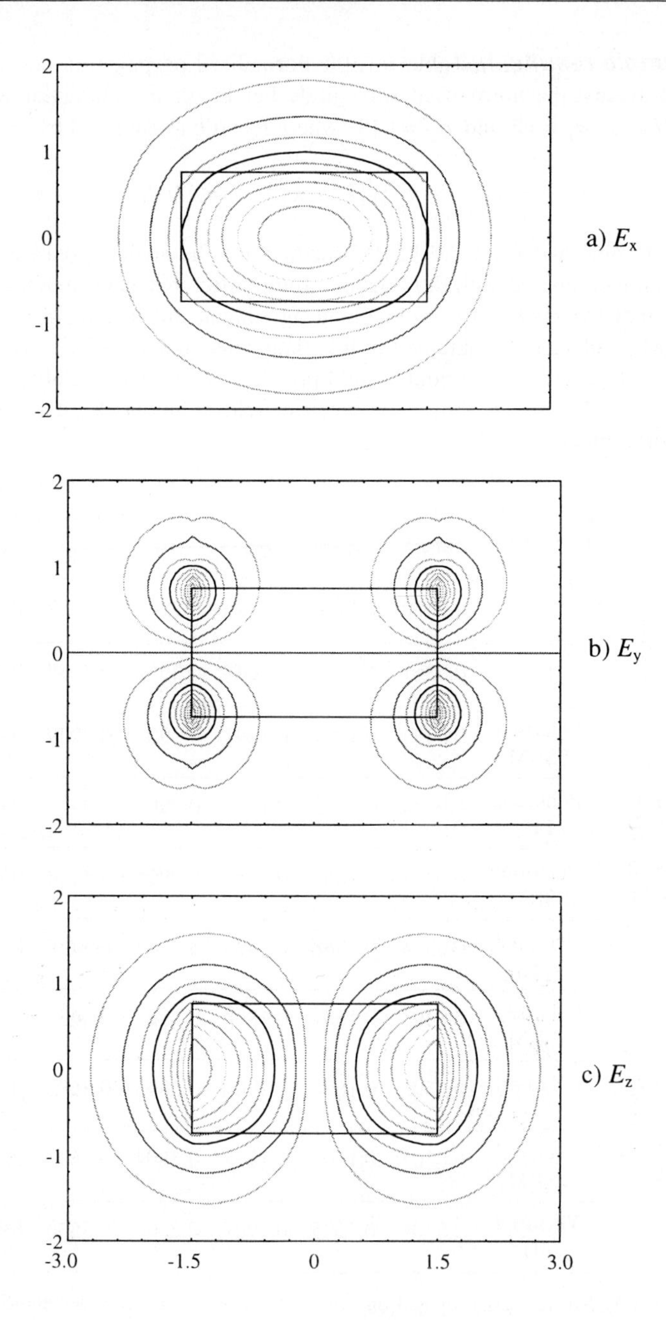

Fig. 1.8. Electric field components of the TE dominated vectorial mode of the rectangular waveguide with $n_1 = 1.5$, $n_2 = 1.45$ and a normalised height of $h = 1.0$.

1.3.2 Waveguide facet reflectivities

The calculation of waveguide facet reflectivity has received much interest due to the application in the design of semiconductor amplifiers. The FSRM method has proved most useful in facet reflection calculations for both angled and normal incidence and anti-reflection (AR) coatings are also easily incorporated into the model. The method was originally applied to the case of a slab waveguide incident upon a normal facet [44] but has since been extended to allow for the two dimensional cross-section of the incident waveguide.

Polarised facet reflectivities considering one-dimensional waveguides. This analysis proceeds as per the modal analysis above, except that the field in the region containing the core consists of incident and reflected guided slab modes and a reflected FSRM radiation spectrum. Angling the facet is allowed for by orientating the direction of propagation of the reflected radiation spectrum perpendicularly to the facet. Imposing continuity at the facet followed by the orthogonality condition yields the following expression for the reflectivity [45]

$$R = \frac{\int_{-\infty}^{\infty} \frac{(\beta \sec\theta + s\tan\theta - f\Gamma)}{(\gamma_{un} + f\Gamma)} \tilde{\psi}_I^{+} \tilde{\psi}_I^{-*} ds}{\int_{-\infty}^{\infty} \frac{(\beta \sec\theta - s\tan\theta + f\Gamma)}{(\gamma_{un} + f\Gamma)} \tilde{\psi}_I^{-} \tilde{\psi}_I^{-*} ds}, \tag{1.39}$$

where $\gamma_{un} = \sqrt{k_{un}^2 - s^2}$, $\Gamma = \sqrt{k_0^2 - s^2}$, $\tilde{\psi}_I$ is the Fourier transformed TE or TM guided slab mode and the $\pm$ indicates either forward or backward going modes projected onto the angled facet. β is the guided slab mode propagation constant, θ is the angle of incidence and $f = 1$ (TE mode) or $(\beta^2 + s^2)/k_0^2$ (TM mode). When AR coatings are deposited on the end of the facet, Γ is replaced by the appropriate plane wave response function [44, 46]. Numerical evaluation of equation (1.39) can be achieved in a matter of seconds on a PC.

Polarised facet reflectivities considering two dimensional waveguides
In this section an expression for the reflectivity of a quasi-TE mode supported by a rectangular waveguide normally incident onto a facet is derived. The expression for TM mode incidence can be obtained in a similar fashion.

The two dimensional Fourier transform of the dominant field components of the incident and reflected TE mode are H_y and E_x and these can be found as described in subsection 1.3.1. The radiation fields are expressed as a superposition of plane waves where s and t are the Fourier variables

$$H_y^{rad} = \begin{cases} f(s,t)\exp(j\gamma_b z), & z \le 0, \\ g(s,t)\exp(-j\Gamma z), & z \ge 0. \end{cases} \tag{1.40}$$

Due to the plane wave representation, the transforms of the radiated electric field are given by

$$E_x^{rad} = \begin{cases} \dfrac{-\omega\mu\gamma_b}{(k_b^2 - s^2)} H_y^{rad} = -Y_B H_y^{rad}, & z \le 0, \\ \dfrac{\omega\mu\Gamma}{(k_0^2 - s^2)} H_y^{rad} = Y_F H_y^{rad}, & z \ge 0. \end{cases} \tag{1.41}$$

The boundary conditions are now applied at the facet, that is, continuity of H_y and E_x is required:

$$(1+R)H_y + f = g\,, \tag{1.42}$$

$$(1-R)E_x - Y_B f = Y_F g\,. \tag{1.43}$$

Eliminating g yields an expression for the reflected radiation field

$$f = \frac{-R(E_x + Y_F H_y) + (E_x - Y_F H_y)}{(Y_B + Y_F)}\,. \tag{1.44}$$

The requirement that this radiation field is orthogonal to the guided slab mode field is

$$\int_{-\infty}^{\infty}\int_{-\infty}^{\infty} H_y^{rad} E_x \, dsdt = 0 \tag{1.45}$$

and upon substituting (1.44) this gives the TE mode reflection coefficient as:

$$R_{TE} = \frac{\displaystyle\int_{-\infty}^{\infty}\int_{-\infty}^{\infty} \frac{(E_x - Y_F H_y)}{(Y_B + Y_F)} E_x d\,sdt}{\displaystyle\int_{-\infty}^{\infty}\int_{-\infty}^{\infty} \frac{(E_x + Y_F H_y)}{(Y_B + Y_F)} E_x d\,sdt}\,. \tag{1.46}$$

For the TM mode the dominant field components are E_y and H_x and a similar analysis gives

$$R_{TM} = \frac{\displaystyle\int_{-\infty}^{\infty}\int_{-\infty}^{\infty} \frac{(H_x + Y_F E_y)}{(Y_B + Y_F)} H_x dsdt}{\displaystyle\int_{-\infty}^{\infty}\int_{-\infty}^{\infty} \frac{(H_x - Y_F E_y)}{(Y_B + Y_F)} H_x dsdt}\,, \tag{1.47}$$

where

$$H_x^{rad} = \begin{cases} \dfrac{\omega\varepsilon_b\gamma_b}{(k_b^2 - s^2)} E_y^{rad} = Y_B E_y^{rad} \ , \ z \le 0, \\ \dfrac{-\omega\varepsilon_o\Gamma}{(k_0^2 - s^2)} E_y^{rad} = -Y_F E_y^{rad} \ , \ z \ge 0 . \end{cases} \tag{1.48}$$

Vectorial facet reflectivities. Evaluation of the reflection coefficients of fully vectorial modes incident upon 2D facets has also been achieved. The approach takes into account the cross-polarisation coupling and higher order mode coupling at angled facets due to the minor field components and loss of symmetry [46].

Example results. Fig. 1.9 shows an example of the variation in reflectivity with facet angle for both the TE and TM cases. The incident waveguide has core and cladding indices of 3.6 and 3.492 respectively, the guide width being 0.1 μm and the operating wavelength 0.86 μm. By comparing with a BPM [47] formulation it can clearly be seen that good agreement is obtained.

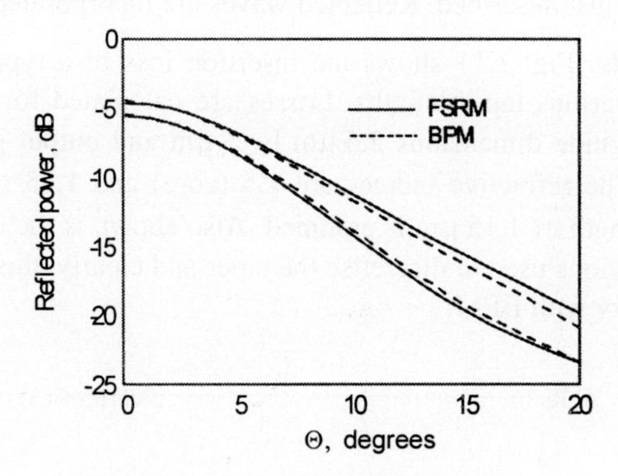

Fig. 1.9. Comparison of FSRM and BPM for an angled facet.

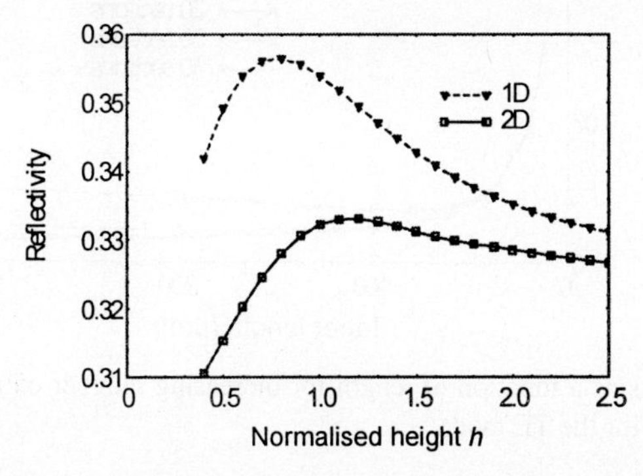

Fig. 1.10. Comparison of 1D and 2D TE mode reflectivity for a buried rectangular waveguide.

Fig. 1.10 demonstrates the need to correctly account for the two dimensional nature of practical waveguide facets. TE mode reflectivity is plotted as a function of normalised height for a buried rectangular waveguide with core and cladding indices 3.6 and 3.492 respectively and $W = 2H$. It can be seen that the results of the 1D formulation differ from those of the more realistic 2D formulation [48].

1.3.3 Propagation in 3D structures

The FSRM method has recently been extended to the analysis of propagation in three dimensional structures. The approach is based upon the two dimensional FSRM propagation method of Smartt et al [49]. The method has successfully been applied to the analysis of 3D step discontinuities, tapers and air gaps for both quasi-TE and quasi-TM polarisations [41,50], and has been shown to be highly efficient in comparison with BPM. The general approach proceeds as a sequence of interacting waveguide steps, each of which is analysed in a similar manner to the facet problem just described. Reflected waves are incorporated in the analysis.

Example results. Fig. 1.11 shows the insertion loss of a typical three dimensional taper for various taper lengths. Losses are calculated for a linear stepped taper with input guide dimensions 2.5 μm by 5 μm and output guide dimensions 1 μm by 2 μm. The refractive indices are 1.5 (core) and 1.45 (cladding) and an operating wavelength of 1.15 μm is assumed. Also shown is the convergence with the number of sections used to discretise the taper and clearly, this is far fewer than would be necessary with BPM.

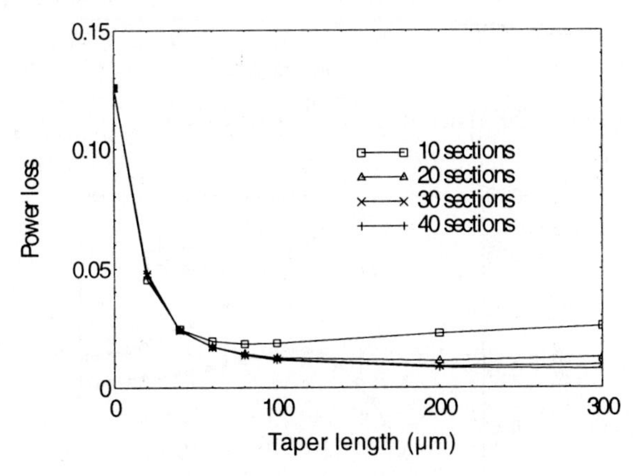

Fig. 1.11. Taper loss as a function of length for increasing number of sections in the staircase approximation for the TE mode.

Conclusions. The FSRM method has been successfully applied to the study of a wide variety of practical buried structures within the field of integrated optics.

Compared with the more intensive numerical techniques, the method has been shown to yield results with a high degree of accuracy with minimum computational effort. Its computational efficiency makes the FSRM method an ideal tool for the design engineer working within an iterative design environment.

References

[1] W. P. Huang, ed., "Methods for Modeling and Simulation of Guided-Wave Optoelectronic Devices Part II: Waves and Interactions". PIER 11 in *Progress in Electromagnetic Research*, EMW Publishing, Cambridge, Massachusetts, USA, 1995.

[2] W. P. Huang, ed., "Methods for Modeling and Simulation of Guided-Wave Optoelectronic Devices: Part I: Modes and Couplings", PIER 10 in *Progress in Electromagnetic Research*, EMW Publishing, Cambridge, Massachusetts, USA, 1995.

[3] T. Itoh, ed., *Numerical Techniques for Microwave and Millimeter Wave Passive Structures*. J. Wiley Publishing, New York, 1989.

[4] C. Vassallo, "1993–1995 Optical mode solvers", *Opt. Quantum Electron.*, vol. 29, No. 2, pp. 95–114, 1997.

[5] R. E. Collin, *Field Theory of Guided Waves*, IEEE Press, 2nd ed., pp. 419, 1990.

[6] G. Kowalski and R. Pregla, "Dispersion Characteristics of Shielded Microstrips with Finite Thickness", *AEÜ*, vol. 25, pp. 193–196, 1971.

[7] E. Kühn, "A Mode Matching Method for Solving Field Problems in Waveguide and Resonator Circuits", *AEÜ*, vol. 27, pp. 511–518, 1973.

[8] R. Pregla, "The Method of Lines for the Unified Analysis of Microstrip and Dielectric Waveguides", *Electromagnetics*, vol. 15, pp. 441–456, 1995.

[9] R. Pregla and W. Pascher, "The Method of Lines", in T. Itoh, ed., *Numerical Techniques for Microwave and Millimeter-Wave Passive Structures*, J. Wiley Publ., New York, pp. 381–446, 1989.

[10] R. Pregla, "The Method of Lines as Generalized Transmission Line Technique for the An alysis of Multilayered Structures", *AEÜ*, vol. 50, No. 5, pp. 293–300, 1996.

[11] R. Sorrentino, "Transverse resonance technique," Ch. 11 in Itoh's book [3].

[12] W. Schlosser and H.G. Unger, "Partially filled waveguides and surface waveguides of rectangular cross-section", in *Advances in Microwaves*, pp. 319–387, Academic Press, New York, 1966.

[13] D. Marcuse, *Theory of Dielectric Optical Waveguides*, second edition, Academic Press, San Diego, 1991.

[14] S. T. Peng and A. A. Oliner, "Guidance and leakage properties of a class of open dielectric waveguides: Part I – Mathematical formulations," *IEEE Trans. Microwave Theory Tech.*, vol. MTT-29, pp. 843–855, 1981.

[15] A. S. Sudbø, "Film mode matching: A versatile method for mode field calculations in dielectric waveguides," *Pure Appl. Opt. (J. Europ. Opt. Soc. A)*, vol. 2, pp. 211–233, 1993.

[16] A. S. Sudbø, "Improved formulation of the film mode matching method for mode field calculations in dielectric waveguides," *Pure Appl. Opt. (J. Europ. Opt. Soc. A)*, vol. 3, pp. 381–388, 1994.

[17] Available from Photon Design, 86 Courtland Road, Oxford OX4 4JB, UK, fax +44 1865 395480, email dfgg@photond.com.
[18] R. Pregla and W. Pascher, *The method of lines*, Ch. 6 in Itoh's book [3].
[19] U. Rogge and R. Pregla, "Method of Lines for the analysis of dielectric waveguides," J. Lightwave Technol., vol. 11, pp. 2015–2020, 1993.
[20] A. S. Sudbø, "Problems in vector mode calculations for dielectric waveguides," *Linear and Nonlinear Integrated Optics*, SPIE Europto Series Proceedings, vol. 2212, pp. 26–35, 1994.
[21] M. S. Stern, P. C. Kendall, and P. W. A. McIlroy, "Analysis of the spectral index method for vector modes of rib waveguides," *IEE Proc. J*, vol. 137, pp. 21–26, 1990.
[22] G. Sztefka and H.P. Nolting, "Bidirectional eigenmode propagation for large refractive index steps," *IEEE Photonics Technol. Lett.*, vol. 5, pp. 554–557, 1993.
[23] J. J. Gerdes, "Bidirectional eigenmode propagation analysis of optical waveguides based on the method of lines," *Electron. Lett.*, vol. 30, pp. 550–551, 1994.
[24] R. Pregla, "MoL-BPM Method of Lines Based Beam Propagation Method, Methods for Modeling and Simulation of Guided-Wave Optoelectronic Devices," W.P. Huang (ed), PIER 11, *Progress in Electromagnetic Research*, EMW Publishing, Cambridge, Massachusetts, USA, pp. 51–102, 1995.
[25] The name 'angular repetency' for k_0 is an ISO standard that has been adopted to get around the conflicting traditions that 'propagation constant' is k_0 in optics, whereas it is jk_0 for microwaves, and 'wave number' is k_0 in optics, whereas it is $k_0/(2\pi)$ in spectroscopy.
[26] A. S. Sudbø and P.I. Jensen, "Stable bidirectional eigenmode propagation of optical fields in waveguide devices," *Integrated Photonics Research*, OSA Topical Meeting, Dana Point, CA, Feb. 23–25, 1995, paper IThB4.
[27] A. S. Sudbø, "Numerically Stable Formulation of the Transverse Resonance Method for Vector Mode-Field Calculations in Dielectric Waveguides," *IEEE Photonics Technol. Lett.*, vol. 5, pp. 342–344, 1993.
[28] J. van Bladel, *Singular electromagnetic fields and sources*, Ch. 4, Clarendon Press, Oxford, 1991.
[29] A. S. Sudbø, "Why are accurate computations of mode fields in rectangular dielectric waveguides difficult?" *J. Lightwave Technol.*, vol. 10, pp. 418–419, 1992.
[30] G. Golub and C.F. Van Loan, *Matrix Computations*, Johns Hopkins University Press, Baltimore 1983.
[31] E. Anderson, Z. Bai, C. Bischof, J. Demmel, J. Dongarra, J. Du Croz, A. Greenbaum, S. Hammarling, A. McKenney, S. Ostrouchov, and D. Sorensen, *LAPACK Users' Guide*, Society for Industrial and Applied Mathemathics (SIAM), Philadelphia 1992.
[32] Matlab® is available from The MathWorks Inc., 24 Prime Park Way, Natick, MA 01760-1500, USA, fax +1 508-647-7001, email info@mathworks.com.
[33] A. Dreher and R. Pregla, "Analysis of planar waveguides with the method of lines and absorbing boundary conditions," *IEEE Microwave Guided Wave Lett.*, vol. 1, pp. 138–140, 1991.
[34] C. Vassallo and J.M. van der Keur, "Comparison of a few transparent boundary conditions for finite-difference optical mode solvers," *J. Lightwave Technol.*, vol. 15, pp. 397–402, 1997

[35] C. Vassallo and F. Collino, "Highly efficient absorbing boundary conditions for the beam propagation method," *J. Lightwave Technol.*, vol. 14, pp. 1570–1577, 1996.
[36] J. S. Gu, P.A. Besse, and H. Melchior, "Method of Lines for the analysis of the propagation characteristics of curved optical rib waveguides," *IEEE J. Quantum Electron.*, vol. 27, pp. 531–537, 1991, with corrections in vol. 28, pp. 1835–1836, 1992
[37] R. S. Burton and T. E. Schlesinger, "Comparative analysis of the Method-of-Lines for three-dimensional curved dielectric waveguides," *J. Lightwave Technol.*, vol. 14, pp. 209–215, 1996.
[38] R. Pregla, E. Ahlers, and S. Helfert, "Efficient and accurate analysis of photonic devices with the Method of Lines," *Progress in Electromagnetic Research Symposium* (PIERS), Hong Kong, Jan. 6–9, 1997.
[39] K. Ramm, P. Lüsse, and H. G. Unger, "Multigrid eigenvalue solver for mode calculation of planar optical waveguides," *IEEE Photonics Technol. Lett.*, vol. 9, pp. 967–969, 1997.
[40] M. Reed, T. M. Benson, P. Sewell, P. C. Kendall, G. M. Berry, and S. V. Dewar: "Free space radiation mode analysis of rectangular dielectric waveguides", *Opt. Quantum Electron.*, vol. 28, pp. 1175–1179, 1996.
[41] M. Reed, PhD Thesis, University of Nottingham, 1998.
[42] P. Sewell, M. Reed, T. M. Benson, P. C. Kendall, and M. Noureddine, "Computationally efficient analysis of buried rectangular and rib waveguides with applications to semiconductor lasers", *IEE Proc. Optoelectron.*, vol. 144, No. 1, pp. 14–18, 1997.
[43] Finite difference results provided by S. Sujecki.
[44] P. C. Kendall, D. A. Roberts, P. N. Robson, M. J. Adams, and M. J. Robertson, "Semiconductor laser facet reflectivities using free space radiation modes" *IEE Proc. J*, vol. 140, No. 1, pp. 49–55, 1993.
[45] M. Reed, T. M. Benson, P. C. Kendall, and P. Sewell, "Antireflection-coated angled facet design" *IEE Proc. Optoelectron.*, vol. 143, No. 4, pp. 214–220, 1996.
[46] P. Sewell, M. Reed, T. M. Benson, and P. C. Kendall, "Full vector analysis of two dimensional angled and coated optical waveguide facets", *IEEE J. Quantum Electron.*, vol. 33, No. 11, 1997.
[47] P. Kaczmarski, R. Baets, G. Franssens, and P. E. Lagasse, "Extension of bidirectional BPM method to TM polarisation and application to laser facet reflectivity", *Electron. Lett.*, No. 25, pp. 716–717, 1989.
[48] M. Reed, P. Sewell, T. M. Benson, and P. C. Kendall, "Limitations of one dimensional models of waveguide facets" *Microwave Opt. Technol. Lett.*, vol. 15, No. 4, pp. 196–198, 1997.
[49] C. J. Smartt, T. M. Benson, and P. C. Kendall, "Free space radiation mode method for the analysis of propagation in optical waveguide devices", *IEE Proc. J*, vol. 140, pp. 56–61, 1993.
[50] M. Reed, P. Sewell, T. M. Benson, and P. C. Kendall, "An efficient propagation algorithm for 3D optical waveguides", IEE Special Issue on Semiconductor Optoelectronics, February 1998.

2 Beam propagation methods

R. Pregla, W. von Reden, H. J. W. M. Hoekstra, H. V. Baghdasaryan

In the research of optical devices and circuits, a principal theoretical problem is to calculate how a lightwave propagates in an optical medium having an arbitrary refractive-index distribution. Optical waveguide devices are usually very long compared to their transversal dimensions: the ratio is typically of the order of a thousand. Therefore, a rigorous analysis is difficult or not possible at all.

In most applications, however, we need not a "marching-in-time" procedure; instead, we are interested only in the final stationary time-harmonic light field imposed on the structure by a boundary condition, namely the stationary input field. This situation can be simulated numerically and the analysis can be performed by the so called *beam propagation methods* (BPM). Assuming a wave is propagating in the longitudinal +z direction of the optical waveguide structure and assuming the stationary field distribution is known in the cross section plane A ($z = z_k$), the task is to determine the stationary field at the plane B ($z = z_k + \Delta z = z_{k+1}$). Repeating this calculation stepwise from the input to the output, the whole device can be approximately analysed.

In the classical BPM version based on the fast Fourier transformation (FFT) [1], [2], [3], two restrictions are assumed: reflected waves can be neglected, and the refractive-index differences in the optical medium are small. Even under conditions where the restricting assumption of small refractive-index differences is not completely satisfied, the classical BPM yields surprisingly accurate results. Because of its inherent computational stability, it allows to use relatively large step sizes Δz which reduces the computation time.

Nevertheless, the classical FFT-based BPM algorithm tends to fail for high-contrast step-index profiles, such that appear, for example, in semiconductor optical circuits based on the InGaAsP/InP material system. Here the size of propagation steps must be considerably reduced and/or the refractive-index profile itself must be smoothed out. Moreover, polarisation effects and reflected optical waves cannot be handled within the framework of the classical BPM. In recent years, a great variety of numerical techniques have been developed in order to overcome these drawbacks [4]. The basic principles of some of them — especially for 2D structures — are summarised and their mutual dependencies are visualised in the diagram in Fig. 2.1. The extension to 3D and also to vectorial algorithms is straightforward. The propagation is generally described by the Helmholtz equation $\overline{\Delta}\phi + n^2(x, y, z)\phi = 0$, where $\phi = E_y$ and $\phi = H_y$ for TE and TM polarisations, respectively. The co-ordinates are normalised with respect to the free space wave number (angular repetency) $k_0 = 2\pi/\lambda$ according to $\bar{\mathbf{r}} = k_0\mathbf{r}$. For this reason also the corresponding transverse differential operators are written with bars. In the TM case the transverse operator is of the Sturm-Liouville type (see also section 2.2) and H_y can also be normalised to get symmetrical matrices in the lossless case. In most algorithms the operators R or Q and the field functions ϕ or ψ (see Fig. 2.1) are discretised resulting in a matrix and a vector, respectively. The formal solution of the Helmholtz equation has the form of a superposition of a forward and a backward propagating fields. Most of the BPM-algorithms can work with only one of these two terms, *e,g.*, with the forward propagation waves only. Therefore, the problem is to extract this part from the general differential equation. The common way is to re-write the Helmholtz equation into the slowly varying envelope (SVE) form by introducing the product ansatz of a basic propagator and the SVE function ψ. This equation can be discretised for forward propagation [5]. However, in the standard FD-BPM the second order differential quotient is neglected, or this differential equation is factorised and the part (factor) describing the forward propagation is extracted. Series expansion of the square root operator up to the first order term leads to the paraxial approximation or Fresnel equation. No higher-order terms should be introduced in the expansion to avoid unphysical solutions. As can be understood from its name, this approximation gives good results only if the field is propagating nearly parallel to the waveguide axis. Wide-angle solutions are obtained by re-writing the SVE equation in such a way that a recursive relation can be constructed [6]. From this relation Padé approximations of the exact Helmholtz forward operator equation are obtained. To obtain the field ψ^{k+1} at $z_{k+1} = z_k + \Delta z$ from the field ψ^k at z_k, the differential operator $d\psi/dz$ is replaced by the two-point difference operator. The result of this operation is valid on an intermediate position between z_k and z_{k+1}. Therefore, the right side should also be constructed at this position. This is approximately fulfilled by using the arithmetic mean value of the discretised fields at z_k and z_{k+1}. This approach leads to the well known Crank-Nicolson discretisation scheme. In the explicit or implicit schemes, on the right side the discretised field at z_k or z_{k+1} is introduced. The explicit scheme results in unstable solutions if the propagation step is too large [7]. The implicit scheme is unconditionally stable but shows numerical loss. In case of the wide-angle scheme the polynomials are factorised. The solutions to the Crank-Nicolson scheme for the nth order Padé approximation can be decomposed into an n-step algorithm. In each partial step a unitary matrix equation with tri-diagonal matrices has to be solved. The algorithm is unconditionally stable [8]. Instead of using finite differences in the cross section, Q can also be discretised by finite elements [9]. A different way to obtain an algebraic approximation for forward propagation is to construct a series expansion of the propagator for the Fresnel approximation [10], or for the forward propagator in the formal solution [11], which should be split into one term in the numerator and another one in the denominator. The algebraic equations can be written in a form analogous to the Crank-Nicolson scheme.

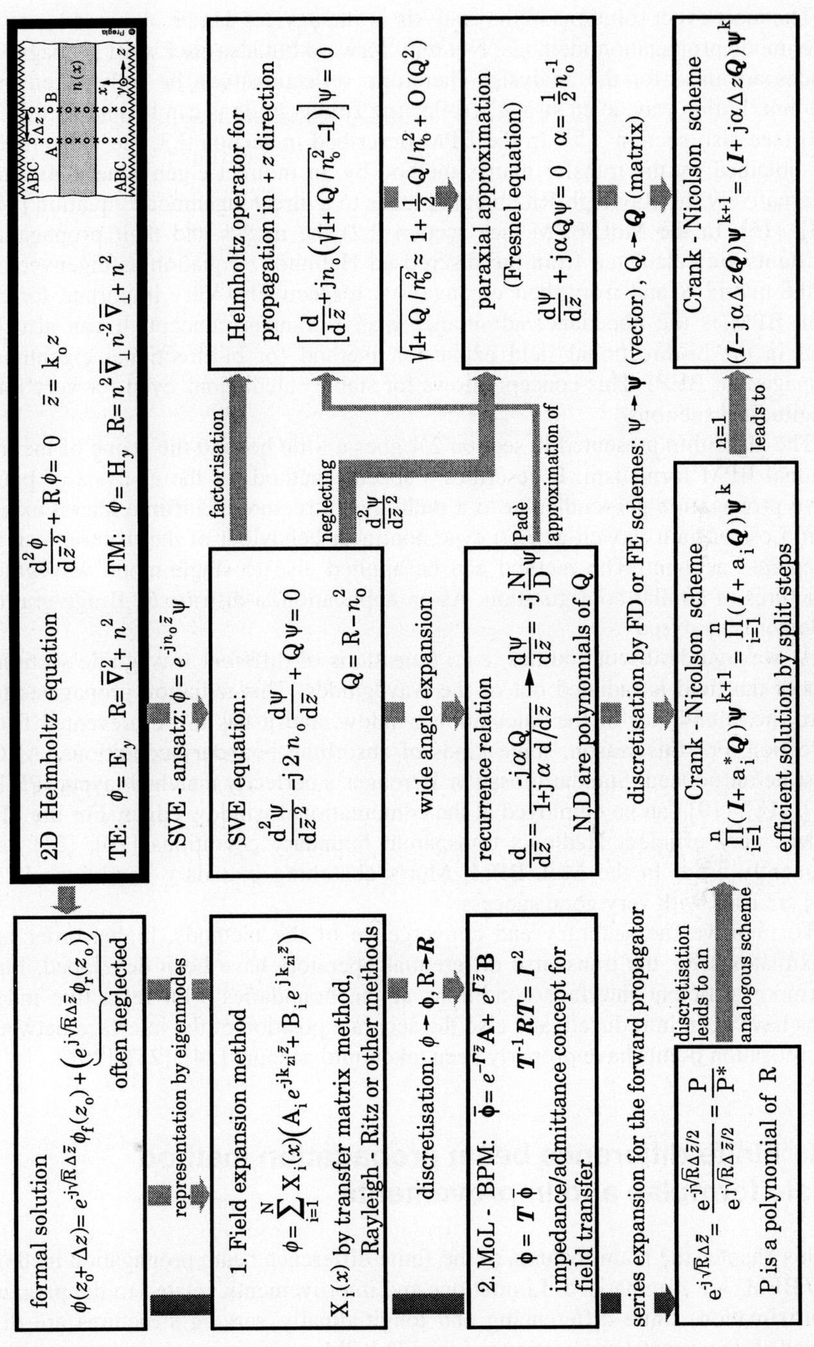

Fig. 2.1. Graphical representation of mutual dependencies of different beam propagation methods.

The most exact solutions in the analysis of the devices can be obtained by using eigenmode propagation methods. Not only forward but also backward propagating modes are used for the analysis. Therefore, reflections can be fully taken into account. In this way, even strongly reflecting Bragg gratings can be analysed [12], [13] (see also section 3.5). In the BPM described in section 2.3, the eigenmodes are obtained by the transfer matrix-method by an indirect eigenvalue algorithm. Alternatively the Rayleigh-Ritz method leads to a direct eigenmode equation [14], [15], [16]. In the MoL-BPM (see section 2.2) the modes and their propagation constants are calculated from the discretised Helmholtz equation as eigenvectors of the matrix $\boldsymbol{R}$ and from their eigenvalues, respectively. Very important for the MoL-BPM is the impedance/admittance matrix transfer concept. It can also be used in the bi-directional field expansion method (or bi-directional eigenmode propagation: BEP). This concept allows for stable calculations even for very long longitudinal sections.

The algorithm presented in section 2.4 goes a little beyond the scope of the traditional BPM formalism. It describes a special method for the analysis of plane wave propagation perpendicular to a multilayer structure of infinite lateral extension. Loss or gain as well as Kerr-type nonlinear behaviour of the material can be taken into account. The method can be applied also to single-mode waveguide structures of similar configuration. As an application, a distributed Bragg-grating reflector is analysed.

At waveguide discontinuities, *e.g.,* transitions of different waveguide sections, part of the field is radiated out of the waveguides. This radiation propagates toward the boundaries of the calculation window, and it has to be prevented from reflection. For this reason, some kinds of absorbing boundary conditions (ABC) must be introduced. Standard lossy or Berenger's perfectly matched layers (PML) [17], [18], [19] can be employed at the computational window edges. For the FD-BPMs, very efficient Hadley's transparent boundary conditions [20], [21], are frequently used. In the MoL-BPM, Mur's absorbing boundary conditions [22], [23] are used with very good success.

To improve the accuracy and convergence of the methods, higher-order approximations for the transverse differential operators have been developed. Furthermore, efficient interface conditions at the boundaries with refractive index steps have been introduced, and also the accurate position of the interface between discretisation points have properly been taken into account [24], [25], [26].

2.1 Finite difference beam propagation method basic formulae and improvements

In this chapter the main features of the finite difference beam propagation method (FD BPM) are summarised. Limitations and improvements, related to the paraxial approximation, finite differencing and longitudinally varying structures are discussed. A few typical applications of the FD BPM are given.

BPM's are aimed at the computing the behaviour of light in structures for which analytical solutions are not known, *e.g.*, structures nonuniform in the propagation direction such as tapers, Y-junctions, and in optical nonlinear devices. For the design of all kinds of optical functions, computation speed is often an important quantity, and for this reason methods like the FD BPM are based on a number of assumptions. Fortunately, these assumptions are often met in good approximation but the designer should always be aware of them. When doing calculations on a certain class of devices the reliability of the method should be tested from time to time, *e.g.*, by calculations of structures similar to those for which the solution is well known.

Recently, a number of improvements have been introduced into the FD BPM, increasing either the applicability or the computational speed, or both. In this chapter the main features of the FD BPM, including recent improvements, are described. We will concentrate mainly on expressions for the continuous wave (CW) case in two dimensions (2D).

2.1.1 Finite difference BPM equations in 2D

Assuming a time dependence of $\exp(j\omega t)$ it follows from Maxwell equations that the wave equations in 2D structures ($\partial/\partial y \equiv 0$) are given by:

$$(\partial^2/\partial z^2 + Q_1^{(\prime)} + k_0^2 n^2)\phi = 0, \tag{2.1}$$

where for TE polarisation $\phi \equiv E_y$, $Q_1 \equiv \partial^2/\partial x^2$, for TM polarisation $\phi \equiv H_y$, $Q_1' \equiv \partial^2/\partial x^2 - (\partial/\partial x)\ln n^2\, \partial/\partial x$.

As uni-directional propagation is considered it is advantageous to introduce the slowly varying envelope (SVE):

$$\begin{aligned} &\text{TE:} \quad \psi = \exp(jk_0 n_0 z) E_y, \\ &\text{TM:} \quad \psi = \exp(jk_0 n_0 z) n^{-1} H_y, \qquad Q_1 \equiv n^{-1} Q_1' n. \end{aligned} \tag{2.2}$$

It can be proved that for square-integrable fields and for real refractive indices the operators Q_1 given above are Hermitean for both TE and TM polarisations, while Q_1' is not Hermitean. The substitution of (2.2) into (2.1) leads to:

$$(\partial^2/\partial z^2 - ja\,\partial/\partial z + Q_1 + Q_2)\psi = 0, \tag{2.3}$$

with $a = 2k_0 n_0$, $\quad Q_2 = k_0^2(n^2 - n_0^2)$,
which may be rewritten, neglecting any z-dependence of the structure, as:

$$\left\{\partial/\partial z - ja\left(1 \pm \sqrt{1 + 4Q/a^2}\right)/2\right\}\psi = 0, \qquad Q = Q_1 + Q_2. \tag{2.4}$$

Equation (2.4) decouples the forward (– sign; positive z-axis) and the backward (+ sign) travelling waves, which is, strictly speaking, only allowed for z-independent structures.

In its standard form the following paraxial approximation, or SVE approximation (SVEA), is used [27]:

$$\sqrt{1+4Q/a^2} = 1+2Q/a^2 + O(Q^2). \tag{2.5}$$

For computations the transversal co-ordinate is discretised. The operator Q is then replaced by a matrix $\mathbf{M}$, where the second order derivative is given in its simplest form by a three point FD operator $(1 \quad -2 \quad 1)$, and the field ψ is from now on treated as a vector consisting of components corresponding to the values of ψ at the discretisation points. Considering propagation along the positive z-axis, it follows that

$$\psi(z_0+\Delta z) = \exp(-j\mathbf{M}\Delta z/a)\psi(z_0). \tag{2.6}$$

For sufficiently small step sizes Δz (2.6) leads to

$$\mathbf{P}_+\psi(z_0+\Delta z) = \mathbf{P}_-\psi(z_0), \quad \mathbf{P}_\pm \equiv \mathbf{I} \pm j\mathbf{M}\Delta z/2a\,. \tag{2.7}$$

Eq. (2.7) is the basic equation for standard FD BPM [27]. In the next section we will discuss its applicability.

In many structures radiation towards the computational window edges may occur. Many different boundary conditions have been suggested and are applied in nowadays computer codes [28, 29]. The most robust one, also applicable in the more sophisticated implementations discussed below, is based on the perfectly matched layers or the Berenger layer boundary condition [30].

2.1.2 Applicability

This section concentrates on the effects of the above assumptions on the applicability of the FD BPM and discusses recent improvements. Considered are the effects of finite differencing, the SVEA, and the assumption of z-independence of the structure for each propagation step, Δz.

For finite differencing one often uses the three-point FD operator

$$\partial^2/\partial x^2 \to \delta_x^2 + O(\Delta x^2) \tag{2.8}$$

with

$$\delta_x^2\psi_p = (\psi_{p+1} - 2\psi_p + \psi_{p-1})/\Delta x^2.$$

For a single plane wave in a uniform part of 2D space this means that

$$\delta_x^2 \exp(jk_x x) = -k_x^2 \exp(jk_x x)\left(1 - k_x^2\Delta x^2/12\right), \tag{2.9}$$

i.e., the error introduced this way may become large for high spatial frequencies along the x-axis. Furthermore, the error becomes severe for discretisation points next to interfaces, as the three-point operator presumes smooth fields. But, at interfaces there are discontinuities for $\partial^2 E_y / \partial x^2$ (TE), $\partial H_y / \partial x$ (TM), and higher-order derivatives. Without any precautions the error here is (*e.g.*, for TE):

$$\partial^2 \psi / \partial x^2 \rightarrow \delta_x^2 \psi_{\text{int}} + O(1). \tag{2.10}$$

As the error thus introduced occurs only for a limited number of points, the corresponding error in the mode index will be $O(\Delta x)$, so we loose one order in accuracy. Taking into account the proper continuity condition (*e.g.*, it can be shown for TM from (2.1) that $(1/n^2)\partial H_y / \partial y$ is continuous), improved (symmetric) FD operators can be formulated that restore the accuracy of $O(\Delta x^2)$ in the mode index [31, 32]. The adapted operator $\delta_{x,\text{int}}^2$ may be derived using Taylor expansions of the field left and right of the interface, taking into account the proper relations between the fields and their derivatives. The application of efficient interface conditions (EIC) results in higher accuracy, almost without any increase in computational effort, and makes the result almost independent of the position of interfaces with respect to the grid points.

Extension of the 3-point FD operator to a 5-point operator would increase the accuracy considerably, without too much extra computational effort. If EICs are used well here the error in the mode index would become $O(\Delta x^4)$.

The SVEA leads to an error in the phase velocity. Considering a modal field ψ_e and its effective index n_e belonging to the matrix **M**, it follows that

$$\mathbf{M}\psi_e = k_0^2\left(n_e^2 - n_0^2\right)\psi_e. \tag{2.11}$$

Using this in the basic equation (2.7), the error in the modal index, Δn_e, can be found:

$$\psi_e(z_0 + \Delta z) = (\mathbf{P}_+ / \mathbf{P}_-)\psi_e(z_0) = \exp(j\theta\,\Delta z)\psi_e(z_0),$$

with $\theta = k_0(n_0 - n_e - \Delta n_e)$. Substitution of (2.11) leads to

$$\Delta n_e / n_0 = \delta^2/2 - 2k'^2\,\delta^3/3 + \ldots \tag{2.12}$$

with $\delta = (n_e - n_0)/n_0$; $k' = k_0 n_0 \Delta z$. So, the spread in the modal indices, or equivalently, the allowed total propagation length, is limited even if $\Delta z \rightarrow 0$. In the above we assumed real indices, leading to a real modal index and to the conservation of the norm, so $\partial|\psi|^2/\partial z = 0$. It is straightforward to show that the expansion above holds also for complex modal indices.

Improvements of the SVEA can be obtained by the extension of the expansions for (2.5) [33, 34, 35, 36] and for the exponent in (2.6). Doing so, the error is strongly reduced For example, if an mth-order correction is used in the scheme

$\mathbf{Q}_{+}\psi(z_0+\Delta z)=\mathbf{Q}_{-}(z_0)$, where $\mathbf{Q}_{\pm}$ contains the matrix $\mathbf{M}$ to the power of m, the error is of the order of $\Delta n_e = n_0 O(\delta^{2m+1}+\ldots)$. Note that the error for the first-order case, according to (2.12), can be improved considerably [36, 37].

The z-dependence for, *e.g.*, a tilted waveguide is usually taken into account by approximating the index with a staircase function, with z-independent segments of the length Δz. In principle this should lead for small enough step sizes to a correct result, as may be shown using arguments from the Green's function theory. The error is introduced by simply taking the end field of a segment as the input for the next segment, neglecting the step index along the z-axis. Benchmark tests [38] on tilted waveguides, in which one mode is launched, show that tilt angles up to 10°–20° can be handled rather well by most BPM's. Here corrections to the SVEA, as discussed above, show a considerable improvement with respect to more paraxial BPM's. This indicates that, in terms of the local (z-independent) waveguide, there is a certain spread in the effective indices. All the methods presented in [38] show at least a small unphysical power loss. Things become even worse if reflection on a tilted interface is considered [39]. Here a severe error in power can be expected as the norm $|\psi|^2$ is conserved and not the power.

In bi-directional BPM's [40], the reflection and transmission at such index steps are usually treated much more accurately, but, on the other hand, such methods are quite time consuming due to either the iterations or the need to handle large matrices. For structures with no or minor back reflection (in the real world) like tilted waveguides or Y-junctions, respectively, an efficient option could be to neglect the multiple reflections, by using a kind of intermediate iterative scheme, such as the Bremmer series [41] up to and including the first order. In other words, at each index step along the z-axis transmission and reflection is calculated correctly, assuming only the forward incoming wave. For a mode in a tilted waveguide or for reflection on a tilted interface this should work well, at least for sufficiently small step sizes Δz and Δx.

The above improvements can also be applied in 3D FD BPM and in time domain BPM's (TD BPM). The 2D BPM can be used not only in 2D cases, but also if the y-dependence of the field does not change too much, such as in Y-junctions and tapers consisting of shallow ridge waveguides. Here the reduction of dimensionality is usually obtained by the well-known effective index method. If the latter is not allowed the 2D version can only be used to find trends in the behaviour of such devices, but for more precise calculations 3D versions are required. Here one may choose among scalar, semi-vectorial or full-vectorial implementations [42] depending on the type of structure under study. Here, in particular, the norm conservation or power conservation is still a topic of research.

The TD BPM has two important advantages if compared to CW implementations. In the first place, uni-directional methods are no longer approximate, as light moves forward in time. Nevertheless, time dependent effects play a role due to dispersion, i.e. a time dependent (linear or nonlinear) susceptibility, and also probably due to moving objects (Doppler effect). The second advantage is that, for a pulse as an input, a certain wavelength *region* is covered by one calculation, at least in the linear case. In the nonlinear case, transient effects can be studied in quite a straightforward way [43]. The price to be paid is the required computation time introduced by the additional dimension.

2.1.3 Applications

In general, BPM's can be applied most effectively to structures for which no analytical solutions exist, or as a rigorous check of approximate analytical results. Numerous applications of all kinds of BPM's can be found in the literature: to structures containing tapers, gratings, Y-junctions *etc.*, to nonlinear materials, *e.g.*, for second harmonic generation (SHG) or to all-optical effects, *e.g.*, soliton propagation. As an example, the result of FD BPM modelling of a waveguiding structure with a sudden transition in the width (from 3 to 20 μm) as used in multimode interference couplers is shown in Fig. 2.2.

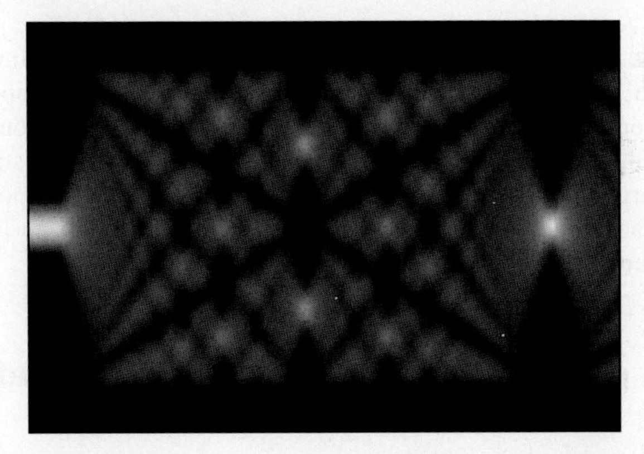

Fig. 2.2. FD BPM calculations on a sudden transition from a monomode to a multimode waveguide.

High accuracy of the phase velocity was obtained by the application of a third-order SVEA correction. For nonlinear materials, among others, the FD BPM is quite suitable. For third-order materials (with light intensity-dependent refractive index) the effect is simply taken into account by adapting the local index at each mesh point. In the case of the second harmonic generation (SHG) the relatively weak coupling between the two frequencies, ω and 2ω, can be accounted for by a split-step method [44]. Hereby, after the linear propagation of the two waves along Δz, the nonlinear coupling is taken into account. As it is usually small, the split-step approximation, having an error $O(\Delta z^3)$, converges rapidly. In Fig. 2.3. we present the result of calculations of almost phase-matched SHG, where the main interaction is between a zero-order mode at ω and a first-order mode at 2ω.

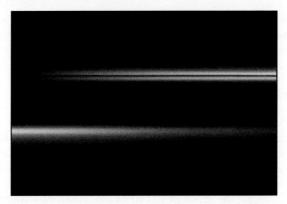

Fig. 2.3. SHG, the coupling is between a zero-order mode at ω (lower part window) and the first-order mode at 2ω (upper part window).

Conclusions. In this chapter we have given a short introduction into the field of FD BPM. The 2D case has been discussed with respect to its applicability and recent improvements in more detail. These improvements have considerably extended the applicability and reliability of the method. The field is still the subject of active research, concentrating on numerical efficiency, on incorporating effects of tilted interfaces like in longitudinally variant structures, and on improvements in finite differencing, in particular close to interfaces.

2.2 Beam propagation method based on the method of lines

In this section the method of lines (MoL) [45] will be described as a versatile tool for the analysis of optical waveguides and components. The MoL is known to have a monotonous convergence behaviour; the phenomenon of a relative convergence and spurious mode solutions are absent. The principles will be demonstrated for the analysis of film waveguides (see Fig. 2.4.).

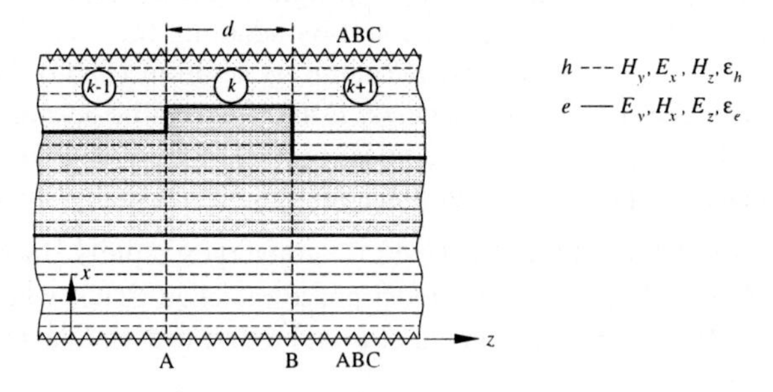

Fig. 2.4. Concatenation of film waveguide sections with discretisation lines.

The MoL has also been used in many other areas of electromagnetic theory, *e.g.*, in a microwave and millimetre wave technique, antenna theory, electronic and photonic devices [45], [46], [47], [48]. Even nonlinear problems have been solved using the MoL [49]. The MoL is closely connected with the mode matching method (MMM) and with the finite difference method (FDM), as it has been explained in Sec. 1.2. As in the MMM, eigensolutions are used to describe the field. These eigensolutions are not obtained by series expansions but by discretisation. In contrast to the FDM the discretisation is not done completely but only as long as necessary. The discretisation is used to transform the partial differential equations to ordinary ones which are then solved analytically. 3D devices can be analysed (even vectorially [50], [51]) by two-dimensional discretisation [45], [52]. They may have an arbitrary composition in the direction of discretisation. The material parameters may have also anisotropic or gyrotropic properties [53], [54]. The MoL has the advantages of the FDM in the discretisation directions. Discontinuities in the analytical direction can be taken into account with high accuracy with the help of a specially developed impedance/admittance transfer. In many photonic components, more or less radiation can take place. By introducing absorbing boundary conditions (ABC) it is easy to take this radiation into account. As yet, radiating structures cannot be accurately modelled by the MMM in a finite domain since the application of Dirichlet or Neumann boundary conditions leads to reflection of radiation from the boundary.

To examine special devices, some extensions have been introduced in the MoL algorithm. Computing along lines of different lengths allows the analysis of sharp bends [55]. Concatenating identical sharp bends with small angles results in an algorithm for examining curved waveguide bends [50], [51], [56], [57]. Recently, the formulas for studying such curved bends using cylindrical coordinates have been derived [58].

When modelling taper structures, introducing cylindrical co-ordinates further reduces the necessary numerical effort in comparison to the calculation in cartesian co-ordinates [59], [60]. An effective method to keep the numerical effort low when analysing periodic structures which consist of a very high ($\approx 10^4$) number of periods is shown in [12], [61]. The extension of the finite difference formulas for the case of thin metal layers is presented in this section. Several COST 240 modelling tasks which will be described in detail in Ch. 3, *e.g.*, waveguide tapers (Sec. 3.3), an electro-optic modulator based on surface plasmons (Sec. 3.4), and waveguide Bragg-grating filter (Sec. 3.5) were analysed using these extensions.

2.2.1 Theory

The film waveguides can be inhomogeneous in the propagation and transverse directions (see Fig. 2.4.). The complete fields are superpositions of the fields of TE and TM modes. All field components are y-independent. The TM and TE modes are obtained from the components H_y and E_y, respectively, which are the solutions of the wave equations (we write the equations for TM (TE) modes on left (right) side).

TM–modes TE–modes

$$\frac{\partial^2 \widetilde{H}_y}{\partial \bar{z}^2} + \varepsilon_r \frac{\partial}{\partial \bar{x}}\left(\varepsilon_r^{-1} \frac{\partial \widetilde{H}_y}{\partial \bar{x}}\right) + \varepsilon_r \widetilde{H}_y = 0, \qquad \frac{\partial^2 E_y}{\partial \bar{z}^2} + \frac{\partial^2 E_y}{\partial \bar{x}^2} + \varepsilon_r E_y = 0. \tag{2.13}$$

The following normalisations have been introduced: $\bar{x} = k_0 x$, $\bar{z} = k_0 z$, $\widetilde{H}_y = \eta_0 H_y$, where $k_0 = \omega\sqrt{\varepsilon_0 \mu_0}$ and $\eta_0 = \sqrt{\mu_0 / \varepsilon_0}$ are the free space wavenumber (angular repetency) and wave impedance, respectively. The remaining field components are obtained by the differentiation,

$$\begin{aligned} \varepsilon_r E_x &= j \frac{\partial}{\partial \bar{z}} \widetilde{H}_y, & \widetilde{H}_x &= -j \frac{\partial}{\partial \bar{z}} E_y, \\ \varepsilon_r E_z &= -j \frac{\partial}{\partial \bar{x}} \widetilde{H}_y, & \widetilde{H}_z &= j \frac{\partial}{\partial \bar{x}} E_y. \end{aligned} \tag{2.14}$$

To solve the partial differential equations (2.13) by the method of lines, the equations are converted to ordinary differential equations by means of discretisation. We assume that the wave propagates in z-direction and therefore we introduce the discretisation in the x-direction. The discretisation lines are sketched in Fig. 2.4. We use two line systems in both TM and TE case [45], [52]. The discretisation yields

$$\begin{aligned} &\widetilde{H}_y \to \widetilde{\mathbf{H}}_y, & &E_y \to \mathbf{E}_y, \\ &\varepsilon_r \to \boldsymbol{\varepsilon}_h, \boldsymbol{\varepsilon}_e, & &\varepsilon_r \to \boldsymbol{\varepsilon}_e, \boldsymbol{\varepsilon}_h, \\ &\frac{\partial \widetilde{H}_y}{\partial \bar{x}} \to \bar{h}^{-1} \boldsymbol{D}_h \mathbf{H}_y = \overline{\boldsymbol{D}}_h \widetilde{\mathbf{H}}_y, & &\frac{\partial E_y}{\partial \bar{x}} \to \bar{h}^{-1} \boldsymbol{D}_e \mathbf{E}_y = \overline{\boldsymbol{D}}_e \mathbf{E}_y. \end{aligned} \tag{2.15}$$

The values of the field components on the discretisation lines are collected in column vectors, the values of discretised permittivities in diagonal matrices. The differential operators are approximated by difference operator matrices $\boldsymbol{D}$. The subscripts correspond to the component (*e.g.*, $\widetilde{H}_y \to$ h-line system: subscript h) from which all the field components are obtained. h is the discretisation step size and $\bar{h} = k_0 h$. For the TM case the permittivity $\varepsilon_r(x)$ must be discretised on both line systems. The difference matrices have the following form [52]:

$$D = \begin{bmatrix} 1 & & & \\ -1 & 1 & & \\ & \ddots & \ddots & \\ & & -1 & 1 \\ & & & -1 \end{bmatrix}, \quad D^{a} = \begin{bmatrix} a_1 & -b_1 & -c_1 & & \\ -1 & 1 & & & \\ & \ddots & \ddots & & \\ & & & -1 & 1 \\ & & c_N & b_N & -a_N \end{bmatrix}. \tag{2.16}$$

The first and last lines depend on the boundary conditions. $\boldsymbol{D}$ is written for the Dirichlet boundary conditions (E_y -component in Fig. 2.1.). If Neumann boundary conditions have to be fulfilled on the lower (upper) side of the structure, the first (last) row in $\boldsymbol{D}$ must be cancelled. $\boldsymbol{D}^{a}$ is the difference operator for absorbing boundary conditions (ABC). This operator is obtained from the operator for Dirichlet boundary conditions by replacing the first and/or last row by the coefficients *a, b, c* [52]. In the discretised form the wave equations (2.13) run

$$\frac{d^2\widetilde{\mathbf{H}}_y}{d\bar{z}^2} - \left(\varepsilon_{\mathrm{h}} \overline{\boldsymbol{D}}_{\mathrm{h}}^{t} \varepsilon_{\mathrm{e}}^{-1} \overline{\boldsymbol{D}}_{\mathrm{h}}^{(\mathrm{a})} - \varepsilon_{\mathrm{h}}\right) \widetilde{\mathbf{H}}_y = 0, \quad \frac{d^2\mathbf{E}_y}{d\bar{z}^2} - \left(\overline{\boldsymbol{D}}_{\mathrm{e}}^{t} \overline{\boldsymbol{D}}_{\mathrm{e}}^{(\mathrm{a})} - \varepsilon_{\mathrm{e}}\right) \mathbf{E}_y = \mathbf{0}. \tag{2.17}$$

Since the difference operators are not diagonal, the transformation to principal axes is necessary: $\mathbf{E}_y = \boldsymbol{T}_e \overline{\mathbf{E}}_y$, $\mathbf{H}_y = \boldsymbol{T}_h \overline{\mathbf{H}}_y$, and

$$T_{\mathrm{h}}^{-1}\left(\varepsilon_{\mathrm{h}} \overline{\boldsymbol{D}}_{\mathrm{h}}^{t} \varepsilon_{\mathrm{e}}^{-1} \overline{\boldsymbol{D}}_{\mathrm{h}}^{(a)} - \varepsilon_{\mathrm{h}}\right) T_{\mathrm{h}} = \Gamma_{\mathrm{h}}^{2}, \quad T_{\mathrm{e}}^{-1}\left(\overline{\boldsymbol{D}}_{\mathrm{e}}^{t} \overline{\boldsymbol{D}}_{\mathrm{e}}^{(a)} - \varepsilon_{\mathrm{e}}\right) T_{\mathrm{e}} = \Gamma_{\mathrm{e}}^{2}, \tag{2.18}$$

where Γ_{h}^2, Γ_{e}^2 are diagonal matrices.

If ABC's are used, $\overline{\boldsymbol{D}}^{t}$ is the transposed matrix of matrix $\overline{\boldsymbol{D}}$ in Eq. (2.16). The solution for the transformed fields may be written as follows

$$\begin{bmatrix} \overline{\mathbf{E}}_{\mathrm{nA}} \\ \overline{\mathbf{E}}_{\mathrm{nB}} \end{bmatrix} = \begin{bmatrix} \bar{z}_1 & \bar{z}_2 \\ \bar{z}_2 & \bar{z}_1 \end{bmatrix}_{|\mathrm{e}} \begin{bmatrix} \overline{\mathbf{H}}_{\mathrm{A}} \\ -\overline{\mathbf{H}}_{\mathrm{B}} \end{bmatrix}, \quad \begin{bmatrix} \overline{\mathbf{E}}_{\mathrm{A}} \\ \overline{\mathbf{E}}_{\mathrm{B}} \end{bmatrix} = \begin{bmatrix} \bar{z}_1 & \bar{z}_2 \\ \bar{z}_2 & \bar{z}_1 \end{bmatrix}_{|\mathrm{h}} \begin{bmatrix} -\overline{\mathbf{H}}_{\mathrm{A}} \\ \overline{\mathbf{H}}_{\mathrm{B}} \end{bmatrix}, \tag{2.19}$$

where $\mathbf{E}_{\mathrm{n}} = \varepsilon_{\mathrm{h}} \mathbf{E}_x$, $\overline{\mathbf{E}}_{\mathrm{n}} = \boldsymbol{T}_{\mathrm{h}}^{-1} \mathbf{E}_n$,

$$\bar{z}_{1\mathrm{e,h}} = \overline{\boldsymbol{Z}}_{0\mathrm{e,h}} \left(\tanh(\Gamma_{\mathrm{e,h}} \bar{d})\right)^{-1}, \quad \bar{z}_{2\mathrm{e,h}} = \overline{\boldsymbol{Z}}_{0\mathrm{e,h}} \left(\sinh(\Gamma_{\mathrm{e,h}} \bar{d})\right)^{-1}, \tag{2.20}$$

and $\bar{d} = k_0 d$ is the normalised length of the waveguide between planes A and B normalised. $\overline{\boldsymbol{Z}}_0$ are the characteristic impedance matrices given by

$$\overline{\boldsymbol{Z}}_{0\mathrm{h}} = -j\Gamma_{\mathrm{h}}, \quad \overline{\boldsymbol{Z}}_{0\mathrm{e}} = j\Gamma_{\mathrm{e}}^{-1}. \tag{2.21}$$

The subscripts A and B in (2.19) denote that the fields are taken at the cross-sections A and B inside the section *k*:

$$\overline{\mathbf{E}}_{\mathrm{nA,B}} = T_{\mathrm{h}}^{-1}\left(\varepsilon_{\mathrm{h}}\mathbf{E}_x(z_{\mathrm{A,B}})\right), \qquad \overline{\mathbf{E}}_{\mathrm{A,B}} = T_{\mathrm{e}}^{-1}\mathbf{E}_y(z_{\mathrm{A,B}}),$$
$$\overline{\mathbf{H}}_{\mathrm{A,B}} = T_{\mathrm{h}}^{-1}\mathbf{H}_y(z_{\mathrm{A,B}}), \qquad \overline{\mathbf{H}}_{\mathrm{A,B}} = T_{\mathrm{e}}^{-1}\mathbf{H}_x(z_{\mathrm{A,B}}). \tag{2.22}$$

If we define impedance matrices in the cross-sections A, B inside section k according to

$$\overline{\mathbf{E}}_{\mathrm{nA,B}} = \overline{\mathbf{Z}}_{\mathrm{A,B}}\overline{\mathbf{H}}_{\mathrm{A,B}}, \qquad \overline{\mathbf{E}}_{\mathrm{A,B}} = \overline{\mathbf{Z}}_{\mathrm{A,B}}(-\overline{\mathbf{H}}_{\mathrm{A,B}}), \tag{2.23}$$

the impedance transfer obtained from Eq. (2.19) is given by

$$\overline{\boldsymbol{Z}}_{\mathrm{A}} = \overline{z}_1 - \overline{z}_2(\overline{z}_1 + \overline{\boldsymbol{Z}}_{\mathrm{B}})^{-1}\overline{z}_2\,. \tag{2.24}$$

The impedance/admittance transfer concept was first described in [45] and the analogous equation to Eq. (2.24) was given in the admittance form in [62]. Later this concept was adapted for the film mode matching (FMM) in [63]. Only by using this concept, the numerical difficulties described in [64] could be avoided. This concept is very important. It allows for the analysis of very complex waveguides [65] and waveguide circuits. For the MoL it was also described in other co-ordinate systems [66], [67]. Eq. (2.24) gives automatically the correct results even for very long waveguide sections. If $d \to \infty$ the matrix $\overline{z}_2$ approaches zero and $\overline{z}_1$ approaches the characteristic impedance matrix $\overline{\boldsymbol{Z}}_0$.

At the concatenation of different sections (*e.g.*, sections k and $k+1$ in plane B) the fields and therefore the impedances must be matched. This yields

$$\overline{\mathbf{Z}}_{\mathrm{B}}^{k} = \overline{\mathbf{T}}_{k}^{-1}\nu_k\mathbf{T}_{k+1}\overline{\mathbf{Z}}_{\mathrm{B}}^{k+1}\mathbf{T}_{k+1}^{-1}\mathbf{T}_k, \tag{2.25}$$

where all the quantities have a sub- or superscript either e or h, $\nu_{\mathrm{h},k} = \varepsilon_{\mathrm{h},k}\varepsilon_{\mathrm{h},k+1}^{-1}$, and $\nu_{\mathrm{e},k} = \boldsymbol{I}$ (identity matrix). Now a complete analysis of a complex structure is possible. We start at the output of the structure and transfer the load impedance through the various sections using Eq. (2.24) and Eq. (2.25) at each concatenation. If we have an infinitely long waveguide at the output (interface B), the load impedance matrix reduces to the characteristic impedance matrix (2.24) of the infinite waveguide.

At the input of the structure we obtain the total field as the reflected and transmitted wave from the input impedance and the incident wave. Using all the calculated impedances the fields can be computed at each cross-section of the structure. The correct procedure is described in [47], [65].

The derived formulas can also be used for a simpler propagation algorithm. From Eqs. (2.17) and (2.18) the solution (in homogeneous sections) for the propagation in a z-direction is given by

$$\mathbf{H}_y(z) = T_{\mathrm{h}}e^{-\Gamma_{\mathrm{h}}\overline{\overline{z}}}\overline{\mathbf{H}}_y(0), \qquad \mathbf{E}_y(z) = T_{\mathrm{e}}e^{-\Gamma_{\mathrm{e}}\overline{\overline{z}}}\overline{\mathbf{E}}_y(0). \tag{2.26}$$

At the concatenations the matching process results in the following transmitted field

$$\overline{\mathbf{H}}_{k+1} = 2\left(\boldsymbol{T}_k^{-1}\boldsymbol{T}_{k+1} + \overline{\boldsymbol{Z}}_{0,k}^{-1}\boldsymbol{T}_k^{-1}\boldsymbol{v}_k\boldsymbol{T}_{k+1}\overline{\boldsymbol{Z}}_{0,k+1}\right)^{-1}\overline{\mathbf{H}}_k\,, \tag{2.27}$$

$$\overline{\mathbf{E}}_{k+1} = 2\left(\boldsymbol{T}_k^{-1}\boldsymbol{T}_{k+1} + \overline{\boldsymbol{Y}}_{0,k}^{-1}\boldsymbol{T}_k^{-1}\boldsymbol{t}_{k+1}\overline{\boldsymbol{Y}}_{0,k+1}\right)^{-1}\overline{\mathbf{E}}_k\,. \tag{2.28}$$

However, multiple reflections are neglected in this simplified algorithm.

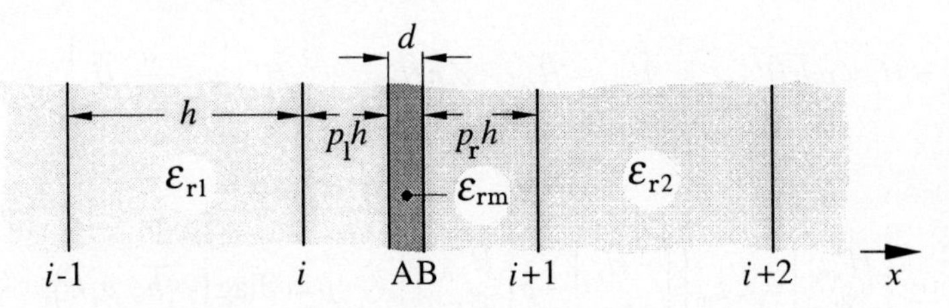

Fig. 2.5. Discretisation in case of thin metallic films.

2.2.2 Special case: very thin layers

Now we consider the case of a metallic film of conductivity κ and finite thickness d smaller than the distance h between the discretisation lines in the structure (see Fig. 2.5), which is important for the analysis of the structure studied in section 3.3. The position of the metallic film with respect to the discretisation lines i and i+1 is defined by p_r and p_l; $(p_\mathrm{r} + p_\mathrm{l} + d/h = 1)$. The field behaviour in the metal is described by the complex permittivity $\varepsilon_\mathrm{rm} = \varepsilon_\mathrm{r} - j\kappa\,\eta_0/k_0$. Therefore the following relation holds between the field and its derivatives at planes A and B inside the metal

$$\frac{d}{d\overline{x}}\begin{bmatrix}\widetilde{H}_{y\mathrm{AM}}\\ \widetilde{H}_{y\mathrm{BM}}\end{bmatrix} = \hat{\mathbf{z}}_\mathrm{m}\begin{bmatrix}-\widetilde{H}_{y\mathrm{A}}\\ \widetilde{H}_{y\mathrm{B}}\end{bmatrix}, \qquad \hat{\mathbf{z}}_\mathrm{m} = \begin{bmatrix} z_{1\mathrm{m}} & z_{2\mathrm{m}}\\ z_{2\mathrm{m}} & z_{1\mathrm{m}}\end{bmatrix}, \tag{2.29}$$

where $z_{1\mathrm{m}} = \gamma_\mathrm{m}\left[\tanh\left(\gamma_\mathrm{m}\overline{d}\right)\right]^{-1}$, $z_{2\mathrm{m}} = \gamma_\mathrm{m}\left[\sinh\left(\gamma_\mathrm{m}\overline{d}\right)\right]^{-1}$, $\gamma_\mathrm{m} = j\sqrt{\varepsilon_\mathrm{rm} - \varepsilon_\mathrm{re}}$ and $\overline{d} = k_0 d$ denotes here the metal film thickness normalised with k_0. ε_re is the effective permittivity describing the propagation in z direction. At the interfaces of both dielectrics with relative permittivities ε_r1 and ε_r2 the conditions run

$$\left.\frac{1}{\varepsilon_\mathrm{r1}}\frac{dH_y}{d\overline{x}}\right|_\mathrm{A} = \left.\frac{1}{\varepsilon_\mathrm{rm}}\frac{dH_y}{d\overline{x}}\right|_\mathrm{AM}, \qquad \left.\frac{1}{\varepsilon_\mathrm{r2}}\frac{dH_y}{d\overline{x}}\right|_\mathrm{B} = \left.\frac{1}{\varepsilon_\mathrm{rm}}\frac{dH_y}{d\overline{x}}\right|_\mathrm{BM}, \tag{2.30}$$

The derivatives on the dielectric sides are therefore given by

$$\frac{d}{d\bar{x}}\begin{bmatrix}\widetilde{H}_{y\mathrm{A}}\\ \widetilde{H}_{y\mathrm{B}}\end{bmatrix}=\hat{\varepsilon}\,\hat{z}_{\mathrm{m}}\begin{bmatrix}\widetilde{H}_{\mathrm{A}}\\ \widetilde{H}_{\mathrm{B}}\end{bmatrix},\qquad \hat{\varepsilon}=\mathrm{diag}\left(\frac{\varepsilon_{\mathrm{r}1}}{\varepsilon_{\mathrm{rm}}},\frac{\varepsilon_{\mathrm{r}2}}{\varepsilon_{\mathrm{rm}}}\right). \tag{2.31}$$

The values of $\widetilde{H}_{y\mathrm{A}}$ and $\widetilde{H}_{y\mathrm{B}}$ have to be calculated from the values of the neighbouring discretisation points i and $i + 1$, respectively. We assume as a first approximation that the derivatives at i and A and at B and $i+1$ have the same value. Therefore we obtain

$$\widetilde{H}_{y\mathrm{A}}\approx\widetilde{H}_i+p_\mathrm{l}\bar{h}\widetilde{H}'_{y\mathrm{A}},\qquad \widetilde{H}_{y\mathrm{B}}\approx\widetilde{H}_{i+1}-p_\mathrm{r}\bar{h}\widetilde{H}'_{y\mathrm{B}},\qquad \widetilde{H}'_{y\mathrm{A,B}}=\left.\frac{d}{d\bar{x}}\widetilde{H}_y\right|_{x_{\mathrm{A,B}}} \tag{2.32}$$

and

$$\begin{bmatrix}\widetilde{H}'_{y\mathrm{A}}\\ \widetilde{H}'_{y\mathrm{B}}\end{bmatrix}=\hat{\varepsilon}\,\hat{z}_{\mathrm{m}}\left(\begin{bmatrix}-\widetilde{H}_i\\ \widetilde{H}_{i+1}\end{bmatrix}-\hat{\boldsymbol{p}}\begin{bmatrix}\widetilde{H}'_{y\mathrm{A}}\\ \widetilde{H}'_{y\mathrm{B}}\end{bmatrix}\right),\qquad \hat{\boldsymbol{p}}=\mathrm{diag}\left(p_\mathrm{l}\bar{h},\,p_\mathrm{r}\bar{h}\right), \tag{2.33}$$

which results in

$$\begin{bmatrix}\widetilde{H}'_{y\mathrm{A}}\\ \widetilde{H}'_{y\mathrm{B}}\end{bmatrix}=\begin{bmatrix}a_1 & b_1\\ b_2 & a_2\end{bmatrix}\begin{bmatrix}\widetilde{H}_i\\ \widetilde{H}_{i+1}\end{bmatrix},\qquad \begin{bmatrix}-a_1 & b_1\\ -b_2 & a_2\end{bmatrix}=\left(\hat{\boldsymbol{p}}+\hat{z}_{\mathrm{m}}^{-1}\hat{\varepsilon}^{-1}\right)^{-1}. \tag{2.34}$$

From the formulas for the second derivatives on discretisation lines i and $i+1$

$$\left.\frac{d^2\widetilde{H}_y}{d\bar{x}^2}\right|_i\approx\left[a_1\widetilde{H}_i+b_i\widetilde{H}_{i+1}-\left(\widetilde{H}_i-\widetilde{H}_{i-1}\right)/\bar{h}\right]\left[\bar{h}(p+1/2)\right]^{-1}, \tag{2.35}$$

$$\left.\frac{d^2\widetilde{H}_y}{d\bar{x}^2}\right|_{i+1}=\left[\left(\widetilde{H}_{i+2}-\widetilde{H}_{i+1}\right)/\bar{h}-\left(b_2\widetilde{H}_i+a_2\widetilde{H}_{i+1}\right)\right]\left[\bar{h}(q+1/2)\right]^{-1}, \tag{2.36}$$

the second order difference operators can easily be obtained.

2.3 Bi-directional eigenmode expansion and propagation method

The bi-directional eigenmode propagation method (BEP) [68], [69] presented here combines the advantages of a multimode transfer-matrix-method for the propagation of a set of forward and backward travelling eigenmodes, and the handling of Maxwell's boundary conditions by the mode matching method (cf. section 1.2) [70], [71]. Because of its accurate treatment of large refractive index steps the method is applicable for lateral and longitudinal discontinuities as are various types of optical passive (*e.g.*, directional couplers, butt coupling, tapers) and active devices (*e.g.*, DBR-, DFB-, VCSE lasers, detectors). Contrary to the beam propagation method, no paraxial approximations are made. The physically interpretable eigenmodes give a much smaller system of equations than the sine and cosine functions of a Fourier expansion [72].

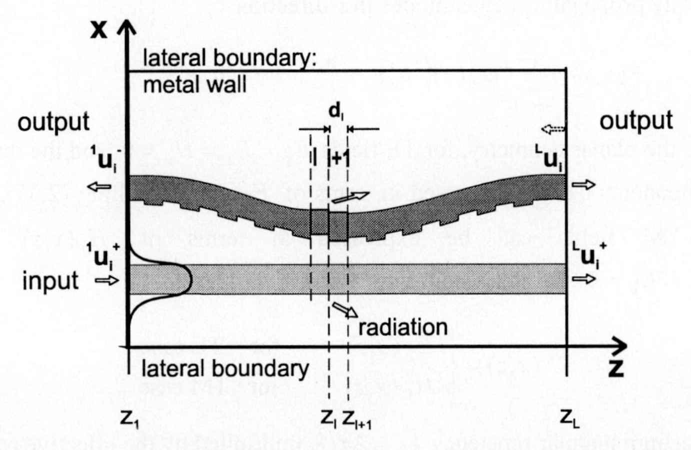

Fig. 2.6. Shown is the ground plan of a device between the first and last co-ordinate. The lateral boundaries on both sides keep the waves within the width w. The bend waveguide structure is divided into steps of constant waveguide sections.

The following considerations are focused on structures in one lateral and the longitudinal dimensions, and hence the modes can be of either TE or TM type. The whole structure in Fig. 2.6. is divided into L sections consisting of a stack of layers. Thus, *e.g.*, a bent waveguide is stepwise approximated by different stacks of layers in each section. For each stack of layers the eigenmodes have to be found [73].

We have to solve Maxwell equations

$$\nabla \times \mathbf{H} = j\omega\varepsilon\varepsilon_0 \mathbf{E}, \qquad \nabla \times \mathbf{E} = -j\omega\mu\mu_0 \mathbf{H}, \tag{2.37}$$

with harmonic time dependence $\exp(j\omega t)$, where ε, μ are the relative electric permittivity and the relative magnetic permeability of the material which are in general space dependent, *i.e.,* they may vary from layer to layer and from section to section but are constant inside the layers. In the following, the time dependence is omitted for brevity. The co-ordinate z_l (Fig. 2.6) denotes the interface between two neighbouring sections. The length of section l is given by $d_l = z_l - z_{l-1}$. Within the BEP formalism the optical field is represented by a superposition of all guided eigenmodes and a set of radiation modes which are discretised by introducing electric or magnetic walls alongside the structure, forcing so the tangential components of the electric or magnetic field at the walls to be zero.

The electromagnetic fields can be expressed as a sum of local eigenfunctions $\varphi_m(x)$ of the different slab waveguides l weighted by the amplitudes ${}^l u_m^+$ for the forward and ${}^l u_m^+$ for the backward travelling waves. The field at any position in the section l is thus a superposition of all (*i.e.*, M) forward and backward (*i.e.*, bi-directionally propagating) eigenmodes in z-direction:

$${}^l F(x,z) = \sum_{m=1}^{M} {}^l\varphi_m(x)\left({}^l u_m^+ e^{-j\,{}^l\beta_m(z-z_{l-1})} + {}^l u_m^- e^{+j\,{}^l\beta_m(z-z_{l-1})}\right), \tag{2.38}$$

Due to the planar geometry, for TE fields $E_x = E_z = H_y = 0$, and the remaining field components can be expressed in terms of $E_y(x,z)$ fulfilling (2.37). Analogously, TM fields can be expressed in terms of $H_y(x,z)$ having $H_x = H_z = E_y = 0$. We thus choose

$${}^l F(x,z) = \begin{cases} {}^l E_y(x,z) & \text{for TE case,} \\ {}^l H_y(x,z) & \text{for TM case.} \end{cases} \tag{2.39}$$

The vacuum angular repetency $k_0 = 2\pi/\lambda$ multiplied by the effective refractive index of the eigenmode m gives the propagation constant of the mode ${}^l\beta_m$.

A plane wave transfer matrix method [73] is applied to calculate the set of eigenmodes ${}^l\varphi_m(x)$ including the discrete set of radiation modes of the slab waveguides automatically. In the TE case, a suitable positioning of ideal metal walls is used to practically eliminate the influence of boundaries on the eigenvalues and the guided mode shapes. In the TM case, ideal "magnetic" walls have to be introduced.

Within the BEP formalism thin layers cause no problems, contrary to the discretisation problems of the finite difference, finite element or MoL formalism, or the local reflection coefficient which uses a plane wave approximation [74], [75].

Boundary conditions require the continuity of tangential components of electric and magnetic fields at the interface z_l between adjacent sections. From (2.39) it follows that

$${}^{l+1}F(x,z)\Big|_{z=z_l} = {}^l F(x,z)\Big|_{z=z_l}, \tag{2.40}$$

and

$$\frac{1}{{}^{l+1}\tau(x)}\frac{\partial}{\partial z}{}^{l+1}F(x,z)\Bigg|_{z=z_l} = \frac{1}{{}^l\tau(x)}\frac{\partial}{\partial z}{}^l F(x,z)\Bigg|_{z=z_l}, \tag{2.41}$$

where in each section l

$$ {}^{l}\tau(x)=\begin{cases} {}^{l}\mu(x) & \text{for TE case,}\\ {}^{l}\varepsilon(x) & \text{for TM case.}\end{cases} \tag{2.42}$$

Generally, the boundary conditions (2.40) and (2.41) can be written in a closed matrix form

$$ {}^{l+1}\Phi(x)\cdot{}^{l+1}\mathbf{P}(z_l)\begin{pmatrix}{}^{l+1}\mathbf{u}^{+}\\ {}^{l+1}\mathbf{u}^{-}\end{pmatrix}\Bigg|_{z=z_l} = {}^{l}\Phi(x)\cdot{}^{l}\mathbf{P}(z_l)\begin{pmatrix}{}^{l}\mathbf{u}^{+}\\ {}^{l}\mathbf{u}^{-}\end{pmatrix}\Bigg|_{z=z_l} \tag{2.43}$$

with the $2\times 2M$ matrix ${}^{l}\Phi(x)$,

$$ {}^{l}\Phi(x)=\begin{pmatrix} {}^{l}\varphi_1(x) & \dots & {}^{l}\varphi_M(x) & 0 & \dots & 0\\ 0 & \dots & 0 & \dfrac{{}^{l}\varphi_1(x)}{{}^{l}\tau(x)} & \dots & \dfrac{{}^{l}\varphi_M(x)}{{}^{l}\tau(x)}\end{pmatrix}, \tag{2.44}$$

the tri-diagonal $2M\times 2M$ propagation matrix ${}^{l}\mathbf{P}(z)$,

$$ {}^{l}\mathbf{P}(z)=\begin{pmatrix}
e^{-j\,{}^{l}\beta_1(z-z_{l-1})} & 0\dots 0 & 0 & e^{j\,{}^{l}\beta_1(z-z_{l-1})} & 0\dots 0 & 0\\
0 & \cdot\;\cdot\;\cdot & 0 & 0 & \cdot\;\cdot\;\cdot & 0\\
\vdots & \ddots & \vdots & \vdots & \ddots & \vdots\\
0 & \cdot\;\cdot\;\cdot & 0 & 0 & \cdot\;\cdot\;\cdot & 0\\
0 & 0\dots 0 & e^{-j\,{}^{l}\beta_M(z-z_{l-1})} & 0 & 0\dots 0 & e^{j\,{}^{l}\beta_M(z-z_{l-1})}\\
j\,{}^{l}\beta_1 e^{-j\,{}^{l}\beta_1(z-z_{l-1})} & 0\dots 0 & 0 & -j\,{}^{l}\beta_1 e^{j\,{}^{l}\beta_1(z-z_{l-1})} & 0\dots 0 & 0\\
0 & \cdot\;\cdot\;\cdot & 0 & 0 & \cdot\;\cdot\;\cdot & 0\\
\vdots & \ddots & \vdots & \vdots & \ddots & \vdots\\
0 & \cdot\;\cdot\;\cdot & 0 & 0 & \cdot\;\cdot\;\cdot & 0\\
0 & 0\dots 0 & j\,{}^{l}\beta_M e^{-j\,{}^{l}\beta_M(z-z_{l-1})} & 0 & 0\dots 0 & -j\,{}^{l}\beta_M e^{j\,{}^{l}\beta_M(z-z_{l-1})}
\end{pmatrix} \tag{2.45}$$

and the $2M$ vector $({}^{l}\mathbf{u}^{\pm})_m={}^{l}u_m^{\pm}$ for $m=1,\dots,M$. There is an inherent numerical problem in the formulation of matrix ${}^{l}\mathbf{P}(z)$ according to (2.45), namely, for complex ${}^{l}\beta_m$ and large $(z-z_{l-1})$ some of the matrix elements for higher order modes become rather large, which should be taken into account in concrete modelling calculations. This problem can be avoided using the impedance transform method described *e.g.*, in [63], instead of the transfer matrix method.

Formulating the orthogonality of the local eigenfunctions ${}^{l}\varphi_m(x)$ as

$$ \int_{x=0}^{x=w}\frac{{}^{l}\varphi_n(x)\cdot{}^{l}\varphi_m(x)}{{}^{l}\tau(x)}\,dx=\delta_{nm}, \tag{2.46}$$

the system of equations (2.43) can be resolved by using the transposed matrix ${}^{l}\Phi^{T}(x)$ for which it holds

$$\int_{x}^{x=w} {}^{l}\Phi(x)\cdot{}^{l}\Phi^{T}(z)\,dx = \mathbf{I},$$

because of the completeness of the set of eigenfunctions ${}^{l}\varphi_m(x)$; here $\mathbf{I}$ is the unity matrix of rank M. Multiplying both sides of (2.43) by ${}^{l}\Phi^{T}(x)$, integrating over the whole x-range and then multiplying by ${}^{l}\mathbf{P}^{-1}(z_l)$ we obtain the solution in the form

$$\begin{pmatrix} {}^{l+1}\mathbf{u}^{+} \\ {}^{l+1}\mathbf{u}^{-} \end{pmatrix} = {}^{l+1,l}\mathbf{T}\cdot\begin{pmatrix} {}^{l}\mathbf{u}^{+} \\ {}^{l}\mathbf{u}^{-} \end{pmatrix}, \qquad {}^{l+1,l}\mathbf{T} = {}^{l+1}\mathbf{P}^{-1}(z_l)\cdot{}^{l+1,l}\mathbf{O}\cdot{}^{l}\mathbf{P}(z_l). \tag{2.47}$$

The matrix ${}^{l+1,l}\mathbf{O}$ in this expression consists of two different overlap integrals of the eigenfunctions, which represent the coefficients of the expansion of the field in terms of eigenfunctions in either interfaces:

$${}^{l+1,l}\mathbf{O} := \int_{x=0}^{x=w} {}^{l+1}\Phi^{T}(x)\cdot{}^{l}\Phi(x)\,dx = \begin{pmatrix} {}^{l,l+1}O_{1,1} & \cdots & {}^{l,l+1}O_{1,M} & 0 & \cdots & 0 \\ \vdots & \ddots & \vdots & \vdots & \ddots & \vdots \\ {}^{l,l+1}O_{M,1} & \cdots & {}^{l,l+1}O_{M,M} & 0 & \cdots & 0 \\ 0 & \cdots & 0 & {}^{l+1,l}O_{1,1} & \cdots & {}^{l+1,l}O_{1,M} \\ \vdots & \ddots & \vdots & \vdots & \ddots & \vdots \\ 0 & \cdots & 0 & {}^{l,l+1}O_{M,1} & \cdots & {}^{l+1,l}O_{M,M} \end{pmatrix} \tag{2.48}$$

In "upward" direction, $l \rightarrow l+1$,

$${}^{l,l+1}O_{m,i} = \int_{x=0}^{x=w} \frac{{}^{l+1}\varphi_m(x)\,{}^{l}\varphi_i(x)}{{}^{l+1}\tau(x)}\,dx, \tag{2.49}$$

in "downward" direction, $l+1 \rightarrow l$,

$${}^{l+1,l}O_{m,i} = \int_{x=0}^{x=w} \frac{{}^{l+1}\varphi_m(x)\,{}^{l}\varphi_i(x)}{{}^{l}\tau(x)}\,dx, \tag{2.50}$$

for $m = 1, \ldots, M$, $i = 1, \ldots, M$.

Both integrals can be evaluated in a closed form. At adjacent sections the waveguides change their refractive indices and/or their thicknesses. The overlap integrals of the modes m and i have to be calculated between the introduced "walls" (the x-range $0 \le x \le w$), but the adjacent stacks of layers define intermediate points for the analytical integration. Hence, the overlap integrals ${}^{l,l+1}O_{m,i}$ and ${}^{l+1,l}O_{m,i}$ are split into a sum of s sub-integrals each. The solutions of the sub-integrals are simple and straightforward, depending on the functions used for ${}^{l}\varphi_m(x)$ (cf. section 2.1). For multilayer waveguides, the mode fields are

$$ {}^{l}\varphi_m(x) = \left({}^{l}\varphi_{s,m}^{+} e^{-j\,{}^{l}k_{x,s,m}(x-x_{s-1})} + {}^{l}\varphi_{s,m}^{-} e^{+j\,{}^{l}k_{x,s,m}(x-x_{s-1})} \right) \tag{2.51} $$

for each layer s together with the separation equation ${}^{l}k_{x,s,m} = \sqrt{{}^{l}k_s^2 - {}^{l}\beta_m^2}$. The material in section l and layer s is characterised by ${}^{l}k_s = k_0\,{}^{l}n_s$. Because of the layered structures, ${}^{l}\mu(x)$ and ${}^{l}\varepsilon(x)$ are piecewise constants, so the integrals in (2.49) and (2.50) can be handled in the same way.

In most optical problems the relative magnetic permeability $\mu(x)$ is equal to 1. Then, the matrix elements ${}^{l,l+1}O_{m,i}$ and ${}^{l+1,l}O_{m,i}$ are equal to each other in the TE case.

In the presence of gain or absorption the modes are no longer power orthogonal. Then the general orthogonality property stated by Bressler et. al. [76] is to be used.

As ${}^{l}\mathbf{P}(z)$ is a sparse tri-diagonal matrix, the matrix inversion and multiplication forming ${}^{l+1,l}\mathbf{T}$ can be worked out analytically. The $2M \times 2M$ elements of the transfer matrix ${}^{l+1,l}\mathbf{T}$ are then

$$ {}^{l+1,l}\mathbf{T} = \frac{1}{2} \begin{pmatrix} & \vdots & & \vdots & \\ \cdots & \left({}^{l,l+1}O_{m,i} + {}^{l+1,l}O_{m,i} \dfrac{{}^{l}\beta_i}{{}^{l+1}\beta_m} \right) e^{-j\,{}^{l}\beta_i d_l} & \cdots & \left({}^{l,l+1}O_{m,i} - {}^{l+1,l}O_{m,i} \dfrac{{}^{l}\beta_i}{{}^{l+1}\beta_m} \right) e^{+j\,{}^{l}\beta_i d_l} & \cdots \\ & \vdots & & \vdots & \\ \cdots & \left({}^{l,l+1}O_{m,i} - {}^{l+1,l}O_{m,i} \dfrac{{}^{l}\beta_i}{{}^{l+1}\beta_m} \right) e^{-j\,{}^{l}\beta_i d_l} & \cdots & \left({}^{l,l+1}O_{m,i} + {}^{l+1,l}O_{m,i} \dfrac{{}^{l}\beta_i}{{}^{l+1}\beta_m} \right) e^{+j\,{}^{l}\beta_i d_l} & \cdots \\ & \vdots & & \vdots & \end{pmatrix}, \tag{2.52} $$

Finally the product of all ${}^{l+1,l}\mathbf{T}$'s describes the wave transmission through the whole structure,

$$ \mathbf{T}_{sys} = \prod_{l=1}^{L-1} {}^{l+1,l}\mathbf{T}. \tag{2.53} $$

The full physical behaviour of the device is stored in the system transfer matrix $\mathbf{T}_{sys}$, and once it is known, the response of any incident field distribution can be calculated. Power conservation (in the absence of absorption) along the device verifies the accurate eigenmode expansion and is attained by including all relevant eigenmodes. The absorber problem for mode propagation methods is discussed in detail in [69].

As an example the front end of a laser waveguide is calculated with $M = 20$ TE modes. An effective index approximation is used; the roots of the eigenvalue problem are searched on the real axis with an adaptive step size control. Counting the nodes of the field distribution ensures that the M lowest eigenmodes are found. Root tracking with reduced functions and checking the orthogonality lead to the roots in the complex plane.

The standing wave pattern in Fig. 2.7 is due to the reflection from the cleaved waveguide facet. Thus the bi-directional character of the method is demonstrated. The total reflected intensity is 39%, and 35% of it is reflected into the fundamental mode. The spot size of the transmitted power into the air spreads with half the radiation angle of 24°. Some more examples are given in [68] and [77].

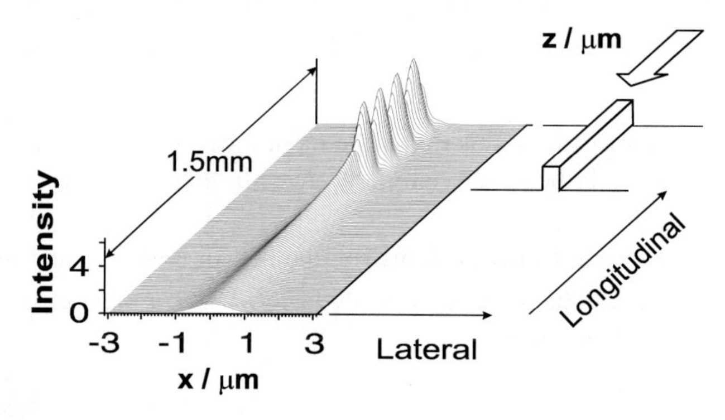

Fig. 2.7. Intensity distribution of a light beam radiated into air and the standing wave pattern in the waveguide with a cleaved facet. Parameters of the simulation: waveguide width $d = 0.2$ µm, refractive indices of the waveguide and of the substrate $n_g = 3.51$ and $n_s = 3.17$, $\lambda = 1.5$ µm.

2.4 Method of backward calculation

In modern optoelectronics, different kinds of optical structures like Bragg gratings, multilayer Fabry-Perot resonators, corrugated waveguides, *etc.* are widely used. For their analysis, a multibeam interference method [78], transfer matrix method [79], recursive method [80,81], and invariant embedding method [82] have been applied. In all these methods the wave in a multilayer medium is described by a superposition of counter-propagating waves. Although this approach is widely adopted for the analysis of linear problems, its application to nonlinear problems is limited to weak nonlinearity.

Here we present another approach in which the field is described by a single quantity. In this case, the boundary problem can be solved for both linear and nonlinear media with an equal ease. The price one has to pay for this is that the medium can have variations only in a longitudinal direction, or the structures should allow for propagation of a single transverse mode only. In this approach, the boundary value problem is solved supposing that at the output plane, only the outgoing wave can propagate. We can then calculate the total field in the backward direction toward the input plane and determine thus the reflection coefficient at the input. In the present section, the principle of the method is briefly described and the method is then applied to a distributed Bragg reflector with loss.

2.4.1 Description of the method

Let us consider the normal incidence of a plane electromagnetic wave on a layered structure (see Fig. 2.8).

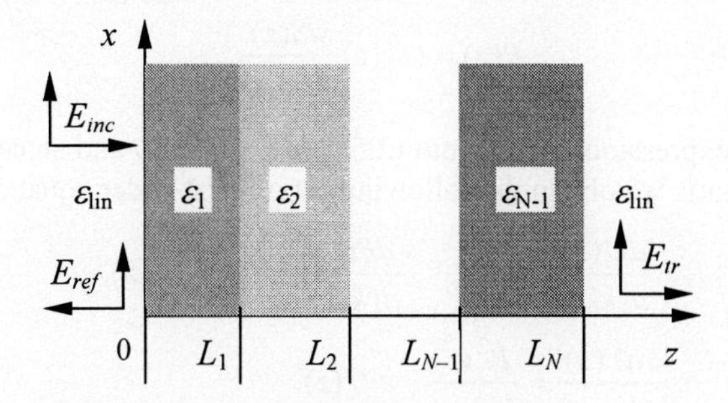

Fig. 2.8. Plane wave interaction with a multilayer structure.

From Maxwell equations for linear polarisation along the **x**-axis one may write

$$\frac{\partial^2 E_x(z,t)}{\partial z^2} + \mu_0 \varepsilon_a \frac{\partial^2 E_x(z,t)}{\partial t^2} = 0, \tag{2.54}$$

where μ_0 is the magnetic permeability of vacuum and ε_a is the permittivity of the medium. This wave equation is valid also for inhomogeneous media, when the condition $\mathbf{E} \cdot \nabla \varepsilon_a = 0$ is satisfied.

We express the field $E_x(z,t)$ in the complex form

$$E_x(z,t) = \frac{1}{2}\left[\dot{E}_x(z)e^{j\omega t} + \dot{E}_x^*(z)e^{-j\omega t}\right], \tag{2.55}$$

where

$$\dot{E}_x(z) = U(z)e^{-jS(z)}. \tag{2.56}$$

Here, $U(z)$ and $S(z)$ are real quantities, the amplitude and phase functions, respectively. Substituting the expression (2.55) into (2.54) we receive the Helmholtz equation for $\dot{E}_x(z)$ in the form

$$\frac{d^2 \dot{E}_x(z)}{dz^2} + \omega^2 \mu_0 \varepsilon_a \dot{E}_x(z) = 0. \tag{2.57}$$

Using the complex notation makes it possible to extend the validity of this equation to media with loss or (linear) gain characterised by a complex permittivity $\varepsilon_a = \varepsilon_a' + j\varepsilon_a'' = \varepsilon_0(\varepsilon' + j\varepsilon'')$. Then, using the expression for the magnetic field obtained from Maxwell equations, the time-averaged power flow density can be expressed as

$$\Pi_z = \frac{1}{2}\mathrm{Re}\left\{\dot{E}_x \dot{H}_y^*\right\} = \frac{1}{2\mu_0 c} U^2(z) \frac{dS(z)}{d(k_0 z)} = \frac{1}{2}\sqrt{\frac{\varepsilon_0}{\mu_0}} P(z), \tag{2.58}$$

where [83]

$$P(z) = U^2(z) \frac{dS(z)}{d(k_0 z)}. \tag{2.59}$$

Substituting expressions (2.56) into (2.57), using (2.59) and separating the real and imaginary parts we obtain the following set of first-order equations:

$$\begin{aligned} &\frac{dU(z)}{d(k_0 z)} = Y(z), \quad \frac{dP(z)}{d(k_0 z)} = \varepsilon''(z) U^2(z), \\ &\frac{dY(z)}{d(k_0 z)} = \frac{P^2(z)}{U^3(z)} - \varepsilon' U(z). \end{aligned} \tag{2.60}$$

This set of equations describes the spatial behaviour of electric field amplitude $U(z)$, its derivative $Y(z)$, and the (normalised) power flow density $P(z)$ in the medium. It allows to take into account loss (or gain) as well as the Kerr-type nonlinearity of a medium.

Now let's consider a boundary problem. We will describe the waves outside the structure in the usual way as an incident, a reflected and an outgoing transmitted waves. Inside the structure we will use the solution given by eq. (2.56). The following relations may be written:

A) In the left half-space ($z < 0$), incident and reflected plane electromagnetic waves can be expressed as follows:

$$E_{xinc} = E_{inc} \exp\left(-jk_0 \sqrt{\varepsilon_{lin}}\, z\right), \quad E_{xref} = E_{ref} \exp\left(+jk_0 \sqrt{\varepsilon_{lin}}\, z\right),$$

where ε_{lin} is the relative permittivity of the linear outside medium.

On the illuminated input surface at $z = 0$, from the continuity of tangential components of electric and magnetic fields the reflection coefficient

$$R = \frac{E_{ref}}{E_{inc}} = \frac{U^2(0)\sqrt{\varepsilon_{lin}} - P(0) - jU(0)Y(0)}{U^2(0)\sqrt{\varepsilon_{lin}} + P(0) + jU(0)Y(0)} \tag{2.61}$$

and the incident field

$$E_{inc} = \left| \frac{U^2(0)\sqrt{\varepsilon_{lin}} + P(0) + jU(0)Y(0)}{2U(0)\sqrt{\varepsilon_{lin}}} \right| \tag{2.62}$$

are obtained as functions of the field parameters at $z = 0$ [84].

B) For adjacent layers of the structure the standard boundary conditions require the continuity of the following values: $U(z)$, $Y(z)$, $S(z)$, $dS(z)/d(k_0 z)$, and $P(z)$.

C) In the right half-space ($z > L$), the transmitted field can be expressed as

$$E_{xtrans} = E_{tr} \exp[-jk_0\sqrt{\varepsilon_{lin}}(z - L)].$$

The boundary conditions at $z = L$ can be written in the form

$$U(z = L) = E_{tr}, \quad S(z = L) = 0,$$
$$Y(z = L) = 0. \quad P(z = L) = U^2(L)\sqrt{\varepsilon_{lin}}.$$

Obviously, starting to solve the boundary problem from the output side of the structure ($z = L$), we reduce it to a Cauchy problem, *i.e.*, to an initial value problem. Setting up the initial values for U, Y, and P, at $z = L$, the numerical integration of the set of differential equations (2.60) is carried out towards the input plane ($z = 0$). Then, the reflection coefficient at the input and the incident field can be easily determined using (2.61) and (2.62).

The presented method has been used to investigate light propagation through nonlinear films and multilayer structures [85, 86, 87, 88].

2.4.2 Application to distributed Bragg reflectors

The method described above is applicable for a distributed Bragg grating reflector also. A well-known coupled-wave theory is now commonly used for theoretical analysis of Bragg reflectors [89]. Contrary to the coupled-wave approach, our method is not limited to shallow permittivity perturbations. The Bragg grating reflector with the following parameters has been analysed by the described method: L is the length of medium, ε_{lin} and ε' are the real parts of the relative permittivities outside and within the perturbed medium, respectively; Λ is a grating period; AM is a relative amplitude of a sinusoidal permittivity modulation of the form $\varepsilon'(z) = \varepsilon'_0[1 + AM\sin(2\pi z/\Lambda)]$. Numerical analysis was carried out for the number of periods $N = 300$, the amplitude of modulation $AM = 0.01$ and. $\varepsilon_{lin} = \varepsilon'_0 = 1$. The small amplitude modulation is chosen for comparison with results obtained by known analytical expressions [89]. Reflection properties of this structure without loss ($\varepsilon'' = 0$) and with loss are presented in Fig. 2.9. The results obtained for the lossless grating, $\varepsilon'' = 0$, coincide with the results obtained by analytical expressions presented in [89].

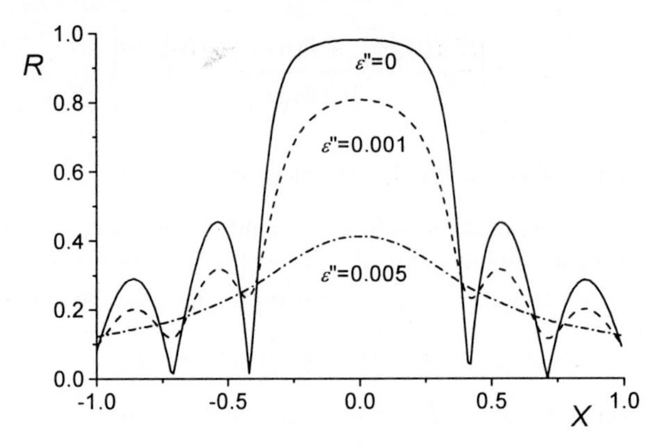

Fig. 2.9. Dependencies of the reflection coefficient of the Bragg grating on the detuning $X = (1/AM)\left(1 - 2\Lambda\sqrt{\varepsilon'_0}/\lambda_0\right)$, λ_0 is a free-space wavelength.

The method of calculation allows also to investigate Bragg gratings with gain; preliminary results have been presented in [90].

References

[1] M. D. Feit and J. A. Fleck, "Light propagation in graded-index optical fibers." *Appl. Opt.*, vol.17, p. 3990, 1978.

[2] D. Marcuse, *Theory of dielectric optical waveguides*, 2nd edn., Academic Press, Boston, 1991.

[3] R. März, *Integrated Optics: Design and Modeling*, Artech House, Boston, 1995

[4] H. J. W. M. Hoekstra, "On beam propagation methods for modelling in integrated optics", *Opt. Quantum Electron.*, vol. 29, pp. 157–171, 1997.

[5] H. E. Hernández-Figueroa, "Simple Nonparaxial Beam-Propagation Method for Integrated Optics," *J. Lightwave Technol.*, vol. 12, p. 644, 1994.

[6] G. R. Hadley, "Wide-angle beam propagation using Padé approximant operators," *Opt. Lett.*, vol. 17, pp. 1426–1428, 1992.

[7] Y. Chung and N. Dagli, "Explicit Finite Difference Beam Propagation Method: Application to Semiconductor Rib Waveguide Y-Junction Analysis," *Electron. Lett.*, vol. 26, No. 11, pp. 711–712, 1990.

[8] G. R. Hadley, "Multistep method for wide-angle beam propagation," *Opt. Lett.*, vol. 17, No. 24, pp. 1743–1745, 1992.

[9] D. Schulz, C. Glingener, M. Bludszuweit, and E. Voges, "Mixed Finite Element Beam Propagation Method", *Proc. European Conference on Integrated Optics*, 1977, pp. 226–229.

[10] A. Splett, M. Majd, and K. Petermann, "A Novel Beam Propagation Method for Large Refractive Index Steps and Large Propagation Distances," *IEEE Photon. Technol. Lett.*, vol. 3, No. 5, pp. 466–468, 1991.

[11] D. Schulz, C. Glingener, and E. Voges, "Novel Generalized Finite-Difference Beam Propagation Method," *IEEE J. Quantum Electron.*, vol. 30, No. 4, pp. 1132–1140, 1994.

[12] S. Helfert and R. Pregla, "Efficient Analysis of Periodic Structures," *J. Lightwave Technol.*, vol. 16, pp.1694–1702, 1998.

[13] J. Čtyroký, S. Helfert, and R. Pregla, "Analysis of a deep waveguide Bragg grating", *Opt. Quantum Electron.*, accepted. 1998.

[14] J. Čtyroký, J. Homola, and M. Skalský, "Modelling of surface plasmon resonance waveguide sensor by complex mode expansion and propagation method," *Opt. Quantum Electron.*, vol. 29, No. 2, pp. 301–311, 1997.

[15] G. Sztefka, and H. P. Nolting, "Bidirectional eigenmode propagation for large refractive index steps," *IEEE Photon. Technol. Lett.*, vol. 5, pp. 554–557, 1993.

[16] J. Willems, J. Haes, and R. Baets, "The bidirectional mode expansion method for two dimensional waveguides: The TM case," *Opt. Quantum Electron.*, Special Issue on Optical Waveguide Theory and Numerical Modeling, vol. 27, no. 10, pp. 995–1108, 1995.

[17] J. P. Bérenger, "A perfectly matched layer for the absorption of electromagnetic waves," *J. Computational Phys.*, vol. 114, pp. 185–200, 1994.

[18] W. P. Huang, C. L. Xu, W. Lui, and K. Yokoyama, "The Perfectly Matched Layer (PML) Boundary Condition for the Beam Propagation Method," *IEEE Photon. Technol. Lett.*, vol. 8, No. 5, pp. 649–651, 1996.

[19] D. S. Katz, E. T. Thiele, and A. Tavlove, "Validation and Extension to Three Dimensions of the Berenger PML Absorbing Boundary Condition for FD-TD Meshes," *IEEE Microwave Guided Wave Lett.*, vol. 4, No. 8, pp. 268–270, 1994.

[20] G. R. Hadley, "Transparent boundary condition for beam propagation," *Opt. Lett.*, vol. 16, No. 9, pp. 624–626, 1991.

[21] G. R. Hadley, "Transparent Boundary Condition for the Beam Propagation Method," *IEEE J. Quantum Electron.*, vol. 28, No. 1, pp. 363–370, 1992.

[22] G. Mur, "Absorbing boundary conditions for the finite-difference approximation of the time-domain electromagnetic field equations," *IEEE Trans. Electromagnetic Compatibility*, vol. EMC-23, pp. 377–382, 1981.

[23] T. G. Moore, J. G. Blachak, A. Taflove, and G.A. Kriegsmann, "Theory and application of radiation boundary operators," *IEEE Trans. Antennas Propagat.*, vol. AP-36, pp. 1797–1812, 1988.

[24] H. J. W. M. Hoekstra, G. J. M. Krijnen, and P. V. Lambeck, "Efficient Interface Conditions for the Finite Difference Beam Propagation Method," *J. Lightwave Technol.*, vol. 10, No. 10, pp. 1352–1355, 1992.

[25] C. Vasallo,"Improvement of finite difference methods for step-index optical waveguides," *IEE Proc. J*, vol. 139, No. 2, pp. 137–142, 1992.

[26] S. Helfert, and R. Pregla, "Finite Difference Expressions for Arbitrarily Positioned Dielectric Steps in Waveguide Structures," *J. Lightwave Technol.*, vol. 14, No. 10, pp. 2414–2421, 1995.

[27] Y. Chung and N. Dagli, "An assessment of finite difference beam propagation method," *IEEE J. Quantum Electron.*, vol. 26, pp. 1335–1339, 1990.

[28] G. R. Hadley, "Transparent boundary condition for the beam propagation method," *IEEE J. Quantum Electron.*, vol. 28, pp. 363–370, 1992.

[29] R. Pregla and D. Kremer, "Method of lines with special absorbing boundary conditions-analysis of weakly guiding optical structures," *IEEE Microwave Guided Wave Lett.*, vol. 2, pp.239–241, 1992.
[30] C. Vassallo and F. Collino, "Highly efficient absorbing boundary conditions for the beam propagation method ," *J. Lightwave Technol.*, vol. 14, pp. 1570–1577, 1996
[31] C. Vassallo, " Improvement of finite difference methods for step-index optical waveguides", *IEE Proc. J*, vol. 139, pp. 137–142, 1992.
[32] H. J. W. M. Hoekstra, G.J.M. Krijnen, and P.V. Lambeck, "Efficient interface conditions for the finite difference beam propagation method," *J. Lightwave Technol.* vol. 10, pp. 1352–1355, 1992.
[33] D. Yevick and M. Glasner, " Forward wide-angle wave propagation in semiconductor rib waveguides," *Opt. Lett.*, vol. 15, pp. 174–176, 1990.
[34] G. R. Hadley, "Multistep method for wide angle propagation," *Opt. Lett.*, vol. 17, pp. 1743–1745, 1992.
[35] H. J. W. M. Hoekstra, G.J.M. Krijnen, and P.V. Lambeck, " New formulation of the BPM based on the SVEA," *Opt. Commun.*, vol. 97, pp. 301–303, 1993.
[36] H. J. W. M. Hoekstra, "On beam propagation methods for modelling in integrated optics," *Opt. Quantum Electron.*, vol. 29, pp. 157–171, 1997.
[37] C. Vassallo, "Reformulation for the beam propagation method," *J.Opt.Soc. Am.* A, vol.10, pp. 2208–2216, 1993.
[38] H.-P. Nolting and R. März, "Results of benchmark test for different numerical BPM algorithms ," *J. Lightwave Techol.*, vol. 13 , pp. 216–224 ,1995.
[39] C. Vassallo, "Wide angle BPM and power conservation," *Electron. Lett.*, vol. 31, pp. 130–131, 1995.
[40] J. Willems, J. Haes, and R. Baets, "The bi-directional mode expansion method for two dimensional waveguides: the TM case," *Opt. Quantum Electron.*, vol. 27, pp. 995–1008, 1995.
[41] M. J. N. van Stralen, H. Blok, and M. V. de Hoop, "Design of sparse matrix representations for the propagator used in the BPM and directional wave field decomposition," *Opt. Quantum. Electron.*, vol. 29, pp. 179–197, 1997.
[42] P. Lüsse and H.-G. Unger, "Stability properties of 3D vectorial and semivectorial BPMs utilizing a multi-grid equation solver," *Opt. Quantum Electron.*, vol 29, pp. 173–178, 1997.
[43] F. Horst, H. J. W. M. Hoekstra, A. Driessen, and T. J. A. Popma, "Simulation of a self-switching nonlinear Bragg reflector using realistic nonlinear parameters," *CLEO*, Hamburg, 1996; F.Horst, thesis, 1997.
[44] G. J. M. Krijnen, W. Torruellas, G. I. Stegeman, H. J. W. M. Hoekstra, and P. V. Lambeck, "Optimization of SHG and nonlinear phase-shifts in the Cerenkov regime," *IEEE J. Quantum Electron.*, vol. 4, pp. 729–738, 1996.
[45] R. Pregla and W. Pascher, "The Method of Lines". In T. Itoh, (ed.), *Numerical Techniques for Microwave and Millimeter Wave Passive Structures*. J. Wiley, New York, pp. 381–446, 1989
[46] R. Pregla, "The Method of Lines for the Unified Analysis of Microstrip and Dielectric Waveguides," *Electromagnetics*, vol.15, pp. 441–456, 1995.
[47] R. Pregla, "The Method of Lines for Modeling of Integrated Optic Structures." *Latsis Symposium*, Zürich, pp. 216–229, 1995.

[48] U. Rogge and R. Pregla, "Method of Lines for the Analysis of Dielectric Waveguides", *J. Lightwave Technol.*, vol. 11, pp. 2015–2020, 1993.

[49] M. Bertolotti, P. Masciulli, and C. Sibilia, "MoL Numerical Analysis of Nonlinear Planar Waveguide," *J. Lightwave Technol.*, vol. 12, pp. 784–789, 1994.

[50] R. Pregla, J. Gerdes, E. Ahlers, and S. Helfert, "Algorithm for Waveguide Bends and Vectorial Fields", in Techn. Digest on *Integrated Photonics Research*, (Optical Society of America, Washington, DC), vol. 9, pp. 32–33, 1992.

[51] J. Gerdes, S. Helfert, and R. Pregla, "Three-dimensional Vectorial Eigenmode Algorithm for Nonparaxial Propagation in Reflecting Optical Waveguide Structures." *Electron. Lett.*, vol. 31, No.1, pp. 65–66, 1995.

[52] R. Pregla, "MoL-BPM Method of Lines Based Beam Propagation Method," Methods for Modeling and Simulation of Guided-Wave Optoelectronic Devices, In W. P. Huang (ed), *PIER 11 in Progress in Electromagnetic Research*, EMW Publishing, Cambridge, Massachusetts, USA, pp. 51–102, 1995.

[53] K. H. Helf, J. Gerdes, and R. Pregla, "Full-Wave Analysis of Traveling-Wave Electrodes with Finite Thickness for Electro-Optic Modulators by the Method of Lines," *J. Lightwave Technol.*, vol. 9, pp. 461–467, 1991.

[54] R. Pregla, "Method of Lines for the Analysis of Multilayered Gyrotropic Waveguide Structures," *IEE Proc. H*, vol. 140, pp. 183–192, 1993.

[55] R. Pregla and E. Ahlers, "The Method of Lines for the Analysis of Discontinuities in Optical Waveguides," *Electron. Lett.* vol. 29, No. 21, pp. 1845–1847, 1993.

[56] W. Pascher and R. Pregla, "Analysis of Curved Optical Waveguides by the Vectorial Method of Lines," *Radio Science* vol. 28, pp. 1229–1233, 1993.

[57] R. Pregla E. Ahlers, "Method of Lines for Analysis of Arbitrarily Curved Waveguide Bends," *Electron. Lett.*, vol. 30, No. 18, pp. 1478–1479, 1994.

[58] S. Helfert, Analysis of Curved Bends in Arbitrary Optical Devices using Cylindrical Coordinates, *Opt. Quantum Electron.*, accepted, 1998.

[59] S. Helfert and R. Pregla, "Modeling of Taper Structures in Cylindrical Coordinates," *Proc. Integrated Photonics Research Topical Meeting*, Dana Point, USA, vol. 7, pp. 30–32, 1995.

[60] S. Helfert and R. Pregla, "New developments of a Beam Propagation Algorithm Based on the Method of Lines," *Opt. Quantum Electron.*, No.25, pp. 943–950, 1993.

[61] S. Helfert and R. Pregla, "A Very Stable and Accurate Algorithm for the Analysis of Periodic Structures with a Finite Number of Periods," *Proc. Progress in Electromagnetics Research (PIERS)*, Hong Kong, vol. 1, p. 105, 1997.

[62] U. Rogge and R. Pregla, "Method of Lines for the Analysis of Dielectric Waveguides." *J. Lightwave Technol.*, vol. 11, pp. 2015–2020, 1993.

[63] A. S. Sudbø, "Improved formulation of the film mode matching method for mode field calculations in dielectric waveguides," *Pure Appl. Opt. (J. Europ. Opt. Soc. A)* vol. 3, pp. 381–388, 1994.

[64] A. S. Sudbø, "Film Mode Matching: A versatile method for mode field calculations in dielectric waveguides," *Pure Appl. Opt. (J. Europ. Opt. Soc. A)* vol. 2 pp. 211–233, 1993.

[65] R. Pregla, "The Method of Lines as Generalized Transmission Line Technique for the Analysis of Multilayered Structures," *AEÜ* vol. 50, pp. 293–300, 1996.

[66] R. Pregla, “General Formulas for the Method of Lines in Cylindrical Coordinates,” *IEEE Trans. Microwave Theory Tech.* vol. 43, pp. 1617–1620, 1995.
[67] R. Pregla, “Concatenations of waveguide sections,” *IEE Proc H*, vol. 144: pp. 119–125, 1997.
[68] G. Stefka and H.-P. Nolting, “Bidirectional Eigenmode Propagation for Large Refractive Index Steps“, *Photon. Technol. Lett.* vol. 5, pp. 554–557, 1993.
[69] J. Willems, J. Haes, and R. Baets, “The Bidirectional Mode Expansion Method for Two Dimensional Waveguides: The TM-Case“, *Opt. Quantum Electron.* Vol. 27, no. 10, p. 995–1007, 1995.
[70] M. C. Amann, "Rigorous Waveguiding Analysis of the Separated Multiclad-Layer Stripe-Geometry Laser", *IEEE J.Quantum Electron.* vol. 22, pp. 1992–1998, 1986.
[71] W. Schlosser and H.-G. Unger, “Partially Filled Waveguides and Surface Waveguides of Rectangular Cross Section“, in L. Young (Ed.), *Advances in Microwaves* vol. 1, pp. 319–387, Academic Press, 1966.
[72] C. H. Henry and Y. Shani., "Analysis of Mode Propagation in Optical Waveguide Devices by Fourier Expansion", *IEEE J. Quantum Electron.* vol. 27 pp. 523–530, 1991.
[73] J. Chilwell and I. Hodgkinson, “Thin-films field-transfer matrix theory of planar multilayer waveguides and reflection prism-loaded waveguides", *J. Opt. Soc. Am.* vol. A-1, pp. 742–753, 1984.
[74] Y. Chung and N. Dagli, *Integrated Photonics Research* ‘92, New Orleans, paper WE4-1, 1992
[75] P. Kaczmarski and P. E. Lagasse, "Bidirectional Beam Propagation Method", *Electron. Lett.* vol. 24, pp. 675–676, 1988.
[76] A. D. Bressler, G. H. Joshi, and N. Marcuvitz, "Orthogonality Properties in Passive and Active Uniform Wave Guides", *J. Appl. Phys.* vol. 29, pp. 794–799, 1958.
[77] J. Willems, J. Haes, R. Baets, G. Sztefka, and H.-P. Nolting, “Eigenmode Propagation Analysis of Radiation Losses in Waveguides with Discontinuities and Grating-Assisted Couplers", *Proc. Integrated Photonics Research Technical Digest Series* vol. 10, pp. 229–232, Palm Springs 1993.
[78] M. Born and E. Wolf, *Principles of Optics*, 5th ed., Pergamon Press, Oxford, 1975.
[79] J. R. Wait, *Electromagnetic Waves in Stratified Media*, Pergamon Press, New-York, 1962.
[80] L. M. Brekhovskikh, *Waves in Layered Media*, Academic Press, New York, 1960.
[81] I. Ohlidal, “Immersion Spectroscopic Reflectometry of Multilayer Systems. I. Theory,” *J. Opt. Soc. Am.* A, vol. 5, No.4, pp. 459–464, 1988.
[82] Y. B. Band, “Optical Bistability in Nonlinear Media: An Exact Method of Calculation,” *J. Appl. Phys.*, vol. 56(3), pp. 656–659, 1984.
[83] F. G. Bass, and Yu. G. Gurevich, *Hot electrons and strong electromagnetic waves in semiconductor plasma and in gas discharge*. Nauka, Moscow, 1975 (in Russian).
[84] O. V. Bagdasaryan and V. A. Permyakov, “Branching of conditions and effect of energy flux limitation of TE mode in medium with ionization nonlinearity.” *Radiophys. Quantum Electron.* vol. 21, pp. 940–947, 1978 (in Russian).
[85] J. H. Marburger and F. S. Felber, “Theory of a lossless nonlinear Fabry-Perot interferometer,” *Phys. Rev. A,* vol. 17, pp. 335–342, 1978.

[86] W. Chen and D. L. Mills, "Optical response of nonlinear multilayer structures: Bilayers and superlattices," *Phys. Rev. B*, vol. 36, pp. 6269–6278, 1987.

[87] U. Langbein, F. Lederer, T. Peschel, and U. Trutschel, "Nonlinear transmission resonances at stratified dielectric media," *Physics reports (Review Section of Physics Letters)* vol. 194, pp. 325–342, 1990.

[88] H. V.Baghdasaryan, A. V. Daryan, T. M. Knyazyan, and N. K. Uzunoglu, "Modeling of Nonlinear Enhanced Performance Fabry-Perot Interferometer Filter," *Microwave Opt. Technol. Lett.*, vol. 14, No.2, pp.105–108, 1997.

[89] A. Yariv and M. Nakamura, "Periodic Structures for Integrated Optics," *J. Quantum Electron.*, vol. QE-13, No.4, pp. 233–253, 1977.

[90] H. V. Baghdasaryan, G. G. Karapetyan, T. M. Knyazyan, S. I. Avagyan, and N. K. Uzunoglu, "Computer Modelling of a Fiber Bragg Grating-Amplifier," *COST 240 Workshop SOA-based Components for Optical Networks*, Prague, Oct. 27–28, 1997, Book of abstracts, pp. 18-1–18-3, 1997.

3 Benchmark tests and modelling tasks

H.-P. Nolting, J. Haes, S. Helfert

In this chapter we describe benchmark tests and modelling tasks to mutually compare the performance of various beam propagation methods and other software developed and/or currently in use in various laboratories. Two sets of tasks for testing the behaviour of beam propagation methods applied to both lossless and lossy (absorbing) waveguide structures are described in the next section. Theoretical background required to understand the latter task is presented in Section 3.2. Then, an important problem of modelling light propagation in waveguide tapers, both symmetric and asymmetric, is discussed in Section 3.3. Comparative modelling of an example of optical waveguide devices containing thin metal films that can support propagation of surface plasmon waves is described in Section 3.4. In the last section, results of truly bi-directional modelling of light propagation in a very deeply etched Bragg waveguide grating filter are presented. The work described in this chapter was organised within the framework of the Action COST 240.

3.1 BPM benchmark tests

Generally, the beam propagation method is an established technique to simulate the distribution of an optical field in integrated optical circuits. For a long time the "classical" BPM based on the fast Fourier transform [1] was the only algorithm used for this type of simulation. The increasing number of simulations carried out for integrated optical circuits in the InGaAsP/InP material system and assessments on the limitations of the method resulted in the conclusion that the FFT-based algorithm becomes unreliable for high-contrast step-index profiles. Over the last years a great variety of numerical techniques have been applied to beam propagation in order to overcome this lack of reliability.

Driven by the current availability of software the tests are restricted to two dimensional beam propagation, *i.e.*, to beam propagation starting from an one dimensional cross section. To indicate which methods meet general user requirements an objective comparison based on well defined test problems is necessary. To support the assessment of algorithms from the user's point of view the following performance parameters were tested:

Accuracy. Since the beam propagation method is continuously evolving into a more and more realistic design tool for integrated optical circuits, the accuracy of different algorithms is an important information for the BPM users. To provide an absolute assessment on the accuracy of the algorithms, all benchmark tests were based on problems possessing a quasi-analytical solution. This is in contrast to an earlier benchmark test for longitudinally invariant waveguides [2]. Of course, this limits considerably the range of suitable problems.

Reliability and robustness of the method. The algorithms should be stable and reliable. This is commonly tested for new algorithms by calculating the propagation of an eigenmode along its waveguide and verifying the stability of the field distribution, but this alone is not enough. To test the stability of an algorithm, the tilted waveguide test described below defines a sequence of tests of increasing severity for BPM algorithms.

Universality. A BPM software package should be universal, *i.e.*, it should be a reliable tool being able to tackle a great variety of problems regardless of the underlying geometry and refractive index profile. Algorithms for very special cases and smoothing algorithms to transform a given refractive index geometry to a simpler form that is easier to be handled numerically do not satisfy this requirement. The ultimate goal should be to tell the algorithm the geometry exactly as it is.

Efficiency. An increasing number of scientists use personal computers and workstations to design and simulate integrated optical structures. A supercomputer environment is thus not considered acceptable. However, if the memory space and/or time complexity of BPM-algorithms tends to exceed the limits of a personal computer environment, its applicability is significantly reduced.

Members of the Working Group 2 of COST 240 have been engaged in the development and investigation of two sets of benchmark tests for the accuracy of BPM algorithms. The first set which contains the "Tilted waveguides" and the "Square meander coupler" problems, was investigated in the years 1992 to 1994 by 25 participating groups using 19 software packages. The second set called "Symmetrical coupler and gain-loss waveguide" was investigated during 1996 to 1997 by a smaller number of groups. The latter task is based on the eigenmode analysis of the gain loss waveguide presented in Sec. 3.2.

The software used in the tests can be divided into two general groups, beam propagation methods and eigenmode expansion methods. The last one serves mainly as a reference for the problems. To give a clear view on the correlation between a special BPM algorithm and the result of the calculations we have focused in the next subchapters only on the family of the algorithm. We have also suppressed information on the individual groups that delivered the results. More detailed information can be found in the referenced original publications.

The family tree of BPM algorithms is discussed in detail in Ch. 2. Here we briefly list all BPMs that have been used in the tests:

- FFT-BPM
- FD-BPM (using 3-point and 5-point FD operator)
- FD/FE-BPM
- series expansion BPM
- Padé approximation BPM
- adaptive grid FE-BPM
- MoL-BPM

3.1.1 Tilted waveguides

The slab multimode waveguide being investigated consists of a single 8.8 μm wide step-index waveguide with a refractive index step of $\Delta n = 0.13$. InP was chosen as the background material. At the wavelength of $\lambda = 1.55$ μm the waveguide supports 11 guided modes for both TE and TM polarisations. The benchmark problem consists of a sequence of simulations in which the fundamental (zero-th), 5th and 10th order mode propagate under different tilt angles of 0°, 5°, 10° and 20° with respect to the z-axis. The waveguide parameters are given in Fig. 3.1. For each of the examples the power transfer to other modes was calculated. Obviously, the test has the advantage of being rather simple. The field distributions and the eigenvalues can be calculated quasi-analytically [3] by standard methods. The exact solution should exclude any power transfer to any other mode.

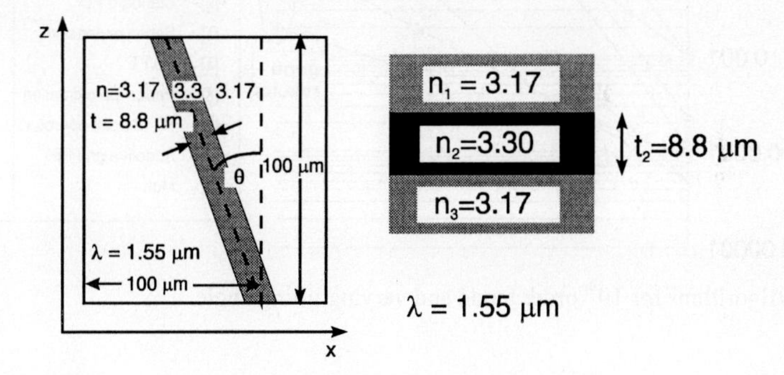

Fig. 3.1. Description of the tilted waveguide problem.

The power transfer $\Delta P = 1 - \left|\langle \phi_0(x - z\tan\theta)\, \phi_z(x)\rangle\right|^2$ for a propagation length of $z = 100$ μm and the remaining power $P_{end} = \int_{-\infty}^{\infty} \left|\varphi_z(x)\right|^2 dx$ has to be calculated.

This numerical experiment measures the sensitivity of the algorithm with respect to the discretisation density. The higher order mode in combination with the tilted propagation direction can be looked at as an approximation to radiation modes. By studying these "radiation-like" modes it is easier to estimate the accuracy of various algorithms for treating radiation modes. Furthermore, the tilt angle gives insight into the behaviour of the algorithms concerning wide propagation angles. But, in fact, it does not give a full characterisation of wide angle behaviour because propagating a single mode at wide angle can always be carried out using paraxial algorithms provided that the reference value of the refractive index profile is chosen properly [4]. The essential advantage of the true wide-angle algorithms is their capability to propagate fields having wider spatial spectral widths.

Originally, the simulations should be carried out for both for TE and TM polarisations. It turned out, however, that a restriction to the TE polarisation and to the most extreme situations, the fundamental mode with the tilt angle of 0° and 10th order mode with tilt angles between 0° and 20°, was useful and sufficient for the comparison.

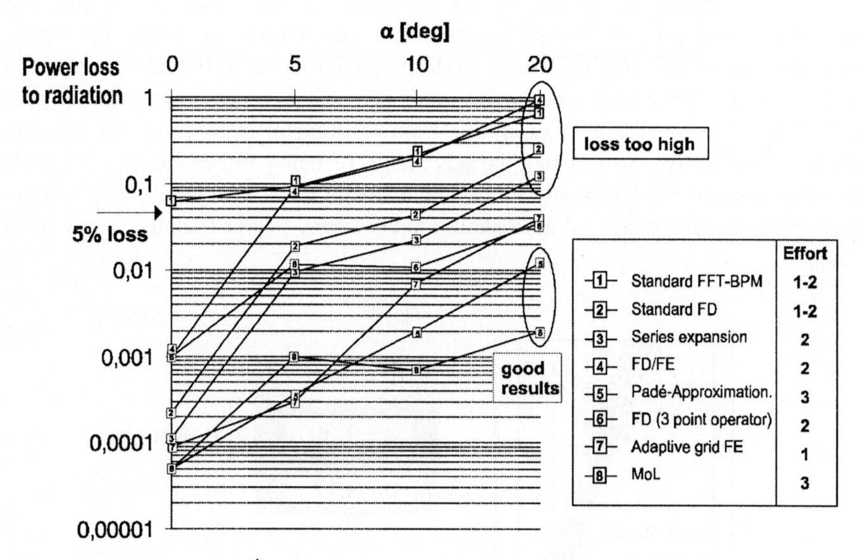

Fig. 3.2. Algorithms for 10th order mode and varying tilting angle.

All BPM algorithms except the 'classical' BPM mastered the propagation of the fundamental mode along the untilted waveguide. The overall power loss which occurred in this test was < –30 dB for all participants. In contrast, the propagation of the 10th-order mode resulted for most of the algorithms in severe problems which increased with increasing tilt angle. Fig. 3.2 illustrates the evolution of power loss with increasing tilt angle. Here we have used mean values for describing the results of standard FFT-BPM, standard FD-BPM and series expansion methods as a group, although the scattering of the results is very large and indicate the breakdown of the corresponding algorithms. As expected, we observe an increasing power loss with increasing tilt angle. Fig. 3.3 showing the power distribution of 10th mode for tilting angles of $\theta = 5°$ and $\theta = 20°$, illustrates that unphysical mode behaviour is the source of the power transfer to radiation modes. Fig. 3.4 shows the power distribution plot calculated with the MoL-BPM. This indicates that the simulation of the tilted wave by using additionally the radiation modes in that method leads to precise results and that the stepwise approximation of a tilted waveguide (see insert) can be used.

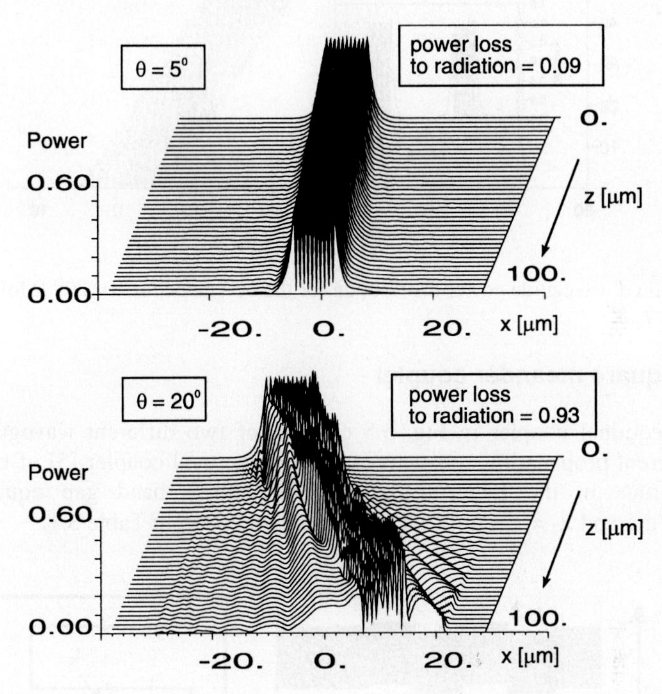

Fig. 3.3. Tilted waveguide power distribution for the 10th mode for $\theta = 5°$ and $\theta = 20°$ (FD/FE method).

It was originally planned to measure the efficiency of the algorithms in terms of the relative CPU-time, i.e. by the CPU-time (t_{CPU}) for the benchmark test related to the CPU-time t_{ref} of a whetstone computer benchmark test. However, the non-uniformity of computers, operating systems and programming languages made it impossible to obtain the desired results. To give at least a rough idea of the performance we introduced the "effort". This number is defined to range from 1 to 3, indicating raising effort which is correlated to the number of discretisation points and time complexity in a more qualitative way. We see in the insert of Fig. 3.2. that the best candidates do not necessarily require the highest effort.

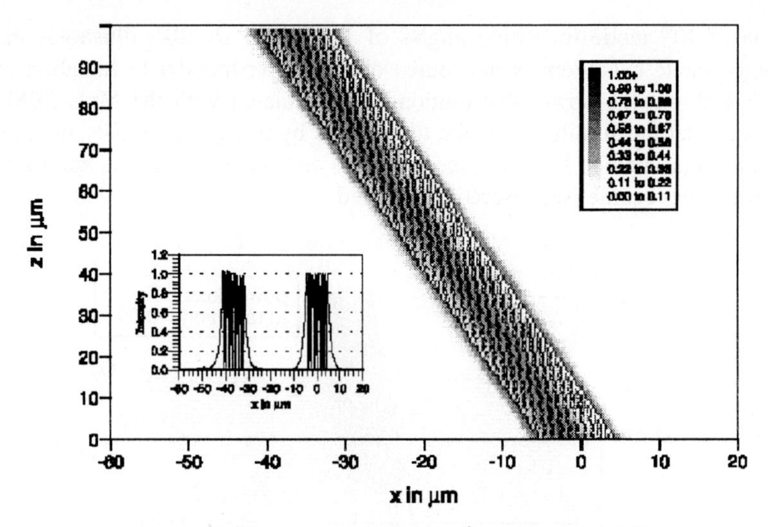

Fig. 3.4. Tilted waveguide power distribution for the 10th mode for $\theta = 20^{0}$ (MoL-BPM).

3.1.2 Square meander coupler

The directional coupler in Fig. 3.5 consists of two different waveguides with very different propagation constants (suppose a vertical coupler [5] of two different materials in the InGaAsP/InP system with a band gap equivalent to $\lambda_1 = 1.05$ µm and $\lambda_2 = 1.3$ µm). Its parameters are given in Table 3.1.

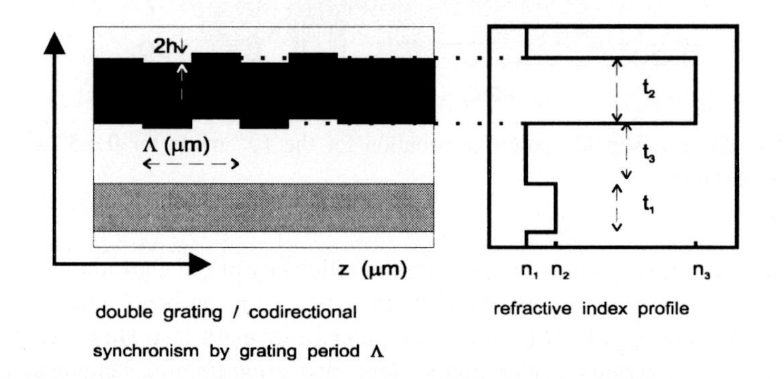

Fig. 3.5. Description of the square meander type problem.

t_1 [µm]	t_2 [µm]	t_3 [µm]	n_1	n_2	n_3
0.6	0.3	2.5	3.20222	3.28098	3.16878

Table 3.1. Parameters of a meander coupler.

The coupling behaviour at a wavelength of $\lambda = 1.555$ µm for a square grating with a height of $2h = 100$ nm was calculated. The following parameters of the coupler had to be determined:

- the period length Λ at resonance for a complete power transfer for the coupling between the two fundamental coupler modes,
- optimum device length L at resonance,
- the dispersion behaviour in terms of the channel separation $\Delta\lambda$, power at resonance P_{res} and the extinction ratio of first sidelobes on both sides, P_{left} and P_{right}, neglecting for simplicity the material dispersion.

This benchmark problem was intended to test the spectral behaviour of the BPM codes, since it is characterised by a finite spatial spectral width. In addition, a fine discretisation is necessary to handle the corrugation depth of 100 nm. The influence of boundary conditions on the results has been studied as well.

A reference solution of high accuracy is obtained by the guided eigenmode expansion propagation method [6] (GEEP) or, if radiation is included, by the bidirectional eigenmode propagation method [7] (BEP) or by the method of lines (MoL-BPM). The following results were obtained:

1. The grating period has been calculated to be $\Lambda = 101.7794$ µm.
2. The number of periods for complete power transfer from symmetric to asymmetric coupler modes (from the fundamental to the first higher mode of the asymmetric coupler) are $N = 42$.
3. The optimum device length is then $L = 4275$ µm.
4. The dispersion behaviour of the device if material dispersion is neglected is shown in Fig. 3.6. In order to ease the calculation the refractive index was kept constant for all wavelengths. The channel separation, *i.e.* the wavelength distance between the maximum and the first zero of the spectral characteristics, is $\Delta\lambda = 14$ nm.
5. In resonance the power transfer is $P_{res} = 0.5\ P_{in}$ between both eigenmodes. We thus have a loss of 3 dB. The extinction ratios at the first side-lobes are $P_{left} = 0.058$ and $P_{right} = 0.045$.

The results of the square meander coupler are summarised in Table 3.2. Five different software packages (including four mode expansion propagation methods (MEP) were used to obtain a reference for the comparison of different BPM calculations. There is only a slight difference between the MoL and the eigenmode propagation techniques. Here we focus only on the period length, device length and channel separation. The discrepancies in the amplitudes of the power transfer function as function of the wavelength (P_{res}, P_{left}, P_{right}), which may be correlated to boundary problems of these methods, will not be discussed here in detail.

Only five BPM results could achieve a sufficient accuracy. The period length has been found by four methods, one of them showing slight deviation (FD/FE). This seems to be the consequence of using absorbing boundary conditions.

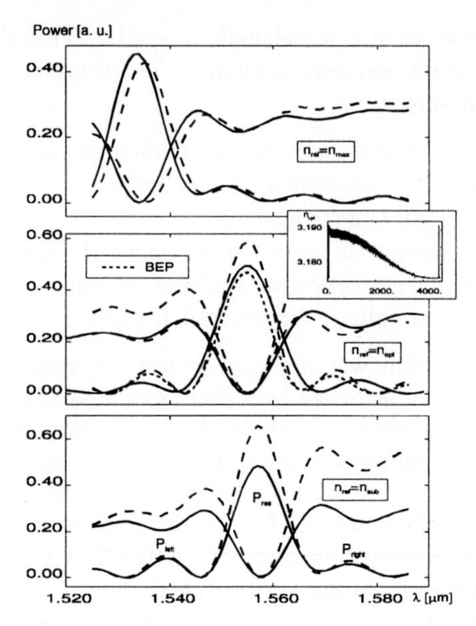

Fig. 3.6. FD/FE dispersion characteristics of a meander coupler for different n_{ref} and boundary conditions. BEP and GEEP results are also shown. The solid lines mean "high" reflection, broken lines "low" reflection at the boundaries, as explained in the text.

Reference (MEP)						
Method	Period length (μm)	Device length (μm)	Channel separation (nm)	P_{res}	P_{left}	P_{right}
GEEP I	101.78	4275	14	0.5	0.058	0.045
BEP	101.78	4275	14	0.5	0.058	0.045
MEP I	101.78	4275	14	0.478	0.046	0.046
GEEP II	101.78	4275	14.3	0.478	0.045	0.052
MEP II	101.65	4269	14	0.563	0.054	0.052
MoL	102.22	4293	14	0.532	0.046	0.055

Results (BPM)						
Method	Period length (μm)	Device length (μm)	Channel separation (nm)	P_{res}	P_{left}	P_{right}
Padé	101.77	4300	14	0.53	0.0457	0.0528
Adaptive mesh	101.78	4275	13.5	0.6	0.08	0.12
FD/FE	102.5	4305	13	0.6	0.06	0.08
FD2BPM	101.78	4280	14.5	0.46	–	–
FFT BPM	88–107	–	–	–	–	–
FD	101.78	4275		no coupling observed		

Table 3.2. Results of BPM and reference calculations for meander coupler as a wavelength sensitive multiplexer. Different implementations of identical methods are distinguished by labels I and II.

Furthermore, for paraxial approximation (without wide-angle enhancement), the choice of the reference refractive index n_{ref} turned out to have a strong influence on the spectral response. In an additional study we have investigated the influence of both the boundary conditions, *i.e.* by artificially introducing low and high reflectivity, and the choice of n_{ref}. We have studied the following three choices:

$$n_{rel} = \begin{cases} n_{\max} \\ n_{\text{substr}} \\ n_{\text{opt}} \end{cases} \qquad \text{with} \qquad n_{\text{opt}}^2 = \frac{-\int \frac{\partial u}{\partial x}\frac{\partial u^*}{\partial x}\,dx + \int n^2(x)\,k_0^2 u\,u^*\,dx}{\int k_0^2\,u\,u^*\,dx}$$

Here $n_{\max}$ is the highest refractive index occurring in the waveguide structure and n_{substr} is the refractive index of the background material (InP). The optimal refractive index n_{opt} [4] has to be calculated at each propagation step as defined in the equation above. The optimum refractive index can be interpreted as the mean value of all contributing modes, weighted with their relative contribution to the total field. Thus, the value changes during the coupling of the meander coupler. This is shown as an insert in Fig. 3.6. The actual choice of the refractive index is indicated by the labels at each dispersion curve. We observe a strong shift of the dispersion curves for the different values of n_{ref}. The FWHM remains more or less unchanged. The boundary conditions (high or low reflections) have a strong influence on the amplitude of the dispersion characteristic. Thus, transparent boundary conditions would be superior in this case.

An illustrative example of a large variety of competitive algorithms is presented in Table 3.2. Note the insufficient accuracy of standard FD BPM and classical FT BPM for calculations of this type.

Discussion and conclusions. The results of the benchmark examples presented here show explicitly that the successful propagation of a fundamental mode along a straight waveguide with smooth dielectric profile is a necessary but by no means also sufficient validation of a BPM-algorithm. The "classical" BPM failed in all of the benchmark examples discussed here. The large scattering in the calculated power for the same class of methods (as in case of the standard FD-BPM) may be probably correlated with the different discretisation densities that have been used. The most important conclusions of the benchmark test can be formulated as follows:

- the MoL-BPM which directly solves the Helmholtz equation yields very accurate results in all three benchmark tests,
- the advantages of wide-angle BPMs for the corresponding BPM problems could be verified,
- the use of non-uniform and especially of adaptive meshes resulted in massive advantages with respect to performance and accuracy,

- the improvement of paraxial BPM-algorithms by properly chosen reference refractive indices could be observed,
- transparent boundary conditions seem to be useful in all applications.

Generally, the standard FD-BPM algorithm can be significantly and rather trivially improved by using Padé approximations [8], [9] to allow steeper propagation angles choosing the right reference refractive index [4] and using transparent boundary conditions[10], [11].

3.1.3 Gain-loss waveguide benchmark test

The gain-loss problem that will be described in detail in the next section was used for defining another benchmark test [12] for the beam propagation method since it possesses a quasi-analytical solution. This test is complementary to the two tests described above in the sense that it is focused on the imaginary part of the refractive index that was always zero in the previous tests.

The definition of the tasks is the following: calculate the wave behaviour of a waveguide structure composed of a waveguide section with a balance of gain and loss that is placed between two lossless (single-mode) input and output waveguides with identical (real) refractive indices (see Fig. 3.7), for various values of the loss coefficient α of the gain-loss section. The existence of a branching point α_{branch} that divides the region of absorption coefficients into two parts is known from the dispersion diagram (see Sec. 3.2).

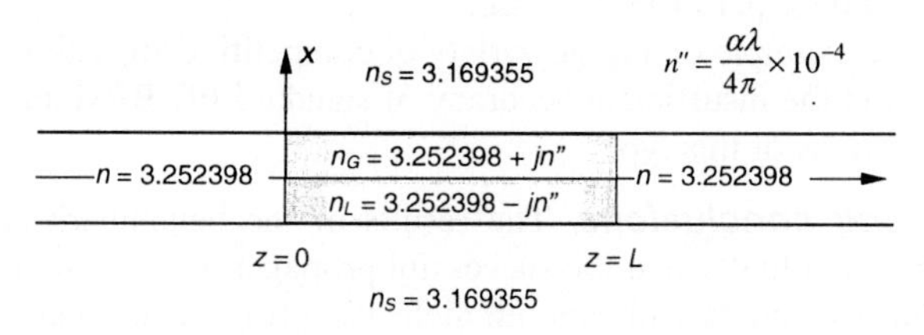

Fig. 3.7. Waveguide structure with complex refractive indices for the BPM benchmark test.

For $\alpha_1 < \alpha < \alpha_{branch}$, the waveguide section with balanced gain and loss supports two *lossless* modes with similar field distributions, and thus a beat length Λ can be observed in the transmitted power. This length is increasing with increasing values of α. The value of α_{branch} can be easily determined from (3.5) (cf. Sec. 3.2) by plotting $1/\Lambda^2$ over α, as it is shown in the left part of Fig. 3.8. For $\alpha > \alpha_{branch}$ we have a superposition of one amplified and one attenuated mode in the gain-loss section, where only the amplified one will survive, and thus the effective gain constant α_{eff} as a function of α from a plot of α_{eff}^2 can be determined. This is shown in the right part of Fig. 3.8. Shown as an example are results from a FD-BPM algorithm for complex refractive indices.

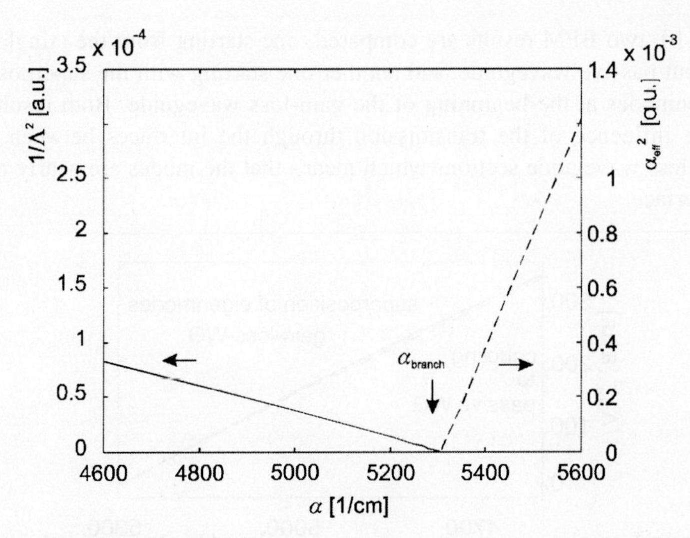

Fig. 3.8. Results of the BPM benchmark test with beat length Λ and α_{eff} as a function of α. The variation of power during propagation along the device is derived in Sec. 3.2.

To utilise the quasi-analytic solution of the eigenmode analysis for this benchmark test as a reference, we have to demonstrate that the generation of radiation modes at the interfaces from the passive waveguide to the gain-loss section is negligible. Radiation modes may influence the results due to the amplifying and attenuating effect of the gain-loss waveguide. In Fig. 3.9 the total power loss due to reflection and excitation of radiation modes at the interfaces between the sections is shown.

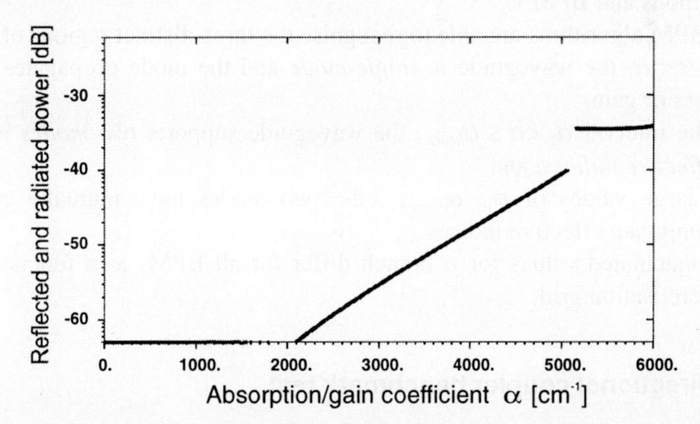

Fig. 3.9. Total power loss due to mode matching of the guided modes at the interface between the passive and the gain-loss waveguide sections. This calculation is based on the BEP (cf. Sec. 2.3).

In Fig. 3.10, two BPM results are compared, one starting from the (single) mode of the input passive waveguide, and another one starting with the superposition of both eigenmodes at the beginning of the gain-loss waveguide. Both results show negligible influence of the transmission through the interfaces between passive and gain-loss waveguide sections which means that the modes are nearly matched at the interface.

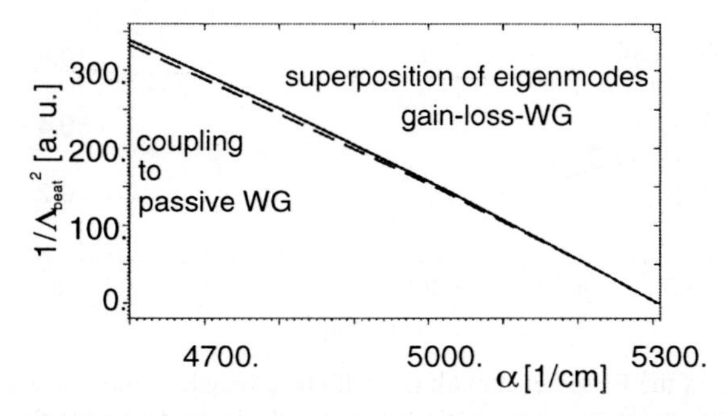

Fig. 3.10. A comparison of two situations: starting at the passive waveguide with the single eigenmode and starting with the superposition of both eigenmodes at the gain-loss waveguide. Both results closely coincide, showing negligible influence from the mode mismatch at the interface.

The conclusions from this BPM test can be expressed by the following statements:

– Radiation modes will not be generated at the interface between passive and gain-loss waveguide. This has been shown using eigenmode expansion methods and BPM´s.
– All BPM algorithms are able to recognise the three distinct regions of α:
 1. for $\alpha < \alpha_l$ the waveguide is *single-mode* and the mode propagates without loss or gain,
 2. in the interval $\alpha_1 < \alpha \leq \alpha_{branch}$ the waveguide supports *two modes with real effective indices*, and
 3. for large values of $\alpha > \alpha_{branch}$, the two modes have mutually complex-conjugate effective indices.
– The calculated values for α-branch differ for all BPMs as a function of the discretisation grid.

3.1.4 Directional coupler benchmark test

The last point of the above conclusions was the main reason why the gain-loss benchmark test was reformulated to address the discretisation and interface problems of BPMs, especially for waveguides with gain or loss. Here, the accuracy of BPM calculations of directional couplers with and — for comparison — also without an imaginary part of the refractive index is investigated. As in all earlier benchmark tests, we restrict ourselves to two-dimensional beam propagation, *i.e.*, to beam propagation in waveguides with one-dimensional cross sections. In the previous benchmark tests we have discussed the strong influence of the choice of the right reference refractive index, the advantages of transparent boundary conditions and of wide angle BPMs. Here we will focus on the investigation of the accuracy of different discretisation schemes which have been developed for waveguide structures with abrupt transitions of the refractive index at arbitrary positions relative to the BPM grid.

As a device under test for this benchmark test, two devices with well known accurate (numerical) solutions were chosen. Two directional couplers will be compared: a passive symmetrical directional coupler (Sym DC), and a waveguide with a balance of loss and gain (GL WG), as they are sketched in Fig. 3.11.

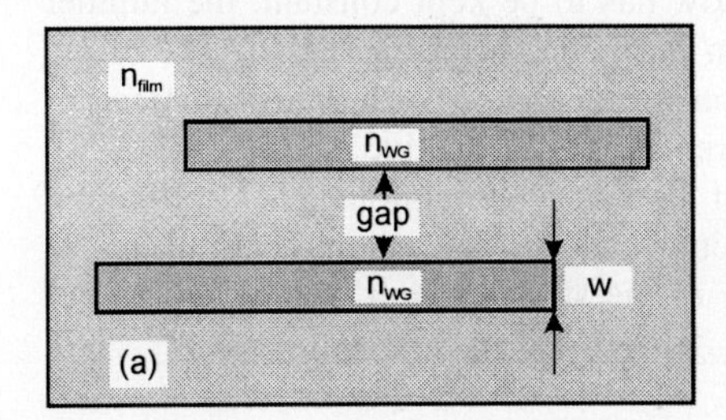

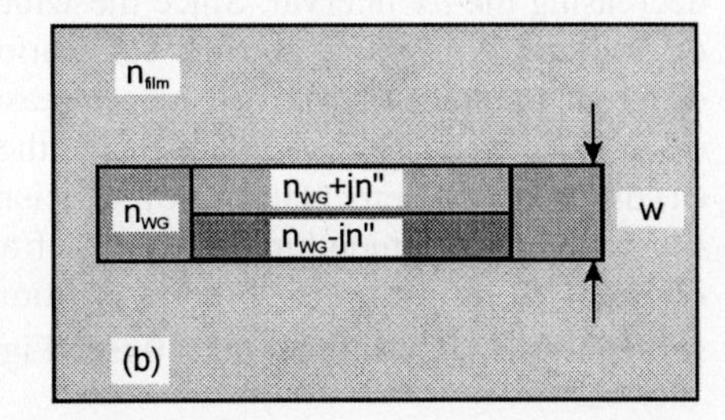

Fig. 3.11. Two devices are under test (a) symmetrical directional coupler and (b) the gain-loss waveguide. The refractive indices $n_{WG} = 3.252398$, $n_{Film} = 3.169355$ and the wavelength $\lambda = 1.550$ µm are common to both structures. In case of the passive symmetrical coupler (a) the widths are $w = 1.02555$ µm and gap = 1.555 µm. The gain-loss waveguide has a total width of w = 1.000 µm and is composed of two layers with mutually complex conjugate refractive indices $n_{WG} \pm jn''$, $n'' = \alpha\lambda/4\pi$. The imaginary parts vary in a very broad range; in terms of α, between $\pm 10^4$ cm^{-1}.

In both cases the spectral behaviour can be calculated semi-analytically using well known eigenmode solvers [3]. We focus on the beat length behaviour of both devices and compare the results of different BPM algorithms with the eigenmode solution.

The calculation of the passive Sym DC is straightforward. Propagating an eigenmode of the input waveguide along the coupler structure leads to a beating of the coupler eigenmodes and the beat length Λ can be determined. Here we investigate the accuracy of the calculation of the beat length Λ.

The dispersion diagram of the GL WG as a function of the imaginary part of the refractive index which is a function of the loss or gain parameter α has been investigated earlier [13], see also section 3.2. The dispersion curves have a particular behaviour at $\alpha_1 = 2725.67$ cm^{-1}, where a second order mode appears, and $\alpha_{branch} = 5226.30$ cm^{-1}, which is a branching point for N_{eff} curves. For values of

$\alpha<\alpha_1$ the waveguide is single-mode and the mode propagates without loss or gain. In the interval $\alpha_1<\alpha<\alpha_{branch}$ the waveguide supports two complex but lossless modes with real effective refractive indices. Finally, for values of $\alpha>\alpha_{\mathrm{branch}}$ the two modes have mutually complex-conjugate effective indices. Here we will focus on the beat length behaviour in the region of $\alpha_1<\alpha<\alpha_{branch}$ which can be used to extrapolate the branching point α_{branch} from the zero of the $1/\Lambda^2(\alpha)$ plot.

Influence of the discretisation scheme on the accuracy. Fig. 3.12 shows the step function of the imaginary refractive index of the GL WG and its digital representation on the equidistant grid x_j. For such simple structure as the GL WG the grid can be chosen in an optimal configuration by putting the step function in the middle between two x_i values. This is not possible for a more complicated geometry like the Sym DC. Thus, an uncertainty of the width w_x in the interval $w - dx/2 < w_x < w + dx/2$ may happen. A more accurate result can be obtained by decreasing the dx interval. Since the window has to be kept constant, the number of discretisation points and the calculation effort has to be increased. Therefore, different methods are in use to get reasonable accuracy with a low number of discretisation points. Among them there are: (i) adaptive mesh size [14], (ii) different implementations of efficient boundary conditions [15], [16], and (iii) rounding or smoothing the step refractive index. All these methods have the objective to introduce the precise position of the step function which may be described by $x_j+q\ dx$ with $1 \geq q > 0$ (see Fig. 3.12b.) into the algorithm.

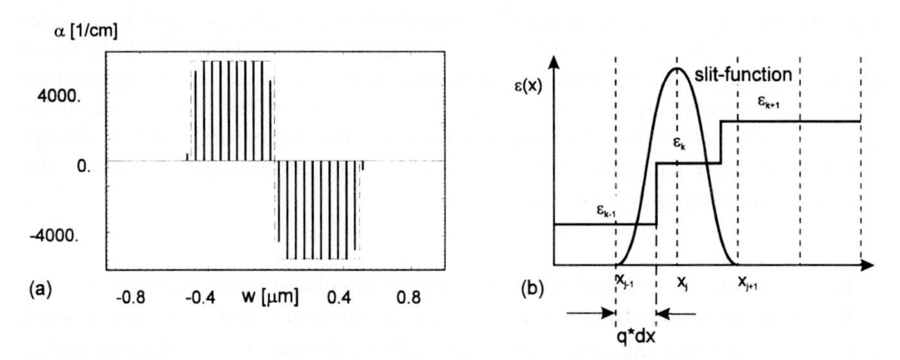

Fig. 3.12. Example of a discretisation method. The original refractive index profile is shown in (a) as a dashed line. The straight lines of (a) shows the imaginary part of the gain-loss waveguide using a convolution with a slit function as is sketched in (b): used is a cosine-slit function for a piecewise constant refractive index waveguides.

The methods having contributed to this investigation can be divided into three classes: Fourier transform BPM, FD BPM, and eigenmode propagation methods like MoL and BEP. The results of Fourier transform BPM are quite far away from the semi-analytic values. This can be attributed to the fact that FFT BPM is generally not suitable for waveguides with large refractive index contrast. Therefore, results from FFT BPM are not considered here. Methods used in the test encompass the finite difference (3-point and 5-point FD operator) BPM with a third order SVEA correction Padé (3,3) and efficient interface conditions [15], the FD/FE BPM with a rounding algorithm for the refractive index profile, and two implementations of the method of lines (MoL, cf. Sec. 2.2). The latter one is, in fact, an eigenmode propagation method. The difference to BEP (cf. Sec. 2.3) is essentially in the way how the eigenmodes are calculated; a discretisation and FD algorithm in the lateral direction is used. Thus, for large values of dx, we should expect a similar behaviour as in case of FD BPM. The BEP results are dropped in the comparison as they are identical with the result of eigenmode solvers that are used as a reference.

The methods used to handle the interface conditions have been all described in the open literature [15], [16] and will not be repeated here. For better understanding of fundamental problems, a very simple method will be described here. A rounding or smoothing algorithm for the refractive index can be used prior to the ordinary FD/FE BPM calculation (see Fig. 3.12a.). Here the convolution-integral

$$\varepsilon(x_j) = \int_{x_j - dx}^{x_j + dx} g(\xi, x_j)\varepsilon(\xi)d\xi$$

of the dielectric constant $\varepsilon = (n_{\mathrm{WG}} + jn'')^2$ and a cosine slit function

$$g(\xi, x_j) = \frac{1}{2dx}\left[1 + \cos\left(\pi\frac{\xi - x_j}{dx}\right)\right]$$

of the width $2dx$ (see Fig. 3.12b) is used.

Results and Discussion. Fig. 3.13 shows the results of BPM methods studying the interface conditions for a moderate grid width of $dx = 50$ nm. To prove the influence on the accuracy, we have shifted the grid relative to the actual device in small steps (variation of q between zero and unity).

The application of interface conditions leads to reasonably accurate results for the Sym DC that are close to the exact value. However, for the GL WG, a variation of the inaccuracy $\Delta\alpha_{branch}$ in the order of ± 10 cm^{-1} for the MoL and between 80 and 92 cm^{-1} for FD/FE BPM was found.

Fig. 3.14 shows the behaviour for decreasing values of dx. All methods show a good convergence to the exact value for very low dx (number of grid points a few thousand). For moderate values of $dx = 50$ nm, the FD BPM with a 5-point operator shows the superior accuracy for both devices. The rounding of the refractive index profile is useful for passive waveguide structures ($dx \cong 50$ nm), but not for waveguides with a complex refractive index. In any case, to ensure the accuracy of calculation, the convergence behaviour has to be proven for low dx, as it is demonstrated in Fig. 3.14. The best approximation can be extrapolated from these curves using a numerical fit for $dx \to 0$.

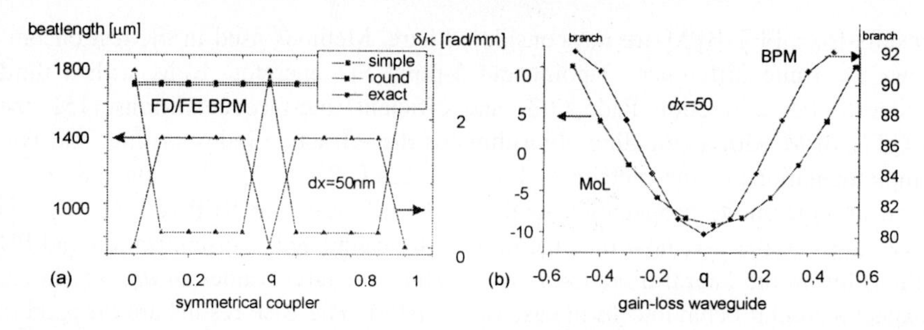

Fig. 3.13. Comparison of results for different relative positions of the test device and the grid (shift q) for $dx = 50$ nm. (a) shows predominantly erroneous results (asymmetrical instead of symmetrical coupler) for the simple BPM and a fairly good coincidence with the exact value for rounded refractive index profile. (b) shows variations of the α_{branch} value as a function of q.

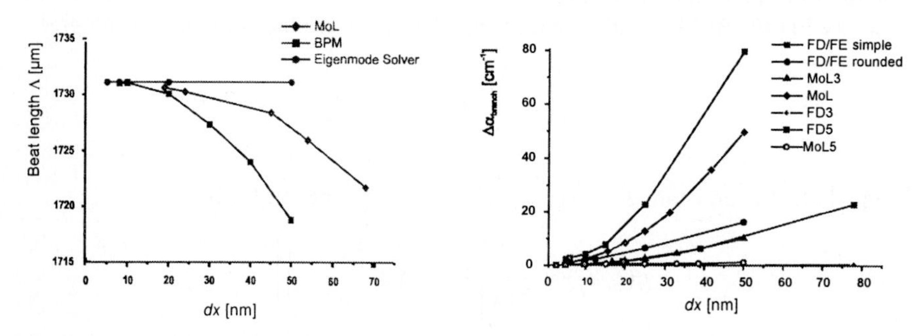

Fig. 3.14. Convergence behaviour for decreasing dx-values. The left figure concerning a symmetrical coupler, whereas the right figure is relating to the gain-loss waveguide. All methods show a good convergence to the exact α_{branch} value.

Conclusions. The accuracy of different BPM algorithms has been investigated with the help of two new benchmark tests based on a symmetrical directional coupler and a waveguide with a balance of gain and loss. In both cases, the spectral behaviour can be calculated semi-analytically as a reference using well known eigenmode solvers. The following conclusions can be drawn from the tests:

- Accurate BPM calculation for waveguides with an *imaginary part of the refractive index* requires a very small grid size dx in the order of a few nm.
- FD operators based on 5-point algorithms have a much higher accuracy than 3-point operators. This is true for both FD BPM and MoL.
- Algorithms which include an efficient interface handling like FD BPM or MoL have higher precision for larger dx values.
- To ensure the accuracy, the convergence behaviour has to be proven for decreasing grid size values dx. The best approximation can be extrapolated from such curves using a numerical fit for dx going to zero (Richardson extrapolation).

3.2 Wave propagation in a waveguide with a balance of gain and loss

A waveguide problem with a balance of loss and gain has been defined to study the mode behaviour in waveguides with complex refractive index and to compare different approaches. This problem is many-layered in both senses of the word. We will not discuss the numerical aspects of finding all roots [17] of the dispersion equation, we will give a quasi-analytic solution for the branching point of the dispersion diagram and discuss the peculiar wave behaviour of this waveguide.

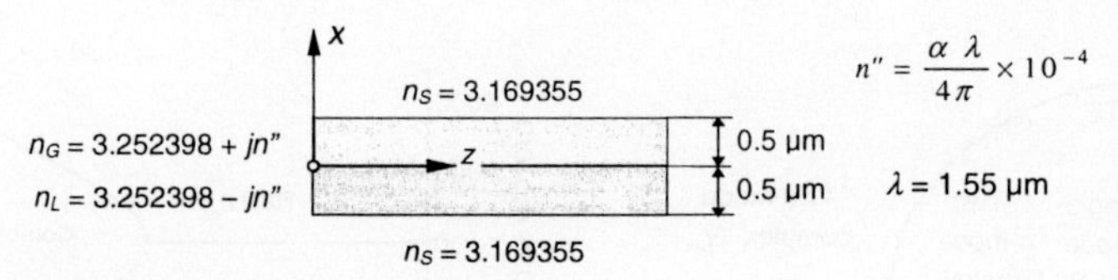

Fig. 3.15. Two-layered waveguide with gain and loss.

A two-layered waveguide with mutually complex conjugate refractive indices n_G, and n_L is surrounded by a substrate with a slightly lower real refractive index n_S. All parameters are given in Fig. 3.15. The imaginary parts of refractive indices of guiding layers vary in a very broad range: in terms of the absorption (gain) coefficient α, between 0 and $\pm 10^4$ cm^{-1}.

3.2.1 Quasi-analytic solution

A large number of well-known methods can be used to find effective refractive indices and mode fields of TE modes for any α in this interval. In the most straightforward method, the electric field distribution of a guided mode in the waveguide is expressed in the form

$$\begin{aligned} &x \le -d:\ E = A\exp\left[k\gamma_S(x+d)\right],\quad -d < x \le 0:\ E = B_1\cos k_0\gamma_L x + B_2 \sin k_0\gamma_L x,\\ &x > d:\ \ E = D\exp\left[-k\gamma_S(x+d)\right],\quad 0 < x \le d:\ E = C_1\cos k_0\gamma_G x + C_2\sin k_0\gamma_G x, \end{aligned} \tag{3.1}$$

where

$$\gamma_S = \sqrt{\varepsilon_{eff} - \varepsilon_S},\ \gamma_L = \sqrt{\varepsilon_L - \varepsilon_{eff}},\ \gamma_G = \sqrt{\varepsilon_G - \varepsilon_{eff}},\ \varepsilon_L = (n - jn'')^2,$$

$$\varepsilon_G = \varepsilon_L^*,\ \varepsilon_{eff} = n_{eff}^2,\quad k_0 = 2\pi/\lambda \text{ is the angular repetency}$$

and the sign of n_{eff} is chosen so that $\mathrm{Re}\{n_{eff}\} > 0$.

The conditions of continuity of the field E and its derivative at $x = 0$ and $x = \pm d$ give a set of homogeneous linear equations for unknown amplitudes A, B_1, B_2, C_1,

C_2, and D. To get a non-trivial solution, the determinant of this set of equations must be equal to zero. It leads to the dispersion equation of the form

$$\begin{aligned}\Phi(\varepsilon_{eff},\alpha) &= \gamma_G(\gamma_S \cos k_0\gamma_G d - \gamma_G \sin k_0\gamma_G d)(\gamma_S \sin k_0\gamma_L d + \gamma_L \cos k_0\gamma_L d) \\ &+ \gamma_L(\gamma_S \cos k_0\gamma_L d - \gamma_L \sin k_0\gamma_L d)(\gamma_S \sin k_0\gamma_G d + \gamma_G \cos k_0\gamma_G d) = 0.\end{aligned} \tag{3.2}$$

Having solved this equation numerically, the field amplitudes A, B_1, B_2, C_1, C_2, D can be calculated, and the field distribution is then explicitly given by (3.1). The calculated dependencies of real and imaginary parts of effective refractive indices versus the absorption coefficient α are plotted in Fig. 3.16.

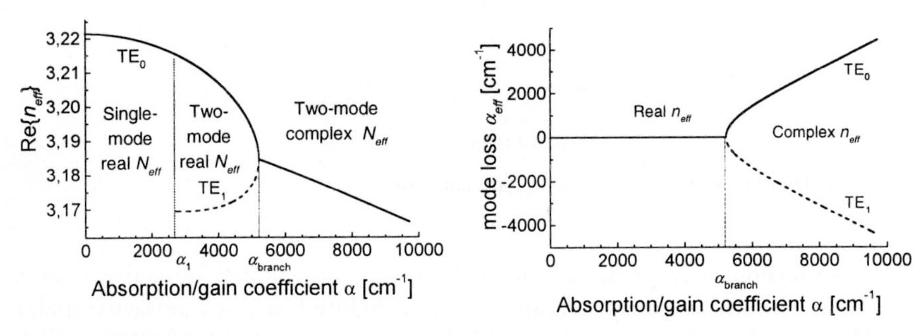

Fig. 3.16. Effective refractive index Re{n_{eff}} and $\alpha_{eff} = (4\pi/\lambda)\,\text{Im}\{n_{eff}\}$ versus α

It is seen that for values of $\alpha < \alpha_1 = 2725\ \text{cm}^{-1}$, the waveguide is single-mode and the mode propagates without loss or gain. In the interval $\alpha_1 < \alpha \le \alpha_{branch} = 5226.3\ \text{cm}^{-1}$, the waveguide supports two modes with real effective indices, and finally, for very large values of $\alpha > \alpha_{branch}$, the two modes have mutually complex-conjugate effective indices. With increasing attenuation and gain in the waveguiding layers above α_1, the effective refractive indices and the mode fields approach to each other, and at $\alpha = \alpha_{branch}$ both effective indices degenerate to a single real value. For larger α, one mode is attenuated while the other one grows, but both propagate with the same velocity. The mode field and phase distributions for $\alpha = 2000\ \text{cm}^{-1}$, $\alpha = 4000\ \text{cm}^{-1}$, $\alpha = 5000\ \text{cm}^{-1}$, $\alpha = 5226\ \text{cm}^{-1}$ ($= \alpha_{branch}$) and $\alpha = 7000\ \text{cm}^{-1}$ are plotted in Fig. 3.17a), b), c) and d) respectively.

The modes fulfil the condition of complex orthogonality (without complex conjugate)

$$\int_{-\infty}^{\infty} E_0(x)E_1(x)dx = 0, \tag{3.3}$$

which follows directly from the wave equation.

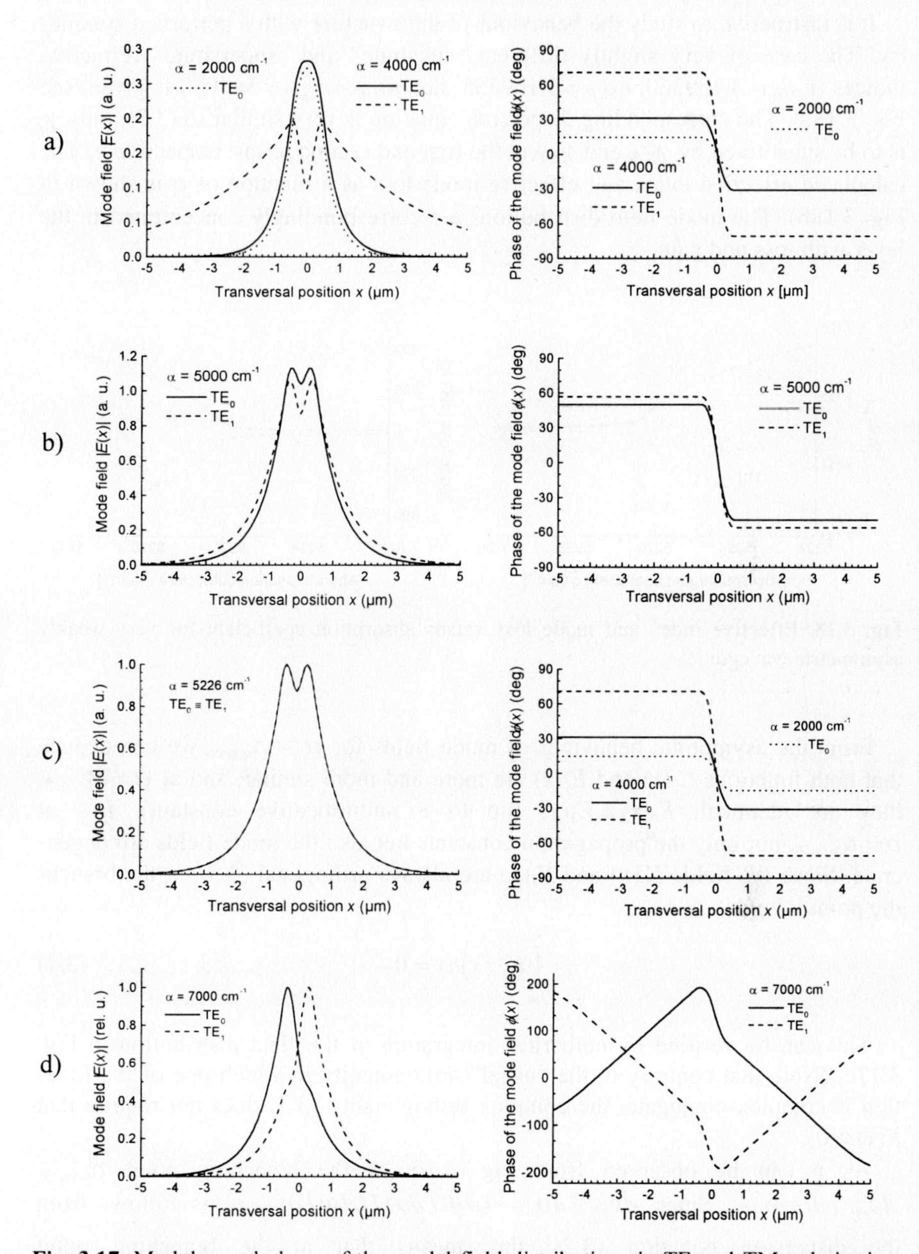

Fig. 3.17. Modulus and phase of the modal field distributions in TE and TM polarisation for a) $\alpha = 2000\ \text{cm}^{-1}$, $\alpha = 4000\ \text{cm}^{-1}$, b) $\alpha = 5000\ \text{cm}^{-1}$, c) $\alpha = 5226\ \text{cm}^{-1}$ (= α_{branch}) and d) $\alpha = 7000\ \text{cm}^{-1}$.

It is instructive to study the behaviour of the structure with a perturbed symmetry. The case of very slightly different "substrate" and "superstrate" refractive indices ($n_{SUB} = 3.169360$, $n_{SUP} = 3.169355$, i.e., $n_{SUB} - n_{SUP} = 5\times10^{-6}$) is given, see Fig. 3.18a). The corresponding dispersion equation is very similar to (3.2), only γ_S is to be substituted by γ_{SUB} and γ_{SUP} in the first and second terms, respectively. The calculated effective index and effective mode loss as a function of α is shown in Fig. 3.18b). The mode field distributions are correspondingly concentrated in the layer with loss and gain.

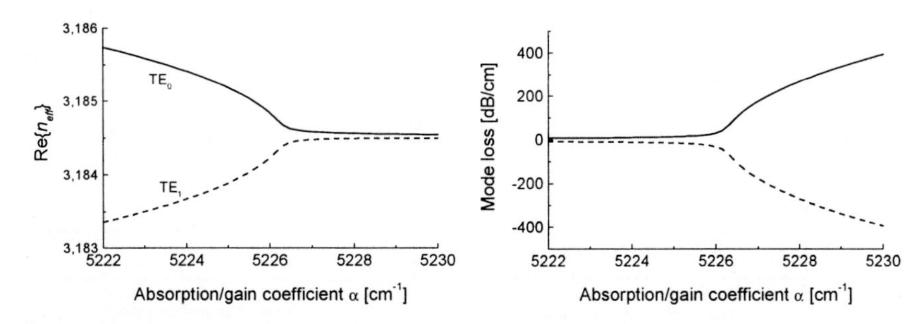

Fig. 3.18. Effective index and mode loss versus absorption coefficient for very weakly asymmetric waveguide.

From the asymptotic behaviour of mode fields for $\alpha \rightarrow \alpha_{\text{branch}}$ we can deduce that both functions $E_0(x)$ and $E_1(x)$ are more and more similar, and at $\alpha = \alpha_{branch}$, they are identical, $E_0(x) \equiv E_1(x)$ (up to a multiplicative constant). *I.e.*, at $\alpha = \alpha_{branch}$, not only the propagation constants but also the mode fields are degenerate. Since the fields $E_0(x)$ and $E_1(x)$ are always orthogonal (3.3), at the branching point it holds

$$\int_{-\infty}^{\infty} E^2(x)dx = 0\,. \tag{3.4}$$

This can be verified by numerical integration of the field distribution in Fig. 3.17c. (Note that contrary to the "usual" orthogonality in which one of the function is complex-conjugate, the complex orthogonality (3.3) does not require that $E(x) \equiv 0$).

As it can be observed from Fig. 3.16, at the branching point α_{branch}, $d\varepsilon_{eff}/d\alpha \rightarrow \infty$. Since $d\varepsilon_{eff}/d\alpha = -(\partial\Phi/\partial\alpha)/(\partial\Phi/\partial\varepsilon_{eff})$, as follows from the dispersion equation (3.2), this means that at the branching point $\partial\Phi/\partial\varepsilon_{eff} = 0$. We can thus get the critical value of α_{branch} by solving a set of two complex transcendental equations

$$\Phi\left(\varepsilon_{eff}, \alpha\right) = 0, \qquad \partial\Phi/\partial\alpha = 0 \tag{3.5}$$

for ε_{eff} and α. The (numerical) solution gives $\alpha_{branch} = 5226.302\ \text{cm}^{-1}$, $\varepsilon_{eff} = 10.1412428\ \text{cm}^{-1}$ ($n_{eff\ branch} = 3.1845318$) The behaviour of the dispersion curves in Fig. 3.16 can easily be explained by the following arguments: Since $\Phi(\varepsilon_{eff}, \alpha)$ is a regular function of the complex variable ε_{eff} and a smooth function of α, it can be expanded into a Taylor (Laurent) series. It follows from (3.2), however, that in the expansion in the vicinity of the branching point the corresponding two terms vanish. The dispersion equation can thus be approximated by

$$\Phi\left(\varepsilon_{eff}, \alpha\right) \approx \Phi'_{\alpha}\left(\alpha - \alpha_{branch}\right) + \frac{1}{2}\Phi''_{\varepsilon_{eff}}\left(\varepsilon_{eff} - \varepsilon_{eff,branch}\right)^2 = 0, \tag{3.6}$$

where

$$\Phi'_{\alpha} = \partial\Phi\left(\varepsilon_{eff,branch}, \alpha_{branch}\right)/\partial\alpha, \quad \Phi''_{\varepsilon_{eff}} = \partial^2\Phi\left(\varepsilon_{eff,branch}, \alpha_{branch}\right)/\partial\varepsilon_{eff}.$$

From (3.6) it follows that the approximate solution of the dispersion equation near α_{branch} is

$$\varepsilon_{eff} \cong \varepsilon_{eff,branch} \pm j\sqrt{(2\Phi'_{\alpha}/\Phi''_{\varepsilon_{eff}})(\alpha - \alpha_{branch})}, \tag{3.7}$$

where $2\Phi'_{\alpha}/\Phi''_{\varepsilon_{eff}} \approx 5.6753 \times 10^{-5}$ cm. The expression (3.7) with $n_{eff} = \sqrt{\varepsilon_{eff}}$ represents very well the behaviour of the dispersion curves in the vicinity of the branching point.

For $\alpha < \alpha_{bramch}$ the gain-loss waveguide supports two modes with *real* effective indices. Both these modes can be excited simultaneously. It leads to a mode beating with the beat length

$$\Lambda = \frac{\lambda}{n_{eff,0} - n_{eff,1}} \approx \frac{\lambda\,(n_{eff,0} + n_{eff,1})}{n_{eff,0}^2 - n_{eff,1}^2} \approx \frac{\lambda\, n_{eff,branch}}{\sqrt{2(\Phi'/\Phi'')(\alpha_{branch} - \alpha)}}, \tag{3.8}$$

It means that $1/\Lambda^2$ is *linearly* proportional to $\alpha_{branch} - \alpha$, with $1/\Lambda^2 \to 0$ for $\alpha \to \alpha_{branch}$.

In Fig. 3.19 the resulting dispersion curves of (3.7) are compared to the curves obtained by accurately numerically solving the dispersion equation (3.2).

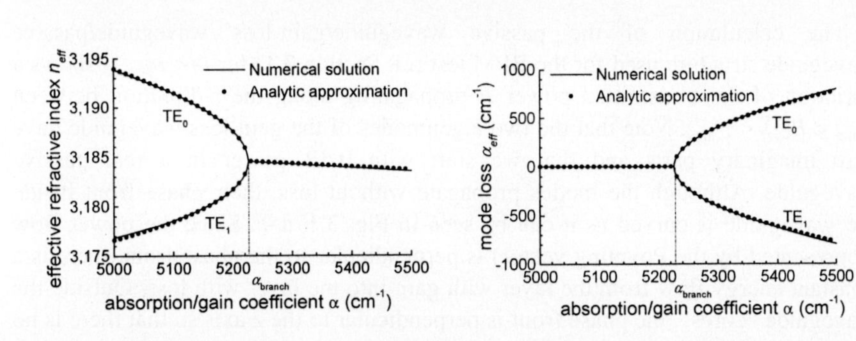

Fig. 3.19. Comparison of the numerically solved exact dispersion equation (3.2) with the approximate (analytic) expression (3.7) near the branching point.

3.2.2 Wave growth in the gain-loss waveguide with lossless eigenmodes

The dispersion behaviour of a mode propagating from a passive waveguide through the gain-loss waveguide and again along a passive waveguide is very interesting (see Fig. 3.20). Using BPM or BEP methods we observe a periodic increase and decrease of the wave amplitude as a function of the length L of the gain-loss waveguide for the two-mode, loss-less mode region of the α values ($\alpha_1 < \alpha \leq \alpha_{branch}$). In this region we observe an interference of both eigenmodes with a beat length Λ depending on the α-value. This characteristic mode behaviour was used to define a BPM benchmark test (cf. Section 3.1). We can easily determine α_{branch} from (3.8) by plotting $1/\Lambda^2$ over α, as is shown in Fig. 3.8, left part. For $\alpha > \alpha_{branch}$ we have a superposition of an amplified and an attenuated mode, where only the amplified will survive, and we can thus determine the effective gain constant α_{eff} as a function α from a plot of α^2_{eff} as is shown in Fig. 3.8, right part. Shown as example are results obtained with a FD-BPM algorithm.

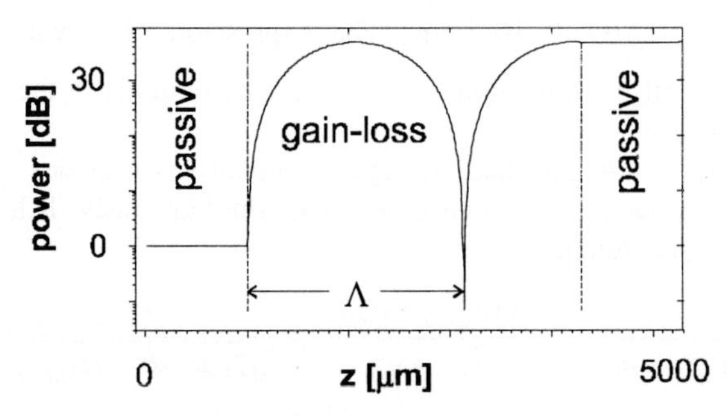

Fig. 3.20. Strong power attenuation/amplification behaviour of the gain-loss waveguide at $\alpha = 5306$ cm^{-1} during propagation. Transmitted power varies from –10 dB to 30 dB relative to the input value (0 dB).

The calculation of the passive waveguide/gain-loss waveguide/passive waveguide structure used for the BPM test (cf. Section 3.1) for $\alpha < \alpha_{branch}$ shows a variation of the transmitted power P propagating along the z-direction between $P_{\min} < P_{start} < P_{\max}$. Note that the two eigenmodes of the gain-loss waveguide have zero imaginary parts and that we start with 0 dB power in a real passive waveguide. Although the modes propagate without loss, their phase front inside the waveguide is curved as it can be seen in Fig. 3.17a–c. Since the power flow (represented by the Poynting vector) is perpendicular to the phase front, there is a constant energy flow from the layer with gain into the layer with loss. Outside the waveguide "cores", the phase front is perpendicular to the z axis so that there is no transversal power flow there. For $\alpha > \alpha_{branch}$, the phase front is curved in such a way that the mode with loss (which also has the field maximum in the lossy layer, see Fig. 3.17d) "absorbs" power from the whole space while the mode with gain "supplies" power into all the space.

Investigation with an eigenmode expansion method like BEP [7] predicts the output power fluctuation as shown in Fig. 3.20, where $P_{\min} \leq -10$ dB and $P_{\max} > 30$ dB.

Following the notation of [18] (with l = number of the section, i, k = number of modes), we can write the amplitude expansion coefficient a_{ik} at the interface of the passive/gain-loss waveguide as

$$^{l+1,l}a_{ik} = \int_{-\infty}^{\infty} {}^{l+1}E_i(x) \cdot {}^{l}E_k(x)dx \Bigg/ \sqrt{\int_{-\infty}^{\infty} {}^{l+1}E_i^2(x)dx \int_{-\infty}^{\infty} {}^{l}E_k^2(x)dx} \,. \tag{3.9}$$

The calculated squared absolute value of the amplitude expansion coefficients a_{ik} at the interface passive/gain-loss waveguide is shown in Fig. 3.21.

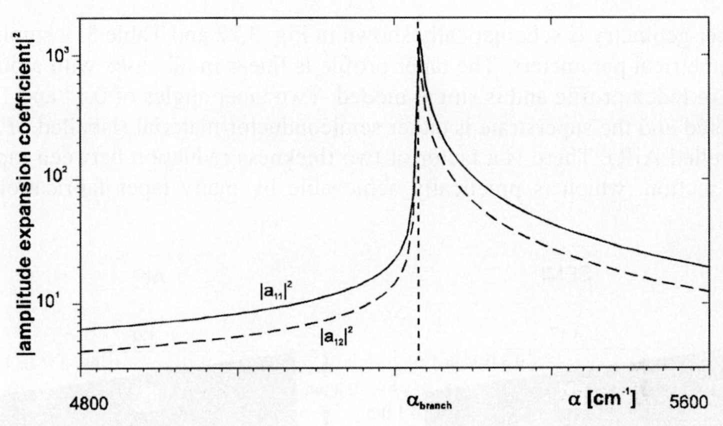

Fig. 3.21. Asymptotic behaviour of the squared absolute value of the amplitude expansion coefficients a_{ik} as a function of α.

A strong increase for α approaching α_{branch} can be observed. As a driving force we easily identify that $^{l+1,l}a_{ik} \to \infty$ for $\alpha \to \alpha_{branch}$ since $\int_{-\infty}^{\infty} {}^{l+1}E_l^2(x)\,dx \to 0$ due to the orthogonality (3.4). We can think of the two modes as of two large numbers $E_1 \cong E_2 = E_1 + \tau$ with $\tau << E_1$. By mode interference we get $P_{\min} \cong \tau^2 << P_{in}$ and $P_{\max} \cong 4|E_1|^2 >> P_{in}$. The asymptotic behaviour of the modes for α approaching α_{branch} is as follows: the beat length Λ approaches to infinity, and for $z \to \infty$ the field amplitude growths to infinity, too.

3.3 Waveguide tapers

The objective of the modelling task described in this section was to investigate the performance and accuracy of a number of propagative algorithms for the simulation of tapered high contrast step index slab waveguides. Both paraxial and non-paraxial formulations of optical field propagation are considered. One may reasonably assume that numerical results obtained by a non-paraxial method like the mode expansion propagation can be considered as reference values. This exercise gives insight in the ability of the more efficient, but less accurate paraxial schemes to model high contrast tapered waveguides. The internal consistency of the various methods is tested using the reciprocity principle.

A concise overview of the main modelling results is given in the following. The interested reader is referred to [19] for a more elaborate discussion.

3.3.1 Modelling task and reciprocity test

The taper geometry is schematically shown in Fig. 3.22 and Table 3.3. summarises the geometrical parameters. The taper profile is linear in all cases with a pure real refractive index profile and is single-moded. Two taper angles of 0.1° and 1.0° are considered and the superstrate is either semiconductor material (labelled SEMI) or air (labelled AIR). There is a factor of two thickness reduction between input and output section, which is practically achievable by many taper fabrication techniques.

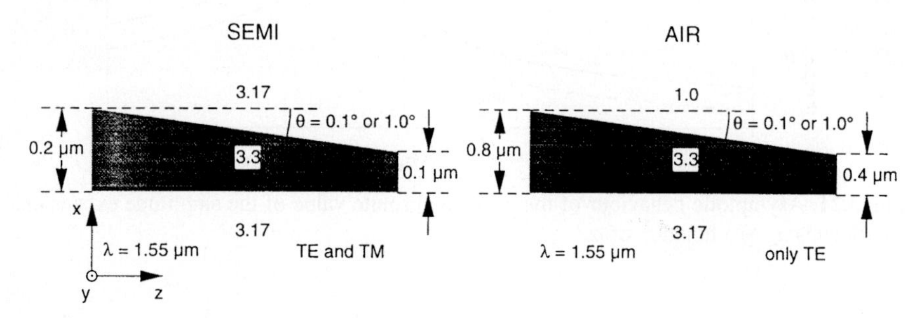

Fig. 3.22. Schematic view of the SEMI and AIR tapers.

	Input thickness [μm]	Output thickness [μm]	Taper angle θ [deg]	Length L [μm]
SEMI	0.2	0.1	0.1	57.3
			1.0	5.73
AIR	0.8	0.4	0.1	229
			1.0	22.9

Table 3.3. Geometrical parameters of two taper structures SEMI and AIR of Fig. 3.22.

The taper is excited by the local fundamental TE mode at the input side at a wavelength of 1.55 μm and carrying unit power. TM modelling results can be found in [19]. The power content of the local fundamental mode at half the taper length and at the output are calculated. To verify the internal consistency of the various methods the principle of reciprocity is tested. If the local fundamental mode is launched with unit power at the left side of the taper in Fig. 3.22, then at the output the local fundamental mode carries a relative power $1 - P_l(L)$, where $P_l(L)$ represents the relative power loss of the guided mode due to propagation through the taper structure. If one excites the same taper at the opposite side with the local fundamental mode, again carrying unit power, then the power content after counter-propagation will again be $1 - P_l(L)$. The total fundamental mode power loss is therefore independent of the propagation direction. It should be stressed that reciprocity alone cannot give any information about the power distribution at any intermediate z position.

3.3.2 Numerical results

An overview of the considered methods is given in Table 3.4. Both paraxial (Fresnel equation) and non-paraxial (Helmholtz) schemes are considered. Most paraxial methods are implementations of the finite difference BPMs. One finite element method is included. The finite difference methods differ by the choice of the reference propagation constant (Univ. Porto and HHI implement a longitudinally varying reference) or by the inclusion of non paraxial corrections (Univ. Twente). The algorithm of Thomson-CSF is an implementation of the basic method. The non-paraxial methods are all mode expansion techniques, whether or not in pure form (IREE Prague and Univ. Gent) or based on a transverse discretisation (Method of lines, Univ. Hagen).

Participant	Equation	Numerical method	Boundary condition (BC)	References
Univ. Twente	Fresnel	FD-BPM	Transparent	Sec. 2.1, [9]
Univ. Porto		FD-BPM	Transparent	[20]
HHI (1)		FD-BPM	Absorber	Sec. 2.1
Thomson-CSF		FD-BPM	Transparent	[21]
AAR-UCL (2)		FE-BPM	Window functions	[22]
Univ. Hagen	Helmholtz	MoL-BPM	Absorbing BC	Sec. 2.2, [23]
IREE Prague		BEP	Absorber	Sec. 2.3
Univ. Gent		BEP	Window functions	Sec. 2.3

(1) Heinrich-Hertz-Institut fur Nachrichtentechnik

(2) Alcatel Alsthom Recherche - University College London

Table 3.4. Overview of all participants in the modelling exercise and details of the implemented methods.

When comparing the power losses of the TE polarised fundamental mode in the SEMI case, it was seen that most results agree very well with each other. The calculated power loss equals $P_1(L)$ = 2.7 % for θ= 0.1° and $P_1(L)$ = 8.8 % for θ= 1.0°. The field evolution for the θ= 0.1° case is plotted in Fig. 3.23. The θ= 1.0° taper is too short to be able to broaden the field profile. More details including TM results and some indication on the discretisation effort are given in [19].

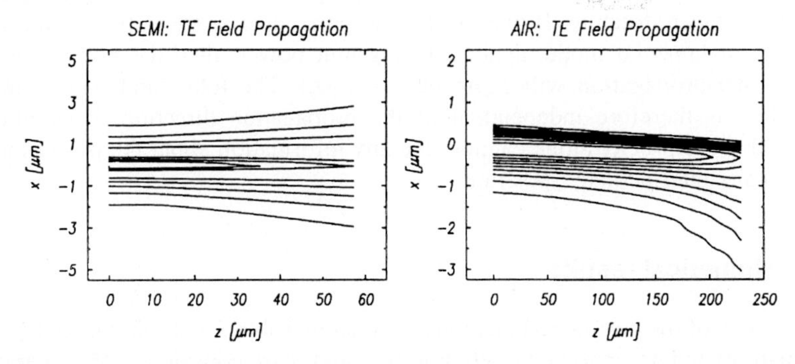

Fig. 3.23. Contour plot of the TE field propagation for the θ= 0.1° case for the SEMI taper (left) and for the AIR taper (right). The propagation direction is in both cases the forward direction. The plots were obtained by the Univ. Gent using 50 propagation steps and 51 radiation modes for the SEMI case and 60 modes for the AIR case. The waveguide/air interface is clearly visible.

AIR – TE	θ= 0.1°, L = 229 µm				θ= 1.0°, L = 22.9 µm			
	Forward		Backward		Forward		Backward	
	$P_1(L/2)$	$P_1(L)$	$P_1(L/2)$	$P_1(L)$	$P_1(L/2)$	$P_1(L)$	$P_1(L/2)$	$P_1(L)$
Twente	0.01	3.19	3.29	3.36	0.90	17.7	20.9	17.7
Porto	0.07	3.30	3.26	3.40	0.97	17.8	20.9	17.6
HHI	0.00	3.30	3.40	3.50	0.90	18.4	21.6	18.5
Thomson	0.03	3.46	3.32	3.39	0.94	17.9	21.0	17.7
AAR-UCL	0.07	2.50	2.05	1.83	0.92	14.7	17.8	14.3
Hagen	0.02	3.64	3.58	3.64	0.92	18.5	21.6	18.6
IREE	0.03	3.40	3.34	3.40	0.90	17.7	20.9	17.7
Gent	0.09	3.48	3.36	3.48	0.95	17.6	20.9	17.6

Table 3.5. Summary of the relative fundamental mode power loss values P_1 (in percents of an input power) as obtained by the different participants for the TE mode propagation in the SEMI tapers.

The TE simulations on the AIR taper are summarised in Table 3.5. There is again a good agreement between the different modelling methods, although the mutual variations are somewhat larger than for the SEMI case. The modal power losses are $P_l(L) = 3.4\%$ for $\theta = 0.1°$ and $P_l(L) = 17.8\%$ for $\theta = 1.0°$. For $\theta = 0.1°$, the paraxial algorithms are not able to fulfil the reciprocity criterion exactly. The loss figures obtained by the HHI and Univ. Hagen seem to give an upper limit for the power losses. The field evolution for $\theta = 0.1°$ is drawn in Fig. 3.23, too.

Table 3.5 further reveals that $P_l(L/2) > P_l(L)$, indicating an oscillatory behaviour of $P_l(z)$, and suggesting a strong coupling between the fundamental mode and the radiation modes. In fact, this coupling length can be obtained from the radiation mode spectrum, see [19].

3.3.3 Discussion

Applicability of paraxial algorithms. As is clear from the numerical results, the paraxial methods give reasonable results, but are not very accurate. In particular, reciprocity is not fulfilled. A major issue in the application of a paraxial propagative scheme to taper simulations is the choice of the reference propagation constant. In a down tapered waveguide the local fundamental mode propagation constant decreases monotonically in the propagation direction. The difference in modal propagation constant between input and output planes of the taper depends on the waveguide cross section. For example, a large core/cladding index contrast can result in a significant decrease in propagation constant and can therefore complicate the choice of the reference value. A method of determining a longitudinally varying reference propagation constant is outlined in [4] and has been applied by the University of Porto.

In case the fundamental mode and the radiation modes couple to and fro, the correct propagation of the radiation modes is of importance. If the effective index difference between the fundamental mode and radiation modes is too high, errors may be introduced as the reference index is mainly adapted to the effective index of the fundamental mode. It can be shown [19] that higher order corrections on the slowly varying envelope approximation do not influence the basic Fresnel result. It might be surprising at the first glance that the same conclusion still holds for the AIR case. However, since the substrate index again equals 3.17, the variation of the propagation constant of the local fundamental guided mode throughout the taper covers almost the same interval as in the SEMI case. The above discussion also reveals that good accuracy can only be achieved if the propagation can be well described by one single mode for which the reference index can be chosen close to the effective index.

Power calculation and reciprocity for paraxial propagation. It is well known [24] that the paraxial wave equation does not conserve the longitudinal power flux per surface unit

$$\frac{1}{2}\mathrm{Re}\left(\int_{-\infty}^{+\infty} \mathbf{E}\times\mathbf{H}^{*}\cdot\mathbf{u}_z\,dx\right) \tag{3.10}$$

but conserves instead the quantities

$$\int_{-\infty}^{+\infty} |\mathbf{E}(x)|^2 dx \quad \text{(TE)}, \qquad \int_{-\infty}^{+\infty} \frac{1}{n^2} |\mathbf{H}(x)|^2 dx \quad \text{(TM)}. \tag{3.11}$$

One should use the integrals (3.11) to calculate the power content of a propagating field at a certain longitudinal position. Especially, the power transfer between input and output sections has to be calculated using (3.11), where $|\mathbf{E}|$ and $|\mathbf{H}|$ stand for the magnitude of the local fundamental mode. This is also the case when doing a unidirectional mode expansion propagation [24]. It should be noted that the slowly varying envelope approximation, which neglects the second order z derivative in the wave equation, excludes backward propagating waves.

Formula (3.10) reduces, in case of the propagation of a single TE polarised mode, to

$$\frac{\beta}{2\omega\mu_0} \int_{-\infty}^{+\infty} |\mathbf{E}(x)|^2 dx = \frac{\beta}{2\omega\mu_0} \int_{-\infty}^{+\infty} |E_y(x)|^2 dx. \tag{3.12}$$

Expression (3.12) is proportional to (3.11) with the propagation constant as waveguide dependent proportionality factor. Comparison of (3.11) (TE) with (3.12) also leads to the insight that (3.11) (TE) can be considered as a paraxial approximation of (3.10) or (3.12) where the variation of the propagation constant is neglected. Calculating the power transfer from input to output using (3.12), the ratio between the propagation constants at input and output section is introduced. It follows directly from (3.12) that

$$\frac{1 - P_l'(L)_{forward}}{1 - P_l'(L)_{backward}} = \left(\frac{\beta_{z=L}}{\beta_{z=0}} \right)^2, \tag{3.13}$$

where the prime denotes power calculation using the Poynting vector. Hence the reciprocity criterion cannot be fulfilled if the power transfer is calculated with (3.10) or (3.12). The property of reciprocity reduces in this case to the observation that the ratio of the forward and backward power transfer depends only on the respective waveguide cross sections at input and output and is independent of the connecting waveguide, as expressed mathematically by (3.13). The ratio (3.13) in case of the SEMI taper equals 0.993.

3.3.4 Conclusion

In this section, the applicability of different propagative algorithms for the modelling of tapered high contrast step index slab waveguides was assessed. Mode expansion results were considered as reference figures to which numerical results of the numerically more efficient paraxial schemes were compared.

All but one paraxial BPM's were implemented using a finite difference based formulation of the scalar Fresnel wave equation. The non-paraxial schemes solved the Helmholtz equation by the method of lines BPM or full mode expansion techniques. The numerical results showed that all methods, both paraxial and non-paraxial, mastered the propagation through the different taper structures. However, the paraxial methods have difficulties to fulfil the reciprocity criterion. This is in contrast to the non-paraxial schemes. The mode expansion methods, which are the natural methods to solve periodic and segmented waveguide problems (see Sections 2.2, 2.3 and 3.5), have proven their usefulness for the simulation of waveguides with continuously varying refractive index profile.

Furthermore, the way the power content of a propagating mode has to be calculated in the case of the paraxial wave equation has been discussed. More in particular, the relation between the choice of power flux expression and its consequences on the principle of reciprocity has been elaborated.

3.4 Electro-optic modulator based on surface plasmons

In this section, different beam propagation methods based on the method of lines (MoL), mode matching (MM) and finite differences (FD) will be compared and studied with respect to their ability to deal with structures having large lateral contrast of complex refractive indices that allow for the propagation of surface plasmons. Surface plasmons are related to plasmon oscillations of the free electrons in metals which may lead to negative values of the real part of the dielectric constant, ε, for a certain optical frequency region. As a consequence, TM polarised guided wave solutions, so called surface plasmons (SP), can exist, which propagate along an interface between a metal and a dielectric. Due to the large field confinement SPs are often used as 'sensing modes' in optical sensors [25]. Another application of SPs, among others, utilising their usually strong damping caused by energy dissipation by electron collisions can be found in electro-optic (EO) modulators [26].

The EO modulator described in [26] has been chosen for a benchmark test of beam propagation methods (BPMs) in the framework of the COST 240 Working group 2 because of its high index contrast occurring at the metal-dielectric interface. In the next sections, the problem will be defined, then the computational methods will be briefly presented together with the parameters used for model calculations. Then, the computational results will be presented and discussed.

3.4.1 Problem definition

The EO modulator which is the subject of the present benchmark test is depicted in Fig. 3.24. The EO part, for $500\,\mu\text{m} < z < 2500\,\mu\text{m}$, contains two silver layers, leading to modal fields which are in principle a combination of, among others, four surface plasmon modes.

A fundamental mode launched into the SiON waveguide at $z = 0$ starts to couple to the strongly attenuated modes of the EO structure at $z = 500\ \mu\text{m}$. The magnitude of this coupling depends critically on the phase-matching between the modes of the isolated waveguides A and B (see Fig. 3.24.). This phase matching can be modulated by a voltage across the EO polymer sandwiched between the two silver electrodes. The benchmark test consists of calculating the throughput of the structure at the wavelength of $\lambda = 1.523$ µm and for a refractive index of a polymer n_p varying within the range 1.58–1.59. The refractive indexes of the used materials and the layer thicknesses are given in Table 3.6.

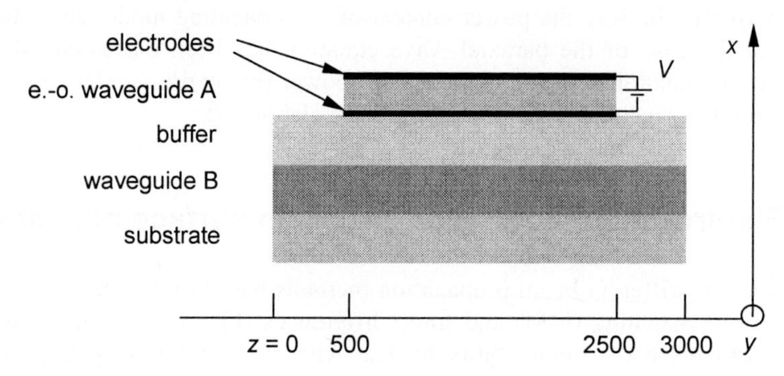

Fig. 3.24. The EO modulator structure for the benchmark test. Details can be found in Table 3.6.

	Layer	index	thickness (µm)
1	air	1.00	∞ /2
2	silver	0.14-i11.0	0.07
3	polymer	1.58-1.59	1.1
4	silver	0.14-i11.0	0.07
5	SiON	1.56	1.3
6	SiON	1.7	0.934
7	SiO_2	1.449	∞ /2

Table 3.6. Parameters for the EO structure given in Fig. 3.24.

Note the extremely large refractive index contrasts, $\Delta n^2 \approx 120$, at the metal-dielectric interfaces. For this reason this structure has been chosen to test the ability of the BPMs to cope with that. In particular, for BPMs based on a discretisation along the x-axis we have to do with a trial by fire, which should give insight into their limitations, convergence properties, and probably the needs for further improvements.

3.4.2 About the computational methods

All the methods applied to the present test structure have been introduced in Ch. 2 of this book. The main features as well as the most important computational parameters are given in Table 3.7.

Institute	Method	step size Δx (nm)	step size Δz(μm)	interface conditions	Nr. of points/ basis functions
IREE	MMM	–	zIS	analytical	14
FU Hagen	MoL	25.9	zIS	analyt./EICs	equidistant
U Rome	MoL	1.0	10	Fermi funct.	300 non-equidistant
U Twente	FD2 BPM	2.0	1	EIC's	4000 equidistant

Table 3.7. Main features of the applied computational methods. Abbreviations used here are: MMM – mode matching method [27]; MoL – method of lines [28]; FD2 BPM – finite difference beam propagation method [29] including a second-order correction for the slowly varying envelope approximation (SVEA). If a non-equidistant grid spacing is used the smallest value for this is given. EIC's – efficient interface conditions [29]. zIS indicates that one step was used for each z-independent section (zIS).

All the methods except the (bi-directional) MoL used by FU Hagen, are uni-directional; for the MMM method, its uni-directional option was used. The validity of the uni-directional approximation will be discussed in the next section. Due to the large index contrast along the transversal direction (x-axis) all methods based on a discretisation in that direction (*i.e.*, MoL and FD BPM) needed special precautions for an accurate treatment of the second-order derivative at the interfaces. The EIC's mentioned above use corrections in the standard three-point FD operator for the second-order derivative close to interfaces which take into account the continuity of H_y, $n^{-2}\,\partial H_y/\partial x$ and the discontinuity of $\partial^2 H_y/\partial x^2$. More or less equivalent expressions for the corrected FD operator can be found in the literature [29]. Besides the EIC's for the dielectric media the MoL of FU Hagen uses analytical functions in the Ag layers. Due to these analytical expressions the field and its first derivative with respect to the transverse co-ordinate x are matched at the metal-dielectric interfaces giving a relation between the fields on the two sides of the metal layer. The finite difference scheme (1 –2 1) for these points was then replaced by expressions obtained by the analytical approach.

In the MoL of U Rome the index close to interfaces is approximated by a Fermi function

$$\varepsilon_{Fermi} = \varepsilon_l + \Delta\varepsilon / \{1 + \exp(-\gamma\, x / \Delta x)\}. \tag{3.14}$$

Here $\Delta\varepsilon = \varepsilon_r - \varepsilon_l$, the subscripts r and l denote right and left of the interface, receptively, Δx is the step size and γ is an adjustable parameter chosen such that the losses are maximum.

The average index used for the SVEA in the FD2 BPM was $n_0 = 1.624$, the window width was so large that reflection from the computational boundaries can be neglected.

3.4.3 Results and discussion

In Fig. 3.25 the propagation of the field is given, for two values of the polymer index, n_p, somewhat off resonance at $n_p = 1.59$ and close to resonant coupling ($n_p = 1.58685$). The fields result from FD2 BPM calculations, the other methods give virtually identical results. From Fig. 3.25 it can be anticipated that (multiple) reflections may play a minor role.

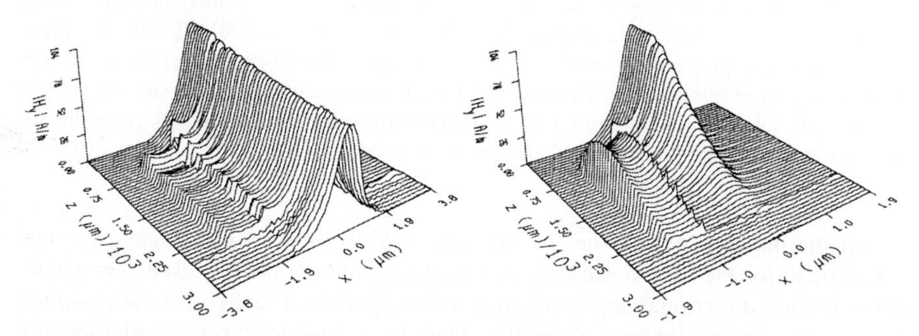

Fig. 3.25. Field propagation for two different polymer indices, left $n_p = 1.59$, right picture $n_p = 1.58685$.

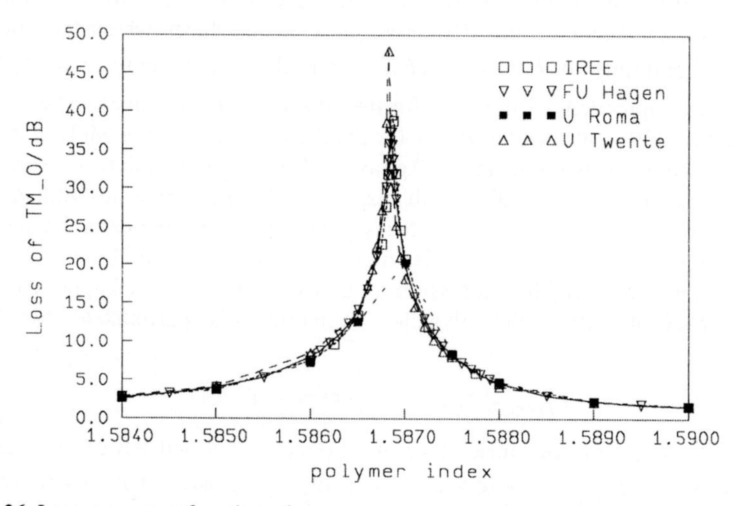

Fig. 3.26. Loss curve as a function of n_p.

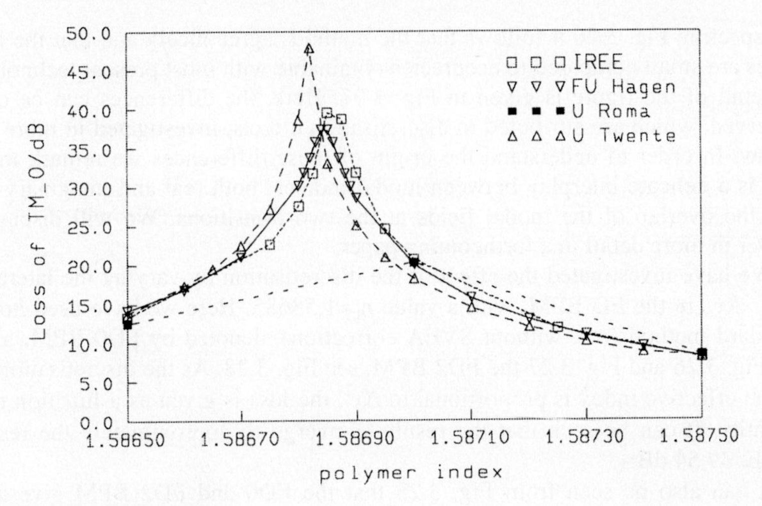

Fig. 3.27. Detail of Fig. 3.26.

However, reflection from the end face of the EO structure is expected not to be completely negligible. A rough estimate, also taking into account that (in terms of local modes) only the strongly damped plasmon modes are partly reflected, leads to only a small error in the throughput of at most a few percent. This will influence the throughput given in Fig. 3.26 not very significantly. This is confirmed by comparison of the results from the bi- and uni-directional MoL by FU Hagen which showed a difference of less than a tenth of a dB.

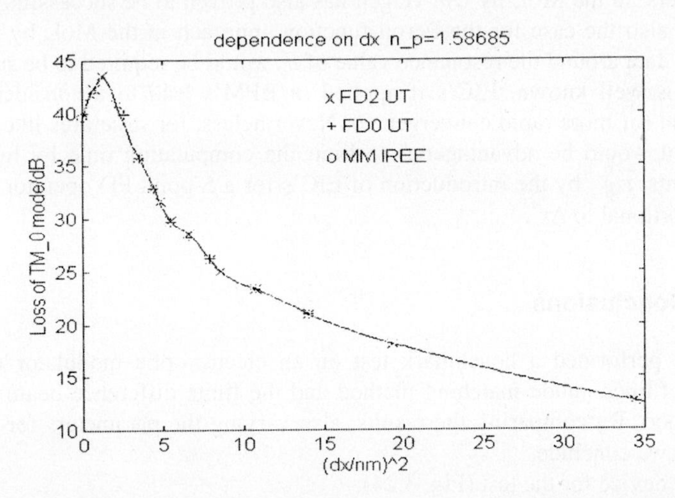

Fig. 3.28. Dependence of the loss in the test structure, calculated with the FD BPM, as a function of the lateral step size, Δx . The result of the MM is given for comparison.

Inspecting Fig. 3.26 it follows that the methods agree nicely and that the differences are small compared to accuracies obtainable with most present technologies. A detail of the figure is given in Fig. 3.27. Here, the differences can be clearly observed, which are attributed to discretisation effects, investigated in more detail below. In order to understand the origin of these differences we remark that the loss is a delicate interplay between modal indices, both real and imaginary parts, and the overlap of the modal fields at the two transitions. We will discuss this matter in more detail in a forthcoming paper.

We have investigated the effect of the discretisation by varying the lateral step size, Δx, in the FD BPM, with a value n_p=1.58685. Here we have used both the standard method (*i.e.*, without SVEA correction), denoted by FD0 BPM, and, as for Fig. 3.26 and Fig. 3.27 the FD2 BPM, see Fig. 3.28. As the discretisation error in the effective index is proportional to Δx^2, the loss is given as a function of this quantity. It can be seen that the results converge to approximately the result of IREE, 39.54 dB.

It can also be seen from Fig. 3.28 that the FD0 and FD2 BPM give almost identical values. This indicates that higher-order modes and radiative modes, which are propagated with small phase-errors by the FD0 BPM, do not play an important role. *I.e.*, these modes excited at the transition at $z = 500$ μm will not re-enter waveguide B at $z = 2500$ μm. This picture is confirmed by MMM calculations which show a rapid convergence as a function of the number of basis functions.

So, based on the above and also on the ability of the MMM to handle large index contrasts, we conclude that the MMM will give the (nearly) correct results for the device of the benchmark test. The other methods agree fairly well. The newly implemented analytical treatment of the high index layers, also using EIC's for the other layers, in the MoL by Uni-Hagen has also proven to be successfully. This is probably also the case for the Fermi-function approach in the MoL by U Rome, but more data around the resonance value of n_p would be required to be sure.

As it is well known, EIC's if applied in BPM's lead to a considerable improvement (or more rapid convergence). Nevertheless, for structures like the present one, it would be advantageous to limit the computation time by further improvements, *e.g.*, by the introduction of EIC's for a 5-point FD operator converging proportional to Δx^4.

3.4.4 Conclusions

We have performed a benchmark test on an electro-optic modulator using the method of lines, mode matching method and the finite difference beam propagation method. By comparing the results, also varying the parameters for the computations we conclude,
about the device for the test (Fig. 3.24):

- the transmission loss is ~2dB, the extinction is at maximum ~40 dB and is >30 dB in an index range of the EO polymer of $\sim 10^{-4}$,

- the large index contrast at silver interfaces requires a careful interface treatment for methods based on a discretisation,
- the MMM seems to be most suitable for the test structure,

about the applied methods (see Table 3.6, Fig. 3.26 to Fig. 3.28):
- the results of the applied methods agree fairly well, *e.g.*, differences in the position of the loss peak correspond to a change of less than 10^{-4} in the EO polymer index,
- small differences are attributed to discretisation errors,
- newly introduced methods for the treatment of the finite difference operator near interfaces, and the field in high-index layers appear to perform well,
- for computational schemes based on a discretisation, further acceleration of the convergence, as a function of lateral step size, would be desirable for structures with large index contrast as in the structure for the benchmark test.

3.5 Waveguide Bragg grating filter

Bragg-gratings are very attractive optical devices. They are currently used as wavelength-selective reflectors with applications in distributed feedback (DFB) and distributed Bragg reflector (DBR) lasers or rejection filters as described, *e.g.* in [30, 31], and are expected to play an important role in sensors, photonic band-gap devices, *etc.*

While the operation of weakly modulated waveguide Bragg gratings is well described by the coupled-mode theory (see, *e.g.*, [32]), the behaviour of "strong", or deeply etched gratings that can be fabricated using up-to-date technology has not yet been analysed in detail. It was one of the reasons why modelling of a deep waveguide grating with large number of periods was chosen as a modelling task in the Action COST 240. Because of a finite but large number of periods of the order of 10^4, the computational methods had to exhibit an extremely high numerical stability. Further problems are related to CPU time and memory requirements. The modelling task is described in the next section. The third section gives a short overview of the applied computational methods with some comments on their implementation. Numerical results are presented in the next section that is followed by a brief summary of results.

3.5.1 Formulation of the problem

The arrangement of the Bragg waveguide grating filter to be analysed is shown in Fig. 3.29. A slab waveguide consists of three layers with refractive indices $n_a = 1$, $n_g = 1.53$, and $n_s = 1.52$, respectively, the thickness of the waveguiding layer being 2.4 μm. The Bragg grating filter with a rectangular groove profile is formed by etching the waveguide (silicon oxy-nitride) layer into a depth of 0.5 μm. The design Bragg free-space wavelength is $\lambda_B = 650$ nm, and the TE polarisation is assumed.

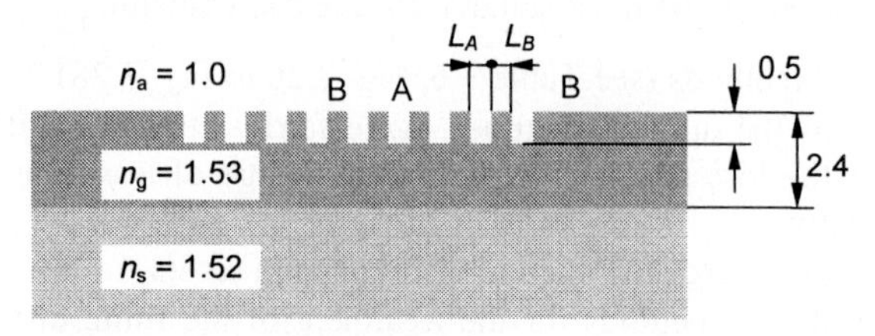

Fig. 3.29. Deeply etched waveguide Bragg grating.

The grating can be considered a periodic sequence of alternating waveguide sections A (etched) and B (unetched). Both sections are single-mode at the operating wavelength. In order to obtain the required Bragg wavelength λ_B, the lengths of the sections A and B were chosen as quarter-wave segments,

$$L_A = \frac{\lambda_B}{4n_{eff,A}}, \qquad L_B = \frac{\lambda_B}{4n_{eff,B}}, \tag{3.15}$$

where $n_{eff,A}$ and $n_{eff,B}$ are the effective refractive indices of the modes in the section A and B, respectively. These values were calculated as follows:

section A: $n_{eff,A} = 1.52506353$, $L_A = 106.553$ nm,
section B: $n_{eff,B} = 1.52645903$, $L_B = 106.456$ nm.

Note that the total period length $L = L_A + L_B = 213$ μm is smaller than one half of the etching depth. Due to the large refractive difference of the refractive index between SiON and air, very high scattering loss and hence low grating reflectivity was expected.

A standard coupled contra-directional mode theory gives a reflectivity of 99% for a device length of 218 μm, *i.e.*, for 1023 periods, and shows a strong dependence of the (symmetrical) spectral response on the choice of the thickness of the "unperturbed" waveguide. The task was to see whether these predictions could be confirmed or refuted with more accurate modelling methods.

3.5.2 Outline of computational methods applied

Let us briefly overview the methods that were used for the analysis of the grating. They are listed in Table 3.8. At the Sandia National Laboratories, a fourth order finite difference scheme was developed [33]. The solution was obtained via direct inversion of the band matrix resulting from the nine-point stencil difference equations [34, 35]. The algorithm was very fast, but required a lot of memory space due to the large numbers of periods.

Institute	Method
Sandia National Laboratories	4th order FDM
IREE Prague	BEP (MMM)
FernUniversität Hagen	MoL, MMM

Table 3.8. Methods used for the analysis of the waveguide Bragg grating.

IREE and Hagen implemented bi-directional eigenmode propagation methods for the analysis of the structure. The methods differ essentially in the way how the eigenmodes were determined. The BEP method [7], [36] uses the transfer matrix approach to find the eigenmodes. To discretise the spectrum of radiation modes, electric walls are used at some distance from the grating, and absorbing layers with complex refractive indexes are used to suppress the reflection from the walls. The dispersion equation is solved numerically using the combination of the interval halving (for real refractive index profiles) and the Newton method (in the complex plane for complex refractive indices). In the application of the Newton method it is taken into account that the dispersion equation is a regular function of a complex variable n_{eff}^2 .

In the Rayleigh-Ritz method (RRM) [37], the eigenmodes are approximated by a series of sine functions. For the analysis with the MoL [23, 38] the wave equation is discretised perpendicular to the direction of propagation. RRM as well as MoL result in a matrix eigenvalue problem from which the eigenmodes can be determined.

Since all methods used are well documented in the literature, they will not be described in detail here. We bring short remarks concerning their implementation, instead.

Because of the large number of periods, the approach based on transfer matrix methods applied to longitudinal direction leads to numerical problems due to the exponentially increasing behaviour of the reflected modes. For this reason all bi-directional eigenmode propagation algorithms were applied with an impedance transfer approach [39]. Impedance transfer can be understood as a generalisation of an impedance transformation by a section of a transmission line in a standard transmission line theory that is known to be very stable for arbitrarily long lines and lines composed of many different line sections.

For the analysis with both MoLs, a special approach was developed [40, 41] which combines the advantage of the impedance transfer with Floquet's theorem resulting in a stable and fast algorithm with a moderate computer memory requirement. Its extension to BEP is described in [42]. In this approach, the grating is considered as a chain of equal periods, and for such a period, its "Floquet modes" and their propagation constant are determined. These "modes" have the property that their transversal field distribution is reproduced up to the phase shift as they propagate by one period. Using the set of Floquet modes as basis functions, it is possible to transfer the transversal impedance from the end of the grating to its input in one step, and then to compute the field distribution along the grating without storing the impedance values at the interfaces between individual "periods". Thus, the whole analysis of the grating can be performed with a very moderate numerical effort.

3.5.3 Numerical results

Fig. 3.30 shows the reflectivity of the grating at the Bragg-wavelength of 650 nm computed by the four methods listed in Table 3.8. A very good agreement can be observed, thus confirming the reliability of the methods. Further examinations showed that absorbing boundaries used for the calculations were not really necessary, although the curves calculated with them are smoother than those obtained with perfectly reflecting walls. It indicates that light scattering out of the waveguide by the grating is rather small. However, the maximum reflectivity at the Bragg wavelength is far less than predicted by the coupled mode theory.

Much higher reflectivity can be reached if the wavelength is shifted to lower values, but far more than 1000 periods are required.

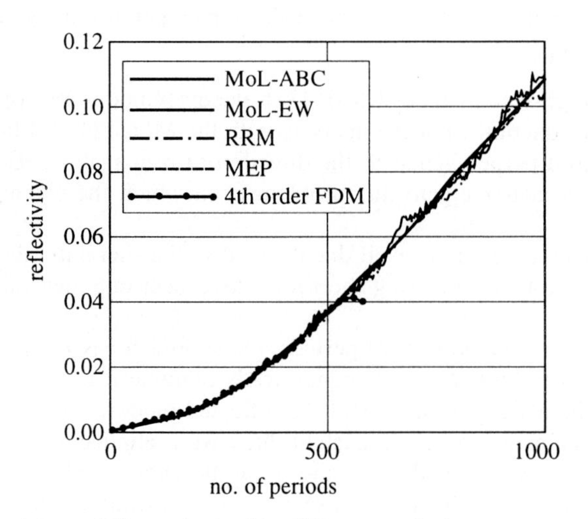

Fig. 3.30. Reflectivity at 650 nm obtained by different methods.

This analysis could not be done with the FDM, due to the high memory space requirement. On the other side, the bi-directional eigenmode algorithms (*i.e.*, BEP, MoL, and RRM) were able to handle this problem with a relative ease.

Fig. 3.31 and Fig. 3.32 show the reflectivity of the grating as a function of the number of periods and the wavelength. In Fig. 3.31, the results obtained with BEP and MoL are mutually compared.

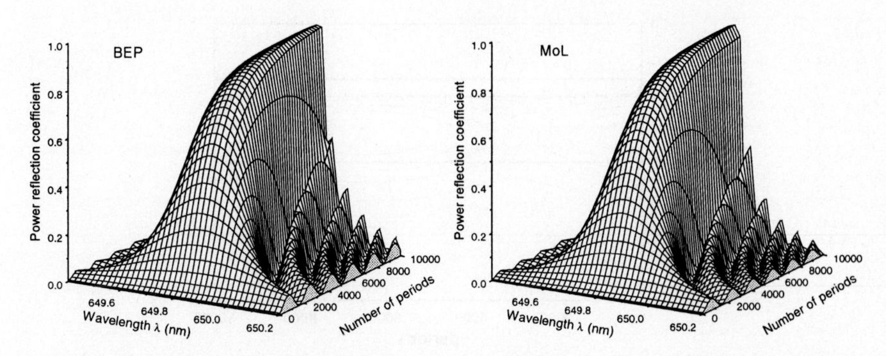

Fig. 3.31. Reflectivity vs. wavelength and number of periods calculated by BEP and MoL.

The results for an infinite number of periods in Fig. 3.32 were obtained by loading the (uniform) waveguide of type B by a wave impedance of one period, which is an impedance that transforms to itself by the transition through one period. The reflectivity value obtained for the wavelength 649.86 nm with an infinite number of periods seems to be the absolute maximum of the reflectivity of the grating (about 91%).

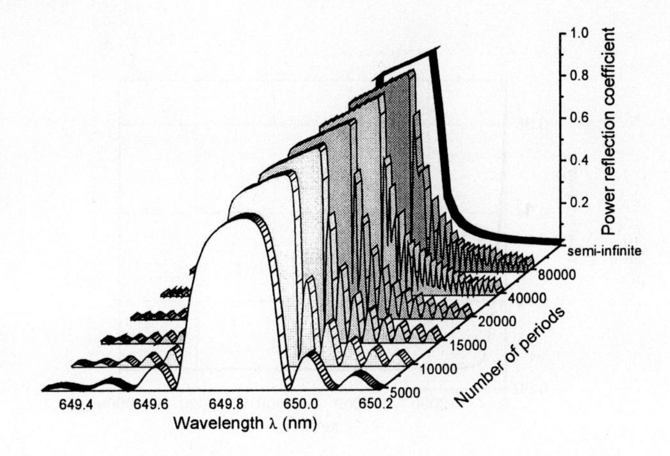

Fig. 3.32. Reflectivity vs. wavelength with the number of periods as parameter.

The behaviour of the electric field inside the grating is plotted in Fig. 3.33 for the wavelength of 649.95 nm. It should be mentioned that this field was computed only in points at the centre of the waveguide sections B. The "periodicity" observed in this field stems from the fact that the standing-wave pattern of the field created by the superposition of the forward and backward waves does not match the grating periodicity exactly, and what we see in the figure is just the beating between these two periodicities (the Moire pattern).

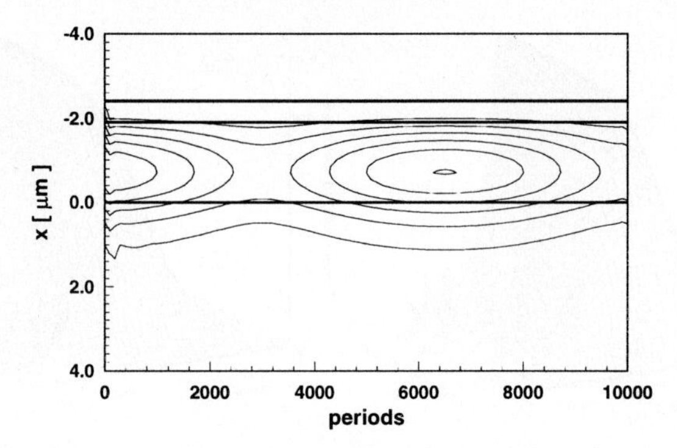

Fig. 3.33. Electric field distribution along the grating with 10000 periods.

The power flow along the grating (normalised to the injected one) is plotted in Fig. 3.34. We see that in spite of the virtual "quasi standing" wave pattern of the field, a rather small continuous power decrease is observed. This figure confirms our expectation that even for a very high number of periods, the loss due to scattering (radiation) out of the waveguide is quite small (power loss less than 5% can be evaluated from Fig. 3.34.).

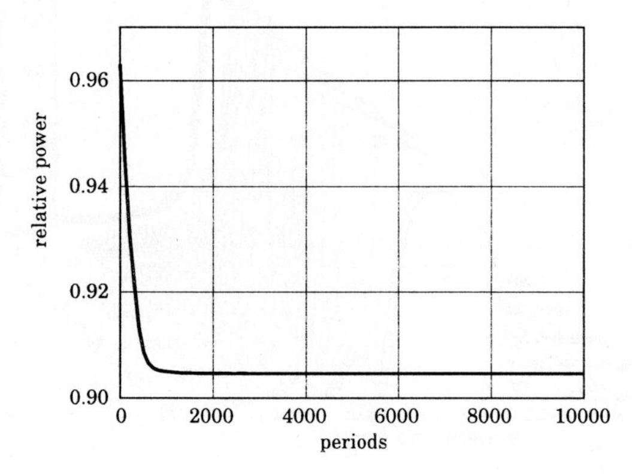

Fig. 3.34. Longitudinal power flow along the grating.

3.5.4 Conclusions

Modelling of a very deep Bragg waveguide grating filter was used as a benchmark test of four numerical methods. All bi-directional eigenmode propagation algorithms were able to handle this task without problems, if an impedance transfer approach was applied. The fourth-order finite difference method was able to analyse relatively high number of periods up to 1000. However, in order to obtain high reflectivity of the grating, about 10 times more periods were necessary.

The examinations showed that a significant part of energy is transferred by higher-order modes. Therefore, a simple coupled mode analysis that takes into account only a fundamental mode of the waveguide fails. The analysis of the grating with accurate numerical algorithms predicts rather unexpected behaviour of the device. The spectral reflectivity curve is asymmetric and shifted against the "Bragg wavelength" determined conventionally towards shorter wavelengths. It can be perhaps understood from (3.15) if we realise that higher-order modes with lower effective refractive indices participate in the power transfer. In spite of the large depth of grooves and the large refractive index contrast, only a small radiation loss of a few percent is observed. For long gratings, the maximum reflectivity was found to be 91%. So, the device can be useful in many applications.

References

[1] M. D. Feit and J. A. Fleck, "Light propagation in graded-index optical fibers", *Appl. Opt.*, vol 17, pp. 3390-3998, 1987.

[2] Working Group I, Cost 216, "Comparison of different modelling techniques for longitudinally invariant integrated optical waveguides", *IEE Proc. J*, vol 136, 1989.

[3] J. Chilwell and I. Hodgkinson, "Thin films field-transfer matrix theory of planar multilayer waveguides and reflection from prism-loaded waveguides", *J. Opt. Soc. Am.*, A-1, pp. 742–753, 1984.

[4] F. Schmidt, "An Adaptive Approach to the Numerical Solution of Fresnel's Wave Equation", *J. Lightwave Technol.*, vol. 11, pp. 1425–1435, 1993.

[5] R. C. Alferness, T. L. Koch, L. L. Buhl, F. Storz, F. Heismann, and M. J. R. Martyak, *Integrated and Guided-Wave Optics*, Houston, Texas, Feb. 6–8, 1989, Technical Digest Series vol. 4, p. 215–218, 1989.

[6] H.-P. Nolting and G. Sztefka, "Eigenmode Matching and Propagation Theory of Square Meander-Type couplers", *IEEE J. Photon. Technol. Lett.* vol. 4, pp. 1386–1389, 1992.

[7] G: Sztefka and H.-P. Nolting, "Bidirectional Eigenmode Propagation for Large Refractive Index Steps", *IEEE J. Photon. Tech. Lett.* vol. 5, pp. 554–557, 1993.

[8] G. R. Hadley, "Multistep method for wide angle beam propagation", *Opt. Lett.*, vol. 17, pp. 1743–1745, 1992.

[9] H. J. W. M. Hoekstra, G. J. M. Krijnen, and P. V. Lambeck, "New Formulation of the BPM based on the slowly varying envelope approximation", *Opt. Commun.* vol. 97, pp. 310-303, 1993.

[10] V. A. Baskakov and A. V. Popov, "Implementation of transparent boundaries for numerical solution of the Schroedinger equation", *Wave Motion*, vol. 14, pp. 123–128, 1991.

[11] G. R. Hadley, "Transparent Boundary Condition for the BPM", *IEEE J. Quantum. Electron.* vol. QE-28 , pp. 363–370, 1992.

[12] H.-P. Nolting and R. März, "Results of Benchmark Tests for Different Numerical BPM Algorithms", *IEEE J. Lightwave Technol.*, vol. 13, pp. 216–224, 1995.

[13] H.-P. Nolting, M. Grawert, G. Sztefka, and J. Čtyroký, "Wave propagation in a waveguide with a balance of gain and loss", *Integrated Photonics Research*, April 29 – May 2, 1996, Boston, Ma, USA, Technical Digest Series vol. 6, pp. 76–79.

[14] F. Schmidt and H.-P. Nolting, "Adaptive multilevel beam propagation method," *IEEE Photon. Technol. Lett.*, vol.4, pp. 1381–1383, 1992.

[15] H. J. W. M Hoekstra, G. J. M. Krijnen, and P. V. Lambeck, "Efficient interface conditions for the FDBPM", *J. Lightwave Technol.*, vol. 10, pp. 1352–1355, 1992.

[16] S. Helfert and R. Pregla, "Finite Difference Expressions for Arbitrarily Positioned Dielectric Steps in Waveguide Structures", *IEEE J. Lightwave Technol.*, vol. 14, pp. 2414–2421, 1996.

[17] R. E. Smith, S. N. Houde-Walter, and G. W. Forbes, "Numerical determination of planar waveguide modes using the analyticity of the dispersion relation", *Opt. Lett.*, vol. 16, pp. 1316–1318, 1991.

[18] H.-P. Nolting and M. Grawert, "A Comparison between Different Methods to Calculate Grating Assisted Asymmetrical Couplers", *Linear and Nonlinear Integrated Optics*, Lindau 94, Germany, Proceedings Europto Series Vol 2212, pp. 328–336, 1994.

[19] J. Haes, R. Baets, C. M. Weinert, M. Gravert, H.-P. Nolting, M. Adelaide Andrade, A. Leite, H. Bissessur, J. B. Davies, R. D. Ettinger, J. Čtyroký, E. Ducloux, F. Ratovelomanana, N. Vodjdani, S. Helfert, R. Pregla, F. H. G. M. Wijnands, H. J. W. M. Hoekstra, and G.J.M. Krijnen, "A comparison between different propagative schemes for the simulation of tapered step index slab waveguides", *IEEE Lightwave Technol.*, vol. 14, pp. 1557–1569, 1996.

[20] Y. Chung and N. Dagli, "An assessment of finite difference beam propagation method", *IEEE J. Quantum Electron.*, vol. 26, pp. 1335–1339, 1990.

[21] R. Accornero, M. Artiglia, G. Coppa, P. Di Vita, G. Lapenza, M. Potenza, and P. Ravetto, "Finite difference methods for the analysis of integrated optical waveguides", *Electron. Lett.*, vol. 26, pp. 1959–1960, 1990.

[22] R. D. Ettinger, F. A. Fernandez, J. B. Davies, and H. Bissessur, "Efficient vectorial analysis of propagation in three-dimensional optical devices", Proc. *Integrated Photonic Research* Topical Meeting, 1993, pp. 392–295.

[23] J. Gerdes and R. Pregla, "Beam propagation algorithm based on the method of lines", *J. Opt. Soc. Am.* B, vol. 8, pp. 389–394, 1991.

[24] J. Haes, J. Willems, and R. Baets, "Study of power conservation at waveguide discontinuities using the mode expansion method", Proc. *Integrated Photonic Research* Topical Meeting, 1995, pp. 115–117.

[25] H. J. M. Kreuwel, P. V. Lambeck, J. M. M. Beltman and T. J. A. Popma, "Mode coupling in multilayered structures applied to a chemical sensor and a wavelength selective directional coupler," Proc. *ECIO,* pp. 217–222, 1987.

[26] A. Driessen, H. M. M. K. Koerkamp, and T.J.A. Popma, "Novel Integrated Optic Intensity Modulator Based on Mode Coupling," *Fiber and Integrated Optics*, vol. 13, pp. 445–461, 1994.

[27] This book, Sec´s. 1.2 and 2.5, and references given there.

[28] This book, Sec. 2.4, and references given there.

[29] This book, Sec. 2.3, and references given there.

[30] K. O. Hill, B. Mało, F. Bilodeau, and D. C. Johnson, "Photosensitivity in Optical Fibers", *Ann. Rev. Mater. Sci.* vol. 23, pp. 125–157, 1993.

[31] W. W. Morey, G. A. Ball, and G.Meltz, "Photoinduced Bragg Gratings in Optical Fibres", *Opt. and Photon. News*, vol. 5, no. 2, pp. 8–14, 1994.

[32] H. Kogelnik, "Theory of Dielectric Optical Waveguides", in: *Guided-wave optoelectronics*, T, Tamir, (ed.) (Springer, Berlin 1988) Ch. 2.

[33] G. R. Hadley, "High-Order-Accurate Finite Difference Equations for Photonics Modeling", in *Proc. Progress in Electromagnetics Research Symp.* (PIERS)}, Hong Kong, Jan. 1997, vol. 1, p. 100.

[34] G. R. Hadley, "Numerical simulation of reflecting structures by solution of the two-dimensional Helmholtz equation", *Opt. Lett.*, vol. 19, no. 2, pp. 84–86, 1994.

[35] G. R Hadley, "Semi-Vectorial Response of Three-Dimensional Reflecting Structures via Iterative Solution of the Helmholtz Equation" in OSA *Integrated Photonics Research.* Tech. Dig., Dana Point, USA, Feb 1995, vol. 7, pp. 194–196.

[36] J. Čtyroký, J. Homola, and M. Skalský, "Modelling of surface plasmon resonance waveguide sensor by complex mode expansion and propagation method", *Opt. Quantum Electron.*, vol. 29, no. 2, pp. 301–311, 1997.

[37] R. E. Collin, *Field Theory of Guided Waves*, Series of Electromagnetic Waves. IEEE press, New York, 2 edition, Chaps 6.2 , and 9.1, pp. 419–428, 1991.

[38] R. Pregla, "MoL-BPM Method of Lines Based Beam Propagation Method", in *Methods for Modeling and Simulation of Guided-Wave Optoelectronic Devices*, W.P. Huang, (Ed.), number PIER 11 in *Progress in Electromagnetic Research*, pp. 51–102. EMW Publishing, Cambridge, Massachusetts, USA, 1995.

[39] R. Pregla, "The Method of Lines as Generalized Transmission Line Technique for the Analysis of Multilayered Structures", *AEÜ*, vol. 50, no. 5, pp. 293–300, 1996.

[40] S. Helfert and R.Pregla, "A Very Stable and Accurate Algorithm for the Analysis of Periodic Structures with a Finite Number of Periods", in *Proc. Progress in Electromagnetics Research Symp.* (PIERS)}, Hong Kong, Jan. 1997, vol.1, pp. 105.

[41] S. Helfert and R. Pregla, "Efficient Analysis of Periodic Devices", *J. Lightwave Technol.*, submitted, 1997

[42] J. Čtyroký, S. Helfert and R.Pregla: "Analysis of a deep waveguide Bragg grating", *Opt. Quantum Electron.*, accepted, 1988.

4 Methods for waveguide characterisation

C. De Bernardi, A. Küng, O. Leminger

Several methods have been devised and used over the years for the determination of waveguide parameters; a few are related to, or derived from, techniques used for optical fibre characterisation, but many are significantly different from their fibre counterparts, or have no counterpart at all in the field of fibres. This difference comes from several reasons, namely from the wide variety of optical materials, refractive index and index differences, dispersion, geometrical shape and symmetry properties, and fabrication techniques which are commonplace in the field of integrated optics. This is in contrast with the far more circumscribed range for fibres, which exhibit (nearly perfect) cylindrical symmetry, negligible attenuation, small refractive index difference, well-known material properties and (obvious but very important) flexibility and availability in long lengths.

Given the range of parameters of specific interest and of their possible values, a reasonably comprehensive treatment of this subject would require a book of its own. On the other hand, a mere list of waveguide parameters with a short description of all the relevant measurement techniques known from the literature would also go beyond the space limits of the present book, without giving a useful picture of the actual problems which challenge who practise the waveguide measurement activity. Additionally, some of the devised methods are applicable only to some particular material or waveguide structure, or work at special wavelengths only, some require unusually complex equipment or special sample preparation, others destroy the sample and thus prevent any possibility of comparison.

Therefore, in the following only a few methods will be described and discussed; these are chosen among the ones more generally applicable, requiring relatively simple equipment, non-destructive, used in many laboratories, and suitable for the measurement of some of the most important waveguide characteristics.

As in the case of fibres, attenuation, refractive index profile, propagation constants and modal characteristics are among the fundamental parameters needed for the practical design of useful waveguide-based devices. Consequently, two methods will be described and discussed for each of the following parameters: i) attenuation; ii) group effective index; iii) guided mode intensity profile; in addition, a method for the determination of modal cut-off wavelength will be presented.

Indeed, the comparison of different methods applicable to a wide range of waveguide structure, shape and composition, is very important to get a better insight of limitations, causes of error, significance and reliability affecting the numeric results of any real measurement activity. The methods described in the following are specially suitable to demonstrate this fact, and moreover have been used in a series of interlaboratory measurements on circulated samples, whose results are discussed in Ch. 5.

4.1 FP resonator method for loss and group index measurement

The analysis of the behaviour of a FP (Fabry-Perot) resonator shows that its transmission characteristics depends on both the round-trip loss (including loss at the mirrors) and the refractive index of the medium filling the cavity. Actually the transmittance T of a Fabry-Perot resonator formed by a lossy medium between two plane parallel mirrors can be expressed as:

$$T = T_0 \frac{e^{-bL}}{1 + R^2 e^{-2bL} - 2R\,e^{-bL} \cos(4\pi L \nu n/c)}, \tag{4.1}$$

where ν is the frequency of the incident optical beam, b is the average loss coefficient and $n = n(\nu)$ the (phase) refractive index of the medium filling the cavity, L its length, R the cavity mirror reflectivity (assumed equal for both mirrors), c the velocity of light *in vacuo,* and T_0 a constant.

In the case of optical waveguides, the refractive index $n(\nu)$ is replaced by the effective index $n_{eff}(\nu)$ of the guided mode(s), L by the geometrical path length between the guide terminal facets and R by the modal reflectivity of these facets.

Provided that there is negligible coupling between different modes in a multimode waveguide, it is possible to write a specific version of equation (4.1) for each mode, and if one can selectively excite a specific mode, this equation provides a way to determine the waveguide behaviour for this mode. However, this is true in principle, but quite difficult to apply to real cases, except for the case of TE and TM polarisations in single-mode waveguides.

4.1.1 Waveguide loss determination

Information about the round-trip loss is provided by the cavity contrast, i.e. the ratio $K=T_{max}/T_{min}$ between maximum and minimum transmission:

$$K = \frac{1 + R^2 e^{-2bL} + 2R\,e^{-bL}}{1 + R^2 e^{-2bL} - 2R\,e^{-bL}} = \frac{(1 + R\,e^{-bL})^2}{(1 - R\,e^{-bL})^2}. \tag{4.2}$$

Propagation loss is easily obtained from equation (4.2) in the form:

$$bL = \ln R + \ln \frac{\sqrt{K} + 1}{\sqrt{K} - 1}, \tag{4.3}$$

from which b is directly calculated (in units of reciprocal length, which can then be converted to the more commonly used dB/cm) from the measured quantities K and L, provided R is known. The latter point is actually rather subtle, in that generally modal reflectivity differs more or less significantly from the plain Fresnel reflectivity of the interface air-core material; however, one of the strong advantages of this measurement technique is indeed its capability to measure directly this parameter too, provided that waveguides of homogeneous features and at least two different lengths are available.

Another important point is that for R<<1 the mirror loss contribution is high, the cavity contrast is low, and b is anyway poorly determined. In such cases (*e.g.*, for guides in low-index materials like silica, most glasses and polymers) the use of external mirrors, contacted to the guide facets, can be helpful, although much care is required, both for the carrying out of the measurement and for the interpretation of the data. In general, the quality of the guide ends is a critical factor to get good measurements: dirt, scratches, chipping or any other facet defect affects strongly the results.

The nomograph shown in Fig. 4.1, calculated directly from equation (4.3), can be useful as a guideline to appreciate the conditions of any specific experimental situation, or to get quickly an estimate of the guide loss from the measured contrast.

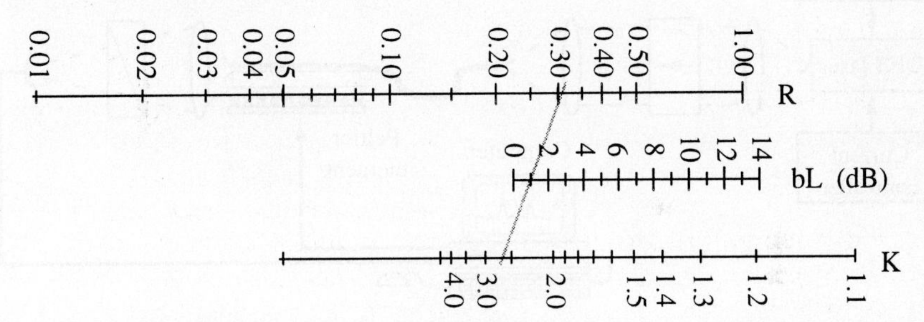

Fig. 4.1. Nomograph for the calculation of waveguide loss, facet reflectivity or cavity contrast from the other two quantities. As an example, a 1 dB loss waveguide with facets reflectivity of 0.3 exhibit a Fabry-Perot contrast of 2.7.

From the experimental point of view, two methods are used in practice to get a scan over at least a full cavity period, and are schematically illustrated in Fig. 4.2. One is based on a fine tunable monochromatic source, by which ν (or equivalently, λ) is swept nearly continuously to cover the range of FP transmission from the minimum to the maximum value; suitable sources include single-mode diode lasers, such as DFB lasers, which can be fine tuned by varying the injection current (rough figures are around 1 GHz/mA in the 1.55 μm range).

The second method uses a wavelength-stabilised source, and carries out a simultaneous scan of L and n_{eff} by changing the temperature of the sample under measurement, by which the same effect of continuously scanning through one or more cavity periods is obtained.

In both cases no precise knowledge is required of the wavelength, the sample temperature or the source power, as only power ratios enter the equations; however, input power stability (for the temperature scan method) or appropriate power normalisation (for the injection current tuning method) are required.

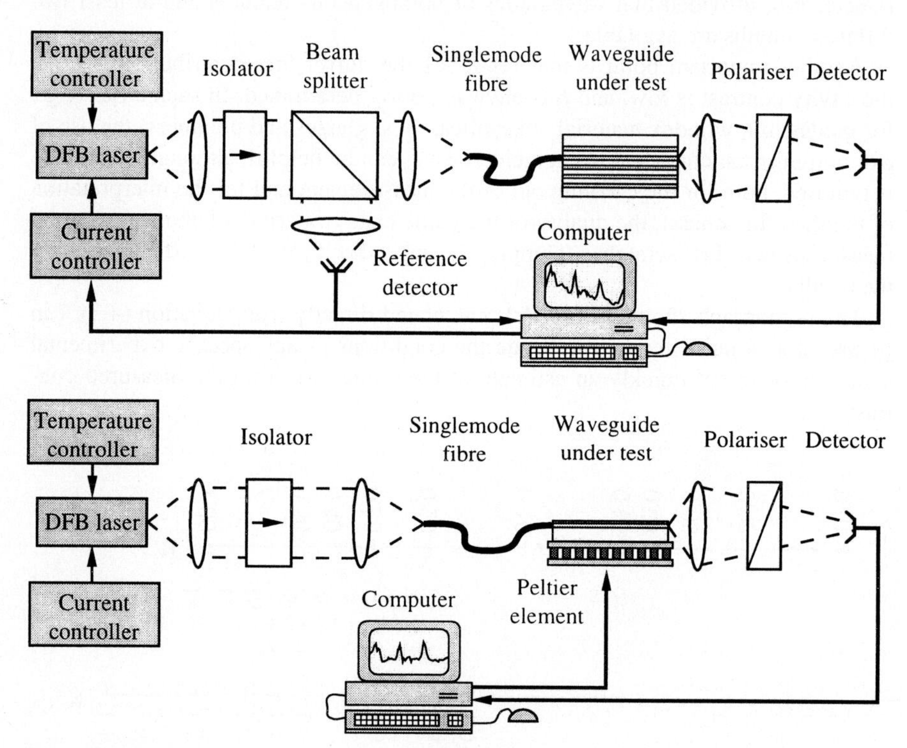

Fig. 4.2. Schematic set-ups of the two method for the measurement of FP contrast of waveguides: a) by current- or temperature-tuning of a single-mode diode laser; b) by temperature scanning of the sample.

Due to the above mentioned effect of facet reflectivity, this technique is most frequently used for semiconductor-based waveguides which exhibit intrinsic reflectivities around 0.3; for them, some results are presented in Ch. 5.

4.1.2 Group effective index determination

The FP cavity transmission T is a periodic function, whose period $p = p(\nu)$ is (provided that b and R are not rapidly varying with ν):

$$p = \frac{c}{2L(n + \nu\, dn/d\nu)}. \tag{4.4}$$

Therefore, the measurement of the spectral periodicity of a waveguide FP resonator and its geometric length yields immediately the group effective index of the guided mode, *i.e.*:

$$n_{g\,eff} = n_{eff} + \nu \frac{dn_{eff}}{d\nu} = n_{eff} - \lambda \frac{dn_{eff}}{d\lambda} = \frac{c}{2Lp}. \tag{4.5}$$

The most effective way of determining the value of p is a Fourier analysis of the cavity transmission characteristics, measured over an adequate range of optical frequencies (wavelengths).

The typical experimental configuration is shown in Fig. 4.3: a narrow linewidth laser source, tunable over a wide wavelength range (about 10 nm or more) is coupled to the guide under measurement, and the transmitted power is measured as function of the wavelength; a wavemeter is used concurrently to measure the emitted wavelength/frequency as precisely as possible for each data point, or alternately a calibrated table of wavelengths for the source settings can be used.

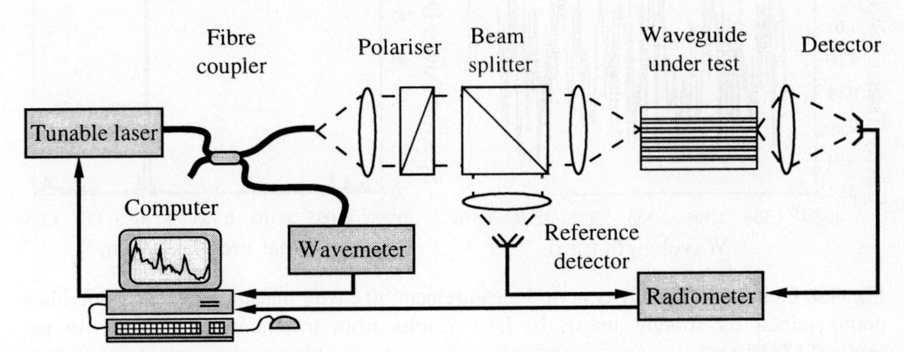

Fig. 4.3. Schematic of a typical set-up for the measurement of the spectral period of a waveguide FP cavity.

The requirements for the accuracy of the wavelength corresponding at each measured transmission value are quite stringent, as the quality of the result of the Fourier analysis is very sensitive to this parameter. On the other hand, there is no need for a dense coverage of the chosen spectral range, or uniform spacing of the available wavelengths/frequencies: an average sampling density of less than one point per period is amply sufficient. Similarly, no particular preparation is needed or limitation applies as to type or characteristics of the sample: even the intrinsic reflectivity of silica waveguides, below 0.04, is adequate. Actually, this technique was tested and validated on plainly cleaved optical fibres [1], demonstrating sensitivities and accuracies to nearly 0.0001 in $n_{g\,eff}$.

Since in actual measurements one cannot guarantee equal spacing of the sampled points, a discrete fourier transform (DFT) algorithm [2] is required to determine p, instead of the commonly used fast Fourier transform (FFT).

The DFT analysis provides the spectral power density function of the cavity transmission, so that peaks appear corresponding to its periodic terms; actually, several peaks may appear, due to the harmonics of the fundamental period, to terms coming from the distribution of sampled points, and to their sums and differences. An example of measured data and corresponding result of the DFT analysis is shown in Fig. 4.4, which shows the spectacular power of this tool. If more than one guided mode is present, peaks corresponding to individual modes may also appear. Since the terms from the sampling distribution can be calculated, or a first approximation of the significant period is already known (for example, from concurrent attenuation measurements by the cavity contrast with a current-tuned laser diode, or from a rough guess of the group index), the peak corresponding to the real cavity term can be univocally determined.

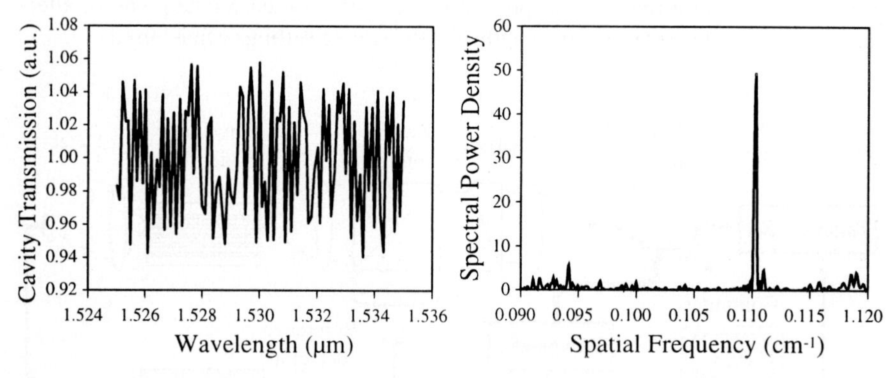

Fig. 4.4. Example of FP cavity period measurement; a) cavity transmission data (individual points joined by straight lines); b) DFT results from transmission data. Cavity period = 0.17418 cm^{-1}.

A reasonable choice of the wavelength span over which to perform the measurements is around 10 nm; wider spectral ranges lead to higher accuracy in the results the Fourier analysis, but then chromatic dispersion in the group index may lead to systematic (although small) errors. Of course, this depends on the dispersive properties of the waveguide material, thus the effect is stronger in III-V based guides. Indeed, if the entire tuning range of about 100 nm provided by currently available tunable lasers is used in smaller subsections (*e.g.*, of 10 nm), a good trade-off between dispersion and Fourier analysis baseline is met, and $P(\lambda)$ and concurrently guide group dispersion can be measured.

For samples of centimetre size, the waveguide length L should be measured to about ±1 μm, in order to avoid additional errors in the calculation of $n_{g\ eff}$.

Curved waveguides can be measured as well, using the appropriate value of L; for silica-on-silicon waveguides, the variation of group effective index with guide curvature has been experimentally evidenced.

4.2 Optical low coherence reflectometry for group index, chromatic dispersion and loss measurements

The design of more and more complex integrated optical circuits (IOC) has shown the need of developing new non-destructive measuring techniques for characterising individually all the functions on an IOC. In 1987, several papers [3], [4], [5] reported the possibility of using optical low coherence reflectometry (OLCR) to characterise the guided wave components with a high spatial resolution. Since then, a lot of work has been carried out on measuring different types of IOC's, mainly glass [6] and $LiNbO_3$ [7], but also on measuring packaged IOC's and all-fibre devices [8]. Nowadays, the OLCR is a well known technique and commercial reflectometers are available on the market.

The use of the OLCR technique can provide a lot of information on the IOC characteristics, but measurements results are not trivial and only a careful analysis of the OLCR scan leads to reliable measurements. Therefore, the theoretical backgrounds and a complete discussion of the physical effects affecting the OLCR scan are presented in this section, in order to discriminate each of these effects from the others and to give a right interpretation of the OLCR scan.

4.2.1 Theory

The output intensity of a Michelson interferometer, as described in Fig. 4.5, is given by

$$I_{\text{det}} = \left\langle \left| E_1 + E_2 \right|^2 \right\rangle = I_1 + I_2 + 2\left\langle \left| E_1 E_2^* \right| \right\rangle, \tag{4.6}$$

where E_1 and E_2 are the electric light fields at the output of each arm of the interferometer, and $I_1 = \left\langle \left| E_1 E_1^* \right| \right\rangle$, $I_2 = \left\langle \left| E_2 E_2^* \right| \right\rangle$ the respective intensities. For polychromatic light with a spectral density $G(\nu)$, the electric fields can be expressed by

$$E_1 = \int_{-\infty}^{\infty} \sqrt{G(\nu)} E_0 \sqrt{TR_1(1-T)} \exp\left[j2\pi\nu\left(t - \frac{2n_1 l_1}{c} \right) \right] d\nu, \tag{4.7}$$

$$E_2 = \int_{-\infty}^{\infty} \sqrt{G(\nu)} E_0 \sqrt{(1-T)R_2 T} \exp\left[j2\pi\nu\left(t - \frac{2n_2 l_2}{c} \right)\right] d\nu, \tag{4.8}$$

where E_0 is the amplitude of the electric field of the input beam, T the intensity transmission factor of the beam splitter, R_1 and R_2 the reflectivity of the mirrors, ν the light frequency, n_1 and n_2 the index of refraction in each arms, l_1 and l_2 the length of each arm and c the velocity of light *in vacuo*.

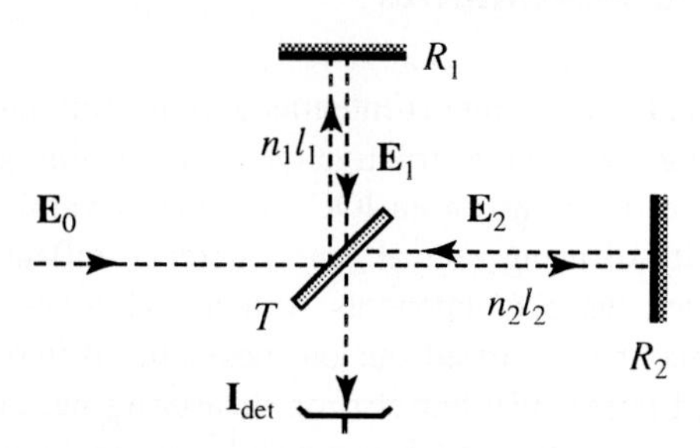

Fig. 4.5. Schematic description of a Michelson interferometer.

The interference term is then given by

$$\begin{aligned} \left\langle \left| E_1 E_2^* \right| \right\rangle &= \int_{-\infty}^{\infty} G(\nu) \exp\left[j2\pi\nu\left(\frac{2n_2 l_2}{c} - \frac{2n_1 l_1}{c} \right)\right] d\nu, \\ &= E_{01} E_{02}\, g\left(\frac{2\pi\nu\Delta l}{c} \right) \cos\left(\frac{2\pi\nu\Delta l}{c} \right), \end{aligned} \tag{4.9}$$

where

$$E_{01} = E_0 \sqrt{TR_1(1-T)}, \qquad E_{02} = E_0 \sqrt{(1-T)R_2 T}\,, \tag{4.10}$$

and where $\Delta l = n_1 l_1 - n_2 l_2$ is the optical path length difference between the two arms and $g(2\pi\nu\Delta l/c)$ (the envelope of the fringes) the Fourier Transform (FT) of the spectral density. This means that interference fringes are obtained only when the two arms of the interferometer are balanced within the coherence length of the light source used. It also means that a broad spectrum light source produces a narrow interference pattern. Therefore, different reflection points spatially separated by more than the light source coherence length give rise to different interference patterns. So using this latter principle, one can built a set-up to identify and measure separately each reflecting site within a waveguide.

4.2.2 OLCR experimental set-up

The OLCR experimental set-up is shown in Fig. 4.6. Its all-fibre configuration is optimised for convenient analysis of the light backscattered from integrated optics components (IOC). The Michelson configuration with very short fibre arms (15 cm) has been chosen in order to obtain a better immunity to external perturbations such as vibrations and temperature variations, but a Mach-Zehnder interferometer configuration [9] could also be used. Since most of the IOC's waveguides are birefringent, polarisation maintaining fibres were chosen in order to allow the analysis of TE and TM modes separately. Launching the light into the sample is performed by positioning the cleaved fibre end in front of the waveguide.

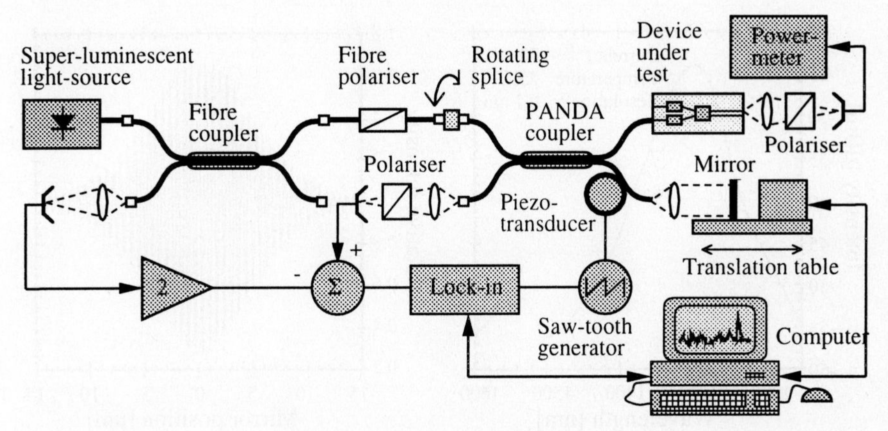

Fig. 4.6. Experimental set-up for optical low coherence reflectometry (OLCR) applied to measurements on integrated optics components (IOC's).

The entire IOC is scanned step by step using a moving mirror in order to change the length of the reference arm. At each position of the mirror, the interference fringe is generated by the light reflected at a given position in the sample and its amplitude is measured using a heterodyne technique: A saw-tooth voltage is applied on the piezoelectric transducer in order to stretch the fibre and obtain a scan of the phase difference between the two arms over 2π. This linear phase modulation over one fringe together with a lock-in detection allows to perform a heterodyne measurement of the amplitude of the fringe. The system is thus able to reconstruct step by step the envelope of the fringe pattern corresponding to reflection at each point inside the IOC. In order to increase further more the dynamic of the system a phase discrimination has been introduced by using two balanced photodetectors, suppressing the residual intensity noise of the light source. All the measurements (mirror displacement, signal processing and recording) are controlled by a computer.

As shown in the theory, the spatial resolution is given by the spectral width of the light source. For the 1.3 μm window, a superluminescent diode (SLD) delivering 50 μW output power into a single-mode fibre and showing a very smooth gaussian shape spectrum (Fig. 4.7) has been chosen. The interference pattern obtained with this light source is presented on Fig. 4.8. and shows a FWMH of 30 μm with residual noise down to -80 dB at a width of 100 μm only. No internal reflections introduced by the pigtailing were detected.

For the 1.55 μm window, the spontaneous emission of a erbium doped fibre amplifier that can easily reach 1 mW is a good alternative. But since the erbium spectrum is made of two main transitions (1536 nm and 1547 nm), it has the side effect of producing a more complex interference pattern.

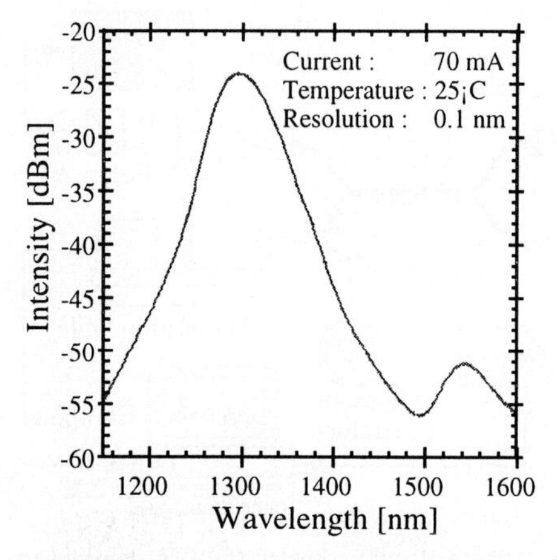

Fig. 4.7. Spectrum of the pigtailed superluminescent diode.

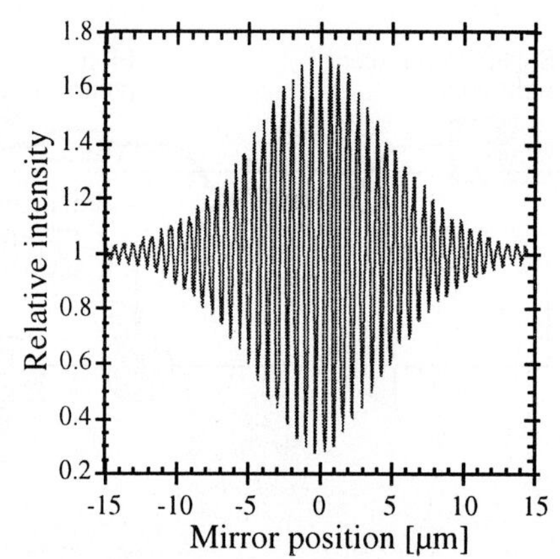

Fig. 4.8. Interference pattern produced by the superluminescent diode.

4.2.3 Calibration and performance of the system

The first measurement is carried out without IOC (see Fig. 4.9). The origin of the mirror position corresponds to the Fresnel reflection at the fibre end. This reflection is used for calibrating the system leading to the reference power level at –14 dB that corresponds to the 3.7% reflection due to the index difference between the cleaved fibre end and the air.

The Rayleigh backscattering gives rise to an interference signal distributed all along the fibre. The dynamic range of the system is mainly limited by the level of this backscattering. It can be estimated by assuming that the fibre losses are only due to the Rayleigh scattering and that the light is uniformly radiated in all directions. PM fibres have losses of 1.1 dB/km at 1.3 μm, thus the amount of light backscattered over the 30 μm of the coherence length of the source is 7.6×10^{-9}. Only 2.5% of this light is coupled back into the numerical aperture of the fibre corresponding to the signal at –97 dB measured for negative mirror positions in

Fig. 4.9. At positive mirror positions, the measured signal corresponds to the Rayleigh backscattering coming from the light reflected at the cleaved fibre end. This light is again reflected at the fibre end and generates the interference signal that limits the overall dynamic of the system. In our system, the dynamic range is therefore limited to –97 dB – (2×14 dB) = –125 dB.

It is to note here that the application of telecommunication fibres would have reduced the Rayleigh backscattering signal, –102 dB at 1.3 μm or –104 dB at 1.55 μm because of their lower losses. Of course, the dynamic should not be limited by the electrical noise coming from the photodetectors and the lock-in amplifier. The power of the light source should be high enough in order to overcome the shot noise limit. The dynamic range and the integration time of the lock-in amplifier defines the minimum input power to be launched in the interferometer. Taking into account a 1 ms lock-in time constant, one gets a calculated minimum SLD power of at least 40 μW.

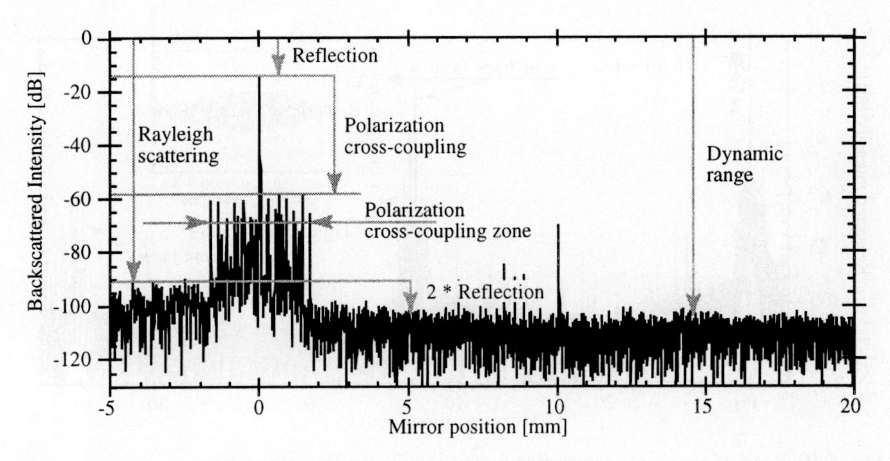

Fig. 4.9. Reference scan of the cleaved fibre end only, performed without IOC for TE polarisation

The use of Hi-Bi fibres allows to control the polarisation state at the fibre outputs but introduces an unwanted signal due to the polarisation cross-coupling along the fibre and in the coupler. Since the two polarisations have different propagation velocities a delay is introduced on the light coupled from one polarisation to the other giving rise to an interferometric signal surrounding each main peak signal coming from the reflections. The width W of this polarisation cross-coupling zone is equal to

$$W = 2\Delta x = 2\frac{l_{fiber}}{l_{beat}}\lambda \tag{4.11}$$

where Δx is the optical path difference introduced by the Hi-Bi fibre, l_{fiber} is the length of the Hi-Bi fibre, l_{beat} is the Hi-Bi fibre beat length and λ is the optical source wavelength. It is therefore recommended to use the shortest PM fibre length possible.

4.2.4 Experimental results on simple IOC's

The following discussion will refer to the example shown in Fig. 4.10 resulting from the scan of a straight Ti:$LiNbO_3$ waveguide. The two main peaks correspond to the Fresnel reflections at the IOC front and back facets. The third peak is generated by the reflection at the front facet of the IOC due to the light that has travelled twice through the sample. The analysis of the interference signal allows to determine several parameters of the IOC.

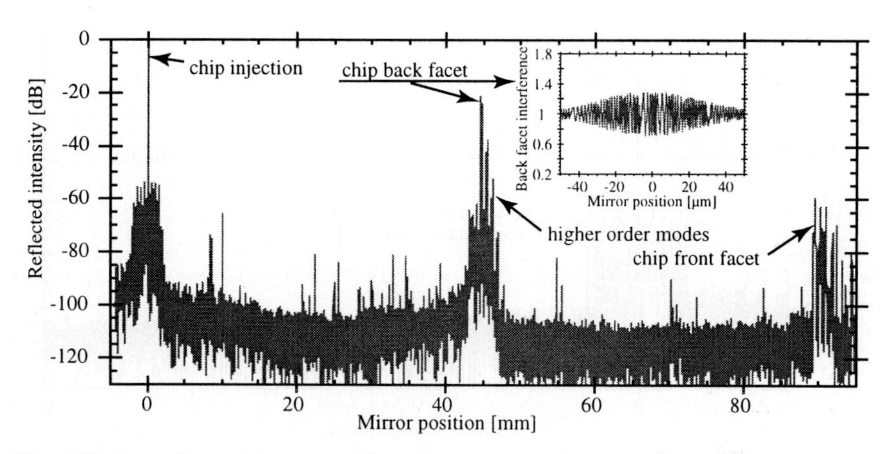

Fig. 4.10. Scan of a straight waveguide in Ti:$LiNbO_3$ for TE polarisation

The group index can be determined by knowing the waveguide length, and by measuring the position of the reflection at the IOC back facet. Looking more carefully at this reflection signal, one can distinguish several peaks. Each of them is generated by the higher order modes having a different propagation constant. Measurement of the group index using this technique is possible for the fundamental mode ($n_{g0} = 2.281$) and also for higher order modes ($n_{g1} = 2.318$, $n_{g2} = 2.358$, $n_{g3} = 2.399$). The knowledge of the different group indices provides information about the index profile of the waveguide.

If one of the arms of the interferometer shows a different chromatic dispersion, the interference term can no longer be represented by equation (4.9). With an additional chromatic dispersion $D(\nu)$ in the second arm, equation (4.8) takes the form

$$E_2 = \int_{-\infty}^{\infty} \sqrt{G(\nu)} E_{02} \exp\left[-j\frac{D(\nu)}{2\pi} 2l_2 \right] \exp\left[j2\pi\nu\left(t - \frac{2n_2 l_2}{c} \right) \right] d\nu. \quad (4.12)$$

By defining a function of dispersion $H(\nu)$, E_2 can be expressed by

$$E_2 = \int_{-\infty}^{\infty} \sqrt{G(\nu)} E_{02} H(\nu) \exp\left[j2\pi\nu\left(t - \frac{2n_2 l_2}{c} \right) \right] d\nu \quad (4.13)$$

and the interference term (4.9) becomes

$$\begin{aligned} \left\langle \left| E_1 E_2^* \right| \right\rangle &= \int_{-\infty}^{\infty} G(\nu) H(\nu) E_{01} E_{02} \exp\left[j2\pi\nu \frac{\Delta l}{c} \right] d\nu \\ &= E_{01} E_{02} \left[g\left(2\pi\nu \frac{\Delta l}{c} \right) * h\left(2\pi\nu \frac{\Delta l}{c} \right) \right] \cos\left(2\pi\nu \frac{\Delta l}{c} \right). \end{aligned} \quad (4.14)$$

The interference pattern is given by the convolution of the FT of the spectral density $g(2\pi\nu\Delta l/c)$ with the FT of the function of dispersion $h(2\pi\nu\Delta l/c)$. The chromatic dispersion always gives rise to a flattening and a broadening of the interference pattern [10]. For a first order chromatic dispersion the fringe pattern is symmetrically broadened. For higher order chromatic dispersion the shape of the interference pattern is modified. By measuring the width of the fringe pattern generated by the reflection at the IOC end facet, it is possible to determine the global chromatic dispersion of both the material ($LiNbO_3$) and the waveguide. Compared to Fig. 4.8, the inset in Fig. 4.10 shows that the interference pattern generated by the light which travelled twice through the sample is homogeneously broadened by a factor $B = 5.2$ corresponding to a signal flattening of –7.4 dB. The total chromatic dispersion D for this IOC waveguide is thus

$$D = \frac{\lambda^2 (B-1)}{c\Delta\lambda^2 2l_{guide}} = 210 \text{ ps/(nm} \cdot \text{km)}, \quad (4.15)$$

where λ is the optical source wavelength, $\Delta\lambda$ the spectral width of the optical source, B the interferogram broadening factor and l_{guide} the length of the waveguide. Interference signals coming from homogeneously distributed effects such as the Rayleigh scattering are not subject to this flattening due to chromatic dispersion, so the noise floor of the system remains at -112 dB.

The losses in optical waveguides are generally determined by measuring the decrease of the Rayleigh backscattering [11]. But Rayleigh backscattering in the IOC's waveguides is generally extremely low and the signal is blurred by the backscattering in the fibre that actually limits the maximum dynamic of our set-up (-112 dB). Losses can then only be determined by measuring the difference between signals coming from the reflections at the back and the front facets of the IOC. But this difference includes the Fresnel reflections and the flattening due to the chromatic dispersion. Therefore, these two contributions have to be accurately determined to reach relevant loss measurements. Using this method, losses of our Ti:$LiNbO_3$ straight waveguides are about 1 dB/cm.

4.2.5 Experimental results on more complex IOC's

Measurements performed on complex IOC's results in more complicated OLCR scans resulting from the numerous reflections sites and/or light travelling in parallel through several waveguides. Nevertheless, in the case of dynamic devices such as Mach-Zehnder intensity modulators it is possible to take advantage of the IOC electrodes to introduce a selective phase modulation to locate the observed reflection site within the IOC.

Another interesting possibility offered by the OLCR technique is the measurement of coupling between TE and TM modes as it often occurs for example in Y junctions.

4.3 Near field imaging for mode profile measurement

The near field distribution in optical waveguides contains an important information on their properties, *e.g.*, the (phase) refractive index profile of the guiding structure and the achievable coupling efficiency to other optical structures. Moreover, the intensity distribution at the output facet of a waveguide is an attractive quantity from the viewpoint of measurements, as the end faces of a guide are inherently accessible and therefore it can be determined generally.

In the general case, the exact description of the field distribution in a two-dimensional guiding structure is given by a full-vectorial wave equation. However, for a weakly-guiding dielectric waveguide, a reasonably approximate description of the transverse field $E(x,y)$ is given by the scalar wave equation

$$\nabla^2 E(x,y) + k_0^2\left[n^2(x,y) - n_{eff}^2\right]\cdot E(x,y) = 0, \tag{4.16}$$

where ∇^2 is the Laplacian operator, $E(x,y)$ is the electric field amplitude proportional to the square root of the intensity $I(x,y)$, $k_0 = 2\pi/\lambda$ is the angular repetency (free-space wavenumber), $n(x,y)$ is the refractive index profile, and n_{eff} is the effective index of the guided mode.

At least for the fundamental mode which has no zero in the field amplitude, and therefore for single-mode waveguides, it could seem easy to get a good measurement of $I(x,y)$ and from it of $E(x,y)$, then to solve equation (4.16) for $n(x,y)$ or to calculate the overlap integral with the field of a fibre or another waveguide. However, the measurement requirements and problems are not trivial, and depend strongly on which type of information is requested.

Two main schemes can be adopted to determine the near-field intensity distribution of a waveguide: i) using an imaging detector to acquire simultaneously the entire intensity map; ii) scanning a single-element detector over a magnified image of the output face. In both cases the image is generated by a high-quality microscope objective, when monochromatic light is coupled to the input face; the choice of the approach depends on several factors that will be briefly discussed.

4.3.1 Near field determination by imaging detectors

For wavelengths shorter than about 1 μm, the most commonly used imaging detectors are silicon-based CCD arrays. These offer excellent quality in terms of linearity, dynamic range, uniformity of response and low cost, and are available with numbers of pixels far exceeding the requirements of this application. Digital CCD cameras offer up to 16 bit signal resolution, and software for basic data processing (e.g. subtraction of dark background, correction for non-uniform responsivity, etc.) is widely available, sometimes free, in particular developed for high-quality astronomic photometry.

For operation at longer wavelengths, and mainly in the 1.3–1.55 μm range useful for single-mode communication fibres, matters are not so favourable; various options are available, *i.e.* IR vidicons, pyroelectric and (more recently) PtSi and InGaAs detector arrays, but none is entirely satisfactory. At present these arrays are rather costly and still suffer from poor uniformity, limited dynamical range (as well as low sensitivity for pyroelectrics) and noise performance, besides providing small numbers of pixels, although things may change in the future for the InGaAs arrays.

Vidicons have also many drawbacks: limited dynamic range (at most about 20 dB), non-uniform response across the sensing surface, rather high dark noise, strong nonlinearity and image distortion. Moreover, some of these unfavourable characteristics (notably responsivity, nonlinearity and dark noise) may change with time and ambient temperature, increasing the difficulty of taking into account their effect in data processing. In spite of that, currently they are by large the most widely used in experimental set-ups like that shown in Fig. 4.11.

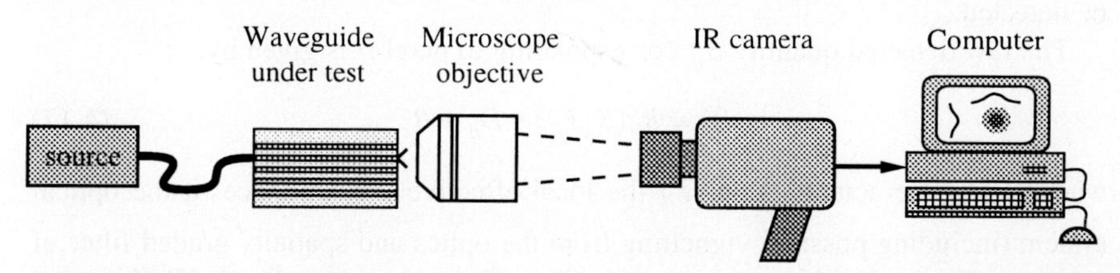

Fig. 4.11 Basic set-up to determine the near field intensity distribution of waveguides by an imaging detector (CCD or IR vidicon camera).

4.3.2 Near field determination by scanning a single-element detector

The use of a single detector obviously overcomes all problems of uniform responsivity; moreover, single-element InGaAs detectors exhibit low noise, high linearity and dynamic range, approaching the characteristics of silicon photodiodes. Given the poorer performance of the available imaging detectors, this technique can provide the best quality for intensity mapping at wavelengths beyond 1 μm. However, this advantage is counterbalanced by the need of complex and costly computer-controlled mechanical scanning equipment: in fact total data acquisition time is relatively long (several thousands of individual points may be required), and mechanical drift during this interval must be very small, to avoid data skewing and systematic errors.

Potential remedies against drift effects come from a suitable choice of the scanning pattern (e.g. a random sequence of pixels, or averaging two scans in reverse order, etc.), but this further increases measurement time. Therefore, this technique is seldom used in practice, also because some sort of imaging device must anyway be available, to check correct focusing, alignment *etc.* of this set-up.

4.3.3 Practical aspects of near field imaging in the 1.3–1.55 μm range

For the above reasons, the following discussion is limited to the case of imaging by an IR vidicon camera.

A convenient way of acquiring the signal from an analog IR camera (as is the case for most available models) is by means of a frame grabber. No need exists for high speed grabbing, or for large pixel counts: although some degree of spatial oversampling is useful, the small size of the guided light spot (not much larger than the diffraction limit of the imaging optics) does not justify more than a few hundred pixels per side. Available grabbers are usually limited to 8-bit sampling, but this is enough, given the limited dynamic range of the detector. A method has been proposed [12] to increase the effective dynamic range, based on a spatially-graded, neutral density filter: this attenuates significantly the central, brighter region of the spot, avoiding saturation, while the unattenuated outer wings can still be detected.

The raw detected quantity S'_{ij} corresponding to pixel ij is given by:

$$S'_{ij} = R_{ij}(S_{ij}F_{ij}) + D_{ij} + B, \tag{4.17}$$

where S_{ij} is the actual signal, F_{ij} the local effective transmittance of the optical system (including possible vignetting from the optics and spatially graded filter, if present), R_{ij} the local responsivity function of the detector, D_{ij} the background signal term, and B a bias term originated from system electronics.

To correct the raw data, first of all a "dark field" must be subtracted, which includes the D_{ij} and B terms and should be acquired with the source off shortly before or after the data.

For further processing, the camera nonlinearity must be known; this can be determined by uniformly illuminating the detecting surface at different known intensity levels spanning the tube dynamic range (preferably shortly before or after the actual measurement, due to the poor time stability of responsivity), acquiring the corresponding signals and fitting them by a suitable function. R_{ij} is well represented by the following form:

$$R_{ij}(w) = a_{ij}(w+b_{ij})^{c_{ij}}, \tag{4.18}$$

where w is the incident intensity and the coefficients $a_{ij}, b_{ij}, c_{ij}, d_{ij}$ in general depend on the camera settings; the exponent c_{ij} is usually in the range 0.45–0.6.

Finally, the quantity F_{ij} can be determined by acquiring an uniformly illuminated field through the same optical system used for the measurements.

It must be noted that the signal S_{ij} obtained from equation (4.17) is not exactly a discrete map of the true near field intensity distribution $I(x,y)$: actually this is blurred by diffraction in the imaging optics, superposed to internal reflections in the optical train (including the camera tube faceplate) and to spurious light. The latter can be residual unguided or partially guided light (e.g. in the slab layer of ridge structures), which may be difficult to remove even by spatial filtering. Practically, indicating by $g(x,y)$ the spatial impulse response of the optics, by $\otimes$ the convolution operator and by $\eta(x,y)$ the superposed noise, one can write:

$$S(x,y) = I(x,y) \otimes g(x,y) + \eta(x,y) \tag{4.19}$$

Diffraction limit problems may be particularly severe in high numerical aperture, small cross section semiconductor waveguides; at a wavelength of 1.5 μm a 0.8 NA microscope objective gives a point-spread function width of about 2 μm, and the highly diverging output beam (primarily in the plane normal to the substrate) can cause diffraction rings from aperture stops in the optics. Finally, defects (like scratches, dirt, edge rounding or chipping) on the output facet perturb the resulting intensity distribution in an unpredictable way.

An example of what can be expected from a good sample when due care is used to minimise perturbations is shown in Fig. 4.12, and compared to examples showing some of the above mentioned effects.

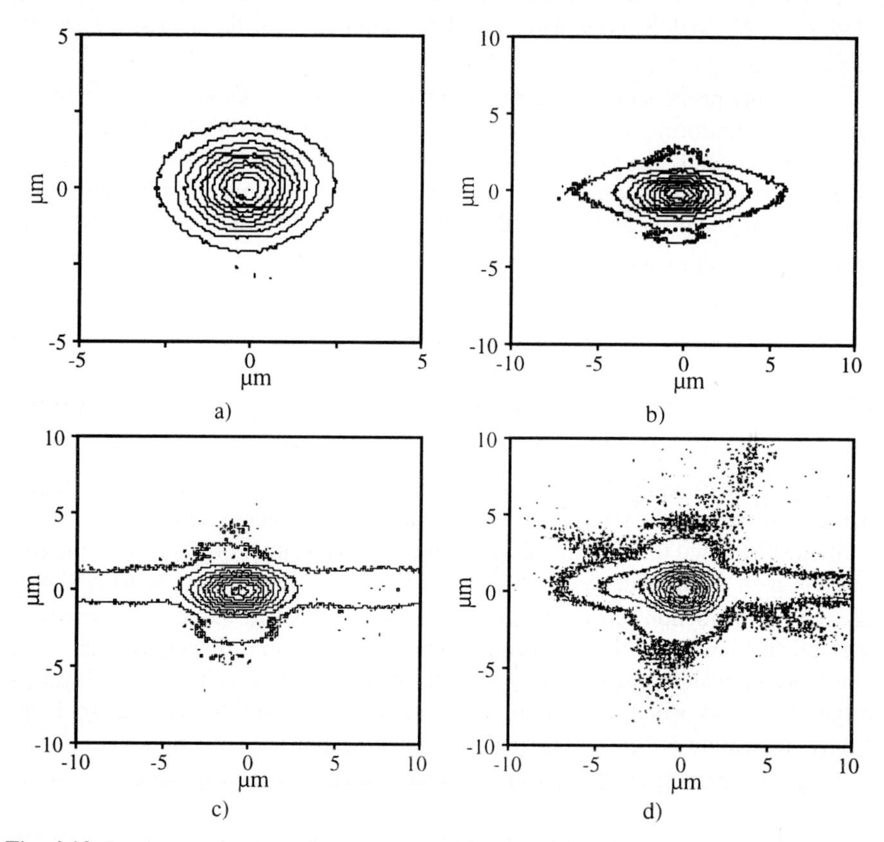

Fig. 4.12. Isophotes of acquired images of semiconductor waveguides output facets; curve spacing is linear in intensity. a) good guide facet, negligible spurious light and diffraction effects; b) significant diffraction effects in the vertical direction; c) high light level in the slab region of a ridge guide, superposed to strong diffraction effects as in b); d) combination of the previous effects with spurious reflections in the tube faceplate (the slanted cross pattern).

The type and degree of further data processing depends on which information is required. If this is the calculated coupling efficiency between waveguides, reasonably good result are obtained by approximating the guided field as

$$E(x,y) \approx \sqrt{S(x,y)}, \tag{4.20}$$

where $S(x,y)$ is an averaged and smoothed function obtained from S_{ij}. Actually the precise form of the field wings contributes little to the overlap integral of the two fields, so that limits from dynamical range, noise and finite resolution are acceptable. In practical cases, such calculations agree within 0.1–0.2 dB with measured values, as demonstrated for both semiconductor waveguides and ion-exchanged glass guides coupled to single-mode fibres.

If refractive index profile reconstruction is the goal, the requirements become much more stringent: the effects of finite resolution of the optics should be taken into account, the field wings become highly important, hence the need of high dynamic range, and low noise. Moreover, the expression for $n(x, y)$ from Eq. (4.16),

$$n^2(x,y) = n_{eff}^2 \left\{ 1 - \left[\frac{\nabla^2 E(x,y)}{k^2 E(x,y)} \right] \right\}, \tag{4.21}$$

is extremely sensitive to noise. This is due to both the Laplacian operator and the division by the field amplitude, which becomes increasingly critical in regions where the measured intensity (and therefore the signal to noise ratio) is low. Additionally, equation (4.16) is not valid at high index-contrast interfaces (e.g. at the air-substrate surface even in diffused waveguides).

This has limited the attempts to calculate $n(x, y)$ mostly to the case of buried, graded-index waveguides; besides relatively small numerical aperture and smooth index profile, they exhibit modal spot sizes large enough to justify neglecting the deconvolution of the spatial impulse response of the optical system. This is really a tough task, since in equation (4.19) in most realistic situations only $S(x, y)$ is known. The case of such "blind deconvolution" has been studied and appropriate algorithms have been devised [13], [14]; however, these enhance the effects of noise, imposing the highest signal quality and appropriate averaging and smoothing of the data.

Limited tests have been performed on the case of buried ion-exchanged guides in glass, comparing the results obtained by either applying or neglecting the deconvolution process. In the case of a detected spot size of about 10×6 μm^2 at $1/e^2$ the deconvolution yields a field only slightly narrower than the measured one, but much noisier and hardly usable for further calculations. Neglecting this step and using a bicubic spline fitted to the (smoothed) detected intensity for further reduction of noise, a qualitatively correct index distribution is obtained, though affected by some calculation artefacts.

Therefore at present the problem of accurately calculating the waveguide index distribution from near field imaging is still open.

4.4 Transverse offset method for mode profile measurement

When measuring the mode profile by near-field imaging, the spatial resolution is determined by the numerical aperture of the microscope objective used and, in principle, cannot resolve structures smaller than the light wavelength λ. This wavelength is often larger than the spot-size diameter of the mode profile, especially for $\lambda = 1.55$ μm, and, in this case, diffraction limits and, in addition to them, lens imperfections and the nonlinear and inhomogeneous sensitivity of the vidicon tube used result in a distorted measured mode profile. In the presence of such constraints, a non-destructive method based on transverse offset scanning of the mode profile field allows better results to be achieved. A simple version of this technique is widely used for measuring the mode field diameter in single-mode optical fibres [15], [16]. In the following, a two-dimensional transverse offset method, suitable for optical waveguides of various kinds, is described. A more detailed presentation can be found in [17].

For the measurement, light is coupled into the investigated waveguide on the input side of the sample by a microscopic objective. The focus of the objective is scanned laterally over a large square area containing the cross section of the waveguide as it is shown in Fig. 4.13. On the output side of the sample, the transmitted light is coupled into a fibre and detected by a photodiode.

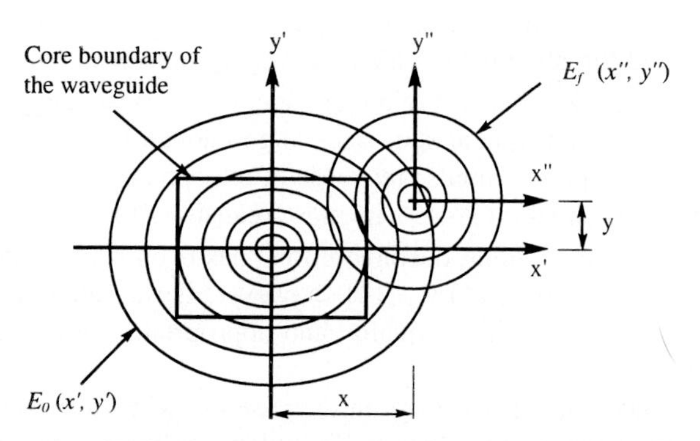

Fig. 4.13. Scanning the front side of the waveguide chip by a microscope objective.

Assuming the investigated dielectric waveguides to be weakly guiding and single-moded, a scalar treatment can be used. The output field of the microscopic objective $E_f(x'', y'')$ set off transversely by (x, y) relative to the field $E_0(x', y')$ of the fundamental waveguide mode (see Fig. 4.13) can be written as

$$E_f(x'-x, y'-y) = c_0 E_0(x', y') + E_R(x', y'), \tag{4.22}$$

where $E_R(x', y')$ sums up the fields of the radiation modes. The total intensity I at the output depends on the offset (x, y) and, using the mode orthogonality, we get

$$I(x, y) = \frac{\left[\iint E_f(x'-x, y'-y) E_0(x', y')\, dx'\, dy'\right]^2}{\iint E_0^2(x', y')\, dx'\, dy'}. \tag{4.23}$$

Here it is tacitly assumed that, going through the sample, $E_R(x', y')$ has been radiated off.

So the square root of the detected intensity $f(x,y)=\sqrt{I(x,y)}$ represents the two-dimensional convolution of the known field distribution of the objective E_f with the unknown mode profile E_0 of the waveguide:

$$f(x,y)=E_f(x,y)\otimes E_0(x,y). \tag{4.24}$$

In a subsequent step, the required mode profile E_0 must be determined from the measured two-dimensional distribution $f(x,y)$ by a deconvolution. A conventional way of calculating it is to take the quotient of both Fourier-transformed fields $\tilde{f}(\omega_x,\omega_y)$, $\tilde{E}_0(\omega_x,\omega_y)$, and perform its inverse Fourier transform. As this procedure is very sensitive to the signal-to-noise ratio, it yields no acceptable results in the presence of usual measurement uncertainties [17].

As an alternative method, the two-dimensional deconvolution can also be carried out by means of iterative convolutions. For the transverse offset method, the optimum algorithm is that published by Gold [18], as it is fast converging and needs no additional parameter assumptions. The iterative scheme of the Gold algorithm runs as follows:

$$E_0^{(0)}(x,y)=f(x,y) \tag{4.25}$$

$$E_0^{(k+1)}(x,y)=E_0^{(k)}(x,y)\frac{f(x,y)}{E_0^{(k)}(x,y)\otimes E_f(x,y)};\quad k=0,1,2,\ldots \tag{4.26}$$

A typical set-up for the measurement is shown in Fig. 4.14. The light source used for the measurement is a 1.55 μm laser diode with a single-mode fibre pigtail. The light beam from its output is collimated by a lens and transmitted through a depolariser, a beam splitter, and a Glan-Thompson polariser to a microscope objective with NA = 0.9 and focused by it with a specified polarisation on the front side of the waveguide chip. The plane of polarisation of the light can be fixed arbitrarily by rotating the polariser. The beam splitter together with an eyepiece makes the adjustment of the equipment easier. The chip is mounted on a three-axis precision translation stage monitored by a computer. It has been found appropriate to scan a 10 μm×10 μm square area with 0.1 μm steps. At the end of the waveguide, the transmitted light is coupled into a single-mode fibre and its optical power is detected by a low-noise germanium photodiode, followed by a transimpedance amplifier. The signal-to-noise ratio is better than 40 dB. The output signal of the amplifier, processed by a 14-bit analog-to-digital converter, arrives finally at the control desktop computer, where the optical power and its square root are recorded in dependence on the position of the chip.

In order to get the unknown mode profile of the investigated single-mode waveguide, the two-dimensional deconvolution described above must be done. The required near-field distribution in the focal plane of the microscope objective was determined in a previous measurement using a far-field method. For the numerical calculation of the deconvolution, the Gold algorithm is used, with 20 iteration steps.

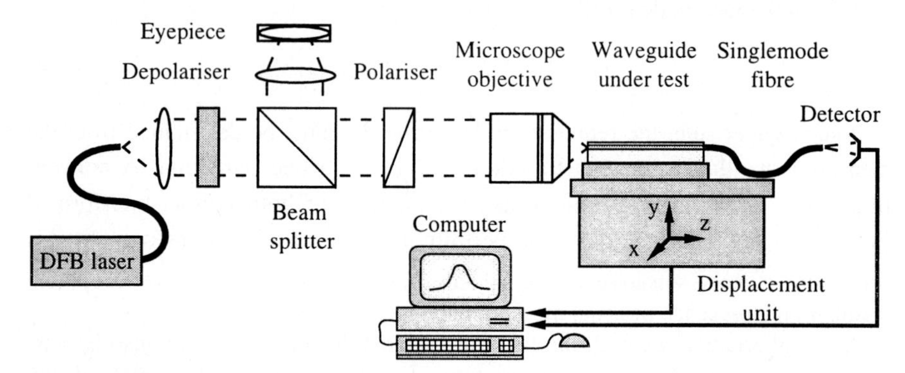

Fig. 4.14. Measurement set-up for the transverse offset method.

As an example, the results obtained for a rectangular semiconductor waveguide fabricated in InP/InGaAsP material are presented. Its width and height are 6 μm and 115 nm, respectively. The refractive indices of the core and substrate are 3.25 and 3.16, respectively. As an eigenmode solver for the stationary mode profile, a scalar calculation was used based on the Rayleigh-Ritz method [19].

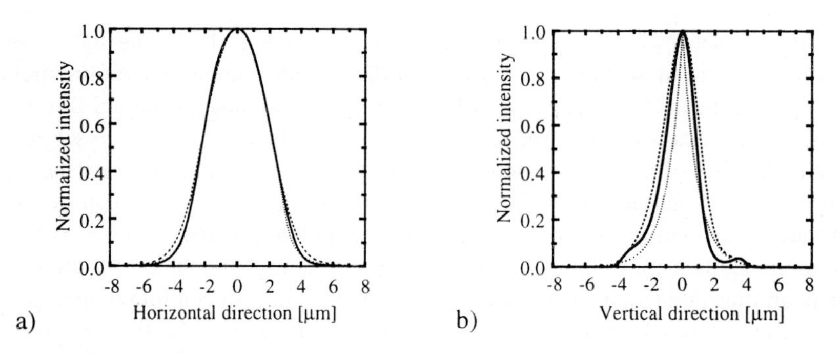

Fig. 4.15. Horizontal (a) and vertical (b) cross-section of the mode intensity profile of a rectangular dielectric waveguide. The solid lines show the experimentally determined (deconvoluted and squared) values, the dotted lines show the calculated values. The dashed lines mark the measured scanned intensity before deconvolution.

For the horizontal direction, Fig. 4.15(a) shows the correspondence between the experimentally determined (deconvoluted and squared) mode intensity profile (solid line) and the calculated one (dotted line). The dashed line marks the measured intensity distribution before the deconvolution. Fig. 4.15(b) shows the same distributions for the vertical direction. Here, the deconvoluted and the calculated distributions differ considerably, due to the peaked narrow mode profile with a double-exponential decay.

To summarise, the presented transverse offset method is an alternative to the conventional near-field measuring technique and allows its resolution limitations to be improved. The main drawbacks are: the method applies only to single-mode waveguides, the cladding and radiation modes must be stripped off completely, and for peaked mode profiles the deconvolution algorithms are inadequate.

4.5 Spectral transmission method for cut-off wavelength measurement

One of the most important parameters in the characterisation of integrated optical devices is a precise determination of the effective cut-off wavelengths of the fundamental and the first higher-order modes, since their difference determines the spectral region of a single-mode operation of a device [20].

Theoretical values for the cut-off wavelengths can be obtained by rigorously solving the wave equations, taking into account the exact refractive index profile and the exact geometrical structure of the waveguide [21]. However, the calculations for arbitrary waveguides are very complicated and only appropriate for straight-channel waveguides. In bent waveguides the effective cut-off wavelength can be shifted considerably to shorter values due to the increasing radiation losses as the wavelength approaches the theoretical cut-off wavelength. Therefore, the experimentally observed effective cut-off wavelengths are of far greater importance for practical use.

For the measurement of the cut-off wavelengths of optical fibres [22], [23], two standard reference test methods based on spectral light transmission techniques are recommended by international standard organisations such as the CCITT [24], namely the single bend and the power step techniques. The latter is also well suited to the measurement of cut-off wavelengths of integrated optical components [21], [25]. An extension of this technique is presented hereafter.

The effective cut-off wavelength for a certain mode is defined as the wavelength where this mode is practically absent at the exit of the waveguide [23]. If all modes at the waveguide entrance are excited, the spectra of the transmitted light will show a drop of power at each wavelength where a mode is "cut off". However, a proper excitation of the waveguides is essential for their correct characterisation. On one hand it has to be assured that all possible modes are excited by an incoherent source with the same power. Therefore, the light spot at the entrance of the waveguide should be larger than the size of the waveguide cross-section. On the other hand, the light spot should be sufficiently well confined to prevent too much light from being injected into the substrate, exciting thus unwanted substrate radiation modes. Furthermore, at the cut-off wavelength of a mode, all the energy of the "cut-off" mode is transferred into substrate in the form of radiation modes. These modes propagate along the waveguide, and owing to its small length their power can be captured by the detector at the output and affect thus the results of the measurements.

4.5.1 Experimental set-up

Fig. 4.16 shows the experimental set-up. The image of a linearly polarised white light source (a halogen lamp) is projected onto the plane of an aperture with a diameter of 100 µm with a pair of lenses. The small aperture was used to select only a small part of the coiled filament of the lamp to ensure a uniform illumination of the endface of the integrated optical device. Using this aperture, the spot size of the launching light beam was sufficiently reduced to 20 µm which ensured an efficient light injection into the waveguide without too much light being lost into the substrate. A high-quality lens such as a microscope objective focuses the image of the source aperture into the waveguide. The numerical aperture of the microscope (NA = 0.35) was chosen to be high enough to excite all the guided modes supported by the waveguide.

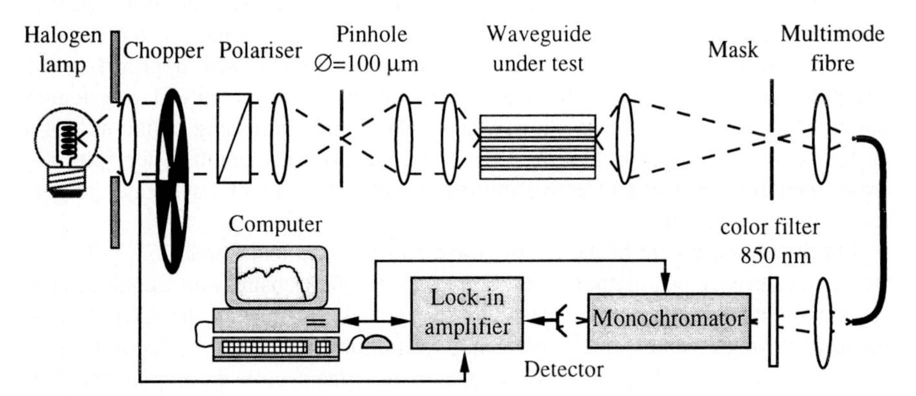

Fig. 4.16. Experimental set-up for the cut-off wavelength measurements

The image of the output light of the waveguide is collimated onto the face of a multimode fibre. A prefixed adjustable aperture mask (spatial filter) enables a precise injection of the light coming from the waveguide into the fibre while eliminating the light coming from the substrate. The fibre output is then directed into a high-resolution grating spectrometer followed by a colour filter to suppress the second order diffraction.

The spectral resolution of the monochromator was chosen to be 8 nm and it was scanned in the desired wavelength region of 900–1600 nm. The spectrally filtered light at the exit slit of the monochromator is detected by an InGaAs detector and the signal is then processed by a lock-in amplifier.

To extract the spectral response of the waveguide iself from the signal containing the spectral desponses of all other components of the system (such as the broadband lamp, the detector, and the monochromator), the spectral output power $P(\lambda)$ had to be normalised. The best results were achieved by measuring the optical output power spectrum $P_r(\lambda)$ of the whole system without the waveguide and using it as a reference. The ratio between the transmission spectra measured with and without the waveguide,

$$r(\lambda) = 10\log\left[P(\lambda)/P_r(\lambda)\right], \tag{4.27}$$

is then plotted as a function of wavelength.

4.5.2 Typical response

Fig. 4.17 gives a typical example of the normalised waveguide output spectrum of $Ti{:}LiNbO_3$ straight waveguides obtained from different Ti-strip widths. The cut-off wavelengths can easily be determined from the responses shown, knowing that the power coupled into each mode of the waveguide increases according to a λ^2 law. The cut-off wavelength for the fundamental mode (λ_{c00}) is defined to be the point at which the transmission spectrum drops by 3 dB after having attained its last maximum (point A for the 6 μm strip waveguide).

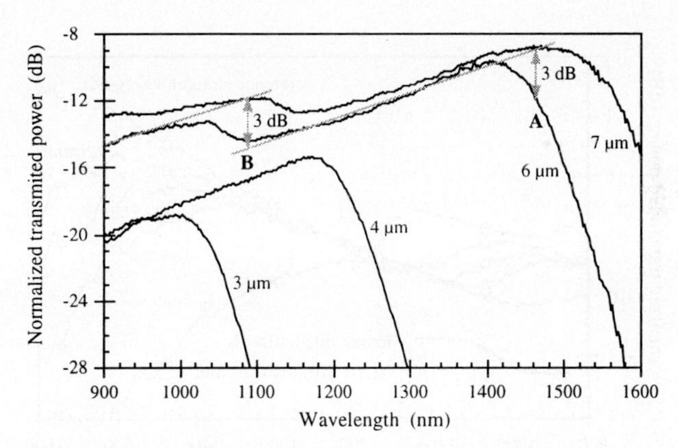

Fig. 4.17. Normalised transmission spectra for different $Ti{:}LiNbO_3$ straight waveguides obtained from different Ti-strip widths.

Since the fundamental and the first order mode transmit the same amount of power because of the uniform excitation, the transmission spectra should drop by 3 dB when the cut-off of the first order mode is reached. The cut-off wavelength for the first order mode (λ_{c01}) is then defined as the point where the normalised transmission spectrum starts to rise again [21-23] (point B for the 6 µm strip waveguide). In Fig. 4.18, the variation of the cut-off wavelengths of the fundamental and first-order modes as a function of the Ti-strip width is summarised for another group of waveguides. From this figure, the single-mode region for each waveguide width can be easily identified with an acuracy adequate for practical purposes.

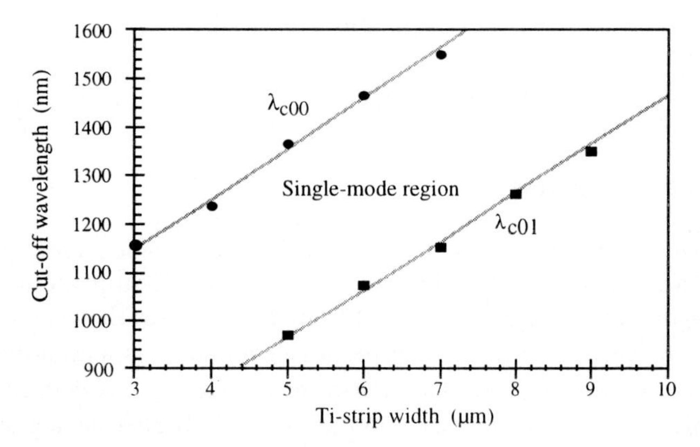

Fig. 4.18. Variation of the cut-off wavelengths of the fundamental (λ_{c00}) and first -order modes (λ_{c01}) as a function of the Ti-strip width

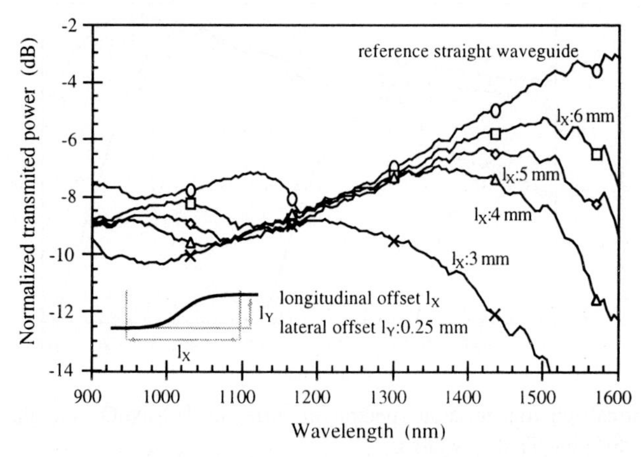

Fig. 4.19. Transmission spectra of bent waveguides with different longitudinal offsets while the lateral offset is kept constant.

The transmission spectra of bent waveguides showed that the induced losses cause a downshift of a cut-off wavelength with decreasing curvature radius. It can also be stated that a decrease of the curvature radius reduces the single-mode region compared to that of a straight waveguide. This can be explained by the additional losses that become more important for the fundamental mode near its cut-off wavelength and for smaller curvature radii. It is thus very desirable to optimise the waveguides in such a way that the cut-off wavelength for the first order mode is very close to the operating wavelength. It can be seen that radiation losses due to bending are lowest just above the cut-off wavelength of the first order mode, due to a very good mode confinement.

4.5.3 Conclusion

Relevant information about the single-mode operation of an integrated optical device can be obtained using the spectral light transmission method for the measurement of cut-off wavelengths. However, in order to get clear and reliable results as those obtained from Ti:$LiNbO_3$ waveguides taken as an example above, a special care is to be taken to excite the waveguide uniformly, without launching too much light into the substrate, and to normalise the transmitted spectra correctly. When measuring semiconductor waveguides, the system can be limited by the numerical aperture of the injection optics (which must be greater than the numerical aperture of the waveguide), or by the material transparency because the transmission window must be wide enough to distinguish the cut-off of the fundamental and first-order modes.

References

[1] F. Pozzi, C. De Bernardi, and S. Morasca, "Group effective indices of different types of optical fibres measured around 1550 nm," *J. Appl. Phys.*, vol. 75, No. 6, pp. 3190–3192, 1994.

[2] E. P. Belserene, "Rhythms of a variable star", *Sky and Telescope*, pp. 288–290, 1988.

[3] R. C. Youngquist, S. Carry, and D. E. N. Davies, "Optical coherence-domain reflectometry: a new optical evaluation technique", *Opt. Lett.*, vol. 12, No. 3, pp. 158–160, 1987.

[4] K. Takada, I. Yokohama, K. Chida, and J. Noda, "New measurement system for fault location in optical waveguide devices based on an interferometric technique", *Appl. Opt.*, vol. 29, No. 9, pp. 1603–1606, 1987.

[5] B. L. Danielson and C. D. Whittenberg, "Guided wave reflectometry with micrometre resolution", *Appl. Opt.*, vol. 26, No. 14, pp. 2836–2842, 1987.

[6] K. Takada, N. Takato, J. Noda, and N. Uchida, "Interferometric optical-time-domain reflectometer to determine backscattering characterisation of silica-based glass waveguides", *J. Opt. Soc. Am. A*, vol. 7, No. 5, pp. 857–867, May 1990.

[7] Ch. Zimmer and H. H. Gilgen, "Optical reflectometry for integrated optical components", *Proceedings ECIO 93*, April 1993, Neuchâtel, Switzerland, pp. 14–16.

[8] P. Lambelet, P. Y. Fonjallaz, H. G. Limberger, R. P. Salaté, Ch. Zimmer, and H. H. Gilgen, "Bragg grating characterisation by optical low-coherence reflectometry", *IEEE Photon. Technol. Letters*, vol. 5, No. 5, pp. 565–567, May 1993.

[9] K. Takada, A. Himeno, and K. Yukimatsu, "Resolution control of low-coherence optical time-domain reflectometrer between 14 and 290 μm", *IEEE Photon. Technol. Letters*, vol. 3, No. 7, pp. 676–678, July 1991.

[10] A. Kohlhaas, C. Frömchen, and E. Brinkmeyer, "High-resolution OCDR for testing integrated-optical waveguides: Dispersion-corrupted experimental data corrected by numerical algorithm", *J. Lightwave Technol.*, vol. 9, No. 11, pp. 1493–1502, 1991.

[11] L.-T. Wang, K. Iiyama, F. Tsukada, N. Yoshida, and K. Hayashi, "Loss measurement in optical waveguide devices by coherent frequency-modulated continuous-wave reflectometry", *Opt. Lett.*, vol. 18, No. 13, pp. 1095–1097, 1993.

[12] D. Brooks and S. Ruschin, "Improved near-field method for refractive index measurement of optical waveguides", *IEEE Photon. Technol. Lett.*, vol. 8 no. 2, pp. 254–256, 1996.

[13] Y. Yang, , N. P. Galatsanos, and H. Stark, "Projection-based blind deconvolution", *J.Opt.Soc.Am. A*, vol. 11, pp. 2401–2409, 1994.

[14] N. F. Law and D. T. Nguyen, "Improved convergence of projection based blind deconvolution", *Electron. Lett.*, vol. 31 no. 20, pp. 1732–1733, 1995.

[15] W. T. Anderson and D. L. Philen, "Spot size measurements for single-mode fibers – a comparison of four techniques". *J.Lightwave Technol.*, vol.LT-1, p. 20, 1983.

[16] M. Artiglia, G. Coppa, P. Di Vita, M. Potenza, and A. Sharma, "Mode field diameter measurements in single-mode optical fibers." *J.Lightwave Technol.*, vol.7, p. 1139, 1989.

[17] M. Halfmann, *Messung der Feldverteilung und des Brechzahlprofils von integriert-optischen Wellenleitern.* Thesis. University of Kaiserslautern, Germany, 1994.

[18] R. Gold, "An iterative unfolding method for response matrices." *AEC Res. and Develop. Rep.* ANL-6984, Argonne National Lab., 1964.

[19] O. Leminger and R. Zengerle, "Eigenmode calculation of dielectric waveguides near cutoff using the Rayleigh-Ritz method with rational basis functions." *Opt. Quantum Electron.,* vol. 27, p. 1009, 1995.

[20] S. I. Najafi, *Introduction to Glass Integrated Optics*, Artech House, Inc., Norwood, MA., 1992.

[21] G. Lamouche and S. I. Najafi, "Scalar Finite-Element Evaluation of Cut-Off Wavelength in Glass Wave Guides and Comparison with Experiment", *Can. J. Phys.* vol. 68, pp. 1251–1256, 1990.

[22] Y. Kitayama and S. Tanaka,: "Length Dependence of LP_{11} Mode Cutoff and its Influence on the Chromatic Dispersion Measurements by Phase Shift Method", *SPIE*, vol. 584. pp. 229–234, 1985.

[23] G. Coppa, B. Costa, P. Di Vita, and U. Rossi, "Cut-Off Wavelength and Mode-Field Diameter Measurements in Single-Mode Fibers", *SPIE*, vol. 584, pp. 210–214, 1985.

[24] CCITT Recommendation G.652 "Characteristics of Single-Mode Optical Fibre Cable", Geneva, 1984.

[25] K. Thyagarajan, A. Enard, P. Kayoun, D. Papillon, and M. Papuchon, "Measurement of Guided Mode Cut-Off Wavelengths in Ti:$LiNbO_3$ Channel Waveguides", *European Conference on Integrated Optics*, ECIO '85, pp. 236–239, 1985.

5 Comparison of experimental results

Ph. Robert

Round robins represent excellent opportunities to compare results obtained by several variations of the same measuring method, and in particular to compare the performances of competing measuring methods. In fact, it is the same measured item which circulates in partners' laboratories, who afterwards meet to discuss the results and try to find the reasons for possible differences and refine experimental procedures. The field of possible measurements relating to waveguides and passive devices is far too vast for us to go into it thoroughly here. We have thus restricted ourselves to some measurements of a basic nature, connected for the most part with theories developed in previous chapters. Three chips have circulated. The first, made of InGaAsP/InP, comprising straight waveguides and integrated mirrors, was supplied by the Institute of Quantum Electronics of the Swiss Federal Institute of Technology in Zurich (ETHZ), the second, a glass chip with straight waveguides obtained by ion exchange, was supplied by CSELT (Centro Studi E Laboratori Telecomunicazioni S.p.A. Torino, Italy), the third, an InP/InGaAsP/InP heterostructure, by the Laboratory of Telecommunication and Remote Sensing Technology of The Delft University of Technology (TUD), The Netherlands.

The participants involved in the round robin measurements on these chips are: ETHZ, CSELT, the Laboratory of Telecommunication and Remote Sensing Technology of TUD (TUD1), Department of Applied Physics of the TUD (TUD2), Alcatel Alsthom Recherche, Marcoussis, France (AAR), Deutsche Telekom, Technologiezentrum Darmstadt, Germany (Telekom), Metrology Laboratory, Swiss Federal Institute of Technology, Lausanne (EPFL), Laboratoire Central de Recherche, Thomson CSF, Orsay, France (Th-CSF).

5.1 Loss measurements

5.1.1 Loss measurements on rib waveguides

A semiconductor chip with rib waveguides supplied by the ETHZ circulated among eight laboratories through Europe within the COST 240 framework, and has been extensively characterised.

This chip fabricated in the InGaAsP/InP material system contains straight waveguides and straight waveguides addressing different types of integrated mirrors as shown in Fig. 5.1. Thanks to the high refractive index of the material, the natural Fresnel reflectivity of the cleaved facets is around 30% and thus provides a good contrast for loss determination using the Fabry-Perot method described in Sec. 4.1 of this book.

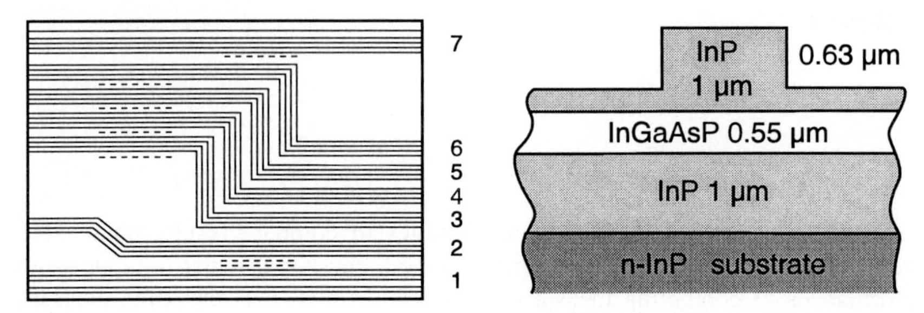

Fig. 5.1. Schematic description of the layout and the waveguide cross-section of the chip supplied by ETHZ. Groups 1 and 7 contain a set of straight waveguides; group 2 contains waveguides with pairs of 45° deviating mirrors; groups 3 to 6 contains waveguides with pairs of 90° deviating mirrors. Each group is composed of 3, 4, 5 and 6 μm wide waveguides and the total chip length is 9 mm.

Two versions of the Fabry-Perot method were used: a) a slow heating or cooling of the sample, by which the optical path length of the cavity is changed under the combined action of thermal expansion and thermo-optic effect, while the wavelength of the coherent light source is kept fixed; b) small change of the light-source wavelength while keeping the temperature of the sample, and thus its optical length, constant. AAR, Telekom, EPFL and Th-CSF used method a), while ETHZ, both TU Delft laboratories and CSELT used method b).

Other slight differences between set-ups consist of the way of coupling light into the waveguide, using either lenses or fibres, or about the way of controlling the polarisation state, either at the input or the output of the waveguide. Light sources were a 1.532 μm HeNe gas laser or DFB semiconductor lasers operating in the 1.55 μm range.

Measurements of the Fabry-Perot cavity contrast were performed on selected waveguides and mirrors on the chips for TE and TM polarisations separately. Values obtained for the total cavity loss are presented in Fig. 5.2.

To get the attenuation for each waveguide, in addition to the measured length of the sample, calculated values of the modal reflectivity were used. These values were computed by a Fourier analysis and Fresnel reflection method [1] applied to the actual waveguide structure, and range according to the waveguide width between 31.5% and 31.9% for TE polarisation, or 24.7% and 25.0% for TM polarisation. This leads to a calculated loss due to the reflection of about 5 dB for TE polarisation and 6 dB for TM polarisation per facet. The sample length was 9 mm, except for measurements performed at CSF Thomson and ETHZ, made after a new cleavage which reduced the sample length down to 5.6 mm, so the facet quality may be different.

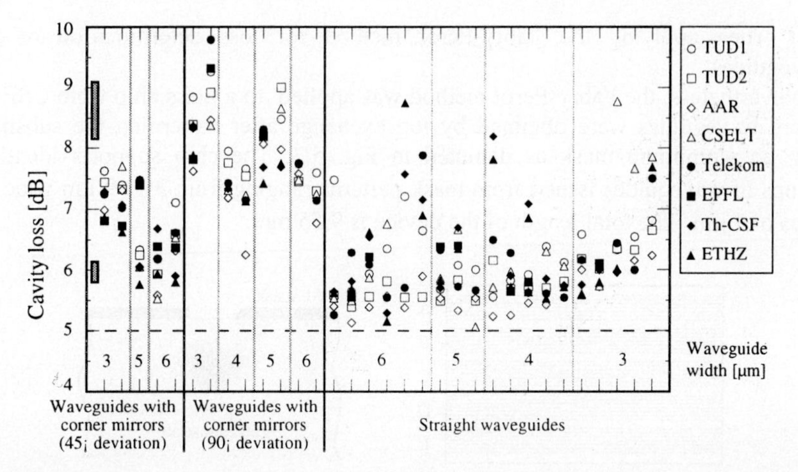

Fig. 5.2. Distribution of the measured total cavity loss (propagation plus reflection loss at facets and mirrors) for TE polarisation [2]. The 5.0 dB level corresponds to the calculated facet reflection loss.

Other discrepancies may result from dust, scratches, but more probably from cleaning the sample which was performed by some laboratories in order to remove the mounting wax they used to hold the chip. These contaminations can alter both the facet reflectivity or the propagation losses if the waveguide surface is altered. The 6 μm waveguides exhibit a multimode behaviour, making the measured cavity contrast rather sensitive to the input coupling conditions.

Measured data from all laboratories are in general agreement: straight waveguides exhibit an attenuation of 1.1±0.3 dB/cm for TE polarisation. It is to notice that the chip length uncertainty of ±10% leads to the uncertainty of ±0.16 dB/cm, taking into account the contrast value of 2.8. An uncertainty of ±0.1% on the facet reflectivity also leads to a ±0.16 dB/cm attenuation error. The results for TM polarisation are not shown here, but they are similar, except for a larger spread due to the lower facet reflectivity. A higher attenuation of 2.4±0.6 dB/cm was observed. The mirror reflectivity was obtained by subtracting the average loss of a corresponding straight waveguide. For TE polarisation, a loss of 0.8±0.2 dB per 90° deviation mirror and a loss of 0.3±0.1 dB per 45° deviation mirror was observed.

5.1.2 Loss measurements on diffused waveguides

In contrast to semiconductor rib waveguides, diffused waveguides are usually built in substrate materials with lower refractive indices: 1.45 for glass or 2.23 for $LiNbO_3$. This results in low-reflection facets, providing possible low insertion loss when pigtailing fibres to the chip, but, on the other hand, it produces a poor contrast when applying the Fabry-Perot method for loss determination of the waveguides.

Nevertheless, the Fabry-Perot method was applied to a glass chip from CSELT where waveguides were obtained by ion exchange after patterning the substrate with an aluminium mask as depicted in Fig. 5.3. The chip supports identical groups of waveguides issued from mask patterns ranging from 3 to 9 μm wide by steps of 1 μm. The total length of the device is 9.55 mm.

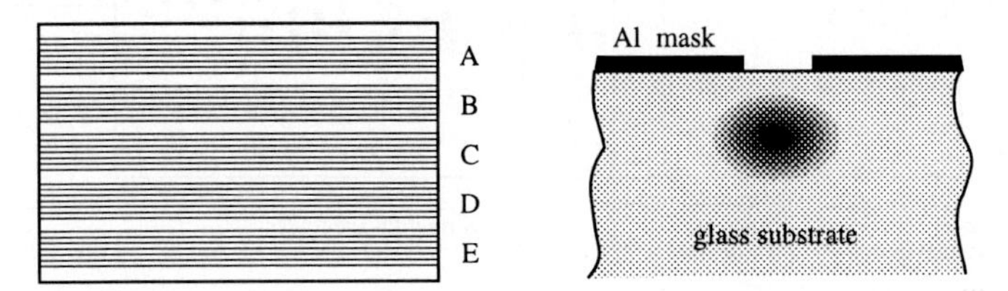

Fig. 5.3. Schematic description of the layout and the waveguide cross-section of the chip – supplied by CSELT. All groups A to E contain a set of straight waveguides issued from mask patterns ranging from 3 to 9 μm wide. The total length of the chip is 9.55 mm.

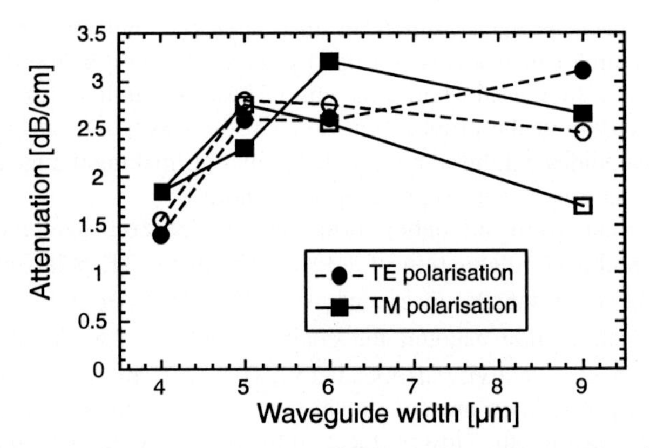

Fig. 5.4. Propagation loss of waveguides issued from mask patterns ranging from 4 to 9 μm wide obtained by the Fabry-Perot method. Circles are for TE polarisation whereas squares are for TM polarisation.

The measured contrast is 1.27 only, and the attenuation was calculated using Eq. (4.3) in Ch. 4. Data obtained by Deutsche Telekom are reported in Fig. 5.4. Reasonable spread of the attenuation values over about 1 dB for each measured mask width is obtained, except for the width of 9 μm, showing a much wider spread. At that width, the waveguides are multimode, making the contrast sensitive to the input coupling conditions.

The attenuation exhibits no significant difference between TE and TM polarisations, but it clearly increases when the pattern width of the mask is increased. One could rapidly conclude that this observed behaviour is due to the larger amount of ions and the better mode confinement which increases the scattered intensity. But this behaviour may also result from a change in the facet reflectivity. For the determination of the waveguide attenuation, a facet reflectivity according to the Fresnel formula was assumed. The facet reflectivity is actually slightly higher and tends to increase a little when the mode size decreases [1]. This results in a decrease of the total cavity loss which is entirely transferred to the attenuation value. This also shows how sensitive this Fabry-Perot method is to the value of the facet reflectivity when measuring waveguides built in a low refractive index substrate.

5.2 Group index measurements

The group effective index $n_{g\,eff}$ of a guided mode, *i.e.*, the ratio between the light velocity *in vacuo* c and the speed of propagation of a guided light pulse, is described by the expression

$$n_{g\,eff} = n_{eff} + \nu \frac{dn_{eff}}{d\nu} = n_{eff} - \lambda \frac{dn_{eff}}{d\lambda}. \tag{5.1}$$

The derivative term contains two different contributions: one depends on the chromatic dispersion of all the materials constituting the guiding structure, and corresponds to the dispersion term of homogeneous media; the second contribution is specific for waveguides, and represents the dependence on wavelength of the guide propagation characteristics, which would exist even for nondispersive media.

Three measurement methods for the group index have been compared by measuring the same sample. The first two are based on the measurement of the free spectral range of the waveguide Fabry-Perot cavity, as described in Sec. 4.1; the first one (FP1) applied at CSELT uses 10 nm scans around the desired wavelength while monitoring each sampled wavelength with a resolution of 1 pm by a wavemeter. The second one (FP2) applied at the EPFL uses 2 nm wide scans, in 5 pm steps, without wavelength monitoring.

The third method, also applied at the EPFL, is based on the technique of a optical low-coherence reflectometry (OLCR), as described in Sec. 4.2, using the superluminescence of an erbium-doped fibre as a broadband light source.

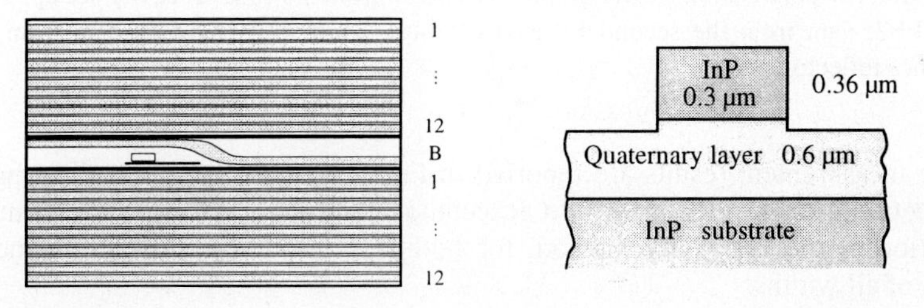

Fig. 5.5. Schematic description of the layout and the cross section of the sample used for group index measurement, fabricated at TU Delft.

The sample was an InP/InGaAsP/InP heterostructure fabricated and supplied by TU Delft. It included 2×12 sets of 10 identically etched ridge guides, as sketched in Fig. 5.5, with ridge widths ranging from 1 to 3.5 μm in steps of 0.5 μm; the chip length was $L = 8.633 \pm 0.001$ mm. The sample also included a set of 20 bends that were not measured.

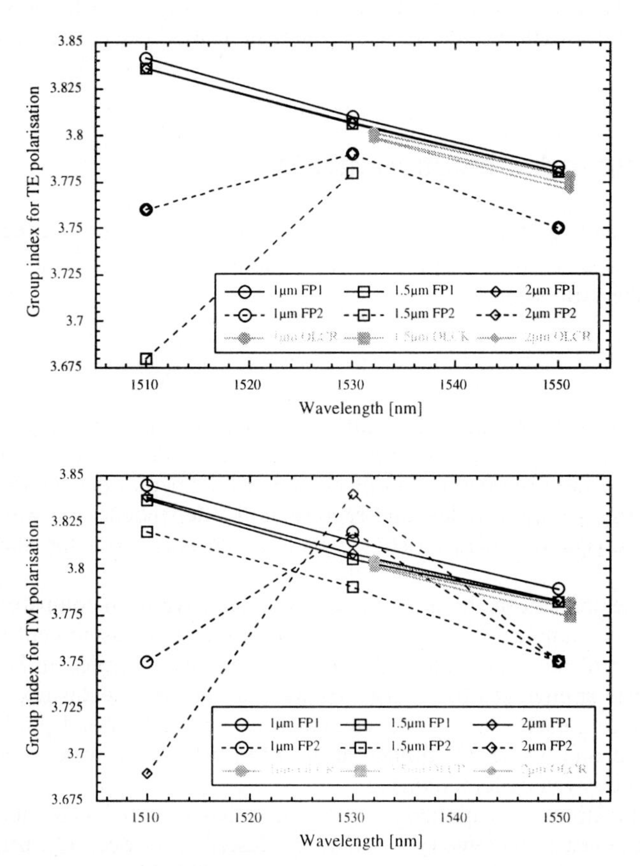

Fig. 5.6. Plot of the data from measurements of group index for TE polarisation (upper graph) and TM polarisation (lower graph). FP1: data from the first FP cavity set-up, 10 nm scans; FP2: data from the second FP cavity set-up, 2 nm scans; OLCR: data from low-coherence reflectometry.

The measurement results are reported in Fig. 5.6. There is an excellent agreement between the results of the OLCR technique and the results of the first implementation of the FP cavity method, for both TE and TM polarisations and for guides of all widths.

The set of data from the second implementation of the FP technique exhibits a large scatter (2 to 3%), specially at the shortest wavelength, which makes it difficult to compare to the other two sets. The lower accuracy may come from two sources: the narrower spectral scan than for the first set (2 nm instead of 10 nm), as accuracy in the cavity FSR is inversely proportional to the spectral base; and wavelength uncertainties at the sampling points, in the absence of a high-accuracy wavemeter.

Neglecting these lower-accuracy data, some general trends can be observed:

- the group index decreases with increasing wavelength, as expected from the decreasing trend of material refractive index and the weaker confinement;
- the group index for TM polarisation is systematically higher than for TE;
- the group index decreases slightly as the guide width increases, in contrast to the usual trend for phase effective index; a detailed analysis of the contributions from material and waveguide dispersion to the derivative term in equation (5.1) is required to interpret this fact.

5.3 Mode profile measurements

The mode profile is a key parameter for the determination of the coupling efficiency when pigtailing the device. It can be computed for semiconductor waveguides, provided that the waveguide geometry and the corresponding refractive index of each layer are accurately known. But this is not the case for diffused or ion-exchanged waveguides, thus measurement of the mode profile is especially crucial since it should provide feedback information to the fabrication process.

Mode profile measurements were performed at a wavelength of 1.55 μm on the ion-exchanged waveguides of the glass chip provided by CSELT. Description of this device was given in the second part of section 5.1. Two different methods were used: i) Deusche Telekom and Universidade do Porto used the near field imaging method described in Sec. 4.3 of the previous chapter; ii) Deutsche Telekom also applied the transverse offset method described in Sec. 4.4. Measurement results are plotted in Fig. 5.7.

Good correlation between the two methods is reported when measuring the mode width at $1/e^2$ of the peak intensity in both horizontal and vertical directions. But a substantial difference is observed when measuring the mode FWHM in both horizontal and vertical directions. This comes from two phenomena: i) concerning the near field imaging method, the vidicon cameras are known to exhibit a highly nonlinear response , not only versus the detected intensity, which can be more or less corrected by a proper linearisation, but also spatially, which causes blurring on the image. Thus the mode profile is generally overestimated when measuring at FWHM because at that point the spatial intensity decay is the steepest; ii) concerning the transverse offset method, the deconvolution algorithms tend to underestimate the mode FWHM of narrow mode profiles with a double-exponential decay.

The data show that the width of the mode profile measured by both methods is constant in the vertical direction, indicating that the ion exchange process is not affected by the pattern width of the mask. The width of the mode profile in the horizontal direction measured by both methods shows an increase of about 1.5 μm when increasing the pattern width from 3 μm to 8 μm.

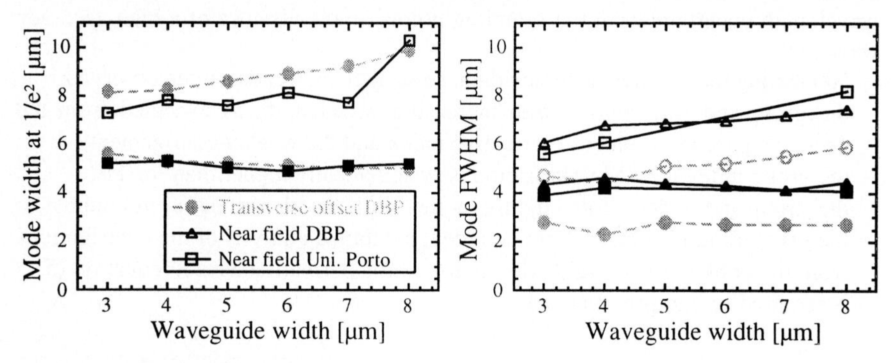

Fig. 5.7. Mode profile measurements performed using the near field imaging method and the transverse offset method. Filled symbols are the width in the vertical direction, whereas empty symbols are for the horizontal direction.

5.4 Cut-off wavelength measurements

Waveguides are often designed to be single-mode at the desired wavelength of operation. This condition can be computed for semiconductor waveguides, provided that the waveguide geometry and the corresponding refractive index of each layer are known, but in the case of diffused or ion-exchanged waveguides, the cut-off wavelengths of each guided mode have to be measured in order to locate this single-mode operation window. Cut-off wavelengths measured for TE and TM polarisation separately can also provide qualitative information about the waveguide birefringence and geometry. All this information is very important for the fabrication process.

Measurements of the cut-off wavelengths were performed by two laboratories, CSELT and EPFL, on the ion-exchanged waveguides of the glass chip provided by CSELT. Description of this device is given in the second part of Sec. 5.1. The two laboratories used the spectral transmission method described in Sec. 4.5.

Since there is no universal definition of the cut-off wavelength that has been adopted, there were slight differences between the two laboratories set-ups. EPFL used the definition given in Section 4.5, considering a drop of -3 dB of the transmitted mode intensity. Thus a uniform excitation of all the modes at the waveguide entrance was created by choosing a pinhole broad enough to illuminate a broader area than the waveguide section. CSELT defined the cut-off to be the wavelength at which the intensity drops down to −10 dB from the last maximum. In order to couple more intensity into higher order modes, the pinhole was smaller and its image was focused slightly aside from the waveguide centre.

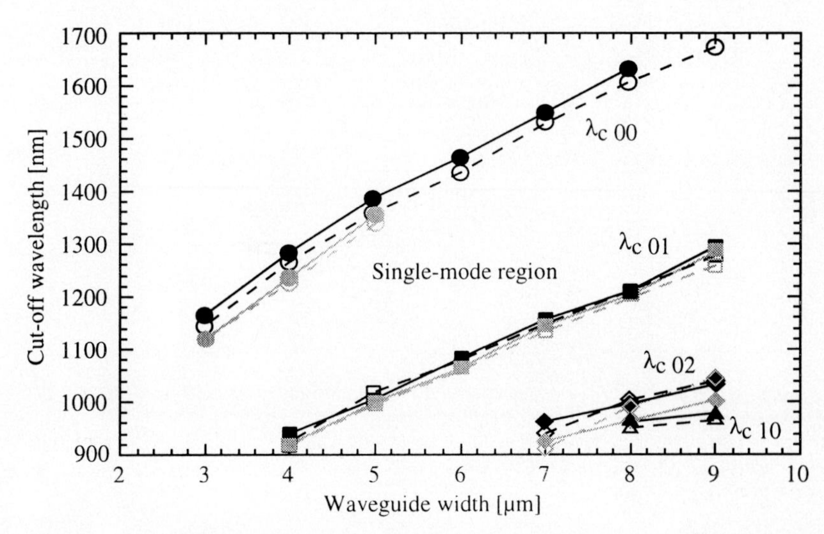

Fig. 5.8. The single-mode region is well determined by the measured cut-off wavelength of the 1st higher-order mode. Solid symbols denote TE polarisation, open symbols TM polarisation. Black curves – CSELT results, grey – EPFL results

The results of measurements are reported in Fig. 5.8. The values agree quite well, and the single-mode window can be easily identified. Due to the difference in the definition of the cut-off, values given by CSELT are systematically higher than values from EPFL. Differences between TE and TM polarisation are very small, indicating that the waveguide birefringence must be small and that the waveguide geometry is almost circular. This last point is corroborated by the mode profile measurements reported in section 5.3 and showing that the waveguides are slightly wider in the horizontal direction than in the vertical direction.

References

[1] P. A. Besse, J. S. Gu, and H. Melchior, "Reflectivity minimization of semiconductor laser amplifiers with coated and angled facets considering two dimentional beam profiles", *IEEE J. Quantum Electron.*, vol. 27, pp. 1830–1836, 1991.

[2] E. Gini, L. H. Spiekman, H. Van Burg, J.-F. Vinchant, S. Morasca, F. Pozzi, C. De Bernardi, R. Zengerle, W. Weiershausen, W. Noell, L. Thévenaz, A. Küng, A. Enard, and N. Vodjani, "Measurements of loss and mirror reflectivity in semiconductor optical waveguides: a European interlaboratory comparison experiment", *Proc. Symp. on Optical Fiber Measurements*, Boulder, USA, Sept. 1994, pp. 112–116.

References

Part II

Semiconductor Distributed Feedback Laser Diodes

Edited by:
Geert Morthier
University of Gent, Belgium

Authors

Geert Morthier	University of Gent, Belgium
Roel Baets	University of Gent, Belgium
Arthur Lowery	University of Melbourne, Australia
Roberto Paoletti	CSELT - Centro Studi e Laboratori Telecomunicazioni, Torino, Italy
Paolo Spano	FUB - Fondazione Ugo Brodoni, Roma, Italy

This part is devoted to the modelling and measurement of single frequency semiconductor light sources such as Distributed FeedBack (DFB) and Distributed Bragg Reflector (DBR) laser diodes; their modelling, the measurement of their characteristics and the combination of both modelling and measurements to derive knowledge about the physical device parameters. In the framework of an interlaboratory co-operation scheme (COST 240) that was operational in Europe in the last years of the nineteen-nineties, a considerable know-how about DFB modelling and characterisation has been developed and it is the goal of this part to summarise the most important and interesting aspects of it. The work has been concentrated on lasers with an emission wavelength of 1.3 or 1.55 μm. Such lasers are the main choice for transmitters in optical communication systems, a choice imposed largely by the minimum optical fibre dispersion at 1.3 μm and by the windows in optical fibre attenuation at both 1.3 and 1.55 μm.

The particular aim of the interlaboratory co-operation was to investigate the accuracy of the modelling and measurement methods. The accuracy of computer models was evaluated by defining specific problems and comparing the results obtained with different models. The accuracy of measurement set-ups was evaluated by circulating devices from one laboratory to another and comparing the results for well-defined measurements. In a later stage, attention was focused more on advanced measurements and parameter extraction from measurements.

While covering the main aspects of the operation of DFB and DBR lasers as well as the corresponding theoretical background, care was taken to obtain an easy-to-read and interesting text. Chapter 6 gives an introduction into the theory of laser diodes and of DFB and DBR laser diodes in particular. In contrast with other books on the same topic, the emphasis here is much more on methods for modelling, measuring and parameter extraction, not on the insights that can be derived from these methods. It has also been tried to incorporate as much as possible practical guidelines. As far as modelling is concerned, the listed examples could even be used as benchmarks for testing other modelling tools.

In the area of modelling, the emphasis has been on longitudinal modelling. An overview of different models as well as an extensive description of the comparison of these models for a number of case studies is given in chapter 7. Several DFB laser types, e.g. lasers with cleaved facets and uniform grating, $\lambda/4$-shifted , AR-coated lasers, multi-electrode lasers and aspects such as instabilities and self-pulsations, linewidth rebroadening, tuning, small signal and large signal response have been considered in these comparisons.

In the area of measurements, round robin exercises in which DFB or DBR lasers were consecutively measured in different institutes or laboratories have been chosen as method. Semiconductor lasers for that purpose have been donated by several European laser manufacturers, namely CNET (F), GEC Marconi (UK), IMC (S), Nortel, BNR Europe (UK) and HP (UK). The round robin exercises have concentrated on those characteristics (such as linewidth, tuning) for which a large spread in the measurement results from different laboratories was originally found or on those characteristics (such as high frequency characteristics) which are known to require precise calibration. The results and conclusions are discussed in detail in chapter 8.

The logical extension of the modelling and measurement activities, the comparison between measurements and modelling, has also been addressed. Obviously, a fair comparison is possible only if most of the laser parameters are well known and therefore quite some effort has been devoted to parameter extraction methods. The optical spectrum measured below threshold is still one of the favoured characteristics for such parameter extraction. In principle many parameters can be extracted from the fitting of the optical spectrum, but the accuracy of the extraction was little known. Effort has however also been devoted to the parameter extraction from the fitting of the RIN (Relative Intensity Noise) and modulation spectra. The results are described in detail in chapter 9.

In addition to the previous activities, a set of standard parameters and symbols that can be used to describe semiconductor lasers unambiguously has been defined. An overview of this parameter set and some details about exact definitions and choices are given at the end of chapter 6. In order to make the exchange of data and results of modelling exercises, measurements, parameter extractions, etc. of opto-electronic devices and structures easier, a standard file-interchange-syntax, the so-called ODIF (or Optoelectronic Data Interchange Format) has been defined. This data interchange format brings the total time needed for comparison, graphing and evaluation of data of different institutes or laboratories down to a minimum. This data interchange format, which is described in much detail in the appendix in this book, is required when electronic media such as floppy disk or tape or e-mail are used to carry the data between different institutes or laboratories.

It is hoped that the selection of material brought together in these four chapters and the appendix forms an easy to read, useful and interesting text for engineers and physicists.

The following scientists have co-operated in the frame of the laser activities of the Action COST 240. Their work is deeply ackowledged:

A. Tsigopoulos, D. Syvridis, Th. Sphicopoulos, C. F. Tsang, M. Marciniak, J. E. Carroll, H. Wenzel, T. Berceli, A. Sapia, P. Correc, S. Hansmann, H. Burkhard, A. Paradisi, R. Hui, I. Montrosset, H. Olesen, B. Johnson, H. E. Lassen, P. Gurney, T. Farrell, J. Buus, S. Balle, S. Lindgren, H. Bissessur, A. Küng, J. Troger, L. Thevenaz, S. Pajarola, A. M. Larsen, B. Christensen, H. Lange, J. G. Provost, H. Walter, T. Marozsak, A. Hilt, S. Mihaly, J. Ladvanszky, M. Puleo, R. Ash, P. Phelan, F. Volpe, Y. C. Chan, M. J. Fice, O. Kjebon, A. Saavedra, J. Nilsson, G. Rossi, G. Magnetti, and S. Van den Bosch.

6 Introductory physics

G. Morthier, R. Baets

In this chapter it is attempted to give the reader an easy-to-follow introduction to the operation of single frequency laser diodes and their characteristics and to provide the basic background that is considered necessary to understand the following chapters on modelling and characterisation of DFB laser diodes. To this end, the structure of a laser diode is first described in section 2 and the fundamental physical laws governing the behaviour of the component are explained in section 3. The subsequent sections then introduce briefly the chapters 7 to 9 and deal with common problems encountered in laser modelling and characterisation (section 4) and general approximations used in laser diode modelling (section 5). Section 6, finally, summarises the work that was done on standardisation of laser parameters in the framework of COST 240.

DFB and DBR lasers differ from other laser diodes only by the presence of a diffraction grating, but their behaviour can be quite different from the more common Fabry-Perot (FP) lasers and is governed by far more complex physics. The grating causes the feedback (necessary for oscillation along with amplification) to be far more frequency selective and gives rise to a larger influence of dispersion. As will become clearer later, the presence of a grating in an active layer also implies a much stronger influence of spatial non-uniformities in carrier and photon density. On the other hand, the fact that single-mode operation can be obtained should allow a better comparison between theory and experiment, because the experimental results do not depend on a stochastically varying distribution of the optical power over a number of longitudinal modes.

6.1 Historical background

It was only around the mid 1970s that the first CW operation of a semiconductor DFB laser was reported [1], although lasing from semiconductor diodes was known already for over 10 years at that time. The interest in DFB or DBR lasers has since then steadily been growing. By the mid 1980s it had already been found that DBR and DFB laser diodes were the only laser diodes among many alternatives (e.g. C3-lasers, injection locked lasers) with sufficient single-mode stability, sufficient modulation bandwidth and sufficiently easy fabrication for use in advanced optical communication systems. The lasers, however, still only showed a modest bandwidth of a few GHz and linewidths of several tens of MHz

[2], while the tuning range of multi-section DBR lasers was typically limited to a few nm [3]. The requirement of emission at 1.3 μm or 1.55 μm has restricted the possible materials to InP and related compounds.

At that time, extensive studies started with the aim to gain a better understanding of the behaviour of these laser diodes and to improve the maximum output power, the modulation bandwidth, and the linewidth and tuning range. It was found that some aspects of the behaviour could only be the result of spatial carrier density variations (the so-called spatial hole burning) and the first truly longitudinal DFB laser models were therefore developed (see e.g. [4]). Using these models, the influence of e.g. active layer dimensions and grating parameters on single-mode stability, modulation speed, etc. was extensively investigated. At the same time, fabrication techniques came to maturity [5] and novel materials (e.g. strained layer quantum wells [6]) with improved behaviour were being used more and more. As the insight into the behaviour and the quality of the fabrication increased, the modulation bandwidth rapidly increased to over 10 GHz [7], the linewidth decreased below 1 MHz and tuning ranges up to 7 nm were being reported [8].

In the meantime, the understanding of DFB and DBR lasers and of their fabrication techniques has improved to such an extent that these lasers now are fully commercially available, be it still at a considerable price. The research in the field of DFB and DBR laser diodes is therefore directed more towards increasing the fabrication efficiency and yield and hence reducing the fabrication cost. How to design a laser diode with predefined characteristics is well known, but the fabrication of this design still requires many fabrication cycles. This is mainly due to the unclear relation between the growth (process) parameters and the physical parameters (such as the coupling coefficient). It is hoped that the work on parameter extraction, as reported in chapter 9, will alleviate this problem. More accurate and efficient characterisation methods however could also help in reducing the development cost.

In laser modelling, there is a tendency towards the development of more complex or more detailed 2D DFB laser models [9-10], but also a tendency towards the development of improved rate equation models taking effects such as spatial hole burning into account. As modelling is focussing more and more on system simulation, parameter extraction and device optimisation, there is indeed a growing need for simple, yet accurate, models that can be used together with or as part of e.g. system simulation or optimisation software.

6.2 Description of DFB and DBR laser diodes

6.2.1 Geometric structure

The fundamental lateral/transverse structure of any laser diode is normally designed to provide carrier confinement and waveguiding or photon confinement. A typical long wavelength structure, the EMBH or Etched Mesa Buried

Heterostructure, is shown in Fig. 6.1. Many variations of this structure are used in practice, but they always consist of a layer (e.g. InGaAsP) with small bandgap and high refractive index surrounded by layers (e.g. InP) with larger bandgap and lower refractive index. The difference in refractive index ensures that the optical power is confined to the waveguiding layer [11]. This waveguiding layer is called active or passive, depending on whether it provides stimulated emission or not. The difference in bandgap ensures that all or nearly all carriers injected from the electrodes reach the active layer (or waveguiding layer in passive sections), where they can recombine via spontaneous or stimulated emission.

The different materials used must have a similar lattice constant to avoid as much as possible the existence of lattice defects at the junctions. Such defects can be a cause of degradation and can reduce the lifetime of a device considerably. Metal contacts are deposited on bottom and top of the device to allow current injection.

Also the longitudinal or axial structure can be of variable nature. In general, one can have multiple sections with the different sections being driven by independent current sources and being active or passive and with or without grating (Fig. 6.2). Gratings can be uniform or with z-dependent amplitude or period (so-called chirped gratings). The dimensions or composition of the waveguide can be different from section to section. This gives rise to discrete reflections at the interface between two sections, reflections which are usually undesired and can e.g. hinder the wavelength tuning.

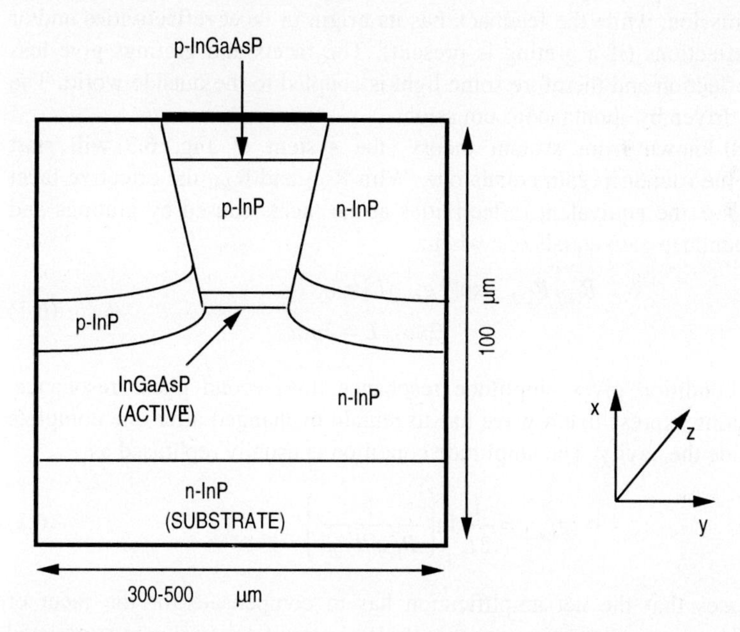

Fig. 6.1. Example of the lateral/transverse structure of a laser diode.

Multi-section lasers are widely applied as wavelength tuneable lasers. If no tuning is desired, single section lasers, which are more stable and easier to control, are generally preferred.

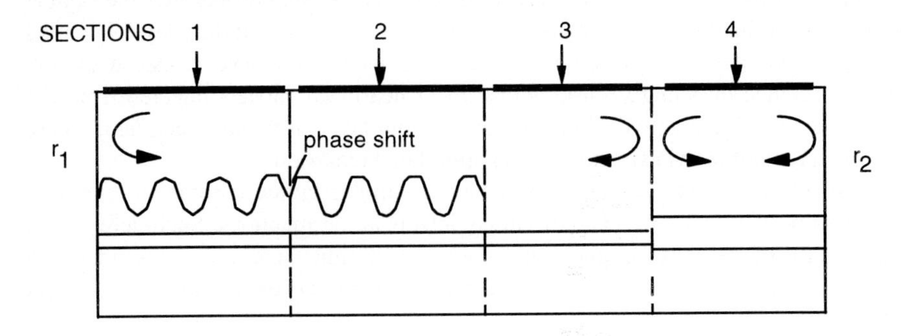

Fig. 6.2. General longitudinal structure of a multi-section laser diode.

6.2.2 Threshold condition

The operation of a laser diode can easily be explained using the systematic representation of Fig. 6.3. In this system, the amplification is provided by the stimulated emission, while the feedback has its origin in facet reflectivities and/or distributed reflections (if a grating is present). The facets and gratings give less than 100% reflection and therefore some light is coupled to the outside world. The oscillation is driven by spontaneous emission.

As is well-known from system theory, the system of Fig. 6.3 will start oscillating if the roundtrip gain equals one. With R_{1eff} and R_{2eff} the effective facet reflectivities, i.e. the equivalent reflectivities at the facets caused by gratings and facets, this roundtrip gain equals one when:

$$R_{1eff} R_{2eff} \exp[2 g_{mod} L] = 1, \qquad 2kn_{eff} L = 2m\pi. \tag{6.1}$$

The first condition gives amplitude resonance, the second phase resonance. These conditions express that a wave has to remain unchanged after one complete roundtrip inside the cavity. The amplitude condition is usually rephrased as:

$$g_{mod} = \frac{1}{2L} \ln\left(\frac{1}{R_{1eff} R_{2eff}} \right) \tag{6.2}$$

which expresses that the net amplification has to compensate for the facet or mirror loss. The net amplification consists itself of amplification due to stimulated emission and attenuation due to internal loss or absorption. The phase resonance condition determines the emission wavelength.

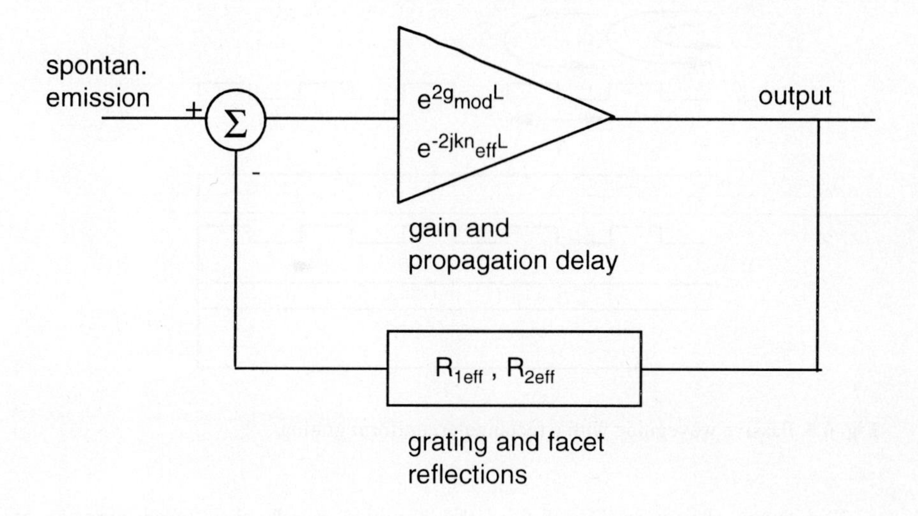

Fig. 6.3. The laser diode as an oscillator.

6.2.3 Distributed reflections

The feedback caused by a grating is not simply equivalent to a constant facet reflectivity, but consists of distributed reflections. This can be seen easily in the case of a uniform, rectangular grating as in Fig. 6.4. If the section is passive (as is the case in a DBR laser), one can replace it by an effective reflectivity at the origin of the grating. This effective reflectivity can be obtained e.g. by carefully adding all the distributed reflections with their appropriate phase delays or, more easily by multiplying the transfer matrices of the different grating sections. This will be worked out further in chapter 7. One finds a wavelength dependent field reflectivity, given by:

$$|r| = \frac{\kappa_i \sinh(\sqrt{\kappa_i^{\,2} - \Delta\beta^2}\,)L}{\sqrt{\kappa_i^{\,2} \cosh(\sqrt{\kappa_i^{\,2} - \Delta\beta^2}\,)L - \Delta\beta^2}}, \qquad \text{with } \Delta\beta = \frac{2\pi}{\lambda} n_{eff} - \frac{\pi}{\Lambda}. \tag{6.3}$$

This reflectivity is maximum for the Bragg wavelength $\lambda_B = 2n_{eff}L$ (with n_{eff} the effective index of the waveguide), but also depends on the coupling coefficient κ_i. For a simple sinusoidal variation of the effective index, this coupling coefficient is simply related to the index variation as follows:

$$\frac{2\pi}{\lambda} \Delta n_{eff} = 2\kappa_i \cos\left(\frac{2\pi z}{\Lambda}\right). \tag{6.4}$$

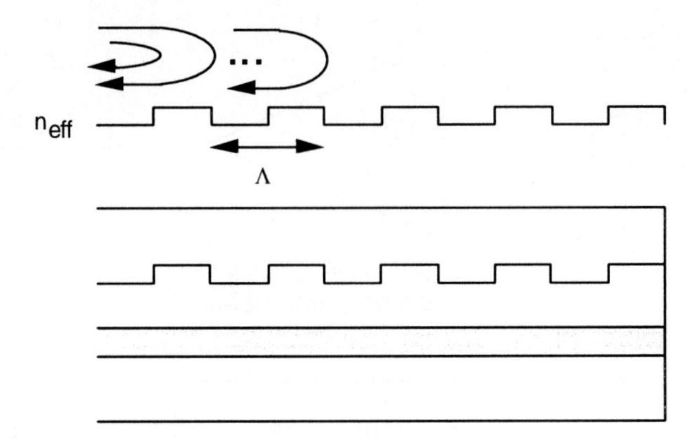

Fig. 6.4. Passive waveguide with a rectangular, uniform grating.

The curve shown in Fig. 6.5 is the complex result of a large number of interferences between the distributed reflections. In a waveguide without gain or loss, all these reflections interfere with a same amplitude since there is only a propagation delay between the different components.

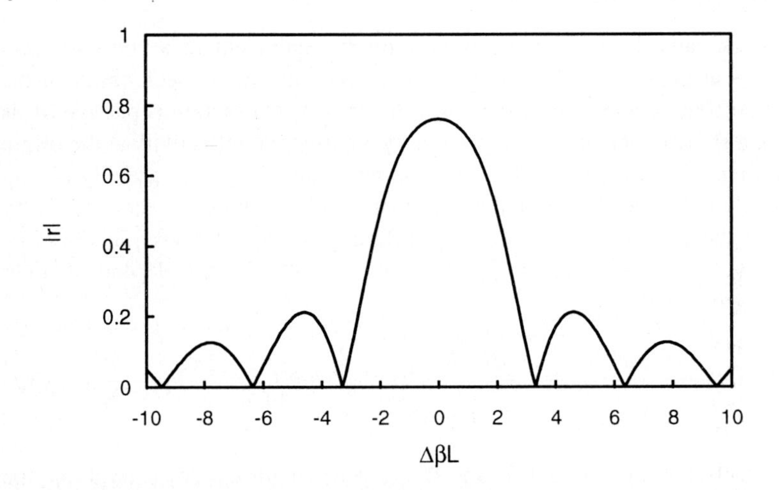

Fig. 6.5. The field reflection characteristic of a Bragg section as a function of the detuning $\Delta\omega L/v_g$ and for $g_{mod}L=0$ ($\kappa L=1$).

In a DFB laser, gratings are situated in active (i.e. gain providing) sections and this makes the analysis of such lasers more complicated. The gain can affect the total reflection of a grating considerably [12]. E.g. the zero value for the reflection on both sides of the main lobe in Fig. 6.5 can eventually change into a peak value for the reflection (or become infinitely large) if the gain is sufficiently large (see Fig. 6.6). This can again be explained from the complex interference. Different reflections can now interfere with different amplitude, with the amplitude strongly depending on the total distance a wave has propagated. And at wavelengths where complete destructive interference occurs without gain, phase delay, gain and distributed reflections can be such that a sort of resonance is obtained.

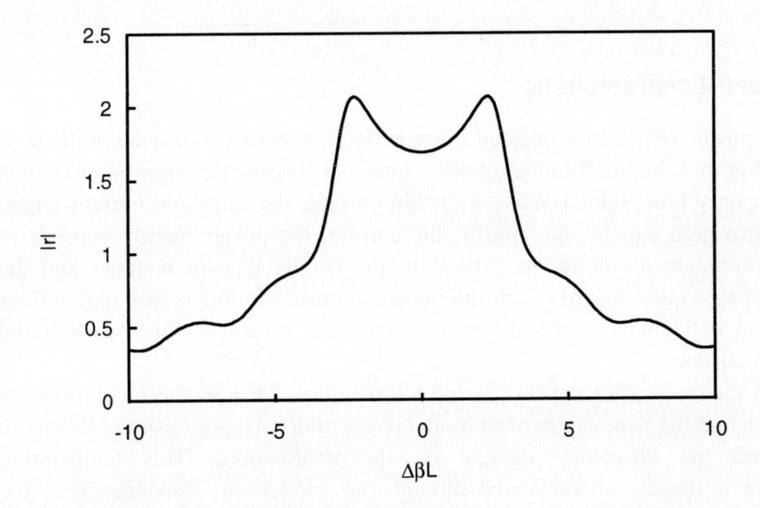

Fig. 6.6. The field reflection characteristic of a Bragg section as a function of the detuning $\Delta\omega L/v_g$ and for $g_{mod}L=2$ ($\kappa L=1$).

Another factor of complexity in DFB laser diodes is usually the interference between distributed reflections and facet reflections. This interference is strongly dependent on the grating phases at the facets and it can cause the behaviour of a DFB laser to change drastically with grating phases at the facets. The current state of the art in technology doesn't allow control of the grating phases at the facets very accurately and, without AR-coating, it is difficult to fabricate laser diodes with predefined behaviour. In DBR lasers, the grating amplitude and section length are usually so large that the field propagating towards the facet is strongly attenuated when it reaches the facet. Interference between facet and Bragg reflections is then less important.

6.3 Introduction to DFB laser characteristics

The term laser characteristics refers to those measurable quantities that are of importance in the design of systems using laser diodes. Since such systems can be of different nature and since their reliability depends on temperature control, power budget, noise influence, speed, etc., a long list of different laser characteristics can be distinguished.

As those laser characteristics will be the subject of the following chapters on modelling, characterisation and parameter extraction (they form the actual subjects of comparison in most cases), we will briefly introduce the most important ones here. They can be divided into static (6.3.1 to 6.3.3), dynamic (6.3.4) and noise (6.3.5) related characteristics.

6.3.1 The P-I characteristic

A typical power (P) versus injected current (I) characteristic of a laser diode is shown in Fig. 6.7. In this P-I characteristic, one can observe that significant optical power can only be obtained above a certain current, the threshold current where the roundtrip gain equals one. Below this current, the power mainly consists of amplified spontaneous emission. Above it, the carrier density is fixed and the power increases quite linearly with the injected current. In this region one defines the external efficiency as the differential increase in power for a differential increase in current.

The P-I characteristics depend on the temperature, because the recombination coefficients and the bandgap depend on the temperature. Hence, also the threshold current and the efficiency depend on the temperature. This temperature dependence is usually characterised through the temperature dependence of the threshold current, which is approximated as $I_{th} = I_{th0} \exp(T/T_0)$. T_0 is called the characteristic temperature.

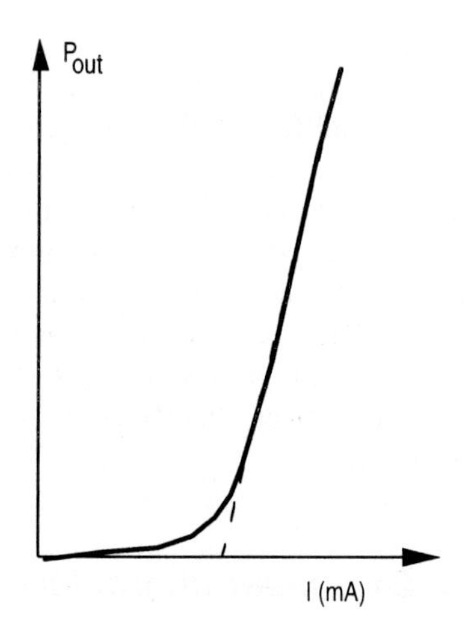

Fig. 6.7. Typical relation between output power (P_{out}) and injected current (I).

6.3.2 The V-I characteristic

The voltage (V) versus injected current characteristic (Fig. 6.8) has two regions. Below the threshold current (around 10 mA in the example), it is not different from a normal diode characteristic. Above the threshold current however, the carrier density barely changes with increasing current and in this case the increase in voltage with current is caused by the series resistance of the cladding layer. The voltage becomes linear with the injected current.

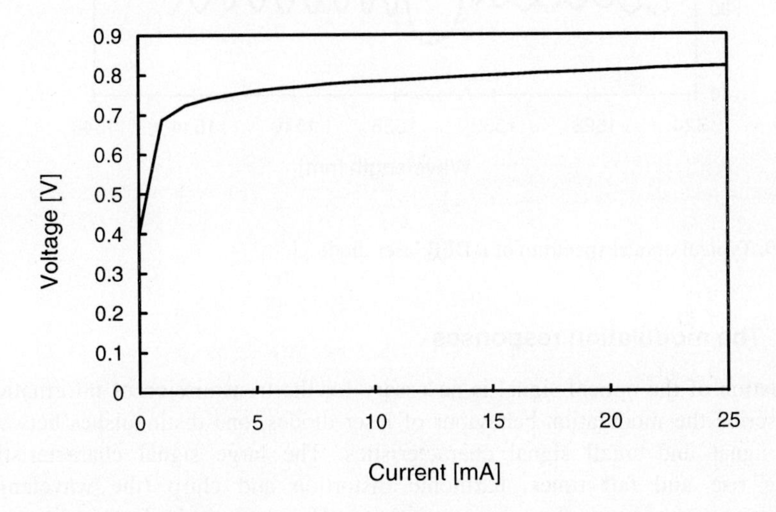

Fig. 6.8. Typical V-I characteristic of a laser diode.

6.3.3 The optical spectrum

The optical spectrum consists of the optical power as a function of the wavelength. A typical spectrum of a single-mode laser is shown in Fig. 6.9. Peaks in the optical power occur at the different longitudinal modes, with the lasing mode having a peak that is 20 to 40 dB higher than the other peaks. The difference in power between the lasing mode and the main side mode is called the Side Mode Suppression Ratio (SMSR) and is usually expressed in dB. Important is also how the wavelength of the lasing mode varies with the injected current, the so-called tuning behaviour. Multi-section lasers are often designed to provide a large tuning range without large accompanying power variations.

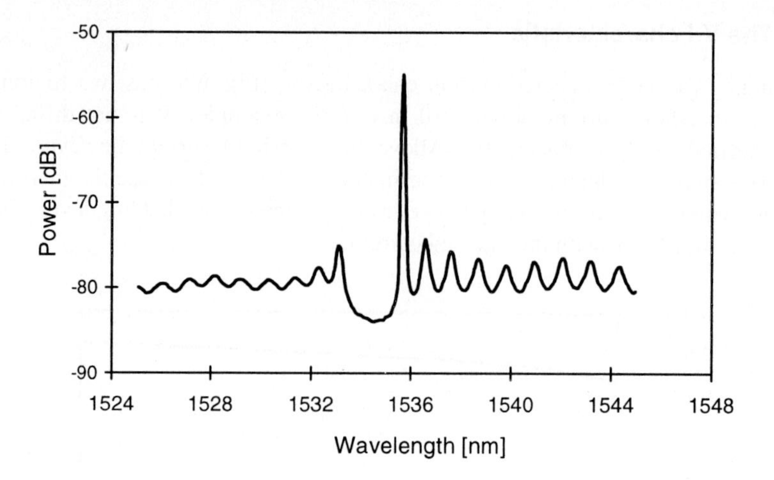

Fig. 6.9. Typical optical spectrum of a DFB laser diode.

6.3.4 The modulation responses

Modulation of the optical signal is necessary for the transmission of information. To describe the modulation behaviour of laser diodes, one distinguishes between large signal and small signal characteristics. The large signal characteristics include rise and fall times, harmonic distortion and chirp (the wavelength variations accompanying the power variations). They will not be further discussed in this book.

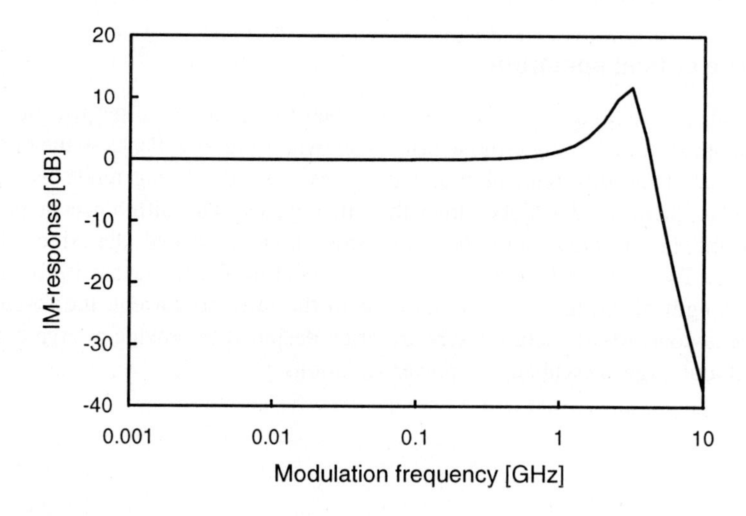

Fig. 6.10. Typical modulation spectrum of a laser diode.

The small signal characteristics are defined as the variations in the optical intensity and the optical frequency for small sinusoidal variations in the current. They are generally displayed as a function of the modulation frequency, as in Fig. 6.10.

6.3.5 The FM- and intensity noise spectra, the linewidth

Both the optical frequency and intensity exhibit small stochastic fluctuations with time which have their origin in the spontaneous emission fluctuations and carrier density shot noise. These fluctuations are called frequency modulation noise (FM-noise) and intensity noise respectively. The Relative Intensity Noise (RIN) is the intensity noise normalised to the average intensity. Both the FM-noise and the RIN are usually characterised by their spectrum as a function of frequency.

The FM-noise determines the 3-dB width of the emission peak in the optical spectrum or the linewidth.

6.4 Problems in modelling and measuring DFB laser diodes

As has been mentioned already in the introduction of this chapter, the presence of a grating in DFB laser diodes causes their behaviour to be strongly dependent on spatial variations in the optical power and the carrier density. In fact, one could say that different spatial parts of the laser more or less can behave differently, even in a geometrically uniform laser diode.

It is well known that spatial effects can be taken into account in the modelling of DFB laser diodes by discretising the structure spatially and solving the spatially-dependent equations self-consistently in this way. However, it is not so well known how fine the discretisation has to be and if such a discretisation is also required in the lateral and transverse directions. Some characteristics such as the P-I and V-I characteristics are not so sensitive to spatial variations, but others like the stability, the FM-response and the harmonic distortion can be extremely sensitive to spatial variations of the carrier density. These carrier density variations, caused by optical power variations, can be negative with respect to the spatially averaged carrier density in some regions of the laser and positive in others. Characteristics like the FM-response are the small net result of different, larger positive and negative contributions and the accuracy of their calculation may vary considerably depending on discretisation.

One also has to bear in mind that the effect of the grating strongly depends on the exact value of the wavelength (or frequency). As a matter of fact, there are cases in which this dispersion of the distributed reflections causes a positive feedback where e.g. frequency variations result in loss variations, which in turn result in carrier density variations (due to gain clamping), which in turn give additional frequency variations in the same direction as the original frequency variations.

Calculations of the modulation responses and the noise spectra are usually based on small signal approximations around a static bias solution. Such small signal calculations can depend on the accuracy of a calculated static solution. This is in particular the case if a laser with unstable or self-pulsating behaviour is considered. Depending on the accuracy of the bias point, values for dynamic characteristics such as linewidth or self-pulsation frequency, but also for current ranges where unstable or self-pulsating behaviour occur can vary over almost an order of magnitude depending on the accuracy of the calculated static solution.

Measurement results, on the other hand, often depend on the specific method and the specific equipment that is used. Interpretation of the results can depend on the specifications of the equipment and on what exactly is being measured. Quantities such as wavelength, linewidth, etc. are not directly measured, but are derived from e.g. reflection characteristics, electrical spectra, etc.. Moreover, the influence of noise introduced by the measurement equipment or current sources has to be taken into account carefully. Comparison of measurements done on a same device, but using different methods or different equipment can help a lot in these interpretations.

Finally, the reliability of theoretical predictions of the characteristics of an actual device really depends on the accuracy of the laser parameters used in the modelling. Some knowledge can usually be gained from the accuracy data of the fabrication equipment, but the exact value of parameters such as the facet reflection coefficients and the coupling coefficients is normally unknown. This inadequate knowledge therefore prevents assessment of the accuracy of laser models from comparison with measurements and, more importantly, optimisation of the performance of a device according to guidelines from modelling. It has therefore been investigated numerically how sensitive laser characteristics are with respect to parameter variations and how parameter values can be extracted from a limited number of measurements.

6.5 General approximations used in laser diode modelling

Since a fully self-consistent analysis of optical fields, carrier and charge densities and current gradients in three dimensions is quite time-consuming, the approximations introduced in a computer simulation are usually manifold. Therefore, many numerical models discussed in the next chapters only consider spatial variations in one dimension, the longitudinal or axial dimension. However, one 3-dimensional numerical model has been used in COST 240 and it has allowed to assess the accuracy of the one-dimensional models to some extent.

In all models, the spontaneous carrier recombination rate is, quite accurately, approximated by a 3rd order polynomial in the carrier density. In particular, $R = AN + BN^2 + CN^3$, with A the coefficient for recombination via traps, B the bimolecular recombination coefficient and C the Auger coefficient. The material gain and the refractive index in the active layer are generally approximated by linear or logarithmic functions of the carrier density, i.e.:

$$g = a(\lambda)(N - N_t) \text{ or } g = a\ln(N / N_t),$$
$$n = n_0(\lambda) + \frac{\partial n}{\partial N}(\lambda)N \tag{6.5}$$

with the functions possibly being wavelength dependent. Linear functions are generally used for the refractive index and the gain in bulk layers. Logarithmic functions are more and more being used to model the gain in Quantum Well lasers.

Also the temperature dependence is usually either ignored or very roughly taken into account. In some models, simple calculations based on one thermal resistance and one thermal capacitance are possible. Non-radiative recombination acts as source of the heating in this case, where one is mainly interested in the thermal tuning through the temperature dependence of refractive index and gain. Calculations taking into account thermal effects however will not be discussed in the following chapters.

In the one-dimensional models, it is often assumed that there is a uniform current density over the stripe electrode and that a constant fraction of the injected carriers reach the active layer where they recombine. An improved model for this carrier injection has however been introduced in many models during the course of COST 240. In this improved model, it is assumed that the potential on the electrodes is constant and the current density injected into the active layer depends on the series resistance R_s of the cladding layers and on the voltage V_{DH} over the double heterojunction. This voltage V_{DH} depends on the carrier density N in the active layer and can be varying in the axial direction. Carrier diffusion is generally neglected. This is justified by the small dimensions of the active layer in the transverse and lateral direction and by the slow variations of the carrier density in the axial direction.

The waveguiding in these one-dimensional models is assumed to be of the index-guiding type, with only the lowest order TE-mode being above cut-off. The profile of this TE-mode is furthermore considered to be independent of the carrier density and wavelength and hence constant along the axial direction. Also the confinement factor Γ (the fraction of the optical power confined to the active layer) is considered a constant, but the propagation constant and effective index vary with wavelength and carrier density. The influence of a complex refractive index variation Δn_c (e.g. including gain, absorption, carrier density dependent real refractive index) on the effective index of the waveguide is then approximated as:

$$n_{eff} = n_{eff,0} + \Gamma \Delta n_c \tag{6.6}$$

with $n_{eff,0}$ the refractive index of the unperturbed waveguide. The modal gain is thus given by:

$$g_{mod} = \Gamma g - \alpha_{int} \tag{6.7}$$

with α_{int} the internal loss.

A different threshold value of g_{mod} is usually required to get lasing of different modes. The difference in threshold value of g_{mod} between the main mode and the most important side mode is in fact a first measure of the single-mode stability of DFB and DBR lasers. The normalised difference ΔgL, with L the device length, is therefore a frequently used characteristic in modelling. The difference in threshold gain has its origin in a difference in facet loss in DFB and DBR lasers and it is zero in FP lasers.

For the coupling coefficients on the other hand, a constant is generally used in cases with uniform grating (but not of course if chirped or tapered gratings exist). To be exact, this coupling coefficient would have to be depending on the mode profile and hence also on the wavelength and carrier density. An exception is normally made for lasers with corrugated active layer, when the carrier density dependence is very strong, or with absorptive grating, where the saturation of the absorptive layer by the optical wave can have a significant impact on the behaviour.

6.6 Standardisation of laser parameters

Crucial to many aspects of the work done within COST 240, i.e. the comparison of modelling or measurement results from different institutes, but certainly also important in the development of a simulation tool (or its user interface) and in making the evaluation of software easier is the ability to describe laser devices by a standard set of parameters. These parameters must have an unambiguous, clear, concise and precise definition which does not rely on assumptions or prior conventions for interpretation. The definitions should also be acceptable to a large majority of practitioners in the field and should bear some allegiance to standard practices and symbol conventions.

Such a standard set of parameters has been defined within WG1 of COST 240 [13] and is presented in the following part. We hope that the results of our efforts to standardise parameters for semiconductor lasers will be of benefit to the photonics community at large. The definitions have certainly proven useful internally and the discussions about them have brought to light some important weaknesses in previous definitions. We also hope that the parameters will form a useful starting point for anyone developing laser models, and will help initiate comparisons between numerical and experimental work, which will help to improve confidence in results and models. An additional benefit is that scientific literature may be easier to read if the symbols are consistent.

The definitions are presented below. Each contains a name, a standard symbol, a unit, and an ASCII equivalent of the symbol to facilitate plain-text electronic data interchange, say by e-mail. Where a parameter is open to misinterpretation, we have included a warning. SI units have been used in all of the definitions (nm, µm, mm, m but not cm).

6.6.1 Derivation of an ASCII equivalent of a symbol

To enable easy conversion from ASCII to normal symbols and back again, the ASCII name for a symbol is derived using the following conventions:

- Roman symbols are written as they appear: a = a (ASCII)
- Subscripts are added using an underscore as a separator: a_b = a_b
- Lower-case Greek symbols are written out in lower-case: λ = lambda
- Upper-case Greek symbols are written out in UPPER-case: Γ= GAMMA

6.6.2 Optical output definitions

These are not laser parameters but are essential in defining the measured performance of a laser, or the output of a simulation program:

A. Power, P, mW, P

The output power in mW from a facet.

B. Field, E, $W^{0.5}$, E

The instantaneous optical field emerging from a facet.

This is strictly a normalised field in that it is normalised such that the optical power equals this field squared. A true electric field would have units of V/m, but the cavity dimensions and impedance would need to be known in order to calculate powers in subsequent models. The photon density can be calculated from a knowledge of this field, the cavity dimensions, and the group velocity of the light in the waveguide.

6.6.3 Cavity dimensions

The following dimension parameters apply to the laser chip:

A. Laser Chip Length, L, μm, L_subscript

The total length, from facet to facet, of the waveguide within the laser chip. The word 'chip' is included in the definition to avoid confusion in cases where the laser has an external cavity.

In cases where the laser model is representing a distinct longitudinal region of a longer device (such as a Bragg region, gain region, phase-shifting region, and absorber region), the length of the model can be denoted by the symbol L with appropriate subscripts. Suitable subscripts are:

a	active	(a region with optical gain)
wg	passive waveguide	(a region with no optical gain)
ph	phase-shift	(a region that provides a variable phase shift)
DBR	Bragg reflector	(a region with a Bragg reflector but no gain)
DFB	Distributed Feedback	(a region with a Bragg reflector and gain)
SA	Saturable Absorber	(a region with a saturable absorber)
mod	modulator	(a region for the purpose of data modulation)
ext	external cavity	(a cavity external to the laser chip)
t	transition	(transition between regions)

In the case of a multi-contact laser, where the waveguide structure is identical under each contact, numerical subscripts can be used to denote the regions under each contact. The left-most region should be labelled '1', e.g. L_1. The ASCII equivalent would be L_1.

Generally capitals are used where the subscript comprises separate words, and lower case where the subscript is an abbreviation of a single word.

This subscript convention can be used for all parameters that vary along the device.

B. Active Region Width, w, μm, w

The nominal width of the active region of the laser. In buried heterostructure devices, the width is simply the width of the active mesa heterostructure. However, in V-groove devices the precise width is difficult to define as the active region tapers towards the edges of the groove. A good choice is to make the active region cross-section $(w.d)$ equal to the area of the active region in the device. Typical values are between 1-μm and 5-μm for single transverse-mode operation.

C. Active Region Thickness, d, μm, d

The thickness of the active region. In multi-quantum well devices the thickness is defined as the sum of the individual well thicknesses. In V-groove devices the average thickness can be used. Typical values are between 0.1 μm and 0.2 μm for bulk devices and 0.003 μm to 0.01 μm per well for quantum-well devices.

6.6.4 Symbols used to describe internal variables

It is necessary to define some terms for the internal variables within the laser in order to define some of the parameters. These are not strictly parameters describing the device, but are essential when formulating models of a device.

A. Photon Density, *S*, m^{-3}, S

This is not a laser parameter, but a symbol convention to be used whenever photon density is used as a parameter or a variable. Subscripts may be used to define, for example, the photon density in a particular mode or model region (transverse, lateral or longitudinal).

B. Photon Number, *S#*, dimensionless, S#

This is not a laser parameter, but a symbol convention to be used whenever the number of photons is used as a parameter or a variable. Subscripts may be used to define, for example, the number of photons in a particular mode or model region.

The Photon Number is obtained by multiplying the Photon Density by the effective cross-section of the optical wave and the length of the cavity. The effective cross-section of the optical mode is defined in terms of the Confinement Factor, Γ (see 6.6.5.E), and the cross-section of the active region. That is:

$$S\# = S\, w.d.L/\Gamma$$

Note that some definitions of Photon Density assume that the cross-section of the optical wave equals the cross-section of the active region for convenience. These should be avoided.

C. Carrier Density, *N*, m^{-3}, N

This is usually the mean carrier density within the active region. Subscripts may be used to define different regions of a laser.

D. Carrier Number, *N#*, dimensionless, N#

This is not a laser parameter, but a symbol convention to be used whenever carrier number is used as a parameter or a variable. Subscripts may be used to define, for example, the Threshold Carrier Number, *N#*th.

The Carrier Number is obtained by multiplying the Carrier Density by the cross-section of the active region and the length of the cavity. That is:

$$N\# = N.w.d.L$$

6.6.5 Optical waveguide parameters

The active region and surrounding layers form an optical waveguide designed to confine light close to the active region in which there is a high level of inversion so that amplification by stimulated emission is probable. The waveguide may be characterised by the following parameters:

A. Waveguide Effective Index, n_{eff} dimensionless, n_eff

The Waveguide Effective Index determines the phase velocity along the laser waveguide. The word 'effective' is used to show that it is the phase velocity within the whole waveguide, and not just the active region. The Waveguide Effective Index also determines the absolute lasing frequency from the length of the resonator or from the grating period. Strictly the wavelength and carrier density should be specified. Typical values vary between 3.5 and 4.3.

Warnings:
The Waveguide Effective Index is wavelength dependent. The Waveguide Group Effective Index describes the linear part of this dependence. A reference wavelength should be specified, *e.g.* λ_{ref}, (lambda_ref).

The Waveguide Effective Index is carrier-density dependent. The Material Linewidth Enhancement Factor describes the linear part of this dependence. A reference carrier density should be specified, *e.g.* N_{ref} (N_ref).

B. Waveguide Group Effective Index, n_g, dimensionless, n_g

The Waveguide Group Effective Index determines the group, or energy, velocity along the laser waveguide. The word 'effective' is used to show that it is the energy velocity within the whole waveguide, and not just the active region. The group index also determines the longitudinal mode spacing within Fabry-Perot lasers over a limited range of frequencies around the lasing frequency.

The energy velocity is important as it enables the optical field to be converted to a photon density for the calculation of the stimulated emission rate.

Typical values of Waveguide Group Effective Index are between 3.2 and 4.0. It is usually determined experimentally from the longitudinal mode spacing of a Fabry-Perot laser. Strictly the wavelength and carrier density should be specified.

C. Material Linewidth Enhancement Factor, α, dimensionless, alpha

This parameter has several names and definitions. Strictly, the linewidth enhancement factor describes the linewidth broadening from the limit given by the modified Schawlow-Townes formula due to phase fluctuations caused by spontaneous emission events. Because the importance of this broadening was first realised by Henry, it is known as Henry's Linewidth Enhancement Factor.

However, this parameter affects many aspects of the laser's performance including [14]: the large-signal chirping of the laser line during modulation, the FM/IM ratio of the device, the effect of spatial-hole-burning on multi-modeness of DFB lasers, and the adiabatic dependence of wavelength on drive current.

A reasonable definition of the linewidth enhancement factor relates the change in index of the active layer with carrier density to the differential gain, d*g*/d*N*, which is equal to the Linear Material Gain Coefficient if the gain is considered to be independent of wavelength. The definition is:

$$\alpha = -\frac{4\pi}{\lambda}\frac{dn}{dN}\frac{dg}{dN}.$$

Typical values fall between 2.5 and 12, though 5.0 is often used for bulk devices operating close to the gain peak, and 3.0 for MQW devices operating close to the gain peak. Strictly, the linewidth enhancement factor should be defined as being negative if the differential refractive index, d*n*/d*N*, is negative, which is in most cases in semiconductor lasers. However, most researchers use a positive value for lasers. This definition implies a positive value for most lasers. Confusion may arise when MQW modulators are defined using the linewidth enhancement factor, as they can have chirp in the opposite sense to that of lasers.

Subscripts may be used to define layers, sections, effective values, values measured by different methods.

Alternative: Differential Refractive Index, *dn/dN*, m^3, dn/dN.

Warning: The linewidth enhancement factor is wavelength dependent.

D. Internal Loss, α_i, m^{-1}, alpha_i

The total power loss factor of the laser waveguide. It is a modal rather than a material parameter in that it accounts for losses in all of the layers that the waveguide mode exists in, rather than the loss of a single layer of material. It does not include facet losses.

The internal loss has carrier-dependent and carrier-independent components:

$$\alpha_i = \alpha_{i0} + N\alpha_{iN}$$

where the components are defined below:

- Fixed Internal Loss, α_{i0}, m^{-1}, alpha_i0

The carrier-density-independent component of the power loss in the laser waveguide. It may include carrier-independent loss mechanisms such as Rayleigh scattering.

• Carrier-Dependent Internal Loss Coefficient, α_{iN}, m^2, alpha_iN

The coefficient for the carrier-density-dependent part of the loss per unit length. Carrier-density-dependent loss can be due to free-carriers (electrons in the conduction band, or holes between valence bands) absorbing photons (intra band absorption). Such absorption is dependent on the carrier density but does not affect the carrier density, as does absorption between the conduction and valence bands. The loss per unit length is the free-carrier absorption factor multiplied by the carrier density.

E. Confinement Factor, Γ, dimensionless, GAMMA

The Confinement Factor describes the proportion of the optical power that travels within the active region's cross section. This is illustrated in Fig. 6.11 which shows a rectangular active region, with the guided optical mode existing within and external to the active region. The confinement factor is the optical power within the active region divided by the total optical power.

Typical values of optical confinement factor are between 0.01 for single-quantum well devices, to 0.4 for bulk devices.

Subscripts may be used to denote confinement factors of other layers (*e.g.* wave guiding layer), or longitudinal sections of the laser. These should be defined by the user if different to those provided in the 'length' definition in 6.6.3.A.

In some situations it may be useful to factorise the confinement factor into vertical and horizontal components, that is:

$$\Gamma = \Gamma_{vertical} \cdot \Gamma_{horizontal}$$

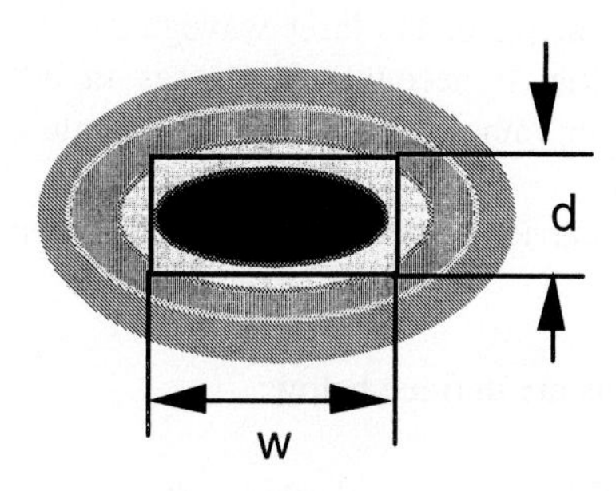

Fig. 6.11. Partial confinement of an optical wave (shaded) in the active region (area wd).

F. Facet Reflectivities, *R*, dimensionless, R

Facet reflectivities define the power reflectivity of the facets of the laser waveguide. For example, $R = 0.3$ means that 30 % of the optical power travelling towards the facet from inside the waveguide will be reflected back into the waveguide; the remaining 70 % will be transmitted to outside of the laser cavity.

The facets can be identified by using subscripts to R. These should be numerical with the left-most facet being labelled '1'. Internal reflectivities within a laser can also use R as a symbol. However, care has to be taken about the phase of the reflection, and it is probably better to use field reflectance for such cases.

G. Facet Reflectances, r, dimensionless, r (NON-PREFERRED)

Facet reflectance is an alternative to facet reflectivity. The reflectance can be determined in terms of the field reflectivity but care has to be taken because there are two conventions for specifying field:

1) The field is normalised such that the optical power is the field squared. In this case the field has units of $W^{0.5}$. The magnitude of the reflectance will be the square-root of the reflectivity. The phase will depend on the direction the wave is imposing on the boundary.

2) The field is normalised such that the optical power density is the field squared over the characteristic impedance of the guide. The units of the field are V/m. Although this definition is strictly correct, it makes the calculation of the reflectivity dependent on the impedances of the waveguide, and/or free space. These impedances are not usually required in laser modelling, and so Definition (1) appears more sensible, and should be used where possible.

6.6.6 Waveguide gratings

Distributed Feedback (DFB) lasers and Distributed Bragg Reflector (DBR) lasers include gratings along their waveguides to provide wavelength-selective feedback to promote lasing at a single frequency. The following parameters can be used to describe such gratings.

A. Nominal Grating Period, Λ, nm, LAMBDA

The (longitudinal) period of the Bragg grating in a DFB/DBR laser. The emission wavelength of the laser will depend on the mean effective index of the waveguide and the grating order.

The Nominal Grating Period is used in preference to the Bragg Wavelength (λ_B, nm, lambda_B), which is the wavelength of the centre of the Bragg grating in a DFB/DBR laser. The Bragg wavelength is equal to $2n_{eff}\Lambda/M$, where n_{eff} is the

Waveguide Effective Index and M is the grating order (defined below). Note that the index is carrier density dependent, and therefore the Bragg Wavelength needs to be defined at some particular carrier density. Thus the Nominal Grating Period is a more fundamental parameter.

B. Grating Order, M, dimensionless, M

The reflection order of the Bragg grating in a DFB/DBR laser. Higher-order gratings have a larger Nominal Grating Wavelength for a given output wavelength.

C. Index Grating Coupling Coefficient, κ_i, m^{-1}, kappa_i

This defines the coupling per unit length of a grating formed solely by index variations along the optical waveguide. Strictly it should be a function of length in order to describe tapered coupling. In cases where the active region is modulated in size (such as gain-coupled lasers), the index coupling becomes a function of carrier density, and should be calculated accounting for the carrier density variation along the laser, which may be time dependent.

D. Gain/Loss Grating Coupling Coefficient, κ_g, m^{-1}, kappa_g

This defines the coupling per unit length of a grating formed by modulating the optical gain (or loss) along the waveguide. In cases where the optical gain is modulated, or the loss is saturable, the gain/loss coupling should be calculated taking into account carrier density variations.

E. Grating Facet Phase, Θ, rad, THETA

The phase of the facet reflections at the end of a grating are critical in determining the spectrum of a laser. Unfortunately, there are at least two definitions currently in use. The definition here has been devised using the following postulates:

1) The definition should not involve the length of the device, because this would require the length to be specified to an accuracy far better than can be measured (6 significant figures), as the device can be thousands of wavelengths long. Similar accuracy for wavelength and π would be required.

2) The definition should be in terms of the physical dimensions of the grating, that is, in terms of its period, rather than the period of the optical wave. This is because the period of the optical wave is unknown until the device is simulated/analysed. 'Grating' in the parameter name implies that the Nominal Grating Period, Λ, is used.

'Quarter-Wave-Shifted' lasers pose a problem as the quarter-wave-shift is in terms of the optical wavelength rather than the grating period. 'Quarter-Wave-Shifted' implies that the laser oscillates exactly at the Bragg wavelength, which is not always the case.

3) The definition should be in terms of the effective index of the modulated waveguide as this is a commonly used parameter. Alternatives could be thickness of the active region or its impedance.

4) The reference planes for zero phase should be defined as being at the peak of the effective index. Defining them as the points where the effective index equals the mean effective index is ambiguous.

5) The phase should be measured as 'outward is positive'. Definitions involving a Cartesian co-ordinate need the facets labelling with respect to this co-ordinate. 'Outward (from the laser centre) is positive' is unambiguous. Also, lasers with say (π, $\pi/3$) facets are likely to have similar spectra to lasers with ($\pi/3$, π) facets using this definition. This may not be the case with other definitions.

6) The phases should be positive.

7) The facet (laser/air interface) is assumed to have a reflection phase of zero degrees (*i.e.* the wave is being reflected by the impedance change from the low-impedance waveguide, to the high-impedance air, giving a positive reflection coefficient of the field.)

Thus the phase is defined as being the phase length from the peak in the index *closest to the facet* to the facet, measured in the outward direction (away from the laser centre) in terms of the grating period.

Fig. 6.12 clarifies the definition. Here, Θ_1 is approximately 1.5 radians and Θ_2 is approximately 3 radians. The facets' phases are distinguished using subscripts to Θ. These should be numerical with the left-most facet being labelled '1'.

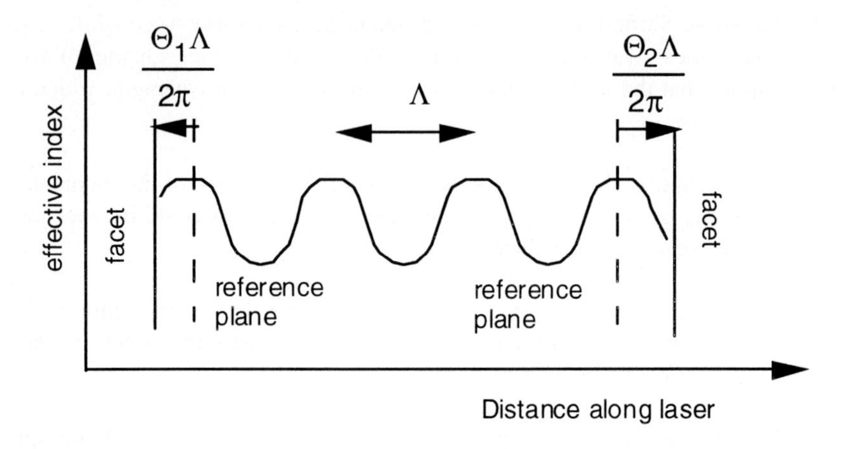

Fig. 6.12. Definition of facet phases relative to index grating.

6.6.7 Stimulated emission parameters

The following parameters relate the optical gain by stimulated emission to the carrier density and the photon density.

A. Nominal Wavelength, λ, nm, lambda

The Nominal Wavelength is used to calculate the photon energy which, in turn, can be used to convert the optical field into a photon density for the calculation of the stimulated emission rate. This parameter affects the efficiency (in watts/amp) of the laser.

B. Linear Material Gain Coefficient, a, m^2, a

The Linear Material Gain Coefficient describes the dependence of the optical power gain of the active region material on the carrier density, N, within the active region; the power gain per unit length in the active region of the waveguide equals the Linear Material Gain Coefficient multiplied by the carrier density in excess of the transparency carrier density. The modal gain per unit length in the overall optical waveguide equals the Linear Material Gain Coefficient multiplied by the carrier density in excess of the transparency carrier density multiplied by the confinement factor of the active layer. Typical values vary between 2.5×10^{-20} m^2 (Bulk InGaAsP) and 25×10^{-20} m^2 (Strained MQW).

Warnings: The Linear Material Gain Coefficient depends on the gain model. The type of gain model should be specified. In particular, the linear material gain coefficient will depend on whether the laser operates at a fixed wavelength (*e.g.*

DFB, external cavity lasers), or at the gain peak (*e.g.* Fabry-Perot laser). This is illustrated in Fig. 6.13, where the gain spectrum is shown for three carrier densities. If a fixed wavelength is used so that the laser follows the line A-A', the differential gain will be small (rate of change of gain with carrier density). Similarly, at fixed wavelength C-C', the differential gain will be large. If the laser is allowed to operate at the peak of the gain spectrum (line B-B'), the differential gain will be moderate, but the gain will be largest for a given carrier density (that is, the threshold carrier density will be the lowest of all cases).

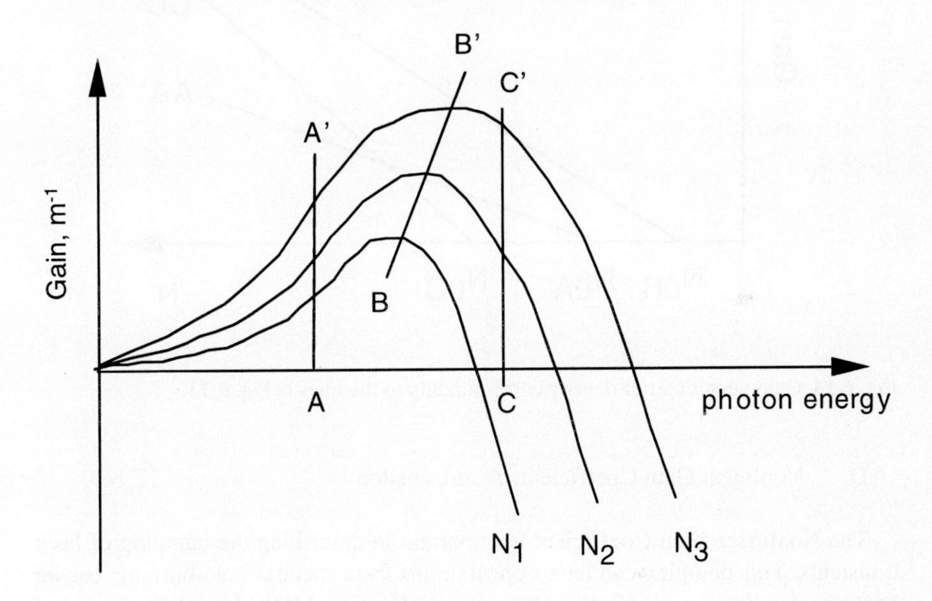

Fig. 6.13. Typical gain spectra for three carrier densities. Lines show constant wavelength operation (A-A', C-C') and operation at the gain peak (B-B').

C. Transparency Carrier Density, N_0, m^{-3}, N_0

The carrier density within the conduction band of the active region at which the material has zero optical gain (or loss) excluding the scattering and other intrinsic losses in the waveguide. Typical values are between 0.8×10^{24} m^{-3} and 1.8×10^{24} m^{-3}.

Warnings: The Transparency Carrier Density depends on the gain model. The type of gain model should be specified. In particular, for a given material the Transparency Carrier Density will depend on whether the laser operates at a fixed wavelength (*e.g.* DFB, external cavity lasers), or at the gain peak (*e.g.* Fabry-Perot laser). This is illustrated in Fig. 6.14, which is plotted from data taken from Fig.

6.13. Note that the operating case for highest differential gain does not correspond to the case for lowest threshold current.

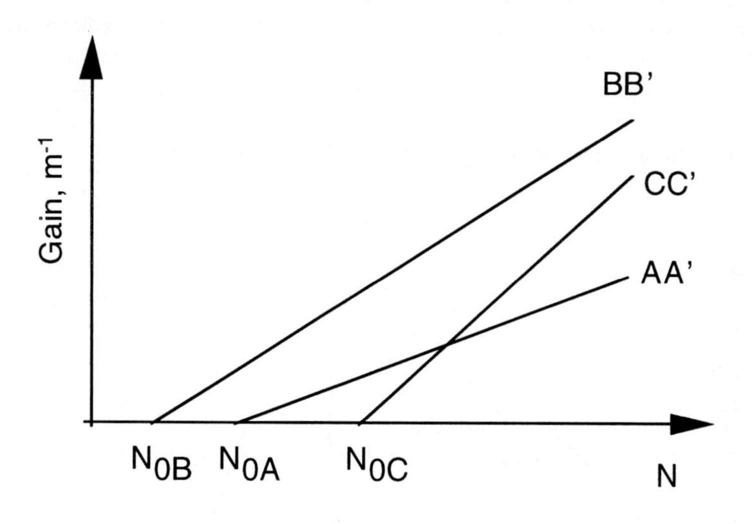

Fig. 6.14. Gain versus carrier density corresponding to the lines in Fig. 6.13.

D. Nonlinear Gain Coefficient, ε, m^3, epsilon

The Nonlinear Gain Coefficient is important in describing the damping of laser transients. This damping can have contributions from spectral hole-burning, carrier heating, standing-wave effects and structural effects in MQW lasers. The value of the Nonlinear Gain Coefficient is fairly well known, and defines the reduction in the optical gain with photon density:

$$\text{Gain} = \text{Gain}_{(S=0)} / (1 + \varepsilon S)$$

where S is the Photon Density, defined above. Note that other definitions of photon density (such as assuming the photons to occupy the same volume as the carriers) give different results for a given Nonlinear Gain Coefficient. Typical values vary between 1×10^{-23} m^3 and 7×10^{-23} m^3.

Warnings: This parameter could be confused with the permittivity of the material. However, the symbol has been chosen in spite of this because it is in common usage. Indeed the symbol's usage is more consistent than any name for the symbol.

6.6.8 Spontaneous recombination parameters

The following parameters describe the probability of spontaneous recombination versus carrier density, and the amount of spontaneous recombination coupled to the lasing mode:

A. Linear Recombination Coefficient, *A*, s^{-1}, A

The recombination rate of carriers can be measured and then described as a function of carrier density by linear, quadratic, and cubic terms. This term represents the linear term of the fit. A typical value is 10^8 s^{-1}.

B. Bimolecular Recombination Coefficient, *B*, m^3s^{-1}, B

The recombination rate of carriers can be measured and then described as a function of carrier density by linear, quadratic, and cubic terms. This term represents the quadratic term of the fit. A typical value is 10^{-16} m^3s^{-1}. Note that the recombination rate associated with this term is not necessarily 'Radiative' alone, because this term is determined by a fit, not by the rate of a fundamental process.

C. Auger Recombination Coefficient, *C*, m^6s^{-1}, C

The recombination rate of carriers can be measured and then described as a function of carrier density by linear, quadratic, and cubic terms. This term represents the cubic term of the fit. A typical value is 8×10^{-41} m^6s^{-1}. Note that the recombination rate associated with this term is not necessarily due to Auger recombination alone.

D. Carrier Lifetime, τ_s, s, tau_s (OPTIONAL: NOT PREFERRED)

Strictly, the carrier lifetime is a carrier-dependent lifetime often described using recombination coefficients for monomolecular, bimolecular and Auger recombination. However, if the carrier density is clamped around a steady value the carrier lifetime can be approximated to being carrier density independent.

$$\tau_s = 1/(A + BN + CN^2)$$

and can be of the order of 1 to 2 ns in semiconductor lasers.

E. Differential Carrier Lifetime, τ_d, s, tau_d

The time for the carrier density to recover to within (1/e) of its steady-state value after being depleted (such as during short pulse amplification). It is given by

$$\tau_d = 1/(A + 2BN + 3CN^2)$$

and can be of the order of 400 ps in semiconductor optical amplifiers with high carrier densities.

F. Population Inversion Parameter, n_{sp}, dimensionless, n_sp

The population inversion parameter is used in the calculation of the spontaneous emission rate into the fundamental transverse and lateral mode of the waveguide. In an ideal ('four-level') laser the power spectral density of the spontaneous emission is related to the gain of the laser, the photon energy. However, in non-ideal lasers in which the transparency carrier density is not zero, the spontaneous emission is increased by a factor called the population inversion parameter [15]. Sometimes this parameter is approximated to $N/(N-N_0)$, the carrier density over the carrier density minus the transparency carrier density. However, often a fixed value is used. Typical values range from 3 down to 1.5, depending on the threshold conditions of the laser. This parameter has little effect on the transient characteristics of the laser, but affects all noise processes within the laser.

G. Spontaneous Emission Coupling Factor, β, dimensionless, beta

This is an alternative to using the population inversion parameter to specify the magnitude of the spontaneous emission. It specifies the proportion of the bimolecular recombination that results in photons coupled to the lasing mode over a frequency range equal to the free-spectral range (FSR) of the laser cavity. That is, the proportion of the photons created by radiative recombination that are coupled to one mode of a Fabry-Perot laser. β is sometimes estimated from the far-field pattern of the laser, as this gives an indication of the acceptance angle of the waveguide to spontaneously-emitted photons [16].

In most cases the use of the Population Inversion Parameter, n_{sp}, is preferable, as it is a more fundamental parameter.

The Spontaneous Emission Coupling Factor has little effect on the transient characteristics of the laser at powers above a milliwatt, as the damping is dominated by the Nonlinear Gain coefficient.

H. Carrier Diffusion Length, L_D, m, L_D

This parameter specifies the diffusion length of carriers diffusing along, or transverse to, the waveguide. It generally has little effect on the performance of the laser if the laser is index-guided. A typical value is 1.0 μm.

6.6.9 Current injection parameters

These describe the injection of carriers into the active region by an external source.

A. Injection Efficiency, η_{inj}, dimensionless, eta_inj

This is the proportion of the injection current which travels to the active region (or into the quantum wells) with regard to the total current injected into the laser. The injection efficiency can be almost 100% in buried heterostructure lasers with good current-blocking regions.

B. Leakage Current, I_1, A, I_l

This is an alternative parameter to the injection efficiency which describes a constant current that leaks around the active region. The leakage current should be close to zero for lasers with good current-blocking regions around the active region.

References

[1] K. Aika, M. Nakamura, J. Umeda, A. Yariv, A. Katzir, and H. W. Yen, "GaAs-GaAlAs Distributed Feedback Diode Lasers with Separate Optical and Carrier Confinement", *Appl. Phys. Lett.,* vol. 27, pp. 145-146, 1975.

[2] M. Kitamura, "Lasing Mode and Spectral Linewidth Control by Phase Tuneable Distributed Feedback Laser Diodes with Double Channel Planar Buried Heterostructure", *IEEE J. Quantum Electron.,* vol. 21, pp. 415-417, 1985.

[3] S. Murata., I. Mito, and K. Kobayashi, "Spectral Characteristics for a 1.5 μm DBR Laser with Frequency-Tuning Region", *IEEE J. Quantum Electron.,* vol. 21, pp. 835-838, 1987.

[4] G. Morthier et al., "Comparison of different DFB laser models within the European COST 240 collaboration", *IEE* Proc. J, vol. 141, pp. 82-99, 1984.

[5] A. Perales, L. Goldstein, A. Accard, B. Fernier, F. Leblond, C. Gourdain, and P. Brosson, "High Performance DFB-MQW Lasers at 1.5 μm grown by GSMBE", *Electron. Lett.,* vol. 26, pp. 236-238, 1990.

[6] W. T. Tsang, M. C. Wu, L. Yang, Y. K. Chen, and A. M. Sergent, "Strained-Layer 1.5 μm wavelength InGaAs/InP Multiple Quantum Well Lasers grown by Chemical Beam Epitaxy", *Electron. Lett.,* vol. 26, pp. 2035-2036, 1990.

[7] E. Meland, R. Holmstrom, J. Schlafer, R.B. Lauer, and W. Powazinik, "Extremely high frequency (24 GHz) InGaAsP diode lasers with excellent modulation efficiency", *Electron. Lett.,* vol. 26, pp. 1827-1829, 1990.

[8] P. I. Kuindersma, W. Scheepers, J.M.H. Cnoops, P.J.A. Thijs, G.L.A. Van Den Hofstad, T. Van Dongen, and J.J. Binsma, "Tuneable Three-Section, Strained MQW, PA DFB's with Large Single Mode Tuning Range (72A) and Narrow Linewidth (around 1 MHz)", Proc. *12th IEEE International Semiconductor Laser Conference,* Davos, Switzerland, pp. 248-249, 1990.

[9] A. D. Sadovnikov and W.-P. Huang, "A Two-Dimensional DFB Laser Model Accounting for Carrier Transport Effects", *IEEE J. Quantum Electron.,* vol. 31, pp. 1856-1862, 1995.

[10] K. Yokoyama, T. Yamanaka, and S. Seki, "Two-Dimensional Numerical Simulator for Multielectrode Distributed Feedback Laser Diodes", *IEEE J. Quantum Electron.*, vol. 29, pp. 856-863, 1993.

[11] M. J. Adams, *An Introduction to Optical Waveguides*, Wiley, Chichester, pp. 157-160, 1981.

[12] S. L. McCall and P.M. Platzman, "An Optimized $\pi/2$ Distributed Feedback Laser", *IEEE J. Quantum Electron.*, vol. 21, pp. 1899-1904, 1985.

[13] A. J. Lowery, H. Olesen, G. Morthier, P. Verhoeve, R. Baets, J. Buus, D. McDonald, and D. D. Marcenac, "A proposal for standardized parameters for semiconductor lasers", *Int. J. Optoelectron.*, vol. 10, pp. 347-355, 1995.

[14] M. Osinski and J. Buus, "Linewidth broadening factor in semiconductor lasers - an overview", *IEEE J. Quantum Electron.*, vol. 23, pp. 9-29, 1987.

[15] C. H. Henry, "Theory of spontaneous emission noise in open resonators and its application to lasers and optical amplifiers", *J. Lightwave Technol.*, vol. 4, pp. 288-297, 1986.

[16] D. Marcuse, "Classical derivation of the laser rate equation", *IEEE J. Quantum Electron.*, vol. 19, pp. 1228-21231, 1983.

7 Modelling of DFB laser diodes

G. Morthier, A. Lowery

The behaviour of DFB lasers is so complex that it cannot be described analytically. Neither can the design of these lasers rely on a number of simple analytical formulae as is the case for FP lasers. The main cause of this complex behaviour is the strong dependence of the distributed feedback, and hence of the facet loss, on local refractive index and carrier density variations. The main consequences of it are changes in side mode rejection or stability and significant contributions to FM-response and harmonic distortion, which are not easy to predict.

An accurate design of DFB lasers therefore has to rely on numerical models in which longitudinal carrier density variations (the so-called spatial hole burning) are self-consistently taken into account. In recent years, several of these longitudinal models have been developed and to a large extent by participants of Working Group 1 (WG1) of the European collaboration framework Action COST 240. The models used by COST 240 participants are described in the next section. The results obtained with the models have been compared extensively on a number of well-defined problems. Those numerical case studies are presented in the last section of this chapter. Although a detailed inclusion of spatial hole burning is claimed for every model, the implementation differs from one model to another (especially if the dynamics are considered) and the different approaches are not *a priori* equivalent. Comparing the models therefore gives valuable information about the relative importance of different dependencies and about the validity of the different approximations. It is a first evaluation of the accuracy of existing laser models. The comparison of numerical and experimental results, as will be described in chapter 9, can of course be regarded as a more thorough evaluation.

7.1 Overview of laser models

Numerical models enable novel devices to be designed without the need for the time-consuming process of fabricating a series of costly prototypes. Furthermore, internal variables, such as carrier density, are easily monitored, allowing a greater understanding of device operation to be gained. Even though advances in electronics have provided us with very powerful computers, it is still not possible to simulate a laser simply by entering the fundamental equations describing a laser's basic physics into a commercial equation solver: approximations have to be made to allow solution in a reasonable time, and skill in formulating numerical solutions and methods can provide huge gains in computational efficiency.

This section describes the development of numerical techniques used to simulate laser diodes, starting from the simplest of laser models, suitable for FP lasers, and progressing to sophisticated and efficient schemes for simulating complex laser diodes, such as DFB lasers with spatial hole-burning.

7.1.1 Desirable characteristics of laser models

The ideal laser diode simulator would:

- contain sufficient detail to simulate all laser characteristics that could affect the performance of the overall circuit or system
- be compatible with circuit or system models
- have interfaces that pass sufficient information between the models so that all possible interactions are identified
- be numerically efficient, with some possibility of trading-off accuracy for speed in the early stages of a design optimisation
- simulate devices over a reasonable spectral bandwidth to allow, for example, wavelength conversion to be simulated

Some typical phenomena that we may wish to simulate include [1]:

- spectra of lasers with frequency-selective elements including gain and index gratings, multiple cavities, external gratings
- laser transient response to large-signal (e.g. digital) stimulation, including the effects of parasitics [2], cavity transit times, carrier hole-burning, timing jitter, and multiple-section cavities or current contacts
- spectral dynamics of lasers including chirping of wavelength during modulation and mode-partition noise [3] and mode-hopping
- interaction of laser output with dispersive elements including filters, fibres, interferometers and gratings
- use of nonlinearities of devices to perform logic operations, clock recovery, switching, wavelength conversion and spectral inversion
- resonances of devices providing millimetre signal generation [4].

7.1.2 Single-mode rate equation laser models

The earliest semiconductor lasers were based on FP cavities formed by cleaving a direct-bandgap semiconductor with a p-n junction grown on it, and selectively injecting current into it. In order to predict the direct-modulation performance of these devices, with a view to their use in optical communications systems, Boers, Vlaardingerbroek, and Danielsen developed the first large-signal model of a semiconductor laser [5]. Their model was based on the laser rate equations originally derived by Statz and de Mars [6], which describe the interaction of the photon and carrier populations within the laser. One rate equation describes the spatially-averaged carrier density, N, within the active region of the laser waveguide, and a second rate equation represents the average photon density, S, of the guided optical wave. The terms in the equations can be derived by accounting for the injection and recombination of carriers, and the emission and loss of photons:

• The carrier density increases due to injected current, and is depleted due to spontaneous recombination and stimulated recombination.

• The photon density increases due to stimulated and spontaneous emission of photons, and is depleted by photon loss through the facets, or by internal scattering and absorption in the waveguide.

The rate equations are coupled, as shown in Fig. 7.1, due to the optoelectronic processes of stimulated and spontaneous *recombination* of carriers producing stimulated and spontaneous *emission* of photons.

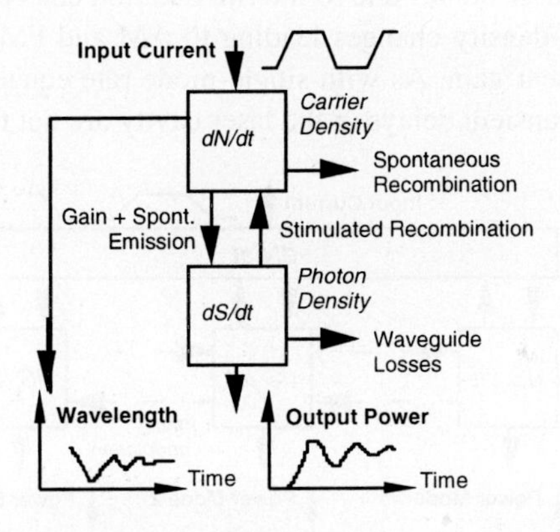

Fig. 7.1. Single-mode rate equation laser model.

Single-mode rate equation models are able to simulate the photon density variation with time, for any injected current waveform. The effects of drive waveform shape, and bias ('0'-symbol current) are easily predicted, as is the maximum modulation frequency. These rate equations can also predict the transient wavelength chirp of the lasing mode [7] from the carrier density for Fabry-Perot devices. However, the wavelength chirp predicted by single-mode equations is inaccurate for DFB devices which have high spatial inhomogeneities, because the effects of carrier spatial-hole burning (SHB) are neglected. SHB in DFB lasers can lead to modal and temporal instabilities [8] and increased chirping [9]. Although the single-mode rate equation model provides a considerable understanding of laser dynamics, undesirable effects such as multimode behaviour and mode-hopping obviously cannot be predicted. Thus, single-mode models are only suitable for system simulations where the laser is known to be stable single-mode. Unfortunately, only a multimode model can be used to predict whether a laser is genuinely single-mode.

7.1.3 Multimode rate equation laser models

Multimode rate equation models [10] extend the single-mode model by using a separate rate equation to represent the photon density within each lasing mode, S_n. The simpler models assume that the modes are only coupled through a common carrier density. As each mode will receive a slightly different gain due to the spectral selectivity of the stimulated emission process, the change in the lasing spectrum with time can be predicted. Furthermore, it can be assumed that each mode undergoes an equal amount of 'chirp' (frequency shift during a transient), which may be calculated from the carrier density. Thus, the narrowing of a Fabry-Perot laser's spectrum during the turn-on of a laser pulse may be predicted. Cross-coupling between the laser modes due to intermodulation can be added [11], which is either due to carrier density changes leading to AM and FM sidebands of each mode, or due to nonlinear gain. As with single-mode rate equation models, spatial inhomogeneities and transient delays in the laser cavity are not treated.

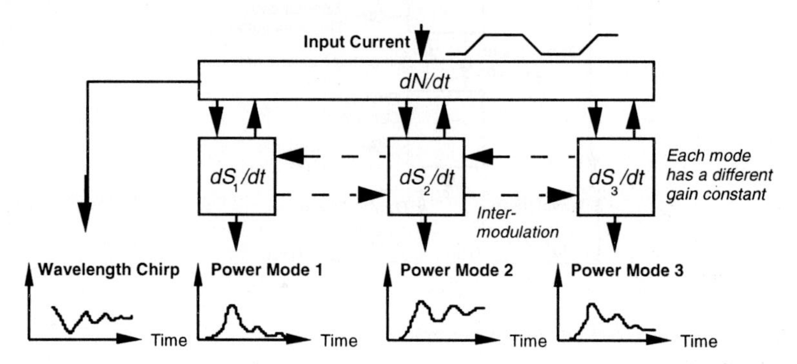

Fig. 7.2. Multimode rate equation laser model and outputs.

7.1.4 Travelling-wave rate equation laser models

In some types of semiconductor laser the transient time of the optical waveguide within the laser cavity is all important to their modulation characteristics. Examples include mode-locked lasers where the laser cavity is extended using a passive section designed to have a microwave resonance close to the drive frequency of the active region [12]. Pulses of light propagate back and forth in the mode-locked laser, and so it is highly dubious to describe the photon density as homogenous throughout the cavity (using a single spatially averaged value) or neglect the delays in the pulse's propagation. Generally, transient time effects must be considered in any laser where the round trip time between facets is in the order of the modulation period. Thus, for a typical laser, these effects should have little importance up to around 20 GHz.

An early approach to modelling time delays was provided by Demokan [13] for simulating mode-locked lasers. The cavity was subdivided longitudinally into sections, each being defined as an iteration delay. Samples of the photon density were passed from section to section at every iteration, similar to a bucket-brigade. Each section also modified the samples according to the local carrier density using modified rate equations. A generalised model based on this approach is shown in Fig. 7.3. Note that carrier diffusion between the sections can also be included, but this is usually negligible.

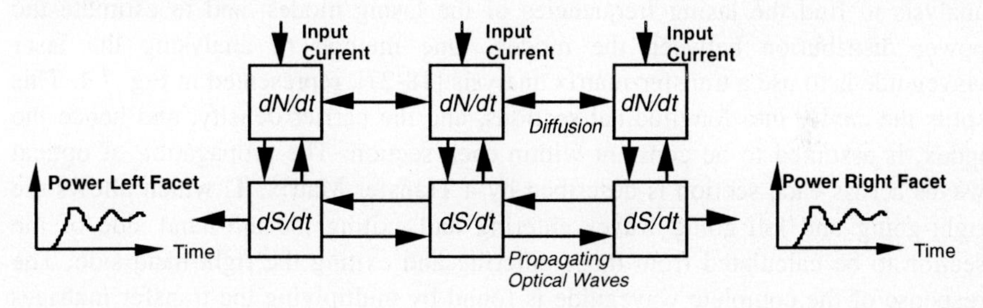

Fig. 7.3. Travelling-wave rate equation model and outputs.

The travelling-wave method of Fig. 7.3 is single-mode. However, it could be extended to multiple modes by replicating the travelling photon wave part of the model for each cavity mode. Only one set of carrier rate equations is required because all modes feed off the same local carriers. Such an extension has been conceived by Hsu [14], however, it suffers from the problem that modal intermodulation is strong in a mode-locked laser, as every mode is modulated with a frequency corresponding roughly to the mode spacing. Thus, this cross-coupling has to be calculated, which can be computationally intensive. Some insight may be gained into reducing the computational task if we consider the correspondence between the time and frequency domains: if we use a representation of the optical waves including optical phase, then the modal spectrum is implicit in this representation without duplicating the photon density rate equations for each and every mode. Furthermore, if the carrier depletion is calculated from the photon density, and the travelling waves are suitably modulated by the carrier density, then the intermodulation between the modes is included automatically, without the need for Fourier transforms. This approach was introduced by Lowery in the Transmission-Line Laser Model [15-17] (TLLM), where rate equations for photon density are replaced by manipulations of the travelling optical electric fields. The TLLM will be discussed in detail in section 7.1.6.

One limitation with all of the above models (excluding the TLLM) is that the phase of internal optical reflections is not treated. In some devices, this can be critical, for example, multi-cavity lasers, or lasers with gratings such as DFB lasers [3]. Thus, some method of considering the effects of optical phase over a wide optical bandwidth is required, and this method must be able to cope with the dynamics of gain and refractive index if it is to form the basis of a time-domain system simulation.

7.1.5 Transfer-matrix models

Lasers with internal reflections, such as DFB lasers, require a sophisticated analysis to find the lasing frequencies of the lasing modes, and to estimate the power distribution between the modes. One method of analysing the laser waveguide is to use a transfer-matrix analysis [18-27], represented in Fig. 7.4. This splits the cavity into longitudinal sections, and the carrier density, and hence the index, is assumed to be constant within each section. The propagation of optical waves across each section is described by a Transfer Matrix, **T**, which allows the right-going and left-going waves entering and exiting the left-hand side of the section to be calculated from those entering and exiting the right-hand side. The response of the complete waveguide is found by multiplying the transfer matrices to find the relation between the waves at the left-hand facet and the right-hand facet. Assuming that the laser will produce output waves from the facets from an infinitesimally small excitation at the facets (or from spontaneous emission within each section), the threshold gains of the waves of the modes, and their frequencies, can be found.

Solution of the transfer matrices is a quick method of finding below-threshold spectra, (sometimes known as Amplified Spontaneous Emission, ASE, spectra), if the carrier density profile can be assumed to be independent of the photon density. However, above threshold, stimulated recombination causes spatial-hole burning of the carrier density profile along the laser. This SHB will perturb the mode positions and frequencies due to changes in the waveguide index, which affects the reflection spectrum of the waveguide grating, making the reflection spectrum position dependent. This change in the grating will, in turn, affect the lasing frequency and the power distribution within the lasing cavity, which will alter the SHB along the cavity. Thus, above-threshold spectra can be found by repeated iteration of the carrier density profile and the waveguide field, as shown in Fig. 7.4, until a stable solution is found.

There are a number of numerical approaches to solving the transfer matrixes consistently with the carrier density (including SHB). Some of those adopted by the members of WG1 include:

• solving for a single longitudinal mode for static characteristics, and warning if the laser is close to becoming multimode.

• using a summation of locally-solved modulation responses to calculate the overall FM and IM modulation responses in the laser (i.e. assuming that the carrier density profile does not change during small-signal modulation).

• approximating the variation in carrier density profile to a function of two carrier-density rate equations for small-signal response modelling [29].

• solving the rate equations for each section self-consistently along the entire cavity using Runge-Kutta methods to find the large-signal response.

• correcting for the modal losses not instantaneously following the gain/refractive index by including a correction term that contains the derivative of the longitudinal Petermann factor [28]

• employing a shooting method to iterate output power and wavelength at one facet, either using Transfer Matrices or a direct solution of the coupled-mode equations.

Some numerical methods can be time-consuming, or even fruitless if the laser does not have a stable mode (i.e. is a self-pulsating device). However, once a stable solution is found, it is easy to apply perturbation analysis to determine small-signal characteristics such as FM response (laser wavelength change with injection current), AM response, and linewidth. Variations of the transfer-matrix method use Green's functions [30] to calculate the lasing spectrum, AM and FM responses, linewidth, and modal instability at high currents.

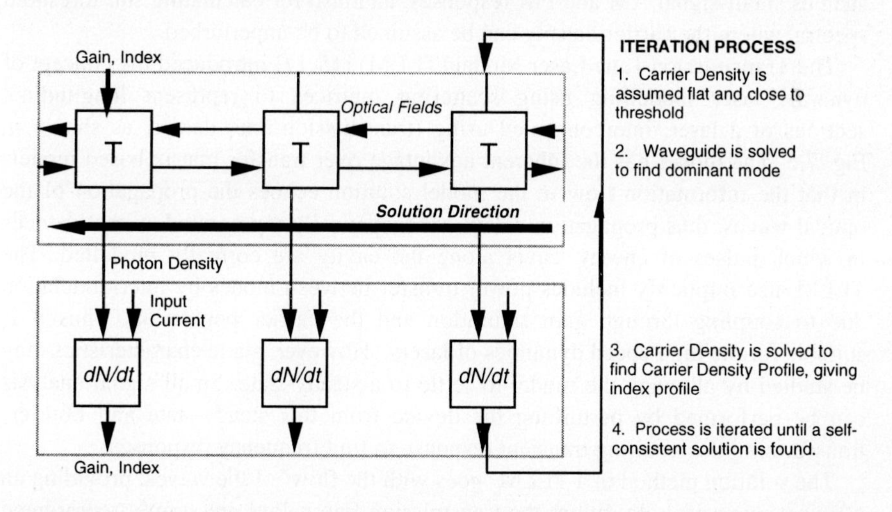

Fig. 7.4. Transfer-matrix model for a DFB laser divided into sections: solution process for one time step.

Time dynamics may be estimated using the transfer-matrix model and solving for a series of steady-states, assuming that the time dynamics are unaffected by the delays along the cavity length [18]. This is computationally very intensive as a self-consistent spectrum has to be calculated for each time step (a time step can be approximately equal to the cavity round-trip time), unless some approximations are made such as assuming a constant carrier density profile. It is again difficult to include the effects of intermodulation between the modes, and this is generally not included in the calculations. One variation, the Power Matrix Method (PMM) [31], increases computational efficiency by following the evolution of a single-mode and using the gain per unit time to estimate the rate of increase of this mode.

7.1.6 Fully time-domain models

The main limitation with all transfer-matrix based approaches is that the optical waveguide equations have to be solved assuming that the power distribution within the cavity at a time step is equal to that in the steady-state, and is unsuitable for devices in which power moves around the cavity during a transient (e.g. laser amplifiers [32], a mode-locked laser, and self-pulsating lasers based on asymmetric mode profiles [8]). This limitation is due to the form of the transfer matrices, which have to be multiplied from left to right, rather than following the propagation of the optical waves by, for example, predicting the waves *out* of a section from the waves *into* a section some time earlier. However, transfer-matrix based models are very useful for analysing characteristics around a steady-state, such as small-signal AM and FM responses, and also for calculating sub-threshold spectra, where the carrier density can be assumed to be unperturbed.

The Transmission-Line Laser Model (TLLM) [15-17] introduced the concept of dynamic laser modelling using scattering matrices to represent longitudinal sections of a laser, interconnected using (transmission-line) delays, as shown in Fig. 7.5. The model has the inherent advantage over transfer-matrix based models in that the information flow in the model solution echoes the propagation of the optical waves, thus propagation delays are realistically represented so that devices in which pulses of energy travel along the cavity are correctly modelled. The TLLM also implicitly includes power transfer between modes by intermodulation due to coupling through gain saturation and the carrier population. Thus it is suitable for the high-speed dynamics of lasers. However, static characteristics may be studied by allowing the model to settle to a steady-state. Small-signal analysis can be performed by perturbing the device from this steady-state and Fourier-transforming the resulting transient response to find frequency responses.

The solution method in a TLLM 'goes with the flow' of the waves, providing an efficient numerical algorithm: the transmission-line delays are simply represented by a storage operation in computer memory, and the scattering matrices can be solved independently as they are separated by these delays. The iteration process essentially follows the propagation of the optical waves back and forth along the laser cavity. At every iteration, and at every section, waves incident upon a scattering matrix are used to calculate waves reflected from the scattering matrix. The reflected waves propagate along the 'transmission-lines' to become incident waves at adjacent sections for the next iteration.

The scattering matrices include terms representing frequency-dependent stimulated material gain (filter and amplifier), attenuation (attenuator), spontaneous emission (noise generators), gain and index gratings (impedance mismatches). The transmission-lines are simply a means of developing a stable algorithm where the numerical errors are apparent as parasitics in the network: the solution is really of Maxwell's equations, and has been shown to be consistent with Quantum theory [33].

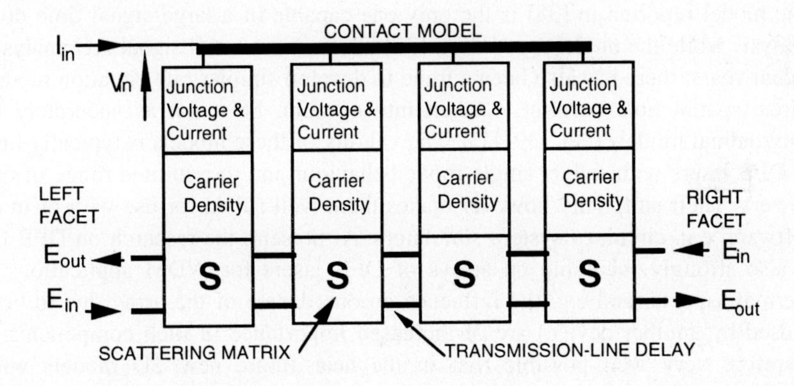

Fig. 7.5. A transmission-line laser model.

The TLLM has some similarity to the travelling-wave models of section 7.1.5 in that the solution follows the propagating waves. The main difference is that optical field, rather than photon density, is passed between the sections. This allows the optical spectrum to be calculated by Fourier-transforming the output field over a number of iterations, with only a small increase in computation over a travelling-wave model. Also, intermodulation between the laser modes via gain nonlinearity or carrier-density changes, is implicit in the model, as all modes share the same carrier density within a section. The change in carrier density at an iteration due to stimulated emission is simply calculated from the sum of the squares of the backward- and forward-propagating fields, and thus is a result of the energy of all modes, and 'beating' between them. These variations in carrier density then amplitude and phase modulate the propagating waves, causing power transfer across the lasing spectrum. Unlike in transfer-matrix models, a steady-state spectrum is not calculated for each time step, and it is not necessary to calculate this power transfer between modes.

The transmission-line laser model has many applications ranging from modelling ultra-short pulse generation in multi-section mode locked lasers to modal instabilities in DFB semiconductor lasers. Because the model is closely related to the real device's topology, it is easily extended beyond the device to include external cavities and interactions between different devices.

7.1.7 Other numerical models

Most of the previously discussed models allow for longitudinal variations, but not for lateral or transverse variations of the carrier density. Since about 1993 however, a few 2-dimensional models have been reported in the literature (e.g. [34-38]). Most of these models are restricted to an analysis of the static behaviour. The model reported in [38] is the only one capable of a large-signal time-domain analysis while the model reported in [36] allows for a small signal AC analysis. In recent years, there has also been a trend to develop simpler rate equation models in which spatial hole burning is taken into account, be it not as accurately as in longitudinal models (e.g. [39-41]). The validity of these models is typically limited to DFB lasers with stable, single-mode behaviour and to a limited range of output powers. Their simplicity however makes them well suited for use with or in other software, e.g. circuit or system simulators. At present, the research on DFB lasers is also strongly focussing on arrays of DFB lasers for WDM applications. The thermal aspects (and e.g. the influence on one device of the array due to heating caused by another device) are of increased importance in such components. It is therefore very well possible that in the near future new 2D models will be presented which allow those thermal effects to be taken into account.

In conclusion, both Transfer-Matrix based models and Scattering-Matrix time-domain models make few assumptions about the operation of complex lasers such as DFB lasers. Both types of model are able to model the small and large signal characteristics of lasers, and can include spatial hole burning. The transfer matrix model is efficient for the small-signal estimation of modulation responses of lasers, and scattering matrices models are more efficient for large signal simulations, especially when modal intermodulation and cavity delay effects are important, for example in mode-locked or pulsating devices.

7.2 Numerical case studies

The numerical case studies were launched in order to compare various laser models in use by university and industry research groups. The case studies produced valuable results as to the accuracy of existing models. Moreover, the studies enabled to adapt some models to make them more attractive for the analysis of practical devices.

7.2.1 Introduction

In total, twelve laser models have been compared. These models belonged to University of Gent (UG), University of Cambridge (UC), University of Melbourne (UM), University of Athens(UA), Humboldt University Berlin (HUB), Royal Institute of Technology Stockholm (KTH), Tele Danmark Research (TDR), Fondazione Ugo Bordoni (FUB), Politecnico di Torino (PT), Deutsche Telekom (DT), Alcatel Alsthom Recherche (AAR) and Centre National d'Etudes des Télécommunications (CNET). As has been outlined in the previous part, most of these models are quite accurate and allow a non-uniform carrier density for the static analysis, but different levels of complexity are used for what concerns the calculation of the dynamic and noise related characteristics. The model of the University of Melbourne is not mentioned in 7.2.2 and 7.2.3. This is due to the fact that the University of Melbourne joined COST 240 at a later stage and does not imply any shortcoming of the transmission line laser model (TLLM, which has since been shown to predict identical characteristics).

The emphasis is on static or low-frequency dynamic results. Only a limited number of high-frequency characteristics (describing effects occurring on a time scale of a few ns at the most or on a frequency scale from 100 MHz to a few tens of GHz) have been calculated since only a limited number of groups had the capability to do so at the time of the case studies. At the end of this chapter, we will also discuss a few large signal, dynamic characteristics.

In all the calculations that will be reported here, a linear and wavelength independent relation between gain, refractive index and carrier density has been assumed and carrier diffusion has been neglected. Thermal effects have also been ignored. For what concerns the details of the current injection, it has been assumed that the current density is uniform over the active layer or in other words that the series resistance is very large.

It should also be remarked that for some calculations inconsistent values for n_{sp} and β_{sp} (both parameters determine the spontaneous emission rate) were defined, which could cause disagreement among the results for the linewidth, the RIN or the side mode suppression. Both parameters are related by the following expression for the spontaneous emission rate:

$$R_{sp} = \beta_{sp} BN^2 V = \Gamma g v_g n_{sp} \tag{7.1}$$

Within COST 240, the first approximation has been used for the calculation of linewidth and RIN in all models. It is only in the calculation of the side mode suppression that both expressions for R_{sp} have been used.

7.2.2 AR-coated, λ/4-shifted DFB lasers

First, we will discuss numerical results obtained for the static or low-frequency dynamic and stochastic characteristics of AR-coated λ/4-shifted DFB lasers. AR-coated, λ/4-shifted DFB lasers possess a grating which undergoes a phase shift of π in the middle, as shown in Fig. 7.6. Such a phase shift lifts the mode degeneracy of the AR-coated DFB laser with uniform grating (see 6.2.3) and gives, at least at low to moderate power levels, a stable single-mode behaviour. Both bulk and MQW lasers with different $\kappa_i L$ values (L=300 μm: $\kappa_i L$=1, 2 and 3; L=1 mm and $\kappa_i L$=1) have been considered. For general reference purposes, the other laser parameters are summarised in Table 7.1. It must be remarked that the choice of n_{eff0} and L gives rise to an unusual emission wavelength around 1.65 μm. Gain suppression has been neglected in all cases.

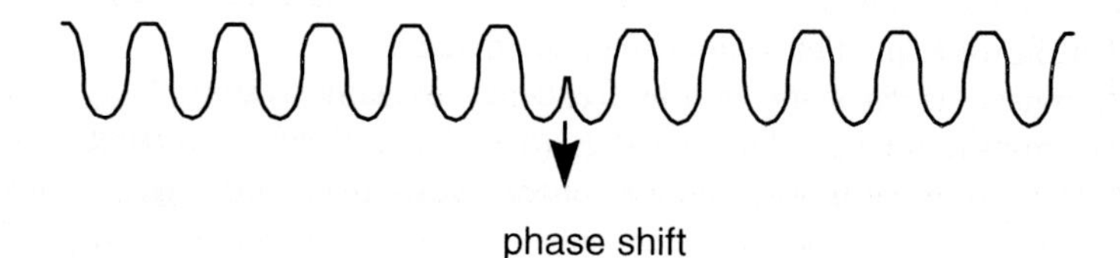

Fig. 7.6. Grating with a quarter wave shift in the middle.

Parameter	bulk	MQW
Grating period Λ [nm]	244.5	244.5
Bragg order M	1	1
Active region width w [μm]	1.5	1.5
Active layer thickness [μm]	0.2	0.1
Confinement factor Γ	0.3	0.15
Effect. refractive index without injection n_{eff}	3.4	3.4
Group index n_g	3.6	3.6
Internal loss α_i [cm^{-1}]	50	40
Linear material gain coefficient a [10^{-16} cm^2]	2.5	7
Transparency carrier dens. N_0 [10^{18} cm^{-3}]	1	1.5
Carrier density dependence of active layer refractive index dn/dN [10^{-20} cm^3]	-1.5	-1.5
Population inversion parameter n_{sp}	2	2
Linear recombination coeff. A [s^{-1}]	10^8	10^8
Bimolecular recombinat. coeff. B [10^{-10} cm^3/s]	1	1
Auger recombination coeff. C [10^{-29} cm^6/s]	7.5	7.5

Table 7.1. Laser parameters used in the modelling exercise.

Since most of the models make use of a similar (transfer-matrix) approach to find the steady state solutions, the results on threshold current I_{th}, emission wavelength λ, threshold gain difference ΔgL and P-I relation show in general excellent agreement. Fig. 7.7 shows the P-I relation, as obtained with eleven models out of the twelve models, for the bulk laser with $\kappa_i L$=2. The dashed lines give the most deviating results, but almost all results coincide with the full line. For this laser, we found I_{th} = 23.75 ± 0.05 mA, λ_{th} = 1.6583 ± 0.0008 μm and ΔgL = 1.46÷1.47. The model of HUB which uses an interpolation scheme gave a slightly higher I_{th} = 24.1 mA. Fig. 7.8 shows the side mode suppression of this laser vs. the output power. Again there seems to be a very good agreement. It can however be noticed that the differences between the results amount to 3 dB (which is a factor of 2). These relatively large differences in absolute value are probably due to the limited accuracy used in the calculations (originating from the approximate solution of the nonlinear system of equations that describe a laser diode) or to different descriptions (based on n_{sp} or β_{sp}) used for the spontaneous emission rate. One can nevertheless observe a good qualitative agreement on the dependence of the side mode suppression on the output power. Fig. 7.9 shows the wavelength deviation from the value at threshold for the same laser as a function of the output power. The small differences between the results are again due to the limited accuracy of the calculations. The deviation of the wavelength from the threshold value is already of the order of 10^{-4} times the absolute wavelength value, which is small when the iteration is on the absolute wavelength value. The wavelength variation obtained by CNET is the largest, but this is probably explained by the inclusion of the dispersion with wavelength in the refractive index in the model of CNET.

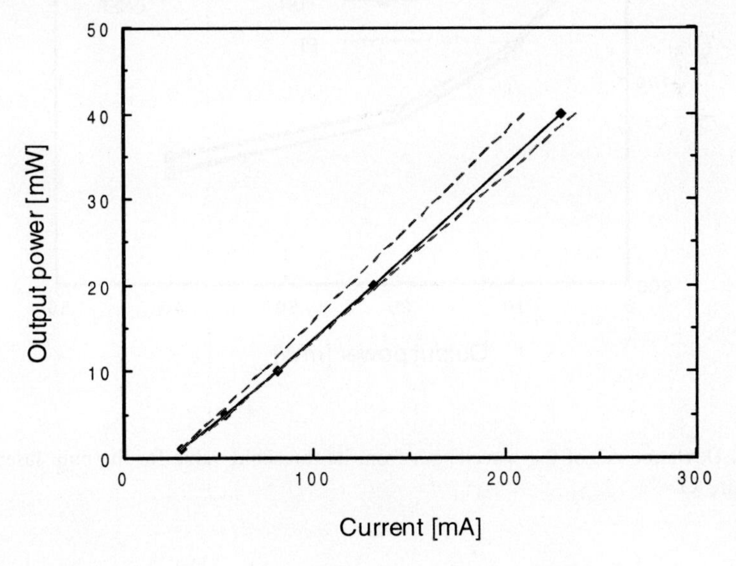

Fig. 7.7. P-I characteristic for the bulk laser with L=300 μm, $\kappa_i L$=2.

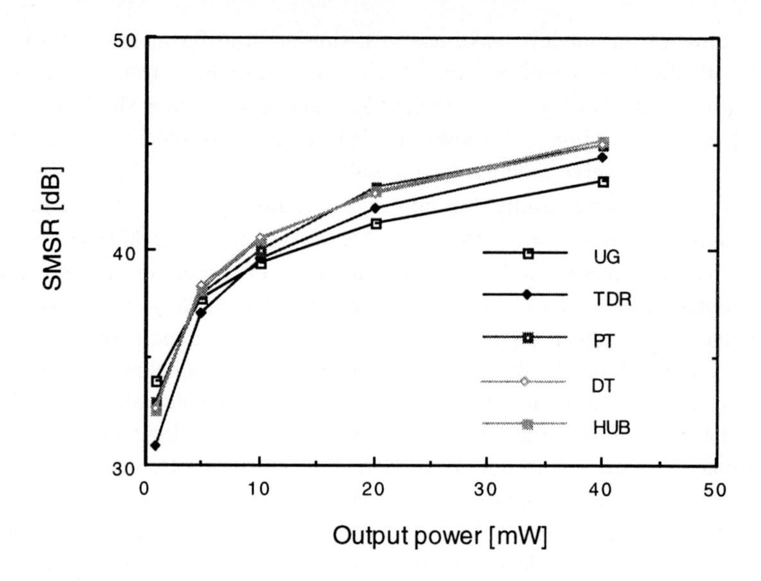

Fig. 7.8. Side mode suppression ratio for the bulk laser with L=300 μm, $\kappa_i L=2$.

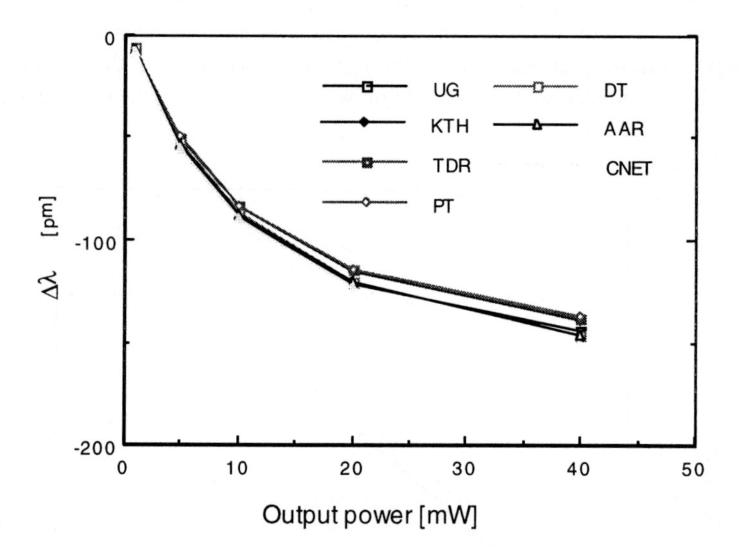

Fig. 7.9. Deviation $\Delta\lambda$ of the wavelength from its threshold value for the bulk laser with L=300 μm, $\kappa_i L=2$.

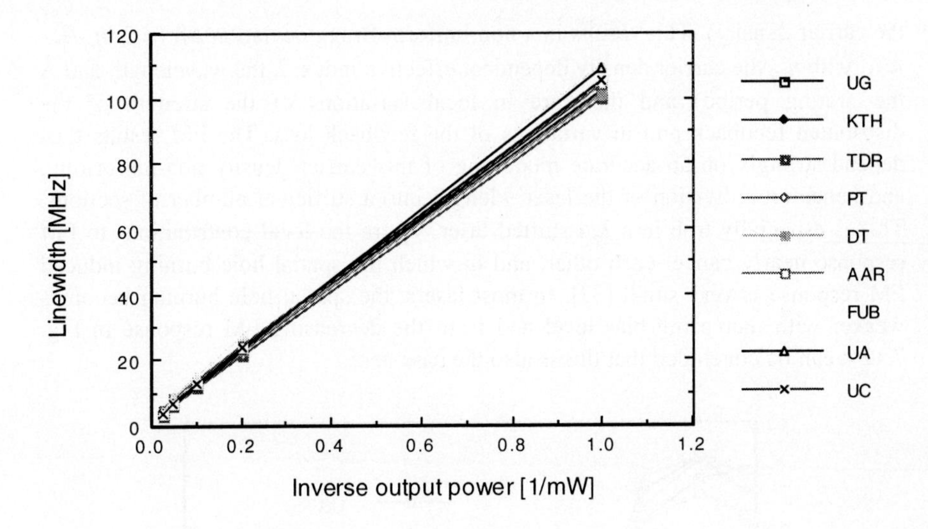

Fig. 7.10. Linewidth for the bulk laser with L=300 μm, kL=2.

The last characteristic of this laser that we will discuss here is the linewidth, which is shown vs. the inverse output power in Fig. 7.10. The results are obtained by various methods such as the Green's function method of Tromborg et al. applied with constant and variable longitudinal power distribution [30], the travelling wave method with z-dependent Langevin functions and the negative conductance oscillator circuit model [19]. The agreement in Fig. 7.10 proves the equivalence of the different methods. The TLLM's results lie on the lower extreme of the results obtained by the other models. The linewidth was calculated from the TLLM deterministically by using a standard formula and calculating: an effective mirror loss; the total number of photons in the cavity; a linewidth correction factor; and an effective linewidth enhancement factor [42]. Linewidth can also be calculated stochastically using the TLLM by simulating the optical field waveform over a very long period, and taking a Fourier transform to find the optical spectrum, or by calculating the near-DC value of the frequency fluctuation spectrum.

A bulk laser similar to the previous one with a $\kappa_i L$-value of 3 becomes single-mode at high output powers. Fig. 7.11 shows the side mode suppression versus the output power of this laser. The variations among the results are of the order of a few dB when the side mode suppression is larger than 30 dB, but they are extremely small when the laser is single-mode (i.e. where the side mode power affects the carrier density and thus the entire laser behaviour). Only the models of UG, TDR and DT could handle real single-mode behaviour though at the time of this case study. Fig. 7.12 shows the FM response of a 300 μm long λ/4-shifted MQW laser with $\kappa_i L$=1. Since gain suppression has been neglected, this FM response is entirely caused by spatial hole burning (i.e. by the non-uniformity in the carrier density). This results in a non-uniform Bragg deviation $\Delta\beta$ (= $2\pi n_{eff}/\lambda$ - π/Λ, with n_{eff} the carrier density dependent effective index, λ the wavelength and Λ the grating period) and therefore in local variations of the strength of the distributed feedback and in variations of the feedback loss. The FM results thus depend strongly on an accurate modelling of this carrier density non-uniformity, and hence on a division of the laser's length into a sufficient number of sections. This is especially true in a $\lambda/4$-shifted laser, where the local contributions to FM response nearly cancel each other, and in which the spatial hole burning induced FM response is very small [32]. In most lasers, the spatial hole burning becomes weaker with increasing bias level and from the decreasing FM response in Fig. 7.12 it can be concluded that this is also the case here.

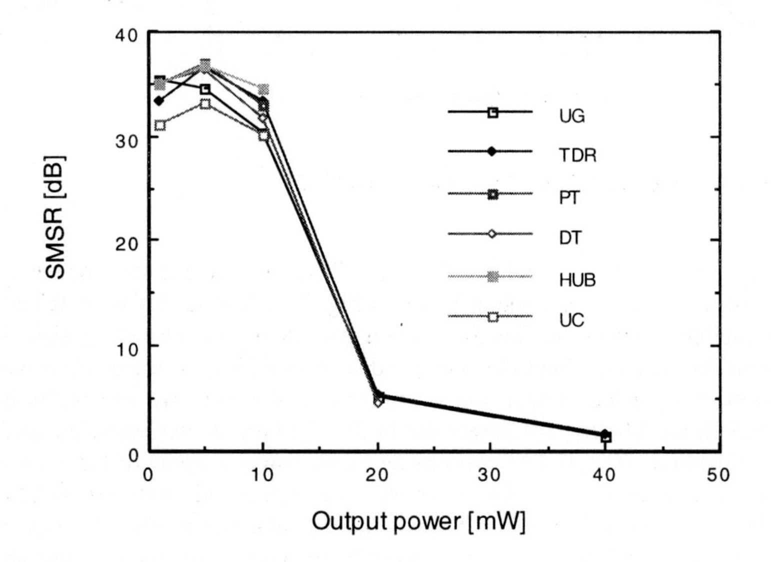

Fig. 7.11. Side mode suppression for the bulk laser with L=300 μm, $\kappa_i L$=3.

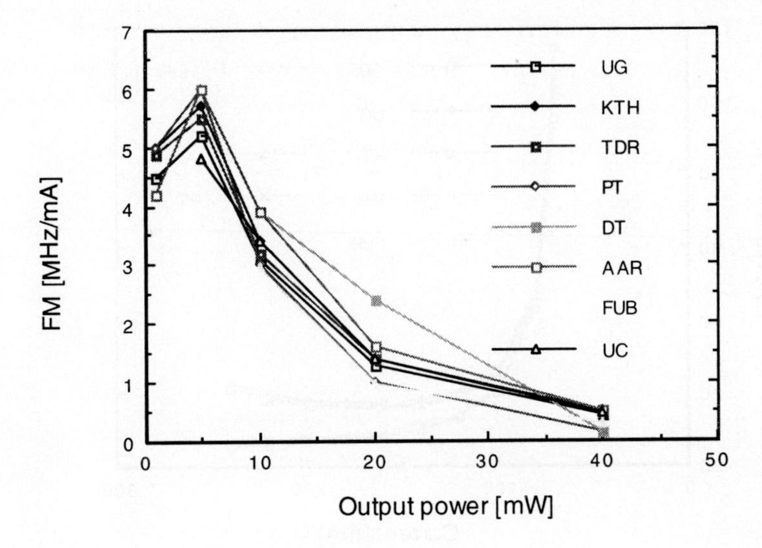

Fig. 7.12. FM response for the MQW laser with L=300 μm, $\kappa_i L$=1.

We finally report on the results for a 1mm long, bulk λ/4-shifted laser with $\kappa_i L$=1; a laser that becomes unstable at high power levels [30]. For this particular laser, a pitchfork bifurcation occurs at 240 mA, where the stable mode splits up into 3 unstable modes, one with symmetric and two with asymmetric longitudinal distributions. While the symmetric solution becomes inherently unstable, the two antisymmetric solutions would be stable if there was no mode competition. However, the large asymmetric perturbation of the cavity will reduce the gain margin and cause the laser to become single-mode. The linewidth results for this laser, which are shown vs. the injected current in Fig. 7.13, are particularly interesting. The linewidth rebroadening which accompanies the instability is found only with the models of UG, PT and TDR, i.e. if small signal fluctuations in the longitudinal mode profile are taken into account. This result is the only one among the results compared within COST 240 where a significant influence of the mode profile fluctuations is seen.

Fig. 7.14 shows the FM-response vs. the injected current. Here, the FM response and thus the spatial hole burning does not decrease with bias level. On the contrary, after a first strong increase with bias level, the FM remains quite constant as a function of the output power and starts to increase again for currents around 240 mA. It also should be noted that the results obtained for the FM response and for the linewidth around 240 mA are strongly dependent on the numerical accuracy obtained in the calculations. E.g. any inaccuracy in the static solution may cause a large inaccuracy in the dynamic, small signal result. Because of the instability of the system, any deviation in the static result is strongly amplified in the small signal result.

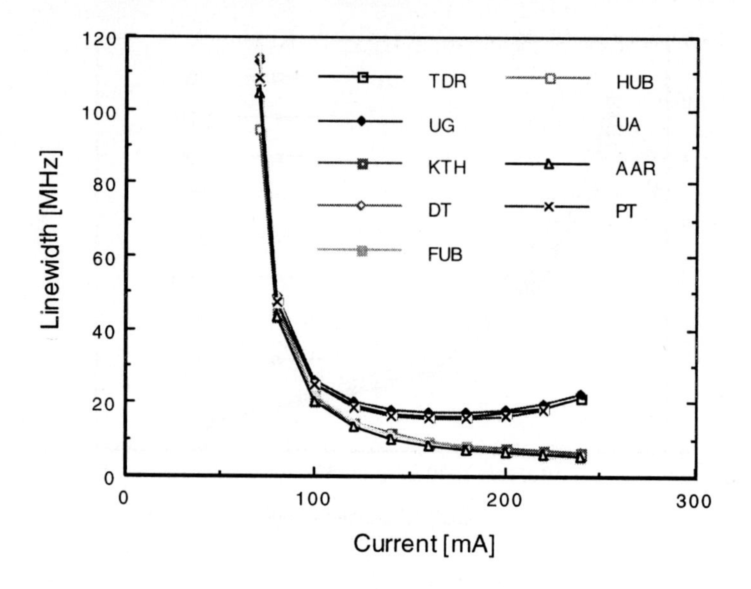

Fig. 7.13. $\Delta\nu$-I characteristic of the bulk laser with L=1 mm, $\kappa_i L$=1.

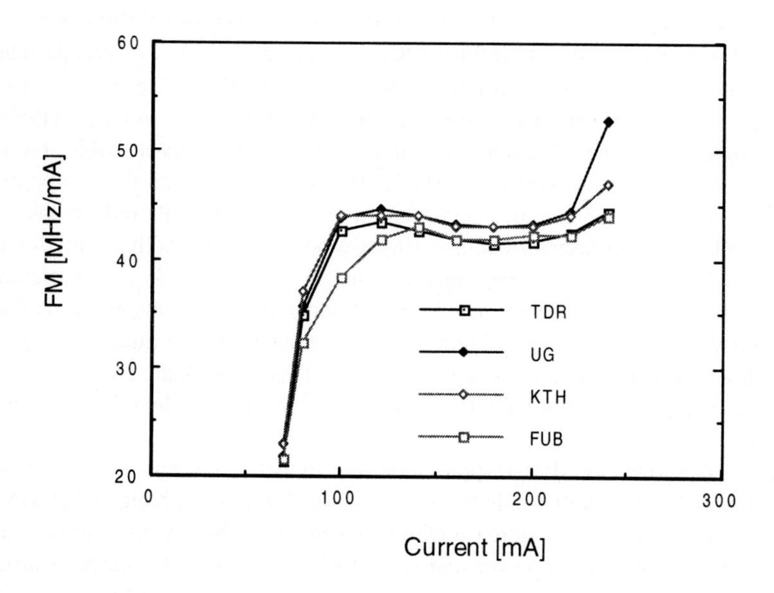

Fig. 7.14. FM-I characteristic of the bulk laser with L=1 mm, $\kappa_i L$=1.

7.2.3 DFB lasers with cleaved facets

As an example of a laser with cleaved facets we will now discuss a 300 μm long MQW laser with uniform grating with period 244 nm and $\kappa_i L$=1.5. For the exact definition of the grating and the facet reflectivities, we have introduced reference planes. The reference plane at the left side (corresponding with z=0) is chosen at the point nearest to the actual facet and where the grating has a phase zero, while the reference plane at the right side corresponds with the point with phase zero nearest to and before z=L=300μm. The grating phase at the right reference plane is given by 2πL/Λ as shown in Fig. 7.15. Our introduction of reference planes means that the reflection coefficients at z=0 and z=L include some phase delay (corresponding with forth and back propagation from reference planes to actual facets). We therefore assume that the facets have complex field reflection coefficients and for our particular choice they are $\rho_1 = j\,0.5657$ and $\rho_2 = 0.5657$.
Table **7.2** displays the threshold characteristics (ΔgL denotes the normalised threshold gain difference between main and side mode [21]).
For the emission wavelength one group (TDR, KTH, PT) finds 1.56093 μm and another group (UG, DT, AAR) finds 1.56084 μm. It is believed that the difference has its origin in two different expressions that are used for the propagation constant (or the effective index):

$$\begin{aligned} \beta &= \frac{\omega_B}{c} n_{eff\,0} + \frac{n_g}{c}(\omega - \omega_B) + \frac{\omega}{c}\Gamma\frac{dn}{dN}N \\ \beta &= \frac{\omega}{c}\left(n_{eff\,0} + \Gamma\frac{dn}{dN}N \right) \end{aligned} \tag{7.2}$$

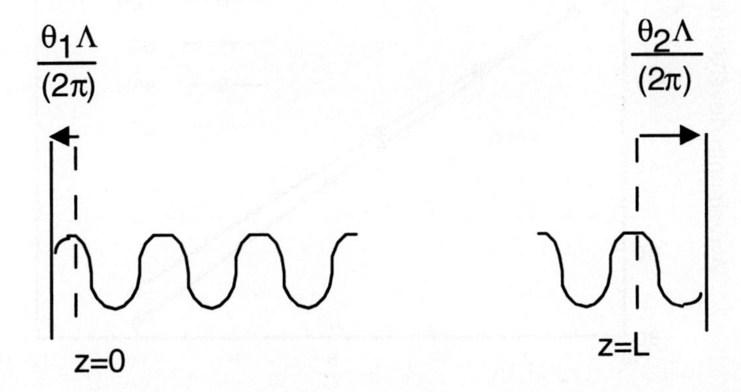

Fig. 7.15. Definition of the grating phases at the facets (with grating phases $\theta_1 = -\pi/2$ and $\theta_2 = \pi$).

	I_{th} [mA]	ΔgL	Wavelength [μm]
TDR	10.39	0.121	1.5609
UG	10.39	0.118	1.5608
UC	10.26	0.123	1.5578
KTH	10.39	0.122	1.5609
DT	10.38	0.124	1.5608
HUB	10.47	0.121	1.5609
AAR	10.37	0.124	1.5608
FUB	10.38	0.124	1.5529
PT	10.39	0.120	1.5581
UA	10.36	0.123	1.5576

Table 7.2. Threshold characteristics of the as-cleaved DFB laser.

The first expression includes the dispersion of the effective index explicitly through the group index, while the second expression assumes that this dispersion is included in a wavelength dependent n_{eff0}. The two expressions are only identical if the group index equals the effective index or for emission at the Bragg frequency (as in λ/4-shifted lasers). The inconsistency lies in the fact that a constant effective index was given and at the same time a group index different from the effective index was used. Fig. 7.16 shows the deviation of the wavelength from its threshold value for this laser, with and without gain suppression taken into account. A value $\varepsilon = 6.36\ 10^{-17}\text{cm}^3$ (6 W^{-1}) has been assumed for the gain suppression coefficient.

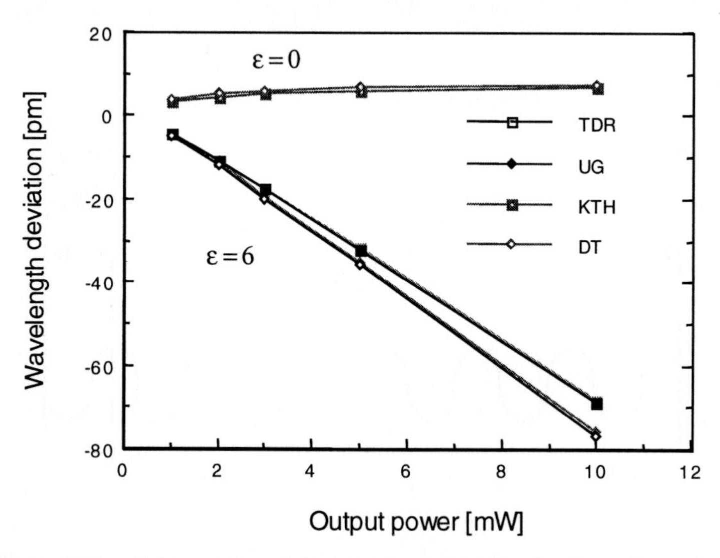

Fig. 7.16. Deviation of the wavelength from its threshold value for the as-cleaved laser with L=300 μm, $\kappa_i L$=1.5.

It can be noticed that the wavelength variation due to spatial hole burning (i.e. with $\varepsilon = 0$) is a lot smaller than in bulk $\lambda/4$-shifted lasers (cf. Fig. 7.9). The spatial hole burning in lasers with cleaved facets, without phase shift and rather small $\kappa_i L$-value is generally weaker than in $\lambda/4$-shifted lasers of which the longitudinal non-uniformity of the optical power is well-known.

Not only static or low frequency characteristics have been modelled for this laser, but FM- and IM-response and spectral density of FM-noise and RIN have been calculated as function of the frequency over a range of 10 GHz. We will show a few examples in which gain suppression has been taken into account. Only a limited number of models were suitable for this analysis, but the agreement between the models is very good. Fig. 7.17 shows the intensity modulation (IM) vs. the modulation frequency at an output power of 3 mW. Fig. 7.18 shows the FM-response and Fig. 7.19 the relative intensity noise spectrum, both for two values of the output power: 1 and 5 mW. The slightly larger IM-response obtained by FUB can be explained by the fact that the carrier recombination is linearised around the threshold value in the model of FUB. As the carrier density increases with increasing power (due to gain suppression), a nonlinear recombination gives a shorter carrier lifetime (and smaller IM-response) at 5 mW than a linear recombination. However, the otherwise excellent agreement between the results of TDR, DT and UG (obtained using a dynamically variable longitudinal power distribution) and the results of KTH and FUB (obtained using a constant amplitude envelope approximation) indicates that variations in the power distribution are again not important for the studied lasers.

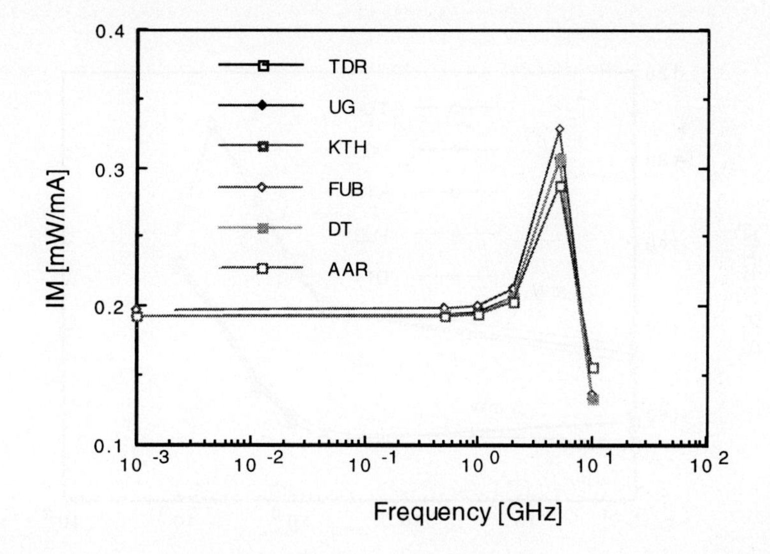

Fig. 7.17. Intensity modulation response vs. modulation frequency for the as-cleaved laser with L=300 μm, $\kappa_i L$=1.5 at 3 mW.

Also a laser identical to the one considered above, but in which a loss grating has been introduced, has been analysed. This loss grating gives rise to extra gain coupling κ_g, but also to an extra absorption loss α_{extra}.

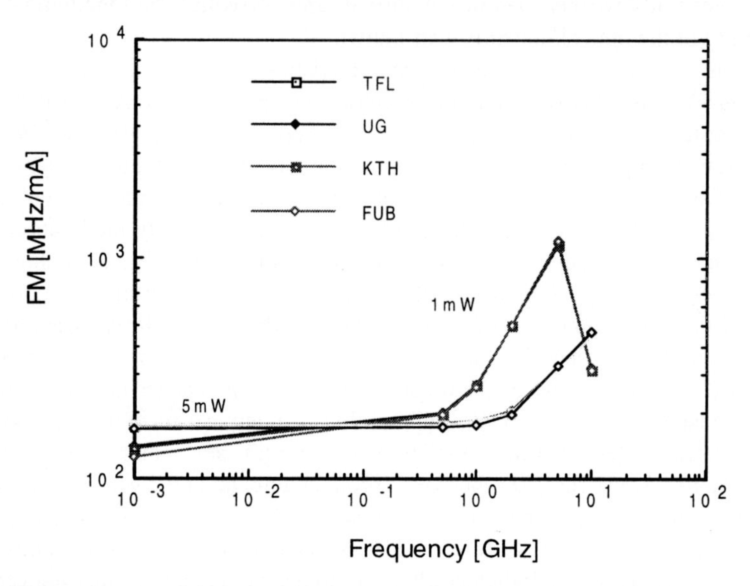

Fig. 7.18. Frequency modulation response vs. modulation frequency for the as-cleaved laser with L=300 µm, $\kappa_i L$=1.5.

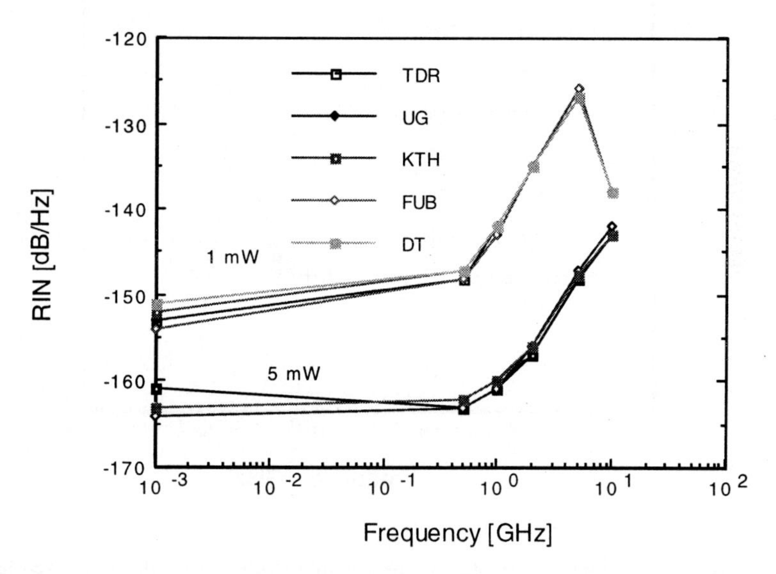

Fig. 7.19. Spectrum of the relative intensity noise for the as-cleaved laser with L=300 µm, $\kappa_i L$=1.5 at 1 and 5 mW.

We assumed a gain coupling coefficient of 10 cm^{-1} and extra losses equal to $2k_g$. This value of the extra loss is the minimum value corresponding to a certain k_g, it is obtained for e.g. a rectangular grating with very small duty cycle. Fig. 7.20 shows results obtained by different groups (UG, KTH and DT) for the linewidth vs. inverse output power.

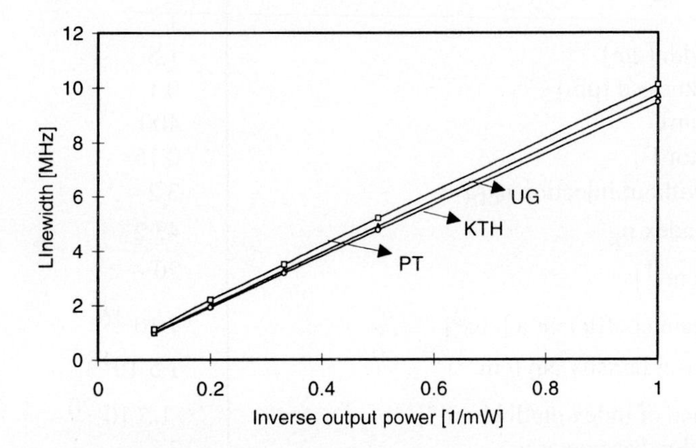

Fig. 7.20. Linewidth vs. inverse output power for the as-cleaved loss coupled laser with L=300 μm, $\kappa_g L$=1.5.

7.2.4 Large signal dynamic behaviour of a λ/4-shifted laser

To compare the large signal, time domain simulation capabilities, the switch-on behaviour of a 400 μm long, AR-coated λ/4-shifted DFB laser with parameters as given in

Table **7.3** has been considered.

Fig. 7.21 shows the results obtained for the time dependence of the optical output power by the University of Melbourne (UM) using the Transmission-Line Laser Model, the University of Cambridge (UC), Deutsche Telekom (DT), the Royal Institute of Technology of Stockholm (KTH) and the University of Gent (UG). The total modelled time has been limited to 1 ns after the current step to allow easier comparison. One can see considerable differences between the results; differences which probably have their origin in the different time steps used by the groups. Accurate modelling of the transient behaviour requires that a sufficiently small time step is used in the calculations.

Parameter	Value
Facet reflectivities R_1, R_2	0
Grating period Λ [nm]	241.5
Normalised coupling strength $\kappa_i L$	2
Bragg order M	1
Active region width [μm]	1.5
Active layer thickness d [μm]	0.1
Laser length L [μm]	400
Confinement factor Γ	0.15
Effective index without injection n_{eff0}	3.2
Effective group index n_g	43.2
Internal loss α_i [cm^{-1}]	20
Linear material gain coefficient a [cm^2]	$7\ 10^{-16}$
Transparency carrier density N_0 [cm^{-3}]	$1.5\ 10^{18}$
Carrier dependence of index dn/dN [cm^3]	$-1.5\ 10^{-20}$
Population inversion parameter n_{sp}	2
Linear recombination coefficient A [s^{-1}]	10^9
Bimolecular recombination coefficient B [$10^{-10}\ cm^3/s$]	1
Auger recombination coefficient C [$10^{-29}\ cm^6/s$]	7.5
Spontaneous emission coefficient β_{sp}	$5\ 10^{-5}$

Table 7.3. Parameters of the $\lambda/4$-shifted laser.

One reason for the apparent delay in the UM result compared with the other result is the use of a stochastic model by UM. The pulses in the stochastic model build up from random spontaneous emission noise. Thus the timing of the pulse is random, as a spontaneous emission event at the appropriate frequency has to occur for a pulse to be initiated. This randomness always introduces an additional delay in the pulse build-up because there has to be enough spontaneous emission at the lasing frequency before a pulse will grow. This contrasts to deterministic models which assume that appropriate spontaneous emission is always present. The average effect of stochastic spontaneous emission is therefore to delay the pulsation. The results presented in Fig. 7.21 by the UM are a box-car average of many transients, and do indeed show a delay compared with other models.

The damping of the transients in Fig. 7.21 is unrealistically small, especially for DFB lasers. This is because a low value of gain nonlinearity was used in order to highlight the pulsation, allowing for more accurate comparison between pulse widths, delays and damping ratios.

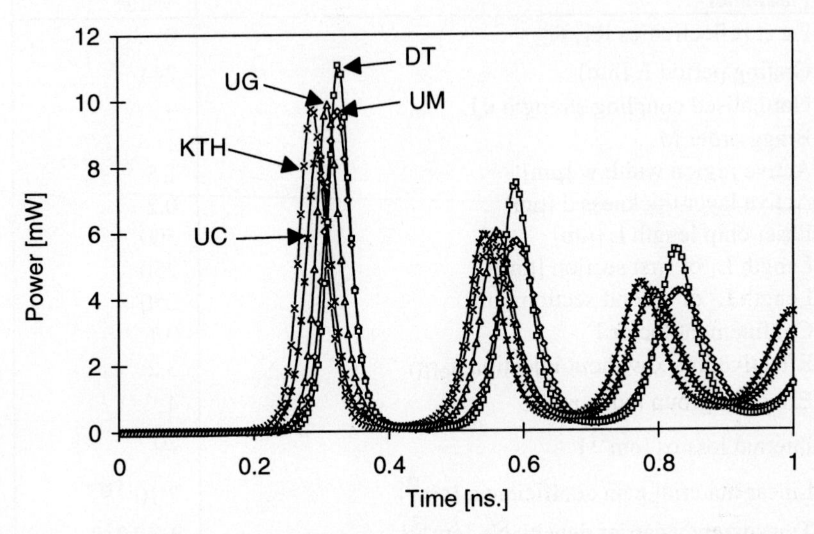

Fig. 7.21. Output power vs. time during switch-on of a 400 μm long, AR-coated λ/4-shifted DFB laser. (diamonds: UM, squares DT, triangles: UG, crosses: KTH, bold crosses: UC).

7.2.5 Self pulsations of a multi-electrode laser

The non-uniform biasing of a two-electrode laser may in some cases lead to self-pulsating behaviour [24]. In fact, it has been found that self-pulsations occur if an increase in the carrier density of one section results in a decrease of the fraction of the optical power in that particular section and in a decrease of the average threshold carrier density. The self-pulsation frequency thereby increases with increasing negative slope of the average threshold carrier density and of the optical power fraction versus the asymmetry in the carrier density.

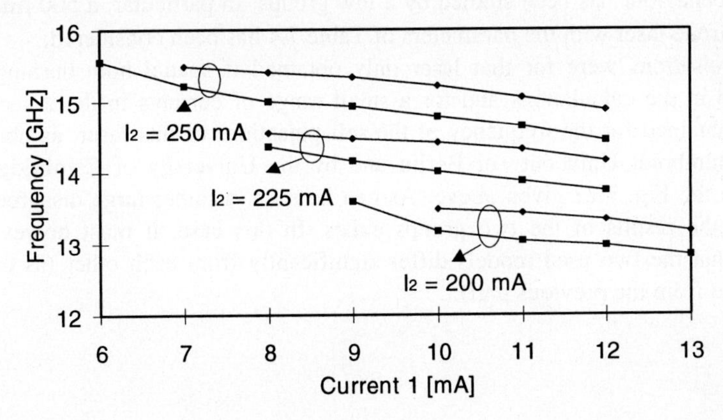

Fig. 7.22. Frequency of the self-pulsations vs. current in the 1st electrode for a self-pulsating 2-section, 500 μm long DFB laser. (■: UC, ♦: HUB).

Parameter	Value
Facet reflectivities R_1, R_2	0
Grating period Λ [nm]	244
Normalised coupling strength $\kappa_i L$	4
Bragg order M	1
Active region width w [μm]	1.5
Active layer thickness d [μm]	0.2
Laser chip length L [μm]	500
Length L_1 of first section [μm]	250
Length L_2 of second section [μm]	250
Confinement factor Γ	0.1
Effective index without injection n_{eff0}	3.2
Effective group index n_g	4.
Internal loss α_i [cm^{-1}]	20
Linear material gain coefficient a [cm^2]	$7\ 10^{-16}$
Transparency carrier density N_0 [cm^{-3}]	$1.8\ 10^{18}$
Carrier dependence of index dn/dN [cm^3]	$-4.371\ 10^{-20}$
Population inversion parameter n_{sp}	2
Linear recombination coefficient A [s^{-1}]	10^9
Bimolecular recombination coefficient B [$10^{-10}\ cm^3/s$]	0
Auger recombination coefficient C [$10^{-29}\ cm^6/s$]	0
Spontaneous emission coefficient β_{sp}	10^{-4}

Table 7.4. Parameters of the self-pulsating laser.

Such behaviour has been studied by a few groups. In particular, a 500 μm long two-electrode laser with the parameters of Table 7.4 has been considered.

Self-pulsations were for that laser only obtained if spatial hole burning was neglected in the calculations and for a small range of currents in the 2 sections. Results obtained for the frequency of the self-pulsations of that laser, as obtained by the Humboldt University of Berlin and by the University of Cambridge are shown in the Fig. 7.22 given above. As can be seen, a rather large disagreement between the results of the two groups exists. In this case, it must however be noticed that the two used models differ significantly from each other (as can be concluded from the previous part).

References

[1] J.Buus, "Principles of semiconductor laser modelling", *IEE Proc. J*, vol. 132, pp. 42-51, 1985.

[2] R.S.Tucker, "High-speed modulation of semiconductor lasers", *J. Lightwave Technol.*, vol. LT-3, pp. 1180-1192, 1985.

[3] S.Sasaki, M.M.Choy, and N.K.Cheung, "Effects of dynamic spectral behaviour and mode-partitioning of 1550 nm distributed feedback lasers on Gbit/s transmission systems", *Electron. Lett.*, vol. 24, pp. 26-28, 1988.

[4] D.Novak and R.S.Tucker, "Millimetre-wave signal generation using pulsed semiconductor lasers", *Electron. Lett.*, vol. 30, pp. 1430-1431, 1994.

[5] P.M.Boers, M.T.Vlaardingerbroek, and M.Danielsen, "Dynamic behaviour of semiconductor lasers", *Electron.Lett.*, vol. 11, pp. 206-208, 1975.

[6] H.Statz and G.de Mars, "Transient and oscillation pulses in masers", pp. 530-537 in C.H. Townes Ed., *Quantum Electronics*, Columbia Press, New York, 1960.

[7] M. Osinski and M.J.Adams, "Transient time-averaged spectra of rapidly-modulated semiconductor lasers", *IEE Proc. J*, vol. 132, pp. 34-37, 1985.

[8] R. Schatz, "Longitudinal spatial instability in symmetric semiconductor lasers due to spatial hole burning", *IEEE J. Quantum Electron.*, vol. 28, pp. 1443-1449, 1992.

[9] J-I. Kinoshita and K. Matsumoto, "Transient chirping in distributed feedback lasers: effect of spatial hole-burning along the laser axis" , *IEEE J.Quantum Electron.*, vol. 24, pp. 2160-2169, 1988.

[10] M.J.Adams and M.Osinski, "Longitudinal mode competition in semiconductor lasers. Rate equations revisited", *IEE Proc. I*, vol. 129, pp. 271-274, 1982.

[11] M.Yamada, "Theory of mode competition noise in semiconductor injection lasers", *IEEE J. Quantum Electron.*, vol. 22, pp. 1052-1059, 1986.

[12] R. S. Tucker, U. Koren, G. Raybon, C. A. Burrus, B. I. Miller, T. L. Koch and G. Eisenstein, "40 GHz active mode-locking in a 1.5 μm monolithic extended-cavity laser," *Electron Lett.*, vol. 25, pp. 621-622, 1989.

[13] M.S.Demokan, "A model of a diode laser actively mode-locked by gain modulation", *Int.J.Electron.,*vol. 60, pp. 67-80, 1986.

[14] K.Hsu, C.M.Verber, and R.Roy, "Pulse fluctuation statistics of an actively mode-locked external-cavity semiconductor laser", *Appl.Phys.Lett.,* vol. 60, pp.307-309, 1992.

[15] A.J.Lowery, "A new dynamic semiconductor laser model based on the transmission-line modelling method", *IEE Proc. J,* vol. 134, pp. 281-289, 1987.

[16] A.J.Lowery, C.N.Murtonen, and A.J.Keating, "Modelling the static and dynamic behavior of quarter-wave-shifted DFB lasers", *IEEE J. Quantum Electron.*, vol. 28, pp. 1874-1883, 1992

[17] A.J.Lowery, "New time-domain model for active mode-locking based on the transmission-line laser model", *IEE Proc. J*, vol. 136, pp. 264-272, 1989.

[18] J.E.A.Whiteaway, A.P.Wright, B.Garrett, G.H.B.Thompson, *et al.*, "Detailed large-signal dymanic modelling of DFB laser structures and comparison with experiment", *Opt. Quantum Electron.*, vol. 26, S817-S842, 1994.

[19] G. Bjork and O. Nilsson, "A new exact and efficient numerical matrix theory of complicated laser structures: Properties of asymmetric phase-shifted DFB lasers", *J. Lightwave Technol.*, vol. 5, pp. 140-146, 1987.

[20] H. Bissessur, "Effects of hole burning, carrier-induced losses and the carrier-dependent differential gain on the static characteristics of DFB lasers", *J. Lightwave Technol*, vol. 11, pp. 1617-1630, 1992.

[21] R. Bonello, I. Montrosset, "Statistical and Dynamical Analysis of Multisection and Multielectrode Semiconductor Lasers", *SPIE* vol. 1787, pp. 151-163, 1992.

[22] P.Vankwikelberge, G. Morthier, R. Baets., "CLADISS - A longitudinal multimode model for the analysis of the static, dynamic, & stochastic behaviour of diode lasers with distributed feedback", *IEEE J. Quantum Electron.*, vol. 26, pp. 1728-41, 1990.

[23] I. Orfanos, T. Sphicopoulos, A. Tsigopoulos, C. Caroubalos, "A Tractable Above-Threshold Model for the Design of DFB and Phase-Shifted DFB Lasers", *IEEE J. Quantum Electron.*, vol. 27, pp. 946-956, 1991.

[24] J.E.Whiteaway, G.H.B.Thompson, A.J.Collar, and C.J. Armistead, "The design and assessment of λ/4 phase-shifted DFB lasers", *IEEE J. Quantum Electron.,* 1989, vol. 25, pp. 1261-1279, 1989.

[25] S.Hansmann, "Transfer matrix analysis of the spectral properties of complex distributed feedback laser structures", *IEEE J. Quantum Electron.*, vol. 28, pp. 2589-2595, 1992.

[26] M.G.Davies and R.F.O'Dowd, "A transfer matrix method based large-signal dynamic model for multielectrode DFB lasers", *IEEE J. Quantum Electron.*, vol. 30, pp. 2458-2466, 1994.

[27] G.Morthier, R. Baets *et al.* (COST 240 Group), "Comparison of different DFB laser models within the European COST-240 collaboration", *IEE Proc. Optoelectron.*, vol. 141, pp. 82-88, 1994.

[28] U. Bandelow, R. Schatz, and H. J. Wunsche, "A correct single-mode photon rate equation for multi-section lasers", *IEEE Photon. Technol. Lett.,* vol. 8, no. 5, pp. 614-616, 1996.

[29] R. Schatz,"Dynamics of Spatial Hole Burning Effects in DFB Lasers", *IEEE J. Quantum Electron.*, vol. 31, no. 11, pp. 1981-1993, 1995

[30] H.Olesen, B.Tromborg, X.Pan and H.E.Lassen, "Stabilities and dynamic properties of multi-electrode laser diodes using a Green's function approach", *IEEE J. Quantum Electron.*, vol. 29, pp. 2282-2301, 1993.

[31] C.F.Tsang, D.D.Marcenac, J.E.Carroll and L.M.Zhang, "Comparison between 'power matrix model' and 'time domain model' in modelling large signal responses of DFB lasers", *IEE Proc. Optoelectron.*, vol. 141, pp. 89-96, 1994.

[32] A.J.Lowery, "Modelling ultra-short pulses (less than the cavity transit time) in semiconductor laser amplifiers", *Int J. Optoelectron.*, vol. 3, pp. 497-508, 1988.

[33] D.D.Marcenac and J.E.Carroll, "Quantum-mechanical model for realistic Fabry-Perot lasers", *IEE Proc. J*, vol. 140, pp. 157-171, 1993.

[34] U. Bandelow, H. J. Wunsche, H. Wenzel, "Theory of Selfpulsations in Two-Section DFB Lasers", IEEE Photon. Technol. Lett., vol. 5, pp. 1176-1179, 1993.

[35] X. Li, W.-P. Huang, "Simulation of DFB Semiconductor Lasers Incorporating Thermal Effects", *IEEE J. Quantum Electron.,* vol. 31, pp. 1848-1855, 1995.

[36] A. D. Sadovnikov, W.-P. Huang, "A Two-Dimensional DFB Laser Model Accounting for Carrier Transport Effects", *IEEE J. Quantum Electron.,* vol. 31, pp. 1856-1862, 1995.

[37] K. Yokoyama, T Yamanaka, S. Seki, "Two-Dimensional Numerical Simulator for Multielectrode Distributed Feedback Laser Diodes", *IEEE J. Quantum Electron.,* vol. 29, pp. 856-863, 1993.

[38] S. F. Yu, R.G.S. Plumb, L.M. Zhang, M.C. Nowell, J.E. Carroll, "Large Signal Dynamic Behaviour of Distributed Feedback Lasers including Lateral Effects", *IEEE J. Quantum Electron.,* vol. 30, pp. 1740-1750, 1994.

[39] J. Kinoshita, "Modeling of high-speed DFB lasers considering the spatial holeburning effect using three rate equations", *IEEE J. Quantum Electron.,* vol. 30, pp. 929-938, 1994.

[40] G. Morthier, "An accurate rate equation description for DFB lasers taking spatial hole burning into account", *IEEE J. Quantum Electron.,* vol. 33, pp. 231-237, 1997.

[41] W. Huang, X. Li, T. Makino, "Analytical formulas for modulation responses of semiconductor DFB lasers", *IEEE J. Quantum Electron.,* vol. 31, pp. 842-851, 1995.

[42] Y. C. Chan, M. Premaratne, and A. J. Lowery, "Semiconductor laser linewidth from the transmission-line laser model", IEE Proc. Optoelectron., vol. 144, pp. 246-252, 1997.

8 Measurements on DFB lasers

R. Paoletti, P. Spano

The measurement of the standard laser performance is generally assumed to be quite easy and to require no particular care. The case of semiconductor lasers is, however, more complicated. Under standard working conditions, semiconductor lasers have a material gain coefficient much higher than the other lasers and a dependence of the refractive index on the gain. These features, together with the low Q factor of a typical semiconductor-laser cavity make these devices very sensitive to external reflections. This reflects on an intrinsic difficulty in performing reliable measurements.

In the European collaboration framework project COST 240, a set of inter-laboratory measurements has been performed in the nineties on different lasers offered by several European manufacturers, namely CNET (France), CSELT (Italy), GEC Marconi (United Kingdom), IMC (Sweden), KTH (Sweden), Nortel (United Kingdom), BNR Europe (United Kingdom), HP (United Kingdom). These lasers have various structures. Quantum well (QW) and bulk, single and multisection, DFB and DBR lasers with both homogeneous and λ-shifted gratings have been investigated. The objective of these measurements was threefold:

- to find, through the comparison of the results obtained using different experimental set-ups and measurement techniques, the critical points that affect the results of the measurements on semiconductor lasers;
- to compare the performance of different lasers;
- to supply a series of data for the modelling exercise, reported in the previous section.

Twenty-two laboratories have been involved. They are:

University of Gent (Belgium), EPFL (Switzerland), ETHZ (Switzerland), Tele Denmark Research (Denmark), CNET (France), Alcatel Alsthom Research (France), Deutsche Telekom (Germany), University of Athens (Greece), University of Budapest (Hungary), CSELT (Italy), FUB (Italy), Politecnico of Turin (Italy), Optronics Ireland (Ireland), University of Porto (Portugal), IMC (Sweden), KTH (Sweden), Nortel, BNR Europe (United Kingdom), HP (United Kingdom), GEC Marconi (United Kingdom), University of Glasgow (United Kingdom), University of Melbourne (Australia).

The measurements covered both the most simple external laser parameters, like emitted power, emission wavelength, side mode suppression ratio (SMSR) and emission spectrum, and the most tricky parameters, like frequency modulation efficiency, relative intensity noise (RIN), frequency of relaxation oscillations, emission linewidth and modulation bandwidth. The analysis of a very large number of results enables to state some general criteria that must be followed to perform reliable measurements of the semiconductor laser parameters.

In the following the results will be reported with some comments that should help those who intend to perform similar measurements on semiconductor lasers.

8.1. Basic measurements

Prior to a discussion on how to perform the measurements of the laser parameters one should underline that a very good optical isolation between the laser and the measurement set-up is always needed to avoid perturbation due to uncontrolled back-reflections from any surface of the measurement set-up. The effect of a given reflection depends on the distance of the reflecting surface [1] (in general, reflections coming from a long distance are more annoying), on the characteristics of the laser under test ("κL" value of the DFB, presence of reflecting facets...) and on its operating conditions (injection current level...). It is commonly recognised that even very low level reflections (-60/-70 dB) can induce some perturbation in the intensity and noise spectrum. Therefore, an optical isolator should be inserted as close as possible to the output facet of the laser, to isolate the laser from as many as possible reflecting surfaces. The level of required isolation depends on several factors; yet, a single stage commercial isolator (usually 30-40 dB isolation), is not enough in many cases, and a double stage isolator (70-75 dB) is strongly recommended. A critical parameter of the optical isolators is the residual reflection of its input facet that, in spite of AR coating, is usually not better than 30 dB (for normal incidence). This means that, if the isolator operates on a collimated beam, it should be slightly tilted (a few degrees, whereas larger values would affect isolation) to misalign the reflected beam and increase the return loss. If the isolator is pigtailed this factor is not controllable; yet, commercial pigtailed isolators claim to have return loss around +60 dB. In the case of pigtailed isolators, direct coupling through the plane-ended pigtail should be avoided; a tapered and rounded fibre end with an AR coating is recommended, which results in a strong reduction of back-coupling into the laser, due to the larger angular spread of the reflections.

When the laser beam is collimated by means of a lens, particular care must be devoted to reflections by the lens itself. Large focal length (and hence working distance) lenses could be used in principle to reduce the amount of reflected light that is coupled back into the laser; yet, due to the correspondingly large beam diameter, this would imply a large aperture of the isolator downstream the lens. Unfortunately, as commercial isolators usually have apertures smaller than 4 mm, only small size lenses can be used.

To check the degree of reflections from the set-up, one should: a) check by a Fabry-Perot Analyser the presence in the spectrum of induced resonance frequencies; b) once these large effects are eliminated, the residual presence of low level reflections can be identified by sweeping the laser current and hence its wavelength. If the laser "breaths", exhibiting periodical fluctuations of the displayed line height, this reveals some perturbing reflection. Yet some short-distance reflections (as from the front facet of a GRIN lens, placed a few hundred μm away) can not be easily identified, since a very large wavelength change would be necessary to shift from an in-phase to the following out-of-phase condition, that is impossible to obtain with realistic current change.

Even the most basic measurements, like the characteristic power/current, can comprise error sources. In Fig. 8.1a and Fig. 8.1b the characteristics P-I are reported for two different lasers.

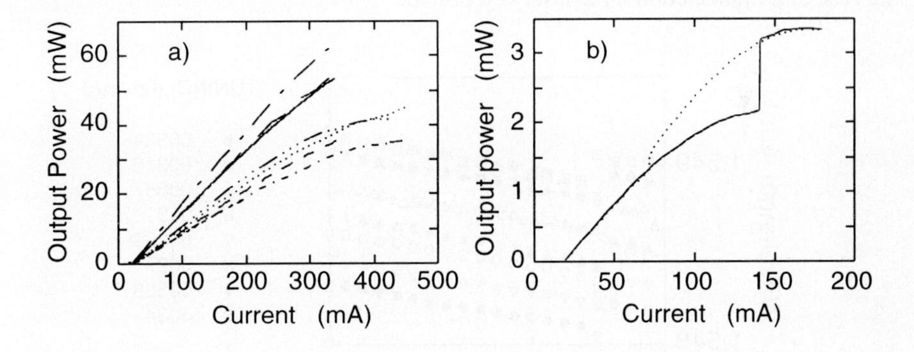

Fig. 8.1. Output power vs. current for: **a)** a DFB laser with a double λ/8 phase shift and an antireflection coated output facet (measurements in different laboratories); **b)** a DBR laser in which tuning and Bragg sections are shorted. In Fig. 8.1b the dotted line has been obtained by increasing the current, the continuous line by decreasing it.

Fig. 8.1a refers to a laser having a double λ/8 phase shift and an antireflection coated output facet. The different curves reported in the plot refer to the results obtained in different laboratories. The spread of the results corresponds to a difference as large as 100% between the upper and the lower curve. This large spread is due to a not optimised temperature control and to the only partial light collection in some cases. The result of this measurement points out that good thermal control and temperature calibration must be assured, as well as a reliable estimate of the amount of power collected by the detector. In the case of lasers, very sensitive to back-reflections from the measurement apparatus, the reflections can represent a further origin of uncertainty. The effect of back-reflections was very evident in two section DFB lasers with as cleaved facets, which are very sensitive to external reflections. The high sensitivity to reflections of these lasers could be ascribed to the presence of the collimating lens in the laser mount.

Fig. 8.1b reports the P-I characteristics for a tuneable three section DBR laser. One can see the bistable behaviour as a function of the bias current for this laser, with a hysteresis loop. In this case the effect of any external disturbance on the characteristics is dramatic, so that a very good care to avoid back reflections must be applied.

The measurements of laser tunability offer in general a lower degree of spread, in particular for lasers having a grating phase shift and AR facet coating. This is not true, however, for the region around the mode jumps, if any, where the external reflections are able to trigger the mode jump. When performing these measurements, differences between the measured wavelengths up to one nanometer can be observed. This is caused by the inaccurate calibration of the spectrum analysers. As an example, in Fig. 8.2 a plot with the results obtained in the case of a multisection DFB laser is reported.

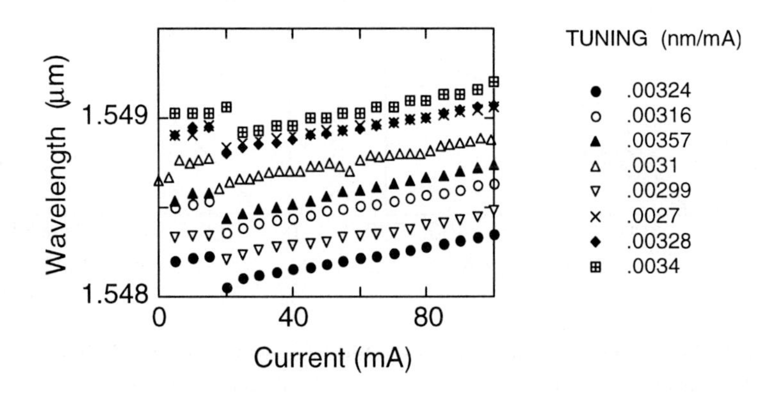

Fig. 8.2. Tuning vs. current in the forward section for a fixed bias current in the central and back section of a three section DFB laser.

The laser of Fig. 8.2 was a device fully packaged with the thermistor and the Peltier element in the laser box. In the case of chip on carrier devices the agreement of the results is much worse. In this case one must pay attention to assure a good thermal contact between the laser, the thermistor and the Peltier element, in order to have reproducible results.

The side mode suppression ratio (SMSR) is a quite simple measurement and the results of the different measurements performed in different laboratories agree quite well. Also in this case good care must be taken to avoid back-reflections, in particular in the case of lasers sensitive to external disturbance. When lasers with SMSR larger than about 30 dB are under test, the use of a spectrum analyser based on a double monochromator is mandatory. The results of the round robin exercise show that single monochromators do not allow to appreciate SMSR higher than about 35 dB. In Fig. 8.3 the examples of the measurement results of SMSR are reported for two different devices.

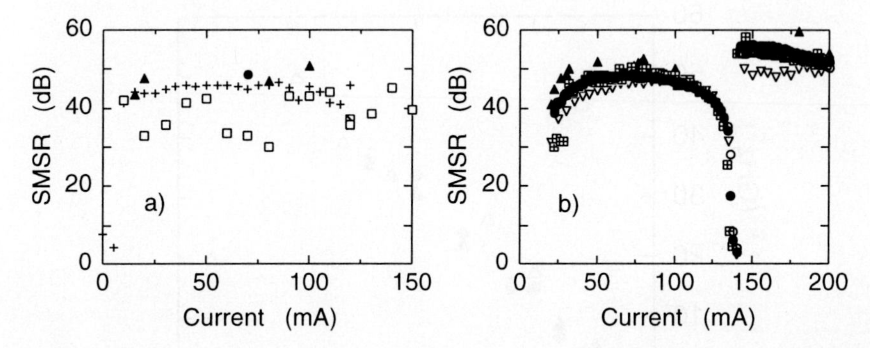

Fig. 8.3. Side mode suppression ratio vs. current in: **a**) a two section DFB laser with no AR coating on the facets; **b**) a three section DBR. In Fig. 8.3b the passive sections are shorted. Measurements in different laboratories.

Fig. 8.3a refers to a DFB laser without AR coating. The scatter of the results is a consequence of the sensitivity of the laser to back reflections. More reproducible results are obtained when measuring a DBR laser, as shown in Fig. 8.3b. In this case SMSR higher than 50 dB have been attained after the mode jump, which is present around 140 mA.

8.2. Emission linewidth and other more specific measurements

Surprisingly, the results from the various measurements performed on the more tricky laser parameters show an agreement that is, in general, much better than for the most basic quantities. An exception is the emission linewidth, as shown in detail later on. The basic reason of the better agreement has to be looked for in the care used to perform this kind of measurements. It is quite obvious that when performing a RIN measurement one must avoid, as much as possible, reflections, the effect of which can be immediately recognised by the generation of resonance in the spectrum at GHz frequencies. The same care is, in general, not taken in the case of the basic measurements.

In Fig. 8.4 a plot of the squared value of the laser resonance frequency f_o, obtained from the RIN spectrum is reported as a function of the total current in a three section DFB laser. The results obtained by the different groups agree quite well with each other. A good agreement is observed also for the result of the RIN spectrum. This is a quite complicated measurement which requires very low noise and high bandwidth electrical amplifiers and refined calibration techniques.

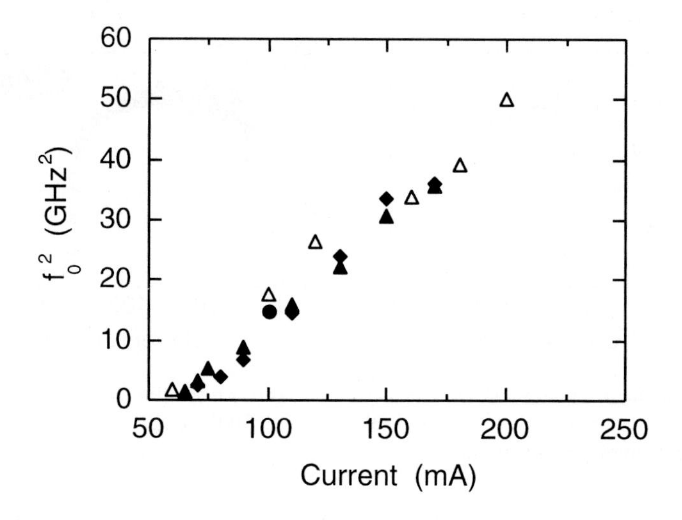

Fig. 8.4. Squared value of the resonance frequency, obtained from the RIN spectrum, vs. total current in a three section DFB laser. Measurements in different laboratories.

In spite of the good agreement of the results of the RIN measurements, the measurement of the linewidth gave very scattered results for all the tested lasers. In lasers with AR coating on the output facets one can in general recognise a trend of the linewidth as a function of the driving current, but with difference among the values obtained by different laboratories as large as two orders of magnitude. In lasers without AR coatings, on the contrary, it is not always obvious to recognise a clear trend common to the results of the various laboratories.

In Fig. 8.5, Fig. 8.6 and Fig. 8.7 the values of the emission linewidth vs. current, obtained by different laboratories, are reported for three different lasers, which represent a sample of the different possible results. We chose to present the results as a function of the current instead of the inverse emitted power, because of the inaccuracy in the evaluation of the emitted power.

In Fig. 8.5 the values obtained by measuring a single contact DFB laser having two $\lambda/8$ shifts in the grating, are reported. With the exception of one set of data, which will be discussed in the following, a maximum deviation at the highest currents of a factor three and a clear common trend can be observed in this case. A smaller deviation is obtained at lower bias currents. Fig. 8.6 refers to a three section tuneable DBR laser. The horizontal axis represents the current in the passive sections, the current in the active section is set to 120 mA. In this case a trend can still be recognised, but the variation of the measured linewidth values ranges from a factor 20 at the highest current to more than two orders of magnitude at low current. Finally in Fig. 8.7 the linewidth values obtained with a two section DFB laser without any grating shift and with as cleaved output facets are reported. Here the maximum deviation of the measured values is a factor 6 but no clear trend is found.

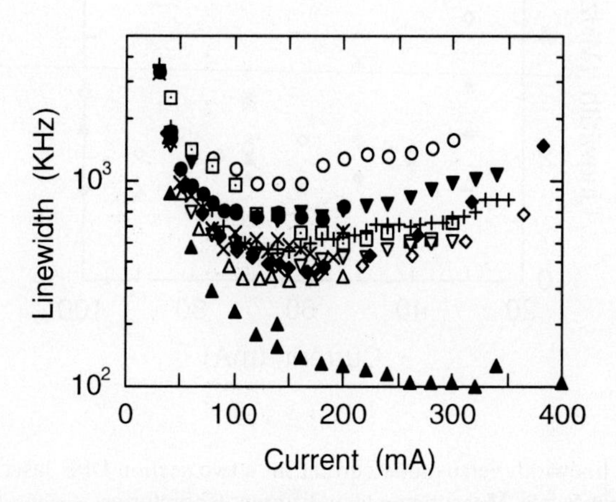

Fig. 8.5. Emission linewidth versus current for a DFB laser having a double λ/8 shift in the grating. Measurements in different laboratories.

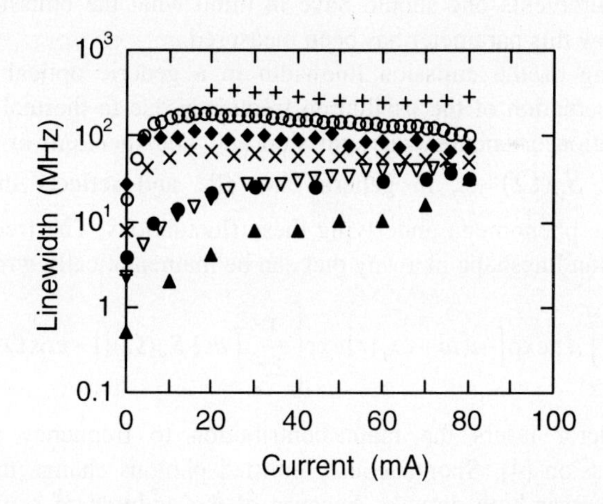

Fig. 8.6. Emission linewidth versus current in a three section DBR laser. Measurements in different laboratories.

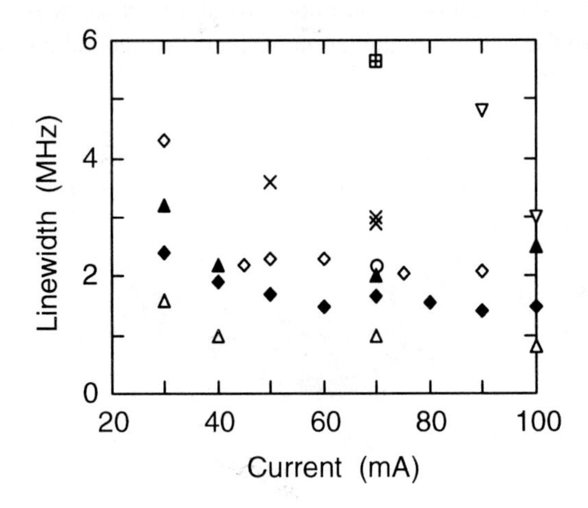

Fig. 8.7. Emission linewidth versus total current in a two section DFB laser having no AR coating on the output facets. Measurements in different laboratories.

To understand the reason for the large spread of the results obtained even in the best set of measurements one should have in mind what the emission linewidth represents and how this parameter has been measured.

The broadening of the emission linewidth in a generic optical oscillator is caused by the fluctuation of the oscillation frequency due to thermal instabilities, mechanical vibration or spontaneous emission [2]. The spectrum of the resulting frequency noise $S_\nu(\Omega)$ is, in general, not flat and reflects the frequency distribution of the phenomena underlying these fluctuations. The frequency noise affects the emission lineshape in a way that can be mathematically written as [3]:

$$S_E(\omega) = I_0 \int_{-\infty}^{\infty} d\tau \exp\left[-i(\omega - \omega_0)\tau\right] \exp\left[\frac{1}{2\tau} \int_{-\infty}^{\infty} d\Omega\, S_\nu(\Omega)\left(1 - \cos\Omega\tau\right)/\Omega^2\right] \quad (8.1)$$

In semiconductor lasers the main contribution to frequency noise is the spontaneous emission [4]. Spontaneously emitted photons change the oscillation frequency of the laser both directly, because of the addition of a photon with a random phase to the coherent field, and indirectly, because the variation of the carrier density following any emission event, modifies the refractive index of the cavity through the linewidth enhancement factor α [5]. This last contribution is, in general, the main contribution to frequency noise in semiconductor lasers. Its spectrum can be assumed to be flat up to frequencies of the order of the resonance frequency, typically some GHz or more. For an ideal flat frequency noise spectrum, eq. (8.1) gives that the emission line has a Lorentzian shape and its width $\Delta\nu$ is unambiguously defined as [6]:

$$\Delta\nu = S_{\nu}(0)/2\pi \tag{8.2}$$

In some cases, however, other phenomena give rise to a non flat contribution at low frequencies with a generic behaviour like $1/f^n$. When this contribution is present the lineshape is no longer Lorentzian and the linewidth cannot be defined [7], because it tends to diverge when one takes into account the low frequency part of $S_{\nu}(\Omega)$. In this case any attempt to measure the linewidth gives a number dependent on the particular method and the parameters used to perform the measurement.

The most common techniques used to perform this kind of measurement are the self-heterodyne [8] and self-homodyne methods. They have similar performance. A sketch of a self-homodyne apparatus is reported in Fig. 8.8.

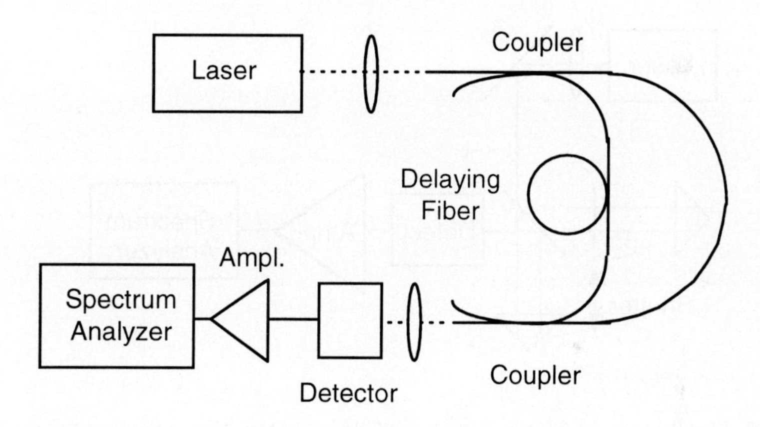

Fig. 8.8. Sketch of a self-homodyne set-up to measure the emission linewidth of a semiconductor laser.

Basically the laser field is superimposed to itself after a delay of some μs, and the beating is detected by a photodiode and displayed on an electrical spectrum analyser. In the hypothesis of a Lorentzian lineshape the beating lineshape is still Lorentzian with a full width at half maximum (FWHM) $\Delta\nu_b$ equal to $2\Delta\nu$ [8]. To apply this technique one needs to overlap two incoherent beams. This means that the delay time must be longer than the inverse of the linewidth. The use of delay fibres of the order of some tens of Km, permits to appreciate linewidth as narrow as 100 kHz. This technique, however, does not provide clear information if the emission lineshape is characterised by a $1/f^n$ contribution. In this case, as stated above, the emission linewidth is not defined and the value of $\Delta\nu_b$ increases as the time delay between the beating fields increases [9]. Really, looking at the results of the linewidth measurements one can find a good correlation between the broader linewidth and the longer delaying fibres. Moreover a fit of the beating lineshape revealed that it is not Lorentzian, but tends to approach a Gaussian curve, which is typical of a frequency noise spectrum characterised by $1/f$ noise contribution at low frequencies.

The lowest values of linewidth reported in Fig. 8.5, Fig. 8.6 and Fig. 8.7 have been obtained by a different technique, that is by the use of an unbalanced Michelson interferometer (MI). A sketch of this interferometer is reported in Fig. 8.9.

This approach exploits the coherence properties of the two beams in the two arms of the interferometer. If the delay between the two beams is much shorter than the coherence time, the interferometer transforms frequency noise in intensity noise, if it is set in quadrature [3].

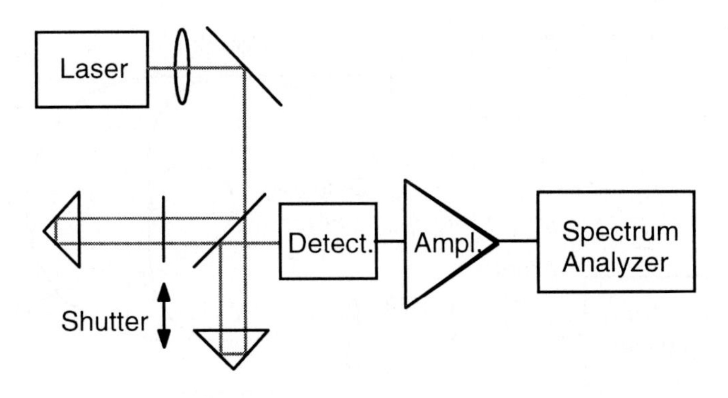

Fig. 8.9. Sketch of a set-up based on an unbalanced Michelson interferometer, used to measure the emission linewidth of a semiconductor laser.

The MI shown in Fig. 8.9 requires a value of isolation lower than that required by a self-homodyne or a self-heterodyne apparatus, because only a small amount of reflections from the apparatus is coupled back into the laser; it allows a careful measurement of the frequency noise spectrum and hence of the lineshape, but the resolution is limited to a few hundred of kHz, and the adjustment and calibration of the apparatus is quite involved. An example of the frequency noise spectrum obtained for the laser of Fig. 8.6, is reported in Fig. 8.10.

Fig. 8.10 was obtained with a zero bias to the passive sections and 100 mA to the active section. It is quite evident that a pronounced tail at low frequency is present. The value of the linewidth in this case cannot be defined. However a number can be still obtained if the noise below a certain frequency is omitted. This is what happens when the self-homodyne or self-heterodyne techniques are used.

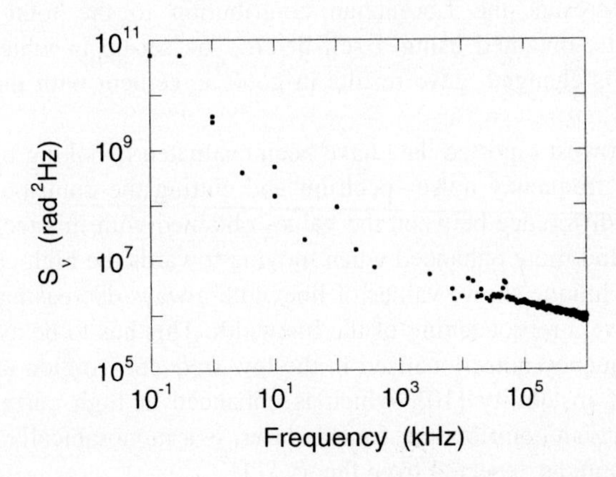

Fig. 8.10. Frequency noise spectrum measured by a Michelson interferometer for the laser of Fig. 8.6. The laser operated with a bias current in the active region equal to 100 mA and no bias in the passive sections.

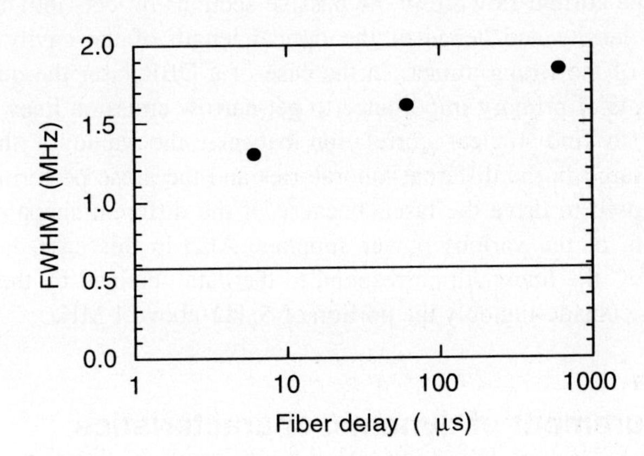

Fig. 8.11. Emission linewidth, expected by a self heterodyne or self homodyne apparatus, as a function of the length of the delay fibre. The data refer to a laser having the frequency noise spectrum reported in Fig.8.10. The straight line represents the linewidth taking into account only the frequency noise contribution above 1 MHz.

The cut off frequency is the inverse of the delay time. As an example in Fig. 8.11 the expected values of the linewidth are reported for the frequency noise spectrum of Fig. 8.10. The data represent the expected linewidth when a self-heterodyne technique is used, as a function of the length of the delay fibre. The continuous line represent the Lorentzian contribution to the total linewidth. Experimental results, obtained using a self-heterodyne set-up in which only the length of the fibre is changed, gave results in good agreement with those of Fig. 8.11.

In Fig. 8.5, the lowest reported data have been evaluated by taking only the flat contribution of the frequency noise spectrum and cutting the contribution below about 1 MHz. The difference between the values obtained with this technique and the others is more and more enhanced when moving towards the high current side. Here the former technique shows values of linewidth always decreasing, while the other techniques give a rebroadening of the linewidth. This has to be ascribed to a contribution of frequency noise localised in the low frequency region and induced from some kind of instability [10], which is enhanced at high current. On the contrary, the Lorentzian contribution, for this laser, is a monotonically decreasing function of the current, as expected from theory [11].

If the intrinsic difference between the various techniques and the parameters of the individual set-ups used to measure the emission linewidth can explain the results of Fig. 8.5, it cannot justify the large spread of data shown in Fig. 8.6. In this case, however, one has to remember that the laser under test is a DBR laser. In DFB lasers the carrier density is clamped above threshold to the threshold value. This is not true for the passive sections of a DBR laser. Any source of noise associated to the current flowing in the passive sections reflects into a fluctuation of the carrier density and hence of the optical length of the cavity and of the grating period of the Bragg mirror. In the case of a DBR laser the quality of the power supplies is of primary importance to get narrow emission linewidth. It was quite difficult to find a clear correlation between the value of the emission linewidth measured in the different laboratories and the noise performance of the power supply used to drive the laser, because of the different shape of the noise power spectrum of the various power supplies. Also in this case, however, the lowest values of the linewidth correspond to the data obtained by the Michelson interferometer, considering only the portion of $S_{\nu}(\Omega)$ above 1 MHz.

8.3. Measurement of dynamic characteristics

8.3.1. Aim of this work

In the previous sections, some critical aspects concerning static measurements of laser devices have been pointed out, showing several difficulties when performing reliable measurements.

The results of this activity suggested to start a new set of measurements, to be performed in several laboratories, dealing with the dynamic characterisation of semiconductor laser devices. The aim of this activity was to define some critical points in the dynamic measurements, as well as to extract additional device parameters for the modelling exercise, suggesting also for this purpose some non usual technique.

Two DFB laser chips have been used for this round robin exercise, provided by two European research centres: KTH (Sweden), and CSELT (Italy). Four laboratories in different countries have been involved in the measurements: KTH, CSELT, ETH Zurich (Switzerland), and University of Gent (Belgium). This small number was justified by the complexity of the measurement set-up, as it will be described in the following sections. However, despite the number of laboratories involved, the comparison of the measurement results was able to identify critical points and special precautions that have to be taken in order to perform reliable dynamic measurements on semiconductor laser devices.

8.3.2. Devices and measurements description

The high frequency measurements on DFB laser require some additional care with respect to the basic continuous-wave measurements. As it has been reported in the previous sections, a very good optical isolation between the laser and the measurement set-up is always needed, to avoid perturbation in the DFB performance introduced by uncontrolled back-reflection.

Moreover, depending on the high frequency involved in the measurements, special cares have to be taken in order to calibrate the measurement set-up. This is particularly true in the case of the laser diode measurements, because of the strongly unmatched load represented by the forward-biased laser diode, whose impedance is usually very low. The large amount of signal power reflected by the unmatched load can therefore affect dramatically the frequency behaviour of the device under test: to avoid these bandwidth distortions, a correct calibration of the equipment (usually a Network Analyser) is always requested.

A considerable problem was to define a "package" for the device under test: commercially available packages for laser diodes, like butterfly modules, are usually limited to 2.5 Gbit/s (or a few GHz of modulation bandwidth). Moreover, the package affects not only the bandwidth behaviour, but also the measurements of the electrical reflection (or S_{11}, in terms of scattering matrix of the electro-optic device), useful for the chip parasitics evaluation [12].

We therefore decided to use unpackaged devices, where the chip laser is mounted on a proper submount, allowing to contact directly the chip by using high frequency (RF) probes.

The devices were two DFB laser diodes: sample "A" was a Multi-Quantum-Well Strained Semi - Insulating Buried Heterostructure, optimised for high frequency operation, with a very low parasitic capacitance (less than 1 pF) obtained by using a very small contact area (40 μm x 150 μm) [13]. Sample "B" was a Bulk Semi - Insulating Buried Heterostructure, in which the contact area was kept quite large (250 μm x 300 μm, of course increasing the parasitic capacitance as well), to make the bonding operation easier [14].

The general measurement set-up used in the dynamic characterisation, is reported in Fig. 8.12: the laser submount is placed on a thermoelectric cooler, to keep the chip temperature to 20 °C. The laser device, biased above threshold, is modulated by the output signal of the Network Analyser. The RF probe assures both the modulation and the bias current, contacting directly the laser chip. The modulated output beam is collected by a lens (selfoc) and an optical isolator into the fibre; this modulated signal is then detected by a fast photodiode connected to the input port of the Network Analyser, assuring the electro-optical signal conversion.

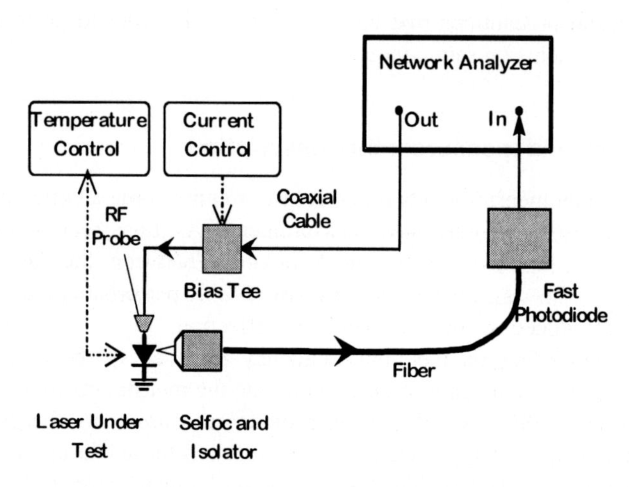

Fig. 8.12. Experimental set-up for bandwidth measurements. The RF probe provides both bias current and modulation signal, mixed by the bias tee; the modulated optical beam, collected by a lens (*selfoc*) and an optical isolator into the fibre, is detected by a fast photodiode, connected to the input port of the Network Analyser.

In order to obtain reliable measurements, the set-up (bias tee, RF probe and photodiode if not included in the network optical unit) should be completely calibrated, allowing the data correction by the network software. In case of electrical reflection measurements, one should also make sure to calibrate the electrical reference plane on the RF probe, by using proper calibration substrate, to be able to extract the correct phase of the load.

Starting from this set-up, some different approaches have been chosen by the laboratories:

- laboratory 1: the set-up described above, with a coplanar (Ground - Signal - Ground, GSG) non-calibrated RF probe, taking into account the transfer function of the fast photodiode (> 40 GHz) with a polynomial interpolation;
- laboratory 2: the set-up provides an optical test set, with a fast photodiode included in the Network Analyser, whose response is taken into account during

the standard calibration procedure; besides, the probes (asymmetrical, in a Ground - Signal configuration, GS) were calibrate with a substrate calibration;

- laboratory 3: similar to the laboratory 1, with an asymmetrical (GS) probe, but extracting the photodiode transfer function by a FFT of the impulse response.
- laboratory 4: similar to the laboratory 1, but without using an optical isolator in the optical path; also in this case the probes (GS) were calibrated with a substrate calibration;

8.3.3. Measurement results: device "A"

The data measured by the laboratories involved in the round- robin have been collected and compared, trying to define the effects of the different measurement set-ups on the results.

The comparison of the results of the modulation response (module) is shown in Fig. 8.13: unfortunately, only the data of the first two laboratories were obtained, since the device was mechanically damaged during shipping. The laboratory 3 and 4 weren't therefore able to perform the measurement session. The comparison shows a very good agreement between the data, indicating that the choice between coplanar or asymmetrical probes is not fundamental up to 20 GHz, and that the external photodiode used by laboratory 1 has been correctly calibrated. Fig. 8.13 shows also the wide bandwidth of the device (more than 20 GHz at higher currents).

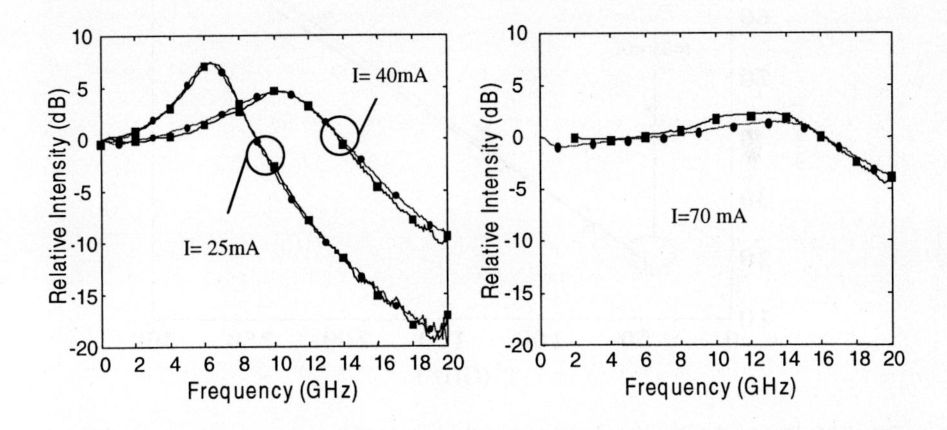

Fig. 8.13. Laboratory 1 (•) and laboratory 2 (■) bandwidth measurements at different bias currents: the results show a very good agreement, despite the differences in the measurement set-up.

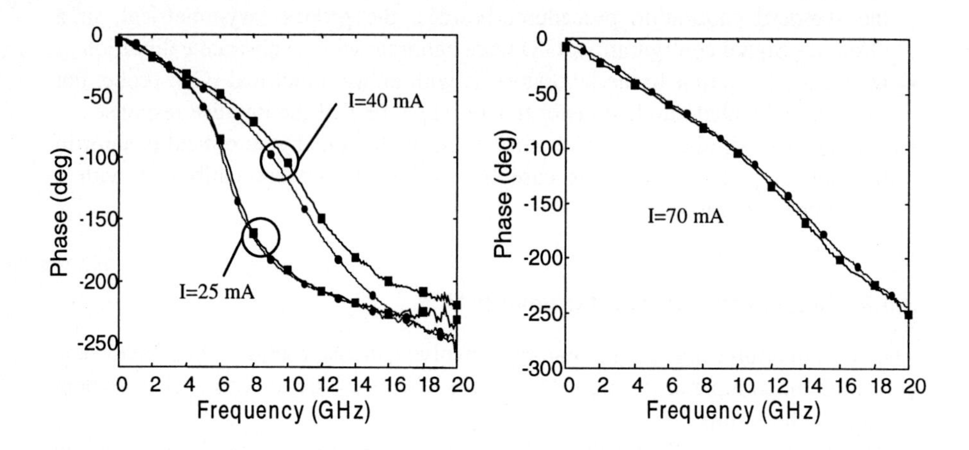

Fig. 8.14. Laboratory 1 (•) and laboratory 2 (■) phase measurements at the same bias currents as in Fig. 8.13; the comparison shows a very good agreement for all the bias currents examined.

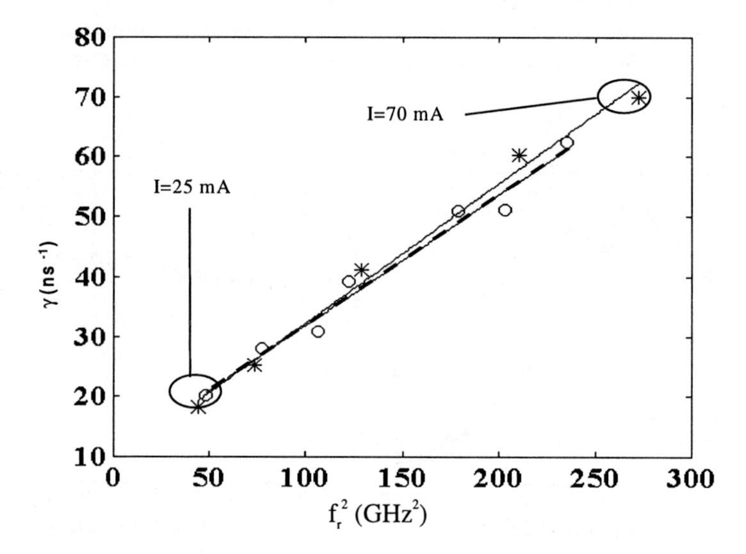

Fig. 8.15. Plot of the damping factor versus the square of resonance frequency, at different bias currents ((*) laboratory 1, (o) laboratory 2). It is also shown the linear interpolation obtained with the data of laboratory 1 (-) and laboratory 2 (-▪).

The results concerning the phase of the frequency response are reported in Fig. 8.14: even in this case the agreement is remarkable.

Fig. 8.15 shows the comparison of the damping factor (γ) versus the square of the resonance frequency (f_r^2), obtained by a 3rd order polynomial interpolation of the frequency responses reported in Fig. 8.13. As described in [15, 18], this plot gives an estimation of the K factor and the gain compression factor ε: the comparison shows therefore a good agreement also for the K factor evaluation (slope of the curves).

8.3.4. Measurement results: device "B"

The device "B" was assumed to be less sensitive to the measurement set-up, due to the limited bandwidth (less than 7 GHz). However, the measurement results indicate that (as reported in Fig. 8.16) large distortion in the bandwidth measurement can occur if no special cares are taken. The errors are mainly due to the optical back-reflection introduced by the set-up, as in the case of the measurements of laboratory 4. In fact, optical reflections have introduced the noticeable ripple in the frequency response, distorting the modulation response with respect to the results of laboratories 1 & 2.

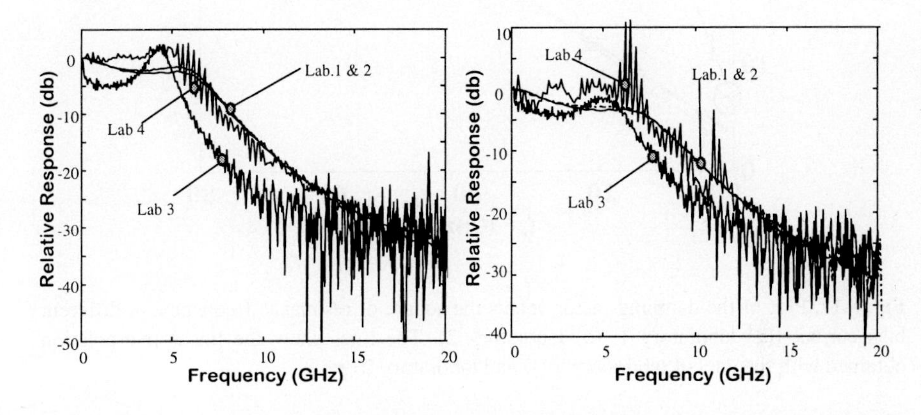

Fig. 8.16. Intensity modulation response at 60 mA (*left*) and 80 mA (*right*) of bias current. The comparison highlight the strong bandwidth dependence on the measurement condition, e.g. the optical reflection in the case of laboratory 4 data.

The data obtained by the laboratory 3 differ slightly from the others: the main reason for this could be some inaccuracy in the temperature control of the device under test, very important at high bias current. The noise that affects this measurement is usually due to a strong optical coupling loss, and can be eventually reduced by enhancing the electrical resolution bandwidth of the Network Analyser. Averaging in the frequency domain, usually called "smoothing", can be sometimes useful, but it can corrupt the response, e.g. the resonance peak of the transfer function; the problem of the noise in the bandwidth measurements will be discussed in detail in the section 9.4.

The comparison of the damping factor (γ) versus the square of the resonance frequency (f_r^2) is reported in Fig. 8.17. γ and f_r^2 are obtained from the interpolation of the frequency responses reported in Fig. 8.16 (for the first two laboratories only). As in the case of the device "A", the results are in good agreement, indicating the good reliability of the measurement performed.

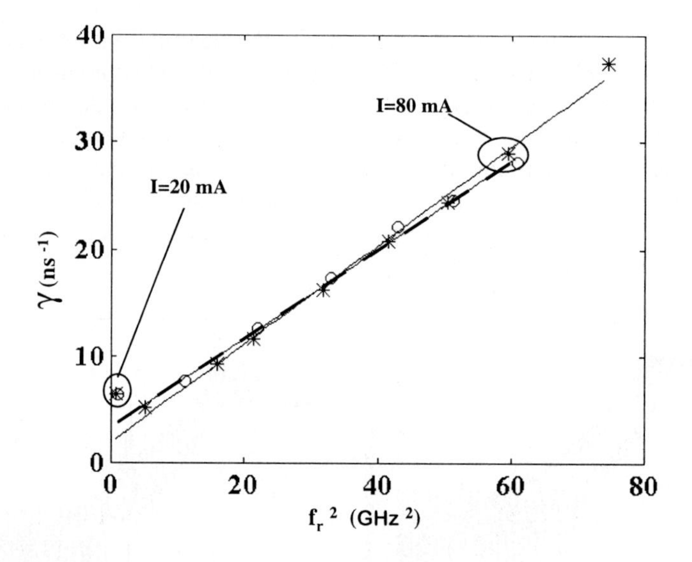

Fig. 8.17. Plot of the damping factor versus the square of resonance frequency, at different bias current ((*) laboratory 1, (o) laboratory 2). The lines show the linear interpolation obtained with the data of laboratory 1 (-) and laboratory 2 (-▪).

8.3.5. Parameter extraction from high frequency measurements

The parameter extraction will be discussed in more detail in the next chapter; however, the round robin exercise has suggested interesting applications of the modulation response measurements in the field of the parameter extraction, for modelling applications.

The development of high speed laser devices has to overcome strong limitations due to the parasitic effects: in the last years, strong reduction of parasitic capacitance has been achieved by using Semi-Insulating blocking layers, which results in a significant improvement in the modulation bandwidth [16, 17]. The correct evaluation of the laser capacitance is therefore extremely important, in the optimisation process involving the parasitics, as well as the dynamics of the active material [18, 19].

An example of parasitics evaluation is reported in Fig. 8.18, in which the S_{11} measurement of the sample "A" is compared with the simulation of its equivalent circuit [12]. A noticeable agreement has been obtained over the entire frequency range (1- 20 GHz).

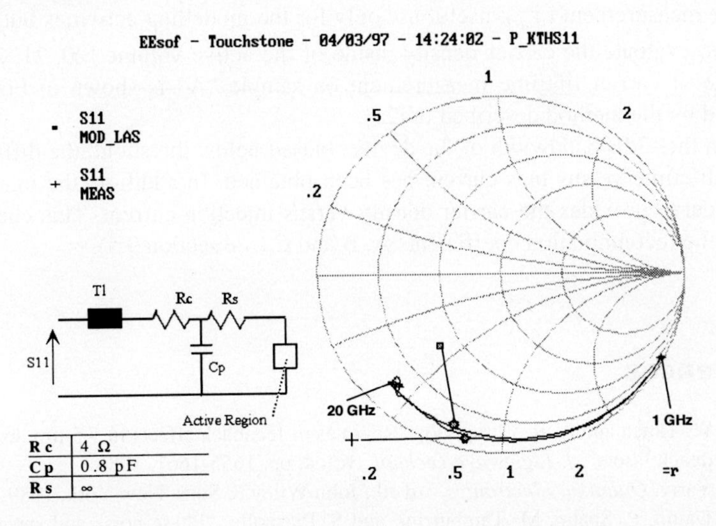

Fig. 8.18. Simulation (■) and measurement (+) comparison of the laser chip reflection coefficient (S_{11}), at zero bias current. The inset shows the parasitic equivalent circuit used for the simulation, whose parameters have been obtained by fitting the measurement response by a least square algorithm.

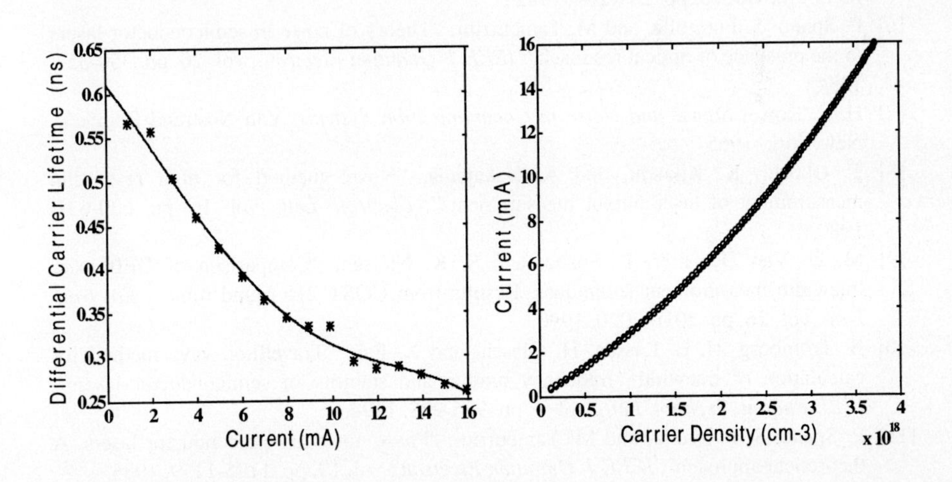

Fig. 8.19. Differential carrier lifetime (*left, *: measured, -: fitted*) and carrier density (*right,* o*: measured, -: fitted)* versus injection current, obtained by measuring sample "B" at different bias current below threshold.

Another important field of the high frequency measurements is related to the parameter extraction of the active material. There are several material parameter that could be evaluated by using dynamic measurements. Among others, experiments are reported in the literature concerning the differential carrier lifetime measurement (τ_d), useful not only for the modelling activities but also as a way to evaluate the carrier density inside of the active volume [20, 21, 22]. An example of carrier lifetime measurement on sample "A" is shown in Fig. 8.19, obtained by the method described in [23].

From the -3dB bandwidth of the device, biased below threshold, the differential carrier lifetime versus bias current has been obtained. In addition, the integration of this curve provides the carrier density versus injection current. This enables to extract the recombination coefficients A, B and C (see section 9.5).

References

[1] R. W. Tkach and A. R. Chraplyvy, "Regimes of feedback effects in 1.5-μm distributed feedback lasers", *J. Lightwave Technol.*, vol. 4, pp. 1655-1661, 1986.

[2] A. Yariv, *Quantum Electronics,* 3rd ed., John Wiley & Sons, New York, 1989.

[3] B. Daino, P. Spano, M. Tamburrini, and S. Piazzolla, "Phase noise and spectral line shape in semiconductor lasers", *IEEE J. Quantum Electron*, vol. 19, pp. 266-270, 1983.

[4] M. W. Fleming and A. Mooradian, "Fundamental line broadening of single-mode AlAs diode lasers", *Appl. Phys. Lett.,* vol. 38, pp. 511-513, 1981.

[5] C. H. Henry, "Theory of the linewidth of semiconductor lasers", *IEEE J. Quantum Electron.*, vol. 18, pp. 259-264, 1982.

[6] P. Spano, S. Piazzolla, and M. Tamburrini, "Theory of noise in semiconductor lasers in the presence of optical feedback", *IEEE J. Quantum Electron.*, vol. 20, pp. 350-357, 1984.

[7] H. E. Rowe, *Signal and Noise in Communication Systems*, Van Nostrand Reinhold, New York, 1965.

[8] T. Okoshi, K. Kikuchi, and A. Nakajama, "Novel method for high resolution measurement of laser output measurements", *Electron. Lett.*, vol. 16, pp. 630-631, 1980.

[9] M. O. Van Deventer, P. Spano, and S. K. Nielsen, "Comparison of DFB laser linewidth measurement techniques. Results from COST 215 round robin", *Electron. Lett.*, vol. 26, pp. 2018-2020, 1990.

[10] B. Tromborg, H. E. Lassen, H. Olesen, and X. Pan, "Travelling wave method for calculation of linewidth, frequency tuning, and stability of semiconductor lasers", *IEEE Photon. Technol. Lett.*, vol. 4, pp. 985-988, 1992.

[11] P. Spano, S. Piazzolla, and M. Tamburrini, "Phase noise in semiconductor lasers: A theoretical approach", *IEEE J. Quantum Electron.*, vol. 19, pp. 1195-1199, 1983.

[12] R. Paoletti, D. Bertone, A. Bricconi, R. Fang, L. Greborio, G. Magnetti, and M. Meliga, "Comparison of Optical and Electrical Modulation Bandwidths in three different 1.55 μm InGaAsP Buried Laser Structures", in *SPIE International Symposia - Photonics West '96*, pp. 296-305, 1996.

[13] O. Kjebon, R. Schatz, S. Lourdudoss, S. Nilsson, and B. Stålnacke, "Modulation response measurements and evaluation of MQW InGaAsP lasers of various designs," in "High-Speed Semiconductor Laser Sources", Paul A. Morton, Deborah L. Crawford, Editors, *Proc. SPIE* 2684, pp. 138-152, 1996.

[14] D. Bertone, A. Bricconi, R.Y. Fang, L. Greborio, G. Magnetti, M. Meliga, and R. Paoletti, "MOCVD Regrowth of Semi-Insulating InP and p-n Junction Blocking Layers Around BRS Laser Stripes", *J. Crystal Growth*, vol. 170, pp. 715 - 718, 1997.

[15] R. Nagarajan, M. Ishikawa, T. Fukushima, R. S. Geels, and J. E. Bowers, "High Speed Quantum-Well Lasers and Carrier Transport Effects", *IEEE J. Quantum Electron.*, vol. 28, No. 10, pp.1990-2008, 1992.

[16] W. H. Cheng, D. Renner, K. L. Hess, and S. W. Zehr, "Dynamic characteristics of semi-insulating current blocking layers: Application to modulation performance of 1.3 μm InGaAsP lasers", *J. Appl. Phys.* , 64 (3), 1988.

[17] P. A. Morton, R. A. Logan, T. Tanbun-Ek, P. F. Sciortino Jr., A. M. Sergent, R. K. Montgomery, and B. T. Lee, "25 GHz Bandwidth 1.55 μm GaInAsP *p*-Doped Strained Multiquantum-Well Lasers", *Electron. Lett.*, vol. 28, No. 23, pp. 2156-2157, 1992.

[18] J. E. Bowers, "High Speed Semiconductor Laser Design and Performance", *Solid State Electron.*, vol. 30, pp. 1-11, 1987.

[19] F.Delpiano, R.Paoletti, P.Audagnotto, and M.Puleo, "High frequency characterisation of high performance DFB laser modules," *IEEE Trans. Comp., Pkg., Mfg. Tech.*, vol. 17, no 3, pp. 412-417, 1994.

[20] R. Olshansky, C.B.Su, J. Manning, and W. Powazinik, "Measurement of radiative and nonradiative recombination rates in InGaAsP and AlGaAs light sources", *IEEE J. Quantum Electron.*, vol. QE_20, pp. 838-854, 1984.

[21] Y. Zou . S. Osinski, P. Grodzinski, P. D. Dapkus, W. Rideaut, W. F. Sharfin, and F. D. Crawford, "Experimental verification of strain benefits in 1.5 μm semiconductor lasers by carrier lifetime and gain measurements", *IEEE Photon. Technol. Lett.*, vol. 4, p 1315-1318, 1992.

[22] G. E. Shtengel, D. A. Ackerman, and P. Morton, "True carrier lifetime measurements of semiconductor lasers", *Electron. Lett.,* vol. 31, p. 1747-1748, 1995.

[23] R.Paoletti, M.Meliga, and I.Montrosset, "Optical modulation technique for carrier lifetime measurement in semiconductor lasers," *IEEE Photon. Technol. Lett.,* vol. 8, no 11, pp. 1447-1449, 1996.

9 Parameter extraction

R. Schatz, D. McDonald, H. Hillmer

"In order to satisfactorily model a system, reliable parameters which describe the system must first be derived from measurement".

This statement is true for any device or plant under test in many engineering domains, be it a chemical process, annealing of materials or the growth of a semiconductor device. This task is usually quite complicated, especially when there is a large number of design details and variables that go into the construction of a desired entity. This is true for semiconductor lasers for telecommunications which are complex devices to grow, process, measure, characterise and model. For this reason, the ability to gather accurate device parameters by suitably modelling and measuring a device or batch of devices can be a demanding task. This is vital for the manufacture of low cost, reliable semiconductor components, the characteristics of which should be reproducible from batch to batch. A more immediate motivation for accurate parameter estimation from measurements is to qualify the models developed in Ch. 7 and to allow these models to intelligently predict how devices will perform in various areas of application.

Following some general remarks on parameter extraction in Sec. 9.1, the sections of this chapter will consider some of the main methods for extracting pertinent laser parameters from measurements, which are quite widely applied by many of the laboratories participating in the Action COST 240. Each method will be discussed in terms of

– performing the measurement, collecting the data, and of the precautions to be adhered to.

– the development of any simplified model required in order that a relatively straightforward parameter extraction procedure can be applied. Methods for obtaining first order approximations for parameters by applying 'rule-of-thumb' methods that can then be used as starting estimates in a search strategy will also be discussed.

– the fitting algorithms used and the benefits of the approaches adopted.

– examples of parameter extraction exercises and estimates of the parameters obtained.

Sections 9.2, 9.3, and 9.4 discuss three of the most widely adopted measurement and characterisation techniques, *i.e.* ASE, RIN and AM modulation. These will be described in detail in terms of the models used to describe the laser characteristics and the algorithms used to extract the various parameters. Section 9.5 will explore characterisation techniques that can be used to check the device parameters obtained from the previous methods. The topic of device yield and the statistical nature of certain device parameters, such as the facet phases in DFB lasers, will be discussed in terms of the effect such parameters have on device performance under large signal modulation conditions.

We hope that by the end of this chapter the reader will have gained an insight into how to extract those parameters for semiconductor lasers from a variety of measurements which will allow accurate and efficient modelling of semiconductor laser diode output characteristics. Indeed, all the parameters discussed in this chapter can be readily applied to the models already developed in Ch. 7.

9.1 General remarks on laser parameter extraction

One of the primary objectives of any parameter extraction strategy is to gather as much information about a laser or laser process from as few measurements and parameter extraction steps as possible. As well as this, measurements in which parameters can be determined in isolation are preferred. However, since the number of parameters required to accurately describe all the traits of a laser diode light output can be quite large (see Table 9.1), those measurements from which a large parameter set can be obtained are preferred.

The models developed in Ch. 7 for DFB devices are numerically quite intensive involving significant computation time depending on the laser characteristic to be modelled and the type model adopted for the calculations. These models are imperative for successful modelling of complex multi-section laser device structures (three electrode DFB, DBR) in which carrier and photon inhomogeneities exist in the laser cavity. However, simplified equations sets can model simpler structures (FP, single section DFB) with good results. In many instants the application of full numerically intensive models to the extraction of laser parameters is not feasible on the basis of the large number of calculations required and the large amount of redundant information generated. This is especially true for those characteristics that are sensitive to parameter values and are influenced by a large parameter set, *e.g.* the Amplified Spontaneous Emission (ASE) spectrum. As such, simplified models are regularly used which allow analytical descriptions of a laser characteristics to be devised and in so doing allow quite accurate determination of certain device parameters with minimal computational effort. The simplified approach to describe the Amplitude Modulation (AM) and Relative Intensity Noise (RIN) spectra are good examples of how this approach is possible for determining, for example, the modal differential gain and the gain compression coefficient of a DFB laser. However, it must be emphasised that this is only possible for the simpler device structures, FP and DFB lasers, which are the laser types treated in this book.

α	material linewidth enhancement factor
α_i	internal loss
R_i	facet power reflectivities
Λ	grating period
M	grating order
κ_i	index grating coupling coefficient
κ_g	gain/loss grating coupling coefficient
Θ_j	grating facet phase
A	linear recombination coefficient
B	bimolecular recombination coefficient
C	Auger recombination coefficient
ε	nonlinear gain coefficient
a	differential material gain coefficient
N_0	transparency carrier density
n_{sp}	population inversion parameter
β_{sp}	spontaneous emission coupling factor
τ_d	differential carrier lifetime
f_r	resonance frequency
Parameters taken from growth conditions or estimated from waveguide calculations	
d	thickness active region
w	width active region
L	length laser chip
Γ	confinement factor

Table 9.1. Parameters used to characterise laser diodes.

Two different aspects of any laser diode need to be addressed before a full understanding of any device is possible, *i.e.* the materials parameters and the parameters of the fully fabricated device. All measurements are performed on fully fabricated devices of the DFB type whose geometry or metallisation have not been altered to allow advanced study of the spontaneous emission or stimulated emission process occurring within the laser. Generally Fabry-Perot (FP) lasers are also fabricated on distributed feedback (DFB) laser wafers to allow certain material parameters to be measured, such as the net material gain spectrum from below threshold measurements. As far as possible, material parameters required by the models of Ch. 7 have been obtained from measurements on finished DFB devices or have been taken from the literature for the InGaAsP/InP, the material system used in the laser devices reported here.

A large suite of measurements is at hand to characterise semiconductor laser diodes. Some are straightforward requiring a minimum of measurement equipment, *e.g.* output power versus input current, while other measurements require sensitive microwave equipment that has to be rigorously calibrated *e.g.* RIN measurement. A selection of the most widely used measurements and the laser parameters that can be obtained from these are set-out in Table 9.2.:

Measurement	Parameters
output power vs. current, voltage vs. current	α_i, η_{int}
ASE spectra below threshold	R_1, R_2, Θ_1, Θ_2, κ_i, κ_g, n_g, dn/dN
AM modulation	γ, f_r, K, a, ε, τ_d
RIN spectra	γ, f_r, K, a
linewidth	γ, f_r, K, a

Table 9.2. List of measurements and corresponding extracted parameters.

Because of the interplay between the various laser parameters, all laser parameters are temperature dependent. All the measurements listed in Table 9.2. are usually performed at a single laser heat-sink temperature, but any measurement that requires the DC bias current to change also involves a temperature rise of the active region due to current heating. To circumvent this, pulsed laser measurement methods are desired at repetition rates of kHz or more and duty cycles of 0.1 % or less. Such measurements are quite demanding and will be discussed in Sec. 9.5.

The models developed in Ch. 7 do not include thermal effects since this requires being able to accurately model the current spreading in the laser structure and the thermal characteristics of the various materials which make up the laser. In addition, the mounting arrangement of the laser on its submount and the current density distribution in multi-contact devices make the development of a thermal model for lasers extremely difficult. The parameters extracted in this section, unless otherwise stated, are not determined for a given temperature but assume that thermal effects are minimal. This assumption is also applied to the simplified models used for some of the parameter extraction algorithms.

9.2 Extraction from the ASE spectrum

The Amplified Spontaneous Emission (ASE) spectrum of a laser biased below lasing threshold is easy to measure and can be used to extract several key parameters that determine the optical properties of a DFB laser, *e.g.*, the coupling coefficient, the effective index, the group index, the modal gain and the magnitude and exact positions of the reflections from the end facets. Especially the coupling coefficient κ has large influence on threshold current, differential quantum efficiency, single-mode yield and light-current linearity of the laser. The coupling coefficient is difficult to control in manufacturing since it depends critically on the geometry and composition of the fabricated grating and surrounding epitaxial layers. It is therefore desirable to measure it on manufactured lasers in order to evaluate and align the manufacturing process and also at an early stage discard components that are unlikely to fulfil the specifications. Accurate parameter values are also needed as input for accurate modelling of manufactured devices and subsequent comparison of modelled and measured data.

A simple and common way to extract the coupling coefficient from the ASE-spectrum is by comparing the wavelength width of the stopband with the mode spacing of the Fabry-Perot modes far away from the stopband [1]. However, this works well only for ordinary (not phase-shifted) DFB lasers with high coupling and small residual end reflections. In practice, the method is too uncertain for measurement on individual lasers and only useful for a large ensemble of lasers from the same batch.

A much more accurate parameter extraction method, which will be treated here, is to manually or numerically fit the measured spectrum to a theoretical formula [2], [3], [4].

9.2.1 Measurement of the ASE spectrum

Although the measurement of the ASE spectrum is quite straightforward, there are some important considerations to keep in mind. External reflections and spontaneous emission from TM-modes can disturb the spectrum. The remedy for this is an optical isolator or, at least, a polariser that is inserted between the laser and the optical spectrum analyser. Another issue is the choice of the spectral range that should be used for the extraction. There is no simple answer to this question but a rule of thumb is that around ten to twenty modes symmetrically spaced around the stopband should be included. The optimum bias should be chosen slightly below the laser threshold in order to give good dynamic range of the measurement without entering the above threshold regime where the clamping process changes the spectrum of the lasing mode and the simple formulas for the ASE below threshold do not apply. Another problem if the laser is biased too close to threshold is that the resolution of the spectrum analyser, typically 0.1 nm, will not be sufficient to fully resolve the sharp peak of the main mode.

It is worthwhile to make a series of measurements at different bias in order to check the accuracy of the extracted parameters [2]. ASE measurements at different bias currents enable also extraction of the material linewidth enhancement factor and the modal gain dependence on bias current. For longitudinally asymmetric laser structures the spontaneous emission spectra from the two facets are different.

Measurement from both the front and the back facet is hence, if the laser mount permits it, another method to check the accuracy of the extracted parameter values.

9.2.2 Theoretical formula for ASE spectrum

The spontaneous emission spectrum for a general laser structure with longitudinally homogeneous inversion and internal losses can be written [2]

$$P_{ASE}(\lambda) = \left(\left|S_{11}(\lambda)\right|^2 + \left|S_{12}(\lambda)\right|^2 - 1\right) K(\lambda) \tag{9.1}$$

Here, λ is the wavelength and the scattering matrix elements S_{11} and S_{12} are the total reflection and transmission coefficients of the laser, $K(\lambda)$ is a factor that depends on the internal efficiency, the inversion factor and the coupling losses between the laser and spectrum analyser. For edge emitting devices, the wavelength dependence of $K(\lambda)$ is usually small compared to the wavelength dependence of the reflection and transmission coefficient and can therefore be neglected or be approximated with a linear function. A linear approximation is in general also sufficient for the wavelength dependence of the gain and of the refractive index for the limited spectral range (typically 10-20 nm) of the measurement.

$$K(\lambda) = K_0 + K_\lambda(\lambda - \lambda_0) \tag{9.2}$$

$$n_{eff}(\lambda) = n_{eff,0} + \frac{n_{eff,0} - n_g}{\lambda_0}(\lambda - \lambda_0) \tag{9.3}$$

$$g_m(\lambda) = g_{m,0} + g_\lambda(\lambda - \lambda_0) \tag{9.4}$$

Here, K_0, $n_{eff,0}$ and $g_{m,0}$ are the amplitude factor, effective index and modal gain at the reference wavelength λ_0, respectively. Furthermore, n_g is the group index and K_λ and g_λ are the wavelength derivatives of the amplitude factor and the modal gain. This reference wavelength can for convenience be chosen to the lasing wavelength or to the nominal Bragg wavelength. For lasers where the assumption of longitudinally homogeneous inversion does not hold, *e.g.*, gain coupled lasers, a slightly more complicated formula has to be used [5].

It is quite straightforward to calculate the scattering matrix for any one-dimensional laser structure by using the transfer matrix method. For an ordinary DFB laser without any phase shift an explicit analytic result can be obtained, using coupled wave theory. The reflectivity and transmission coefficients for a pure sinusoidal grating are [6]

$$r_g = -j\frac{\kappa^* \tanh(\sigma L_g)}{\sigma + j\delta \tanh(\sigma L_g)} \quad \text{and} \quad t_g = \frac{\sigma\ \mathrm{sech}(\sigma L_g)}{\sigma + j\delta \tanh(\sigma L_g)} e^{-j\frac{\pi}{\Lambda}L_g} \tag{9.5}$$

with

$$\sigma^2 = \kappa^2 - \delta^2 \tag{9.6}$$

$$\delta = \frac{2\pi n_{eff}(\lambda)}{\lambda} + j\frac{g_m(\lambda)}{2} - \frac{\pi}{\Lambda} \tag{9.7}$$

If the laser is not anti-reflection coated, the reflectivity and the transmitivity of the total structure, including end reflections have to be used instead

$$r_{1g2} = r_{g2} + \frac{t_{g2}^2 r_1}{1 - r_{g2} r_1} \quad \text{and} \quad t_{1g2} = \frac{t_{g2} t_1}{1 - r_{g2} r_1} \tag{9.8}$$

where

$$r_{g2} = r_g + \frac{t_g^2 r_2}{1 - r_g r_2} \quad \text{and} \quad t_{g2} = \frac{t_g t_2}{1 - r_g r_2} \tag{9.9}$$

Whereas the grating length L_g and grating period Λ are usually known, the amplitude factor $K(\lambda)$, the modal effective index $n_{eff}(\lambda)$, the modal gain $g_m(\lambda)$, and the complex-valued facet reflectivities r_1 and r_2, often need to be extracted simultaneously with the coupling coefficient κ. If both phase and amplitude of the facet reflectivities are unknown, this leads to a total of eleven unknown parameters. The large number of parameters makes the extraction procedure difficult and gives large uncertainty in the extracted values. However, in many cases it is possible to reduce the number of unknown parameters. If the laser facets are cleaved, the amplitude of the reflection coefficient can approximately be determined from the effective index using the Fresnel formula. The facet reflections can be neglected altogether if the facets have a good anti-reflection coating and the coupling coefficient is large. A rule of thumb for this to be valid is that the reflection coefficient divided by the normalised grating coupling coefficient κL_g should be less than 10^{-3}. If the laser is lasing close to the gain peak, the gain derivative g_λ can be neglected. Alternatively, for DFB lasers with as-cleaved facets, it is possible to determine $g_m(\lambda)$ separately by measuring the ASE-spectrum for a broader wavelength range. For wavelengths well away from the stopband the laser will behave as a Fabry-Perot device and the gain can be determined either by using the Hakki-Paoli-method [7] or fitting it to a theoretical Fabry-Perot spectrum. From these measurements it is possible to estimate the modal gain at the reference wavelength and its wavelength derivative.

The qualitative influence of the different parameters on the DFB ASE spectrum is as follows:

- The factor K_0 determines the amplitude level of the spectrum and K_λ its slope.
- The stopband wavelength is determined by the effective index n_{eff}, together with the grating period which is assumed to be known.

- Variation of the group index n_g shrinks or expands the spectrum in wavelength.
- The normalised net modal gain $g_m L$ determines the peakness of the spectrum, especially of the modes with low loss.
- The normalised grating coupling κL_g affects the mode spacing and amplitude of the modes close to the Bragg wavelength.
- The amplitude of the facet reflections affects mainly the amplitude of the modes far away from the Bragg wavelength.
- The asymmetry of the spectrum, *i.e.*, the amplitude difference between the modes on the long wavelength side of the Bragg wavelength as compared with the short wavelength side is dependent on the phase of the facet reflections.

9.2.3 Fitting technique

The fitting of the theoretical formula to the measured spectrum is done by minimising a cost function by variation of the unknown parameters. The cost function gives a quantitative measure of the difference between the two spectra. The most common cost function is the least mean-square estimator, *i.e.*, simply the sum of the squared errors

$$E_{LMS} = \sum_i \left[P_{ASE,i} - P_{ASE}(\lambda_i) \right]^2 \tag{9.10}$$

Here, $P_{ASE,i}$ is the measured ASE level at wavelength λ_i. This is also a maximum likelihood estimator if the noise is gaussian with the same variance for the different measurement points. However, in reality the minimised error is often not mainly due to noise but to model error, *i.e.*, the theoretical model does not fully describe the real system due to imperfections in the laser structure or the measurement set-up. This means that there may be more efficient cost functions that give less error in the extracted parameters. We have found it better to use a least mean-square estimator of the logarithmic quantities, *i.e.*,

$$E_{LMS} = \sum_i \left\{ \ln\left(P_{ASE,i}\right) - \ln\left[P_{ASE}(\lambda_i)\right] \right\}^2 \tag{9.11}$$

Often there exist à priori knowledge of probable approximative values of the laser parameters. This knowledge is not capitalised if a simple least mean-square estimator is used. A more advanced cost function that utilises this à priori knowledge may also avoid local minima that are far away from the global minimum in the parameter space [8].

There exists several numerical methods to find the global minimum of the multi-dimensional cost function [9]. Common to these methods is that they need good start values as input in order to avoid local minima. Approximate start values for K_0, $n_{eff,0}$, and n_g can be determined from the average ASE level, the approximate position of the Bragg stopband and the spacing of the side modes away from the stopband. A tentative start value of the modal gain can be around half of the threshold gain for the nominal structure and the wavelength derivatives, K_λ and g_λ, can be put to zero. The main problem is to find good start values of the complex reflectivities, especially the phase. A robust but calculation intensive way of doing this is to systematically scan the possible reflectivity amplitude and phase and do a minimisation for each set. Another way to reduce the probability of finding local minima is to use adaptive simulated annealing at start.

A simply implemented multi-dimensional minimisation method is to cyclically minimise the cost function with respect to one parameter at a time using, *e.g.*, Brent's method [9], for one-dimensional minimisation (a combination of the secant method and the bisection method). This technique is, however, very inefficient if the parameter directions are not conjugate, *i.e.*, if their effects on the spectrum are not independent so that minimisation in one direction to a large degree spoils the previous minimisation in the other directions. A simple remedy for this is to determine, at the end of each minimisation cycle, the resulting change in the parameters and replace one of the previous minimisation directions with that new direction in parameter space. The old direction to be discarded should be the most successful one in the previous minimisation cycle since this is most parallel with the new direction. This method, called Powell's method [9], works quite well but it does not guarantee that two search directions do not become parallel. A way to avoid this is to reinitialise the original directions after a number of iteration cycles.

Another method that is usually less efficient than the Powell's method but somewhat easier to implement, is the down-hill simplex method [9]. The cost function is evaluated in N+1 points (N is the number of parameters) around the starting point creating a N-dimensional simplex structure. The point with the maximum function value is thereafter reflected through the (N–1)-dimensional plane that is put up by the other points or the entire simplex structure is shrank towards the point with minimum value.

The numerically most efficient parameter extraction method, especially designed for least mean-square fitting, is the Levenberg-Marquardt method [9]. The only drawback with this method is that not only the theoretical cost function but also its derivatives with respect to the different parameters are needed. This makes the method more cumbersome to implement, especially if it should work for a general DFB structure with incorporated phase shifts or chirped grating. It is of course possible to numerically calculate these derivatives but it is then an open question if this technique is still competitive in comparison to Powell's method.

9.2.4 Comparative experimental results

Several parameter extraction exercises has been defined within the COST 240 Action in order to compare parameter extraction methods from different laboratories [8]. We present below the results for one of the lasers that was studied. The following information was given by the manufacturer. The laser is a DFB laser without any phase-shifts and with one cleaved facet and one AR-coated facet.

The period of the second order grating is 475 nm. The active layer is 280 μm long, 1.5 μm wide and 0.14 μm thick. The normalised coupling coefficient and the effective index were expected from design to be around 2 and 3.2, respectively. The spontaneous emission was measured at two bias currents, 25 mA and 26 mA, in order to be able to extract also the linewidth enhancement factor.

Five groups participated in this exercise, Royal Institute of Technology, Stockholm (KTH), University College Dublin (UCD), Tele Denmark Research (TDR), University of Gent (UG) and Ericsson Components, Stockholm (Ericsson).

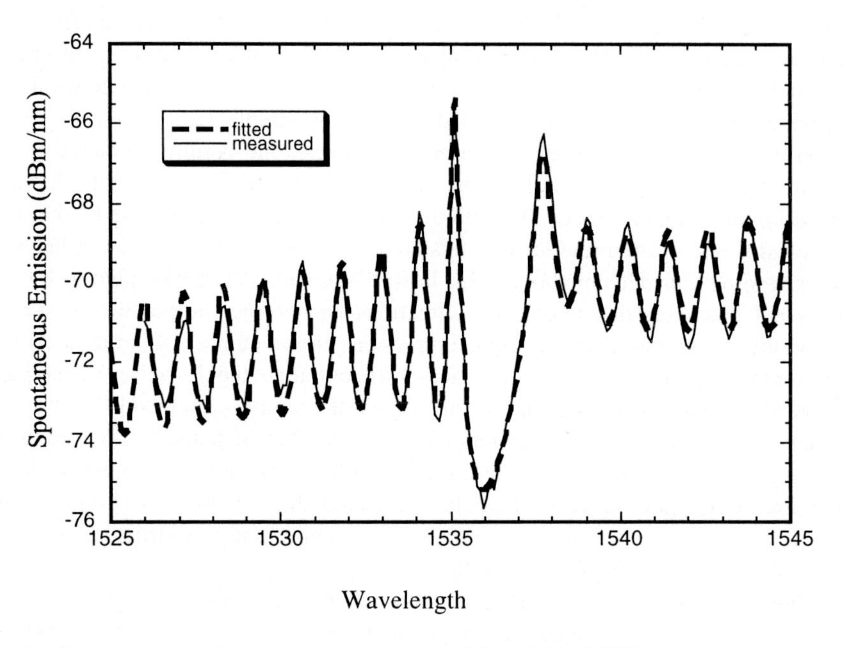

Fig. 9.1. Example of fit to the ASE spectrum at 25 mA bias (KTH)

There was considerable difference in approach between the different laboratories. In order to reduce the number of parameters, the power reflectivity of the cleaved facet was calculated from the effective index using the Fresnel formula and the imaginary part of the coupling coefficient was neglected by KTH and Ericsson. KTH used Powell's method to fit the spectrum, TDR used a manual method, Ericsson used an automatic method for determining the start values and thereafter Levenberg-Marquardt with numerical derivatives, UG used analytical Levenberg-Marquardt and UCD used a combination of adaptive simulated annealing and the downhill simplex method. The measured and fitted spectra at 25 mA and the corresponding extracted parameter values can be found in Fig. 9.1 and Table 9.3.

Parameter [unit]	*KTH*	*TDR*	*UCD*	*UG*	*Ericsson*
power refl. R_1 [%]	27.84	32	15.25	21.03	27.85
power refl. R_2[%]	5.08	0.5	9.79	6.63	4.90
refl. phase Θ_1	1.032	1.257	2.817	1.745	1.012
refl. phase Θ_2	2.755	4.869	2.838	3.620	2.754
index coupl. $\kappa_i L$	2.075	1.96	2.09	2.07	2.04
gain coupl. $\kappa_g L$	not used	0.056	0.34	0.034	not used
group index n_g	3.5489	3.6	3.55313	3.55225	3.5487
effective index n_{eff}@ 25 mA	3.23446 @1.536 μm	3.2471	3.23594 @ 1.535 μm	3.23472 @ 1.535 μm	3.23452 @1.536 μm
linew. enh. fact.	4.141	4.394	-	3.538	3.809

Table 9.3. Extracted parameters from the spectrum shown in Fig. 9.1.

By comparing the results it can be concluded that the coupling coefficient, group index and effective index can be easily determined whereas the other parameters, *e.g.*, the extracted phase and magnitude of facet reflectivities, differ indicating that several sets of these parameters can give almost identical spectra. Similar conclusions have recently been drawn in [4] where a comparison of extraction results on DFB lasers of different types and lengths but with the same coupling coefficient was made.

9.3 Extraction from the RIN spectrum

As with any other electronic or optoelectronic component a thorough knowledge of device parameters is required if a device is to be modelled accurately for sub-system development [10].

Many of the more recent models for semiconductor lasers are quite numerically intensive and are usually concerned more with modelling the effects of nonlinear phenomena in the 'internals' of the laser and whether and how these affect the laser output characteristics. This makes their use as fast efficient parameter estimators from measured data rather poor, also because of the added difficulty of relating model parameters to physically measurable parameters. Early models involved with simulating the dynamic response of semiconductor laser diodes have proven convenient both in terms of their speed of analysis and the direct relation that these models have with the physical processes occurring within the laser, *e.g.* recombination, spontaneous emission, gain saturation etc. In addition to this, analytical solutions for various laser characteristics for the simplified cases can be derived which yield greatly simplified laser diode characterisation. This latter approach has been adopted in the present section concerning the characterisation of diode lasers using the relative intensity noise (RIN) spectrum or the intensity modulation response.

Other methods exist for the estimation of intrinsic and extrinsic laser dynamic parameters some of which are: laser gain switching and observing the relaxation oscillations of the transient up to tenths of GHz; fully implemented optical S-parameter measurements as well as S_{21} measurement of extrinsic laser performance and circuit model parameter extraction; FM modulation and FM noise characterisation [11]; using the terminal electrical noise from the laser contact; parasitic free active layer photomixing; the injection optical probe method; frequency response subtraction method [12]. Other more exotic methods such as Four Wave Mixing for estimating certain parameters are good with regard to the investigation of basic mechanisms but seem rather impractical as a general purpose characterisation method.

9.3.1 Measurement of RIN

Relative intensity noise indicates the maximum possible Signal-to-Noise ratio (SNR) in a lightwave transmission system where the dominant noise source is laser intensity noise. Unfortunately, noise from a large number of sources is actually measured in any real system where the extraneous sources come from feedback induced noise, interferometric conversion of laser phase noise to intensity noise, fibre dispersion, as well as thermal noise of the receiver. With careful optical design such as using optical isolators and tilting the optical interfaces, using index matching gel with fibre adapters and keeping fibre lengths short (especially with multimode lasers), all but the receiver noise can be eliminated in the laboratory. The receiver noise and the laser shot noise may be subtracted from the measured system noise according to the equation:

$$RIN_{laser} = RIN_{system} - \underbrace{\frac{N_{thermal}}{R_L (rP_{opt(AVG)})^2}}_{Thermal} - \underbrace{\frac{2e}{rP_{opt(AVG)}}}_{Shot} \tag{9.12}$$

leaving only the laser excess intensity noise, where $N_{thermal}$ is the thermal noise power spectral density, R_L is the photodetector load resistance, r is the photodetector responsitivity and $P_{opt(AVG)}$ is the average received optical power. Instabilities can arise in the subtraction process when the noise sources are of almost equal magnitude, especially in the case of very low noise lasers at large received powers. These operations are routinely performed by an optical analyser using precise calibration data for the photodiode (by heterodyning two tunable diode pumped YAG lasers on the photodiode), post low noise amplifier and associated passive component frequency responses. Alternative approaches for RIN measurement exist [18] such as using a broadband photodiode and noise figure meter in conjunction with a power meter and a tunable filter.

9.3.2 Theoretical formula for RIN

Due to spontaneous emission events in a laser cavity and the subsequent coupling of these into the lasing mode, the optical amplitude and phase vary in a random way. This results in a AM and FM noise component in the output optical emission which are conventionally expressed as the RIN or FM noise spectra at a specified output power. Both spectra are parasitic free and are extremely valuable at estimating a laser's modulation performance under purely DC conditions. Only the former is employed as an analytic tool here.

In the simplified two level homogeneously broadened laser system, the single-mode spatially averaged carrier and photon density rates, neglecting lateral carrier diffusion, can be expressed by the stochastic rate equations:

$$\frac{dN}{dt} = \frac{I}{eV} - v_g g(1-\varepsilon S)S - \frac{N}{\tau_n} + F_N \tag{9.13}$$

$$\frac{dS}{dt} = \Gamma v_g g(1-\varepsilon S)S - \frac{S}{\tau_p} + R_{sp}(N) + F_S \tag{9.14}$$

where the effects of spontaneous carrier lifetime, spontaneous emission density into the lasing mode and phenomenological gain compression have been included via the carrier lifetime τ_n, the spontaneous emission rate R_{sp} and the nonlinear gain coefficient ε. The parameters v_g, g, S and N are the group velocity, material gain, volume (V) photon and carrier densities, respectively, while I is the DC bias current. The $1-eS$ phenomenological form of nonlinear gain has been used in this noise analysis. The fluctuations arising from spontaneous emission events are included via the Langevin noise terms F_S (photon) and F_N (carrier) which under the Markovian approximation (memoryless system) for these zero mean Gaussian processes satisfy

$$\langle F_i(t)\rangle = 0, \qquad \langle F_i(t)F_j(t)\rangle = 2D_{ij}\delta(t-t'), \tag{9.15}$$

where $\langle\ \rangle$ represents ensemble averaging and D_{ij} (i = S, N; j = S, N) are the diffusion coefficients of the process associated with the corresponding noise source. The use of these single-mode equations to model the noise processes in Fabry-Perot lasers is valid around the resonance frequency where single-mode photon noise is the dominant noise source. At low frequencies the above set of equations is invalid and a full multimode solution is required, which includes spontaneous effects in order to model mode partition noise.

Performing a small signal analysis on these equations but neglecting the small contribution to photon noise from the carrier shot noise (F_N), and applying the Fourier transform of equations, the fluctuating photon component is given by

$$s(\omega) = \frac{(\gamma_N + j\omega)F_s(\omega)}{\Omega_r^2 - \omega^2 + j\omega\gamma} \tag{9.16}$$

where $F_s(\omega)$ is the Fourier transform of the diffusion constant and γ_N is the inverse effective differential lifetime of the carriers, taking into account both the stimulated and spontaneous recombination. Then the damping rate of the resonance in the spectrum is

$$\gamma = 4\pi^2 \left(\frac{\varepsilon}{v_g\, a} + \tau_p \right) f_r^2 + \frac{1}{\tau_d} + \frac{R_{sp}}{S} \tag{9.17}$$

where τ_d is the differential carrier lifetime and the resonance frequency is given by

$$f_r^2 = \frac{1}{4\pi^2} \frac{\Gamma v_g a S}{\tau_p} \tag{9.18}$$

Here, the effect of the gain compression on the resonance frequency has been neglected since it is small at moderate photon densities.

If a linear power current relationship is assumed, which is quite valid close to threshold where heating effects are small, the cavity photon density can be expressed in terms of the bias current

$$S = \frac{(I - I_{th})\tau_p}{eV} \tag{9.19}$$

where e is the electronic charge, $V = dwL$ is the volume of the active region and $I - I_{th}$ is the bias above threshold. Using this relationship and the previous equation then [13]:

$$\frac{f_r^2}{I - I_{th}} = \frac{1}{4\pi^2 e} \frac{\Gamma v_g a}{V} \tag{9.20}$$

The slope of the plot of f_r^2 vs. $(I - I_{th})$ is equal to the right hand side of the equation. Therefore, if the volume of the active region V and the confinement factor Γ are known, the differential gain a can be found.

The laser amplitude intensity noise at a given frequency is characterised by the Relative Intensity Noise, which is defined as

$$RIN(\omega) = \frac{\langle S_{sd}(\omega) \rangle^2}{\langle S_o \rangle^2} \tag{9.21}$$

where $\langle S_{sd}(\omega) \rangle$ is the photon noise spectral density and $\langle S_o \rangle$ is the mean photon density. Under moderate bias conditions the excess RIN can be expressed as:

$$RIN(f) = A \frac{f^2 + (\gamma/2\pi)^2}{\left(f^2 - f_r^2\right)^2 + (\gamma/2\pi)^2 f^2}, \tag{9.22}$$

where parameter A has been related to the Schawlow-Townes linewidth [14], and f is the frequency.

Recent analyses [15], [16] of the effect of carrier transport in quantum well lasers on the intensity noise spectrum have indicated that equation (9.22) has to be expanded to include an additional pedestal low pass filter or a frequency dependent damping factor in the numerator. These forms of RIN explicitly include the effects of carrier transport to the quantum wells even under DC bias conditions, which results in an effective RC roll-off in the noise response. The analysis above is valid in the case of small effective capture times, as is expected for the large well number, small separate confinement region devices.

From the analysis by Olshansky et al. [17] for bulk semiconductor lasers, the damping parameter γ and the resonance frequency f_r are related, for moderate to high photon densities where spontaneous emission is negligible, by

$$\gamma = Kf_r^2 + \gamma_0 \tag{9.23}$$

where

$$K = 4\pi^2 \left[\tau_p + \frac{\varepsilon}{v_g a} \right] \tag{9.24}$$

and can be related to the 3 dB modulation bandwidth by

$$f_{3dB} = \frac{2\pi\sqrt{2}}{K}, \tag{9.25}$$

which for conventional bulk lasers is independent of laser dimensions and facet reflectivity. This allows for an efficient means of estimating a device's ultimate frequency bandwidth performance from purely DC conditions. The case is somewhat complicated for quantum well lasers where structural effects have been seen to influence the modulation bandwidth, where the gain compression parameter becomes a function of the transverse structure of the laser active region and the operating photon density. In practice the maximum modulation bandwidth of semiconductor lasers is often substantially lower than the value predicted by equation (9.25) due to electrical parasitics or thermal effects.

9.3.3 Parameter extraction example

In the subsection 9.3.2 the formula for RIN of a general class of semiconductor lasers has been developed from the spatially averaged single-mode rate equations. The present subsection involves the extraction of the dynamic parameters that govern the small signal response, linewidth and to a large degree the large signal behaviour of semiconductor devices. This is a nonlinear curve fitting problem of measured RIN data to equation (9.22)

The problem is stated as: given a data set (x_i, independent variable; y_i, dependent variable) of dimension n ($i = 1$ to n), we wish to compute the regression coefficients of a describing function which give an error minimum to measured data according to some selection criterion. Mathematically stated, given

$$y_1 = g\left(x, a_1, a_2, \ldots, a_j, \ldots, a_m\right), \tag{9.26}$$

where $a_1, a_2, \ldots, a_j, \ldots, a_m$ are the m regression coefficients of the describing function g, an optimum fit is reached when the quantity E (the χ^2 function) defined by

$$E = \sum_{i=1}^{n} \left[\frac{y_1 - g\left(x, a_1, a_2, \ldots, a_j, \ldots, a_m\right)}{\sigma_i} \right]^2 \tag{9.27}$$

is minimised. Here, σ_i is the standard deviation of the measured data value y_i at x_i. For the least squares criterion, as used below, σ_i is assumed to equal 1 for all data points.

In the case of curve fitting to equation 9.22 the a_i's are the parameters to be extracted (resonance frequency, damping rate etc.), σ_i is the error on the RIN measurements, x_i is the frequency and y_i the RIN value at frequency x_i.

One point of caution when performing this curve fit is to limit the curve fit to that frequency range in which the resonance frequency appears in the noise spectrum. This is particularly true of FP devices where, due to mode partition noise, the RIN spectra amplitude increases rapidly at low frequencies and the analysis of subsection 9.3.2 breaks down.

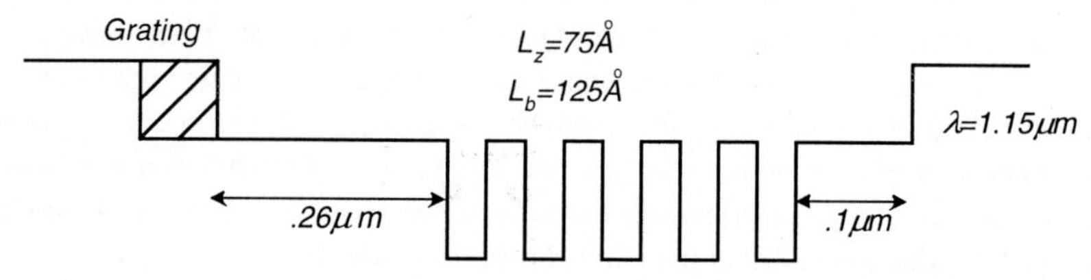

Fig. 9.2. Conduction band diagrams of the DFB laser. L_z is the well width and L_b is the barrier width.

By way of example, a DFB laser that has been circulated within COST 240 for round robin exercises has been characterised in terms of its RIN. Fig. 9.2 schematically illustrates the quantum well structure of the device under flat band conditions. Briefly, the device consists of five quantum wells, 7.5 nm wide, separated by 12.5 nm barriers sandwiched between asymmetric SCH (Separate Confinement Heterostructure) guiding layers. A first order frequency selective grating above the active region is the distributed feedback of the structure. The device is 500 μm in length and has a threshold current of 13 mA at 25°C under CW conditions.

Relative intensity noise measurements were performed using a HP71400C instrument that has been calibrated from 100 kHz to 22 GHz. Measurement results are summarised in Fig. 9.3 for a range of DC bias currents above threshold for frequencies up to 15 GHz. Applying the Levenberg-Marquardt curve fitting algorithm to equation 9.22 where the cost function (E) for the fitting has been formed in accordance with equation 9.27, the resonance frequency (f_r) and the resonance frequency squared (f_r^2), have been extracted for this range of DC bias conditions.

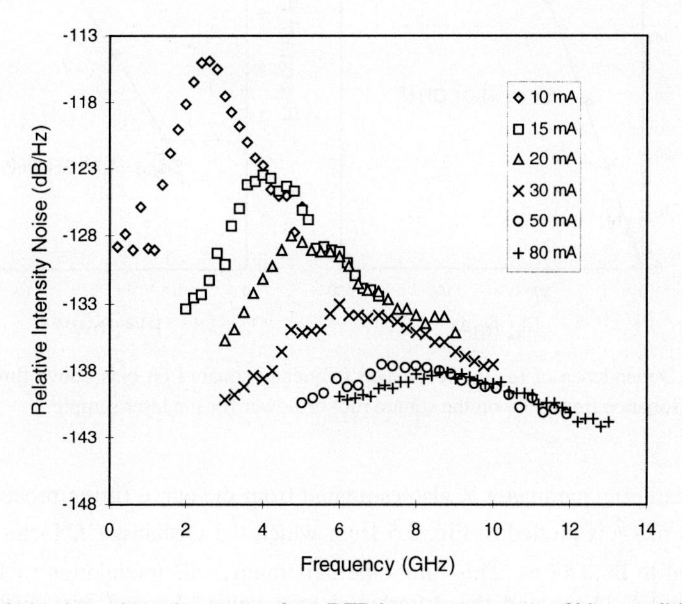

Fig. 9.3. Relative intensity noise spectra for a DFB laser over a range of bias conditions.

The evolution of f_r and f_r^2 with bias above threshold and $P^{1/2}$, respectively, are plotted in Fig. 9.4. Although both curves show good linear performance at low currents/power, both characteristics tend to saturate at higher drive conditions. This is primarily due to thermal effects. This laser shows good bandwidth efficiency (GHz/mW$^{1/2}$), primarily due to the lower barrier material bandgap wavelength of 1.15 μm. The bandwidth efficiency of devices having the more usual 1.3 μm barrier material, and having nominally similar lateral layer dimensions, would tend to be two to three times less than this at comparable output power levels. Using the measured slope of these curves and equation 9.20 the extracted differential gain value for the device is $3.574 \cdot 10^{-20}$ m^2. This compares well with the differential gain ($2.4 \cdot 10^{-20}$ m^2) estimated from sub-threshold measurements [19] for the same device.

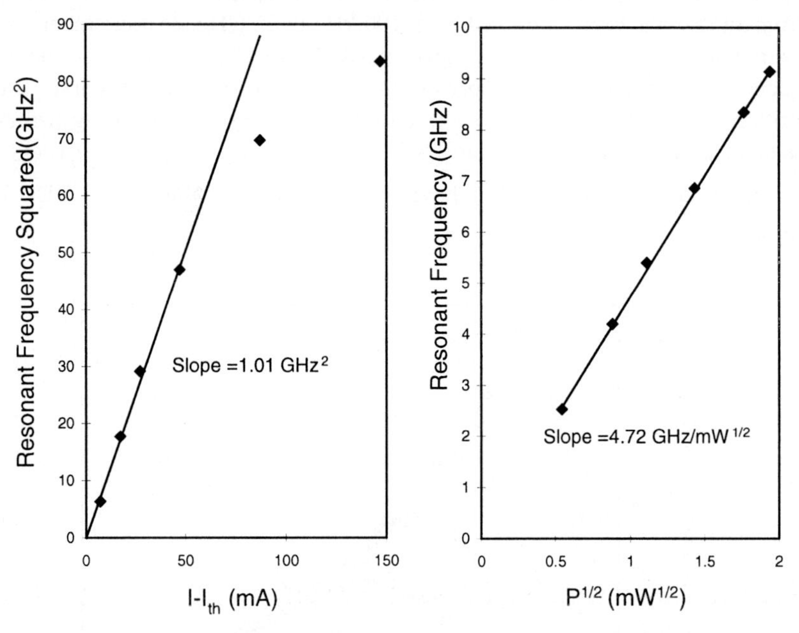

Fig. 9.4. Dependence of (a) the resonance frequency squared on bias above threshold and (b) the resonance frequency on the square root of power for the laser sample.

The damping parameter γ, also estimated from the curve fitting procedure, as a function of f_r^2 is plotted in Fig. 9.5 from which the Olshansky K factor has been estimated to be 0.48 ns. This limits the maximum 3 dB modulation to 18.5 GHz. Using this K factor and the differential gain value obtained previously, a gain compression coefficient ε of $2.17 \cdot 10^{-23}$ m^3 is derived, if the photon lifetime contribution to the K factor is neglected (see equation 9.24). This implies that although the DFB, which has a high effective quantum well barrier, has a large differential gain and is accompanied by a significant gain saturation coefficient. This limits the K factor value to 0.48 ns and the maximum modulation bandwidth accordingly.

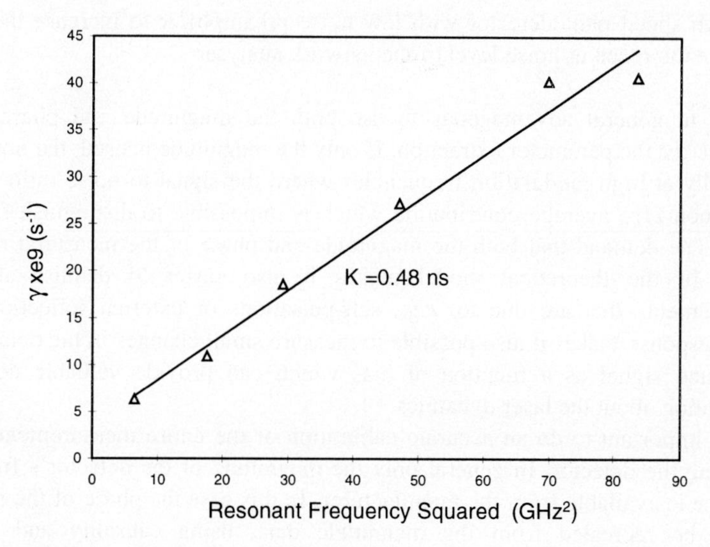

Fig. 9.5. Damping rate variation with the square of the resonance frequency. From the slope, the K factor is 0.48 ns.

9.4 Extraction from modulation response measurements

The dynamic parameters of the laser can, instead of using the RIN spectrum, alternatively be extracted from the small signal modulation response of the laser. Although this demands a more complicated measurement set-up it gives better dynamic range and also access to both magnitude and phase of the response, which make measurements at frequencies far above the relaxation peak possible. This technique enables also, apart from the quantities that can be determined from the RIN spectrum, extraction of the parasitic cut-off frequency of the device.

9.4.1 Measurement of modulation response

Measurements of the modulation response have been treated in detail in Sec. 8.3 and we will here only discuss issues that are important for parameter extraction. The measurement set-up for modulation measurements consists typically of

- a broadband network analyser which supplies a frequency swept electrical modulation signal
- a coaxial probe for applying the modulation signal directly to the laser chip in order to avoid the external parasitics of a bonding wire
- an optical isolator for avoiding that reflected light is coupled back to the laser

- a high speed photodetector with low noise preamplifier to increase the signal above the receiver noise level of the network analyser

It is in general advantageous to use both the magnitude and phase of the response for the parameter extraction. If only the magnitude is used, the noise will, especially at high modulation frequencies where the signal-to-noise ratio is poor, give a non-zero average contribution which is impossible to distinguish from the signal. The demand that both the magnitude and phase of the measured response should fit the theoretical model makes it also easier to dismiss abnormal measurements that are due to, *e.g.*, self-pulsations or external reflections. The phase response makes it also possible to measure small changes in the delay of the modulated signal as a function of bias which can provide valuable additional information about the laser dynamics.

It is important to do an accurate calibration of the entire measurement system including the detector. In general only the magnitude of the detector's frequency response is available from the manufacturer. In this case the phase of the response has to be recreated from the magnitude data, using causality and stability constraints of linear systems. For accurate parameter extraction it is also important to have a large signal-to-noise ratio. This can be achieved by using large temporal averaging factors during measurements. Averaging in the frequency domain, so called smoothing, is not recommended; it will reduce the frequency resolution of the signal and often corrupt the phase information, especially if the delay time is large.

A good rule is to measure the modulation response at five to ten different bias currents until a maximum current is reached where the 3-dB bandwidth saturates or starts to drop. The procedure for the parameter extraction is essentially the same as in the case of RIN measurements: fitting the measured response to a theoretical formula.

9.4.2 Theoretical formula for modulation response

It is simple to show, by a small signal expansion of the standard rate equations, that the intrinsic modulation response will be governed by the same resonant denominator as the RIN spectrum. The measured response will often, in addition to the two complex conjugated poles of this intrinsic response, also contain a third real pole. This parasitic-like roll-off is in general due to the electrical capacitance of the metallic contact and the high conductive layers below it. However, also slow diffusion limited carrier transport through undoped separate confinement layers can give rise to parasitic-like effects since the charge pile-up on either side of the active layer will essentially act as a plate capacitor. Diffusion-limited carrier transport and the finite capture time into the well will contribute to the damping of the resonance frequency [20], [21]. There is also a delay in the measurement system which is difficult to fully calibrate which is necessary to include in the theoretical model.

From the above discussion we consider the following theoretical transfer function

$$H(\omega)=\frac{C}{\left(-\omega^2+j\omega\gamma+\Omega_R^2\right)\left(j\omega+\Omega_p\right)}e^{j\omega\tau_D} \tag{9.28}$$

Here C is a coefficient that depends on the coupling efficiency and the quantum efficiency, Ω_R and γ are the resonance frequency and damping rate of the relaxation peak, Ω_p is the parasitic-like roll-off frequency and τ_D is the delay time. The above formula fits surprisingly well the real response of most lasers.

9.4.3 Fitting procedure and extraction example

The parameters in equation (9.28) can be extracted by fitting it to the experimental data. Note, that the denominator of $H(\omega)$ is a polynomial in $j\omega$ with *real* coefficients where all zeros should have a negative real part. These constraints make it possible to detect unrealistic measurements and calibration errors. The fitting can be done by either using the χ^2-criterion or the least mean-square criterion:

$$\chi^2=\sum_{i=1}^{N}\frac{\left|H_i-H(\omega_i)\right|^2}{\sigma_i^2} \tag{9.29}$$

The above χ^2-criterion, where the data points are weighted down proportional to their standard deviation, is to be preferred if the model error is smaller than the error due to noise, and if the system noise is strongly frequency dependent. This demands a well-behaved laser and accurately calibrated measurement system, otherwise the least mean-square criterion with $\sigma_i^2=1$ is to be preferred. The standard deviation used in the χ^2-formula can in practice often be estimated from a high order numerical derivative with respect to frequency of the measured data. However, this technique works only if the frequency dependence of the noise spectrum and the transfer function is slow compared to the frequency spacing of the measured points. It also assumes that the noise in different measurement points, corresponding to different frequencies, is uncorrelated.

The minimisation of χ^2 is conveniently done using the computationally effective Levenberg-Marquardt method which solves the problem in an iterative manner. There are, however, some issues to keep in mind.

The first issue is to determine how many and which measurement points should be used in the extraction. As a general rule of thumb the measurement points should be spaced closely enough so that the difference in value between two neighbouring points is mainly governed by noise. Hence, it is not necessary to have many points in regions where the signal is noisy or lacks frequency dependent features.

At low frequencies, below 1 GHz, a slight peak or dip in the modulation response is often seen. The occurrence of this peak or dip is in most, but not all, cases correlated with reduced side mode suppression. Since such effects are not included in the theoretical model, it is better to omit this part of the signal during extraction in order to get a lower model error and more accurate extraction parameters. At high frequencies the signal-to-noise ratio is often poor. This noise in the high frequency region is in general not only caused by ordinary amplifier noise but also noise due to calibration problems of the measurement set-up. Even slight movement of the measurement cables will change the phase of distant reflections which can give a noisy appearance of the measurement curve. This type of noise is not uncorrelated at different measurement points and its standard deviation to use in the χ^2-formula cannot easily be determined. Hence, it is preferred to omit the highest frequency part of the curve where the signal-to-noise ratio is very poor.

Another issue is local minima which can occur especially due to the time delay factor. A way to circumvent this is to first only fit the magnitude squared, $|H(\omega)|^2$, of the response to get approximate values of the parameters. For this first step it is possible to use a non-iterative method by linearising above equation by multiplying with the denominator of $|H(\omega)|^2$. Setting the derivative of χ^2 with respect to each one of the parameters to zero gives a linear equation system that can be solved with usual techniques. As an alternative one can use singular value decomposition to solve the overdetermined equation system $\chi^2 = 0$ (one equation for every point). Knowing approximate values of the poles one can thereafter extract an approximate delay time. The last step is polishing the obtained results by minimising the χ^2-value with respect to all parameters simultaneously using the Levenberg-Marquardt method.

The minimum value of χ^2 can be divided into two parts, error due to noise and error caused by discrepancies between the measured signal and the model, $\chi^2 = \chi^2_{\text{noise}} + \chi^2_{\text{model}}$. If the standard deviation of the measurement points is correctly determined, the expectation value of χ^2 due to the noise contribution, assuming Gaussian noise, will equal the number of degrees of freedom, $2N$–5. Here N is the number of measurement points, the factor 2 appears because there is noise in both quadratures and 5 stems from the number of extracted parameters. From this it is possible to independently monitor how the error due to noise and model error change if a measurement point is omitted. If the model error is several times larger than the noise error, it is a sign that either the measurement set-up or the laser do not behave as expected and that the extracted parameter values are not reliable.

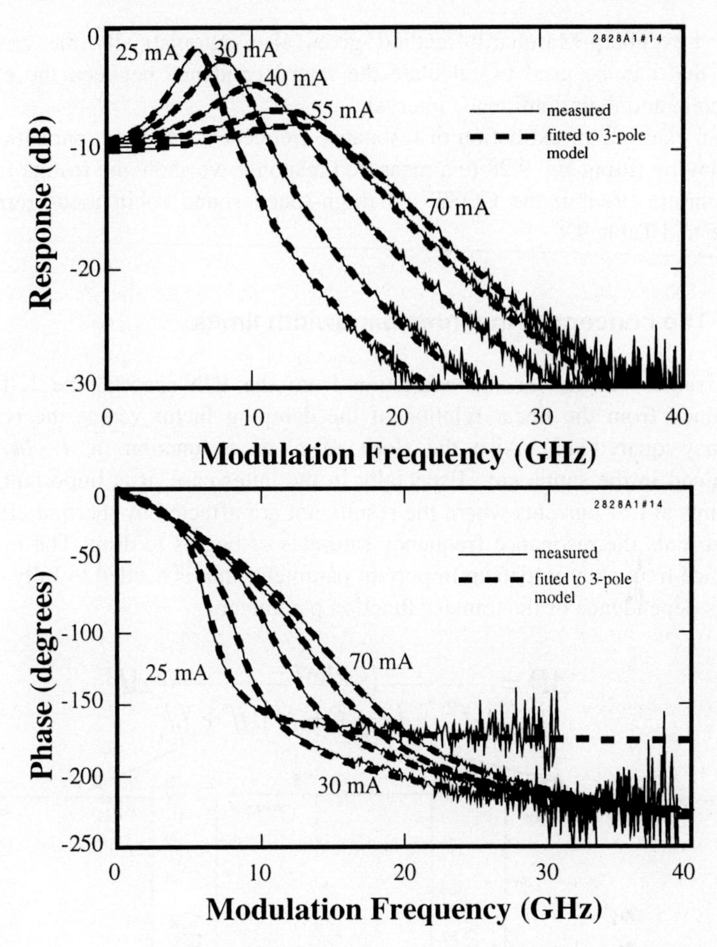

Fig. 9.6. Example of a measured (solid line) and fitted (dashed line) modulation response from the laser used for high speed measurements.

I (mA)	f_r(GHz)	γ (ns^{-1})	f_p(GHz)	τ_D(ns)	H(0)	f_{3dB}(GHz)
25	6.557	17.601	>40	-0.23280	0.34920	9.857
30	8.530	24.448	28.514	-0.22790	0.34730	12.403
40	11.273	40.113	28.167	-0.22810	0.33880	15.736
55	14.307	58.192	19.185	-0.22740	0.33320	18.269
70	15.939	69.335	19.222	-0.22760	0.31500	19.577
90	17.000	80.628	17.316	-0.2275	0.28398	19.650
110	16.793	87.290	18.216	-0.22770	0.24620	19.014
130	15.383	79.550	21.851	-0.22780	0.21660	18.453

Table 9.4. Extracted parameters from the laser in Fig. 9.6.

The Levenberg-Marquardt method gives also automatically the covariance matrix that can be used to calculate the interdependence between the extracted parameters and their confidence interval.

As an example of extraction of resonance frequency, damping, parasitic cut-off and delay by fitting Eq. 9.28 to a measured response we show the results from one of the lasers used in the COST 240 high-speed round robin measurements in Fig. 9.6 and Table 9.4.

9.4.4 The concept of the three bandwidth limits

In the same way as for the extraction from the RIN-spectra, the K-factor is determined from the linear relation of the damping factor versus the resonance frequency squared [17]. Also the slope of f_r^2 as a function of $I - I_{th}$ can be determined in the same way. Especially in the latter case, it is important to only use points at low currents where the results not are affected by thermal effects. At high currents the resonance frequency saturates or begins to drop. The maximum resonance frequency is also an important parameter that is needed to fully describe the bias dependence of the transfer function parameters

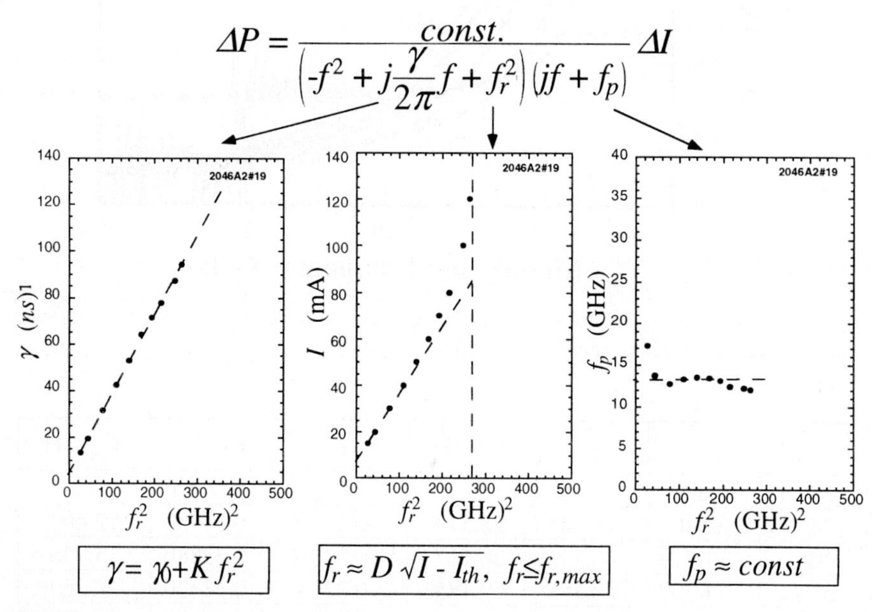

Fig. 9.7. Picture that visualises the parametrisation of the bias dependence of the extracted damping factor, resonance frequency and parasitic-like roll-off frequency.

From these relations it is possible to distinguish three separate reasons that can limit the laser bandwidth [22]:

- The laser can be limited by the parasitic-like real pole which gives a 3-dB *parasitic limit:*

$$f_{3dB,par} = f_p \tag{9.30}$$

Theory: $$f_{3dB,par} \approx \frac{1}{2\pi} \frac{1}{\tau_{diff} + \tau_{cap,e} + \tau_{RC}} \tag{9.31}$$

- At high relaxation frequencies the linear relation between γ and f_r^2 implies a maximum possible bandwidth, the *damping limit* given by [17]

$$f_{3dB,damp} = \frac{2\sqrt{2}\pi}{K} \tag{9.32}$$

Theory: $$f_{3dB,par} \approx \frac{\sqrt{2}}{2\pi} \frac{1}{\tau_p + \dfrac{\varepsilon}{\left(v_g a/\chi\right)}} \tag{9.33}$$

- Often the laser is instead limited by thermal effects that give a maximum relaxation frequency $f_{r,max}$ that occurs at a current I_{max} before the damping limit is reached. This *thermal limit* is given by

$$f_{3dB,therm} = \sqrt{1+\sqrt{2}}\, f_{r,\max}, \tag{9.34}$$

Theory: $$f_{3dB,therm} = \frac{\sqrt{1+\sqrt{2}}}{2\pi} \sqrt{\Gamma v_g \frac{a}{\chi} \frac{I_{\max} - I_{th}}{eV}} \tag{9.35}$$

where a is used to denote the differential material gain. The above theoretical results are obtained from an ambipolar rate equation model with three carrier reservoirs: one for the carriers in the SCH layers, one for the carriers in the gateway states just above the quantum well and one for the optically active carriers in the wells [22]–[24]. The transport effects can then for moderate frequencies be taken into account with a correction factor $\chi = 1 + \tau_{cap,e} / \tau_{esc}$. Here, $\tau_{cap,e} = \tau_{cap}(1 + \tau_{diff} / \tau_{diff(well)})$, where τ_{cap} is the local capture time from the gateway state into the well, $\tau_{cap,e}$ is the escape time out of the well, τ_{diff} is the classical diffusion time from the SCH-reservoir to the gateway state and $\tau_{diff(well)}$ describes the diffusion in opposite direction. Note that above equations are limits, *i.e.*, each equation gives the true maximum 3-dB bandwidth only if the other two limits are significantly higher.

For lasers where the contact parasitics are small it is often found that the parasitic-like cut-off frequency is not constant but decreases with increasing bias [22]. This effect can possibly be explained with transport effects [25], [26]. In these cases the value of the parasitic limit can be compared with the other two limits at the bias which gives the maximum overall bandwidth.

For the laser response in Fig. 9.6 the damping limit was found to be 32.9 GHz, the thermal limit 26.4 GHz and the parasitic limit 17.3 GHz yielding an overall bandwidth of 19.6 GHz. In this case the resonance peak lifts the response at high frequencies so that the resulting bandwidth becomes larger than the parasitic limit.

By comparing the above limits it is possible to identify the physical mechanism that is the main bandwidth limiting factor.

9.5 Other methods and the role of facet properties

As demonstrated in the previous sections any laser parameter can be extracted from experimental data of, *e.g.* ASE spectra, modulation studies and RIN profiles. Not all these different methods are equally sensitive to individual parameters. A high precision is obtained, *e.g.*, (i) by RIN and modulation studies in the determination especially of the differential gain $a=dg/dN$, (ii) by ASE spectra, in particular, in the extraction of the coupling coefficient κ, the effective refractive index n_{eff} and the facet phases, (iii) by wavelength tunability studies especially the differential refractive index dn_{eff}/dN, and (iv) by the recombination coefficients from dynamic measurements. The parameter extraction, however, is an iterative process intended to simultaneously describe all measured properties of a laser device or even better a larger number of devices originating from the same wafer. This iterative process based on different measurements also allows for cross-checks. By using linewidth measurements and simulations, one can check the parameters differential gain dg/dN and refractive index dn_{eff}/dN extracted, *e.g.*, from RIN profiles and wavelength tunability studies, respectively. Note that the experimental problems in linewidth measurements may be considerable (see Ch. 8.2). On the other hand, the quality of the linewidth measurements can be cross-checked by RIN studies and tunability measurements combined with corresponding simulations and parameter extractions, involving the material linewidth enhancement factor α. Finally, the recombination coefficients A (linear), B (square) and C (cubic) are extracted from dynamic studies which have already been described in subsection 8.3.5.

Subsection 9.5.2 is devoted to the crucial influence of DFB laser facet properties. DFB laser parameters such as threshold current, output power, spectral mode positions, wavelength tunability, static linewidths, chirp under high bit rate modulation, single-mode yield and the stability of single-mode oscillation vary considerably between nominally identical devices, due to the influence of the statistics in the facet phases [27]–[31]. Facet coatings and the balance between real and imaginary coupling governs the amount of variation between the lasers which are nominally identical but differ in facet phases. Finally, some aspects are summarised which may be important for both basic research and industrial DFB laser production to enable a separation of the above mentioned variations from technological fluctuations.

9.5.1 Cross-check of extracted parameters by different methods

In the following example a DFB laser of L = 165 μm length and 1.52 μm emission wavelength is considered including 5 lattice-matched $In_{0.53}Ga_{0.47}As$ quantum wells of 7.5nm widths and $In_{0.53}Al_{0.19}Ga_{0.28}As$ barriers/confinement layers. Two $\lambda/8$ phase-shifts are intentionally placed at 45 μm and 75 μm from facet 1 (both facets are as cleaved). The facet phases are unknown due to natural fluctuations. Using transfer matrix model calculations [28], [30], [31] and ASE spectra, the facet phases are extracted yielding Θ_1=1.56π and Θ_2=1.42π, by varying the facet phases continuously but independently from each other between 0 and 2π. The accuracy in the extraction is about $\lambda/30$. The extracted index grating coupling coefficient is κ_i=114cm^{-1}. Fig. 9.8 shows the mode position of the oscillating mode λ (I) versus current. Many effects contribute to the absolute wavelength values of the spectral mode positions. The wavelength critically depends on the effective refractive index n_{eff}, influenced by geometric or physical parameters (nonlinear gain coefficient ε, transparency carrier density N_0, coupling coefficient κ_i, facet phases, lifetime, current heating). n_{eff} has been first determined by waveguiding calculations to about 3.24 yielding the coarse value. Fine-tuning was performed by fitting mainly the calculated spectra to the measured ones, while also considering the other experimental data to be described by the same parameter set. Spectra were measured for different currents and heat sink temperatures to determine $\lambda(I)$ and λ(T). For the calculation of $\lambda(I)$ the variations $n_{eff}(T, N)$, dN/dI and dT/dI are required. dT/dI is an experimental and device-specific property varying, *e.g.*, with L and soldering conditions. dn_{eff}/dT and dn_{eff}/dN are material properties to be extracted. The wavelength differences measured between pulsed and continuous current are automatically described correctly for different lasers of the same wafer which is regarded as an additional proof of self-consistency in the simulations.

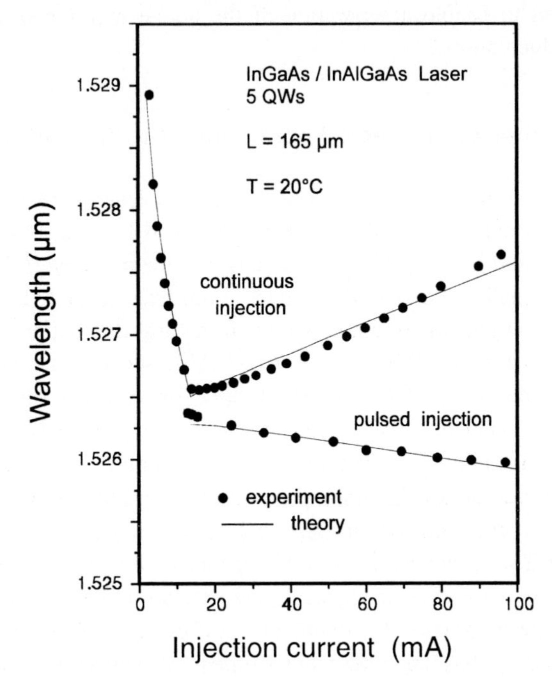

Fig. 9.8. Spectral variation of the oscillating mode versus continuous and pulsed currents for a 1.52 µm DFB laser.

For pulsed injection below and above threshold, $\lambda(I)$ is very sensitive to dn_{act}/dN (act=active layers) yielding $2 \cdot 10^{-20}\text{cm}^{-3}$. This value was also extracted from other lasers of the same wafer, although, under pulsed injection, $\lambda(I)$ is totally different from the values in Fig. 9.8, depending on L and mounting. This demonstrates that a precise parameter extraction is possible without being influenced by differences in L and heat transfer to the heat sink. Note that under pulsed bias, $\lambda(I)$ varies considerably even for lasers of identical L, κ_i and dn_{act}/dN due to individual longitudinal carrier density variations and individual spatial hole burning, both caused by different facet phases.

Fig. 9.9 (a) depicts the frequency f_r for which the maximum RIN has been measured versus continuous current. Fig. 9.9 (b) displays the RIN power at f_r versus current. The experimental data (dots) are very well described by the calculated results (full lines) with $dg/dN = 3.3 \cdot 10^{-16}\text{cm}^2$ (obtained for 150 µm < L < 200 µm and $\lambda \sim 1.52$µm). The values of dn_{act}/dN and dg/dN are now cross-checked by comparing the experimental linewidth (dots) versus current, Fig. 9.9 (c), with a calculated profile (full line), showing good agreement. For high currents we obtain in the experiment a linewidth saturation. Several additions to the linewidth theory were reported in the literature which describe these deviations but involve additional parameters and do not yield a higher accuracy in the verification of dg/dN and dn_{act}/dN.

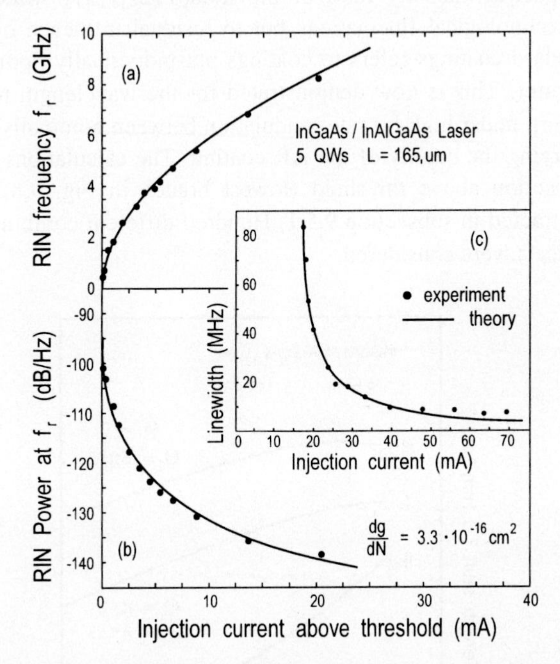

Fig. 9.9. (a) RIN resonance frequency f_r, (b) RIN power at f_r, (c) linewidth versus current.

As already demonstrated in subsection 8.3.5, several important laser properties such as the differential carrier lifetime or the carrier density, both versus current, can be determined from dynamic measurements [32]. Fits to the latter curve (see Figure 8.19) provide the recombination coefficients: A (excitonic and non radiative), B (bimolecular free carrier) and C (Auger) giving $A=1.4\cdot10^{9}\,s^{-1}$, $B=2.7\cdot10^{-10}\,cm^{3}s^{-1}$, $C=0.98\cdot10^{-29}\,cm^{6}s^{-1}$ for the sample "A" studied in subsection 8.3.5. The accuracy obtained by this method is extremely high for A and good for B and C. The excellent agreement between the experimental and theoretical data, fundamental to extract reliable carrier density and recombination parameters, was found to be typical of this technique. This is true especially at very low bias current and close to the device threshold, where more conventional techniques give worse accuracy in the parameter extraction and experimental reliability.

9.5.2 The role of facet properties

This subsection will focus on explanations of the fact that DFB lasers originating from industrial device production from a single wafer can differ considerably in many laser properties. Although standard antireflection (AR) coatings were deposited, the devices behave like individuals concerning variations in threshold current and spectral intensity ratio of the modes [27]–[31] which cannot be attributed to technological fluctuations but to residual influence of facet phases. The term standard coatings refers to coatings not individually optimised for each laser of the wafer. This is now demonstrated for the wavelength tuning $\lambda(I)$ and wavelength chirp under high bit rate modulation between nominally identical laser devices by varying the quality of the AR coating. The calculations are performed for pulsed injection above threshold (lowest branch in Fig. 9.8) based on the parameters extracted in subsection 9.5.1. Hundred different combinations (Θ_1, Θ_2) of uncoated facets were considered.

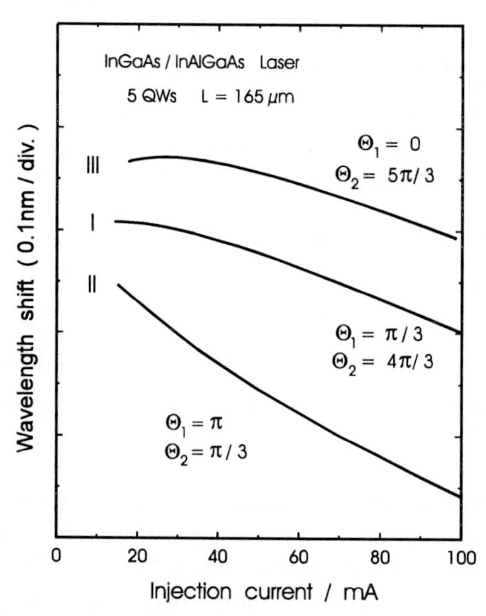

Fig. 9.10. Calculated wavelength variation of the oscillating mode versus current for three different combinations (Θ_1, Θ_2) of the facet phases.

Depending on the facet phases combinations and other physical conditions, either the Bragg mode or a side mode oscillates and individual longitudinal variations of the carrier and photon densities are revealed. Fig. 9.10 displays $\lambda(I)$ of the oscillating mode for 3 lasers differing only in the facet phases combinations. The different slopes are due to the individual longitudinal carrier density profiles [30], [31]. The results of wavelength shift and dynamic chirp for various facet coatings are depicted in Fig. 9.11: (i) lasers with both sides symmetrically AR coated, denoted by AR/AR (ii) an asymmetrically coated laser revealing a high reflection (HR) coating at facet 1 (power reflectivity =1.0) and an antireflection (AR) coating with a power reflection of 10^{-3} at facet 2, denoted by HR/AR.

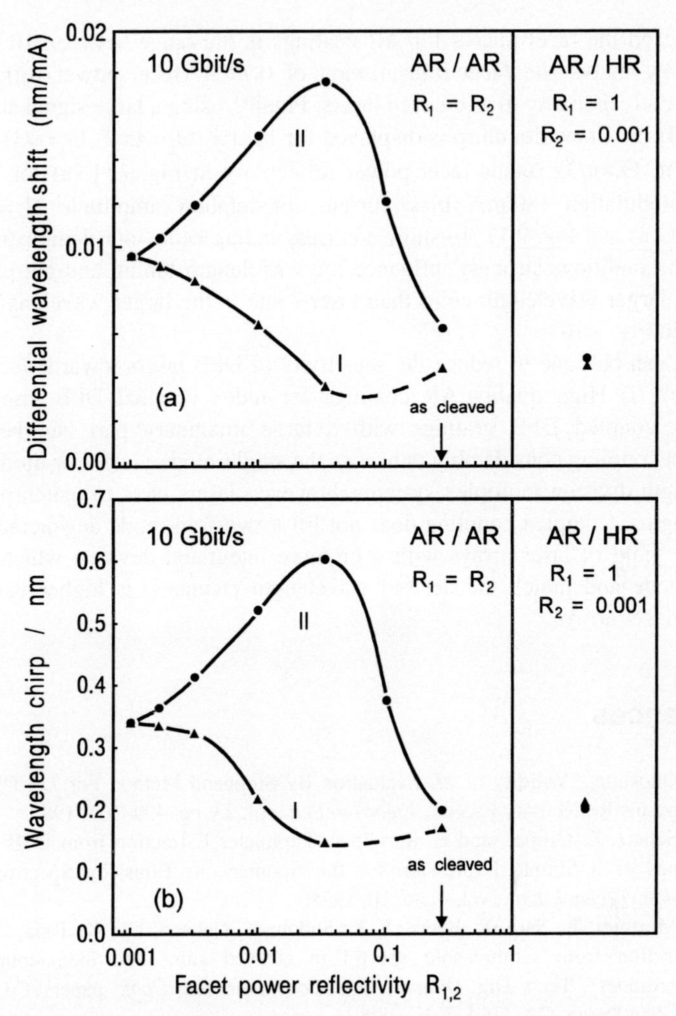

Fig. 9.11. (a) Differential wavelength shift versus the facet power reflectivities for two different facet phases combinations (laser I and II), HR = high reflection, AR = antireflection, (b) Wavelength chirp at 10 Gbit/s direct modulation (bias current I_o= 80 mA, ΔI_{pp} = 40 mA peak to peak).

To enable a condensed display of the calculated wavelength shifts of the different lasers the parameter "differential wavelength shift" is created, being a linear fit to the $\lambda(I)$ variation between 60 mA and 100 mA. Even for moderately bent $\lambda(I)$ variations, the differential wavelength provides a good impression of the strength of the $\lambda(I)$ variation. For both AR/AR and HR/AR coated devices the influence of the facet phases on the wavelength shift tends to disappear for AR coatings better than 10^{-3}. However, there exists an unexpectedly high residual influence on the facet phases for AR coatings in the range between 10^{-1} and 10^{-2}. The arrow marks the facet transmission of 0.7204 (facet power reflectivity = 0.2796) corresponding to as cleaved lasers. Finally, using a large signal analysis [31], [33] the linewidth chirp is displayed for laser I ($\Theta_1 = 4\pi/3$, $\Theta_2 = \pi/3$) and laser II ($\Theta_1 = \pi$, $\Theta_2 = \pi/3$) versus facet power reflectivity in Fig. 9.11 (a) for 10 Gbit/s direct modulation (80 mA bias current, modulation amplitude (I_{pp}=40 mA)). Fig. 9.11 (a) and Fig. 9.11 (b) show a corresponding behaviour demonstrating that the facet conditions strongly influence the wavelength tuning and chirp. Laser II shows a larger wavelength chirp than laser I due to the larger wavelength shift of laser II in Fig. 9.10.

What can be done to reduce the sensitivity of DFB lasers towards facet phases statistics? (i) High quality AR coatings for index coupled DFB lasers or (ii) complex coupled DFB gratings with a large imaginary part can be applied. Complex coupling considerably enhances the single-mode yield. For modern dense wavelength division multiplex systems, however, lasers have to match predefined wavelengths. Complex coupling does not lift a two-fold mode degeneracy. In this case the yield of laser arrays with 4 or more integrated devices, which are both single-mode and match the desired wavelength channels, is higher using index coupled laser arrays.

References

[1] J. Kinoshita, "Validity of κL Evaluation By Stopband Method For λ/4 DFB Lasers with Low Reflectivity Facets", *Electron. Lett.*, vol. 23, pp. 499–501, 1987.

[2] R. Schatz, L. Gillner, and E. Berglind, "Parameter Extraction from DFB Lasers by Means of a Simple Expression for the Spontaneous Emission Spectrum", *IEEE Photon. Technol. Lett.*, vol. 6, no. 10, 1994.

[3] G. Morthier, K. Sato, R. Baets, T. K. Sudoh, Y. Nakano, and K. Tada, "Parameter extraction from subthreshold spectra in cleaved gain- and index-coupled DFB Laserdiodes", Tech. Dig. *Optical Fiber Communication Conf.*, paper FC3, pp. 309–310, San Diego, CA, USA, Feb. 1995.

[4] F. Pusa, J. Skagerlund, L. Gillner, O. Sahlén, R. Schatz, P. Granestrand, L. Lundqvist, B. Stoltz, J. Terlecki, F. Wahlin, A. Mörner, J. Wallin, and O. Öberg, "Evaluation of an automatic method to extract the grating coupling coefficient in different types of DFB lasers," *IEEE J. Quantum Electron.*, vol. 34, no. 1, pp.141–146, 1998.

[5] T. Makino and J. Glinski , "Transfer Matrix Analysis of the Amplified Spontaneous Emission of DFB Semiconductor Amplifiers", *IEEE J. Quantum Electron.*, vol 24, no. 8, pp.1507–1518, 1988.

[6] L. A. Coldren and S. W. Corzine, *Diode Lasers and Photonic Integrated Circuits*, John Wiley & Sons, Inc., 1995.

[7] B.W. Hakki and T.L. Paoli, "CW degradation at 300 K of GaAs double heterostructure junction lasers. II Electronic gain", *J. Appl. Phys.*, vol. 44, no. 9, pp. 4113–4119, 1973.

[8] G. Morthier, P. Verhoeve, R. Baets, and R. Schatz, "Extraction of a large set of laser parameters from different measurements", *15th IEEE Intern. Semicond. Laser Conf.*, Haifa, Israel, paper Th. 2.5 pp. 175–176, Oct. 13-18, 1996.

[9] W. Press, B. P. Flannery, S. A. Teukolsky, and W. T. Vetterling, *Numerical Recipies*, Cambridge University Press, 1986.

[10] "Optical Source Module for Fiber in the Loop (FITL) systems", *Bell Communications Research,* Technical Advisory, TA-NMT-000786, Issue 2, Dec. 1991.

[11] J.M. Osterwalder and B.J. Rickett, "Frequency Modulation of GaAlAs Injection Lasers at Microwave Frequency Rates", *IEEE J. Quantum Electron.*, vol. 16, pp. 250–252, 1980.

[12] P.A. Morton, T. Tanbun-Ek, R.A. Logan, A.M. Sergent, P.F. Sciortino, and D.L. Coblentz, "Frequency Response Subtraction for Simple Measurement of Intrinsic Laser Dynamic Properties", *IEEE Photon. Technol. Lett.,* vol. 4, pp. 133–136, 1992.

[13] J. Eom, C.B. Su, J. S. LaCourse, and R. B. Lauer, "The relation of doping level to K Factor and the effect on Ultimate Modulation Performance of Semiconductor Lasers", *IEEE Photon. Technol. Lett.,* vol. 2, no. 10, 1990, pp. 692–694.

[14] B. Zhao, T.R. Chen, S. Wu, Y.H. Zhuang, Y. Yamada, and A. Yariv, "Direct Measurement Of Linewidth Enhancement Factors In Quantum Well Lasers Of Different Quantum Well Barrier Heights", *Appl. Phys. Lett.*, vol. 62, pp. 1591–1593, 1993.

[15] R. Nagarajan, M. Ishikawa, T. Fukushima, R.S. Geels, and J.E. Bowers, "High Speed Quantum Well Lasers and Carrier Transport Effects", *IEEE J. Quantum Electron.,* vol. 28, pp. 1990–2008, 1992.

[16] A. Hangleiter, A. Grabmaier, and G. Fuchs, "Damping of The Relaxation Resonance In Multiple Quantum Well Lasers By Slow Interwell Transport", *Appl. Phys. Lett.,* vol. 62, pp. 2316–2318, 1993.

[17] R. Olshansky, P. Hill, V. Lanzisera, and W. Powazinik, "Frequency response of 1.3 μm InGaAsP high speed semiconductor lasers", *IEEE J. Quantum Electron.*, vol. 23, no. 10, pp. 1410–1418, 1987.

[18] M.C. Tatham, I.F. Lealman, C.P. Seltzer, L.D. Westbrook, and D.M. Cooper, "Resonance Frequency, Damping And Differential Gain In 1.5 μm Multiple Quantum Well Lasers", *IEEE J. Quantum Electron.*, vol. 28, pp. 408–414, 1992.

[19] L.D. Westbrook, "Measurements of ∂G/∂N and ∂n/∂N And Their Dependence On Photon Energy In λ=1.55 μm InGaAsP Laser Diodes", *IEE Proc. J,* vol. 133, pp. 135–142,1986.

[20] W. Rideout, W. F. Sharfin, E. S. Koteles, M. O. Vassell, and B. Elman, " Well-Barrier Hole Burning in Quantum Well Lasers", *IEEE Photon. Technol. Lett.,* vol. 3, pp. 784–786, 1991.

[21] R. Nagarajan, T. Fukushima, M. Ishikawa, J. E. Bowers, Randall S. Geels, and Larry A. Coldren, "Transport Limits in High-Speed Quantum-Well Lasers: Experiment and Theory", *IEEE Photon. Technol. Lett.*, vol. 4, pp. 121–123, 1992.

[22] O. Kjebon, R. Schatz, S. Lourdudoss , S. Nilsson, and B. Stålnacke, "Modulation response measurements and evaluation of MQW InGaAsP lasers of various designs", Invited talk at Photonics West (High Speed Semiconductor Laser Sources), *SPIE proc.*, vol. 2684–25, San Jose Cal. USA, 27 Jan. –2 Feb. 1996.

[23] T. Ishikawa, R. Nagarajan, and J. E. Bowers, "Analysis of the effects of doping and barrier design on the small-signal modulation characteristics of long-wavelength

multiple quantum well lasers", *Opt. Quantum Electron.,* vol. 26, no. 7, pp. S805–S816, 1994.

[24] D. McDonald and R. F. O'Dowd, "Comparison of two- and three-level rate equations in the modeling of quantum-well lasers", *IEEE J. Quantum Electron.,* vol. 31, pp. 1927–1934, 1995.

[25] N. Tessler and G. Eisenstein, "Modelling carrier dynamics and small-signal modulation response in quantum-well lasers", *Opt. Quantum Electron.*, vol. 26, no. 7, pp. S767–787, 1994.

[26] M. Ishikawa, R. Nagarajan, T. Fukushima, J. G. Wasserbauer, and J. E. Bowers, "Long wavelength high-speed semiconductor lasers with carrier transport effects", *IEEE J. Quantum Electron.,* vol. 28, no. 10, pp. 2230–2241, 1992.

[27] T. Matsuoka, Y. Yoshikuni, and H. Nagai, "Verification of the light phase effect at the facet of DFB laser properties", *IEEE J. Quantum Electron.* vol. 21, pp. 1880–1886, 1985.

[28] S.Hansmann, "Transfer matrix analysis of the spectral properties of complex distributed feedback laser structures", *IEEE J. Quantum Electron.* , vol 28, pp. 2589–2595, 1992.

[29] T. Kjellberg, S. Nilsson, T. Klinga, B. Broberg, and R. Schatz, "Investigation on the spectral characteristics of DFB lasers with different grating configurations made by electron-beam lithography", *J. Lightwave Technol.*, vol. 11, no. 9, 1405–1415, 1993.

[30] H. Hillmer, S. Hansmann, H. Walter, and H. Burkhard, "Experimental and theoretical study of the facet phase influence on the wavelength shift in InGaAs/InAlGaAs quantum well distributed feedback lasers", *Appl. Phys. Lett.*, vol. 64, pp. 698–670, 1994.

[31] H. Hillmer, S. Hansmann, H. Walter, H. Burkhard, A. Krost, and D. Bimberg, "Study of wavelength shift in InGaAs/InAlGaAs QW DFB lasers based on laser parameters from a comparison of experiment and theory", *IEEE J. Quantum Electron.,* vol. 30, p. 2251, 1994.

[32] R. Paoletti, M. Meliga, and I. Montrosset, "Optical modulation technique for carrier lifetime measurement in semiconductor lasers", *IEEE Photon. Technol. Lett.*, vol. 8, p. 1447, 1996.

[33] S. Hansmann, H. Burkhard, H. Walter, and H. Hillmer, "A tractable large-signa dynamic model - application to strongly coupled distributed feedback lasers", *IEEE J. Quantum Electron.*, vol. 12, pp. 952–956, 1994.

Part III

Nonlinear Effects in Semiconductor Optical Amplifiers: Four-Wave Mixing

Edited by:
François Girardin
Swiss Federal Institute of Technology, Switzerland

Authors

Stefan Diez	Heinrich-Hertz-Institut für Nachrichtentechnik, Berlin, Germany
Thomas Ducellier	Alcatel Alsthom Recherche, Marcoussis, France
François Girardin	Swiss Federal Institute of Technology Zurich, Switzerland
Dominique Marcenac	British Telecom Research Labs, Ipswich, UK
Antonio Mecozzi	Fondazione Ugo Bordoni, Roma, Italy
Jesper Mørk	Technical University of Denmark, Lyngby, Denmark
Kristof Obermann	Technical University Berlin, Germany
Simona Scotti	Fondazione Ugo Bordoni, Roma, Italy

Further contributors

Béatrice Dagens	Alcatel Alsthom Recherche, Marcoussis, France
Roberto Dall'Ara	Opto Speed SA, Mezzovico, Switzerland
Alessandro D'Ottavi	Fondazione Ugo Bordoni, Roma, Italy
Laura Graziani	Fondazione Ugo Bordoni, Roma, Italy
Antony Kelly	British Telecom Research Labs, Ipswich, UK
Stefan Kindt	Technical University Berlin, Germany
Igor Koltchanov	Technical University Berlin, Germany
Reinhold Ludwig	Heinrich-Hertz-Institut für Nachrichtentechnik, Berlin, Germany
Faustino Martelli	Fondazione Ugo Bordoni, Italy
Carsten Schmidt	Heinrich-Hertz-Institut für Nachrichtentechnik, Berlin, Germany
Paolo Spano	Fondazione Ugo Bordoni, Roma, Italy
Jörg Troger	Swiss Federal Institute of Technology Lausanne, Switzerland

What do we designate as *nonlinear effects* in semiconductor optical amplifiers? We basically mean all the observable effects which result from a *nonlinear behaviour of the device.* The origins of this behaviour can be either purely nonlinear properties of the semiconductor material or nonlinearities due to the structure of the device, or both of them. The former are for example, the spectral hole burning, the Kerr effect or the two photon absorption, the latter can be, for example, the carrier density saturation or the longitudinal non-uniform distribution of the carriers in the cavity.

How can we use these nonlinear effects in semiconductor devices? Semiconductor materials and devices are of great interest because of the strength of their nonlinear properties. Among many examples we can mention the use of the excitonic absorption effects in quantum-well materials for modulators and four-wave mixing in passive waveguides. Furthermore, the use of stimulated emission in semiconductor active devices significantly enhances the field of applications. For this reason, a world-wide interest in semiconductor optical amplifiers has recently developed.

In the previous Parts of this book passive waveguide structures and linear properties of active semiconductor devices have been discussed. The topic of this last Part is the nonlinearities in bulk semiconductor optical amplifiers, with the example of four-wave mixing. Many aspects of this example are investigated in order to give a full idea of it, and, furthermore, the reader should find here theoretical approaches and detailed experimental methods the relevance of which spreads on all aspects of the treatment of bulk nonlinear devices.

The reader should be aware that the notation slightly differs from the one used in the previous chapters, but that all these changes are mentioned in the text. In particular, the physicists agreement is chosen for the Fourier transformation and the letter " i " is used to designate the imaginary unity number.

10 Why and how to study four-wave mixing?

S. Diez

Photonic networks are the backbone of todays global communication systems. In order to satisfy the ever growing demand for more bandwidth, the capacity of present days core networks has to be increased continuously. Optical fibres possess excellent transmission properties and serve to transmit high data capacity over long distances. In order to use the bandwidth of an optical fibre effectively, optical multiplexing techniques, namely wavelength division multiplexing (WDM) and time division multiplexing (TDM) techniques, are applied. However, wherever optical data are to be routed or switched, i.e. where the respective channels have to be converted in wavelength (WDM-systems) or to be (de)multiplexed in the temporal domain (TDM-systems) signal processors are needed. In fact, signal processing, which is currently mainly performed via the optic-electronic-optic pathway, is the bottleneck in the current high capacity communication networks. In order to overcome the disadvantages that are imposed by the limited speed of electronics, all-optical solutions are of interest. One method for high capacity all-optical data processing is four-wave mixing (FWM) in semiconductor optical amplifiers (SOAs). FWM in SOAs can be used for frequency conversion of optical data signals [1], dispersion compensation using mid-span spectral inversion [2], and time division demultiplexing of high bit rate optical data signals [3, 4]. Besides that, FWM can be used as spectroscopic tool to extract the dynamic SOA parameters [5].

10.1 Applications of semiconductor optical amplifiers

During recent years, semiconductor-optical amplifiers (SOAs) have become promising components for optical communication systems. Owing to tremendous improvements of the single-pass gain and the polarisation insensitivity, SOAs can nowadays successfully compete with rare-earth doped fibre amplifiers in several applications. Providing a single pass gain of up to 30 dB with a polarisation sensitivity of less than 2 dB [6–9] SOAs find increasing interest as booster- and preamplifiers [10]. Although fibre amplifiers are still preferable for inline amplification (where the constraints on the polarisation sensitivity are much tighter) in the 1.5 μm window, SOAs find also access in these applications at 1.3 μm due to less advanced fibre amplifiers for this wavelength range [11].

However, today the most attractive applications of SOAs are associated with their nonlinear optical properties. The injection of light into a SOA changes the carrier density and carrier distribution within the energy bands which results in a modulation of gain and refractive index in the active region of the amplifier. In cross-gain modulation schemes, the homogeneous gain saturation effect can be used to realise frequency conversion of data signals. There, the carrier density is bleached by an intensity modulated signal at the input wavelength, while the gain is probed by a continuous wave at the desired output wavelength [12]. Since a change of the gain in a SOA is always accompanied by a change of the refractive index and vice versa, also cross-phase modulation can be used for signal processing. Interferometric structures, such as Sagnac [13–15], Mach-Zehnder [16–18] or Michelson [19] interferometers have been used to convert cross-phase modulation into an intensity modulated signal in various temporal and spectral switching operations. Another nonlinear process in SOAs is four-wave mixing. The underlying principles of FWM in SOAs, which have also been described in detail in [20–24] will be discussed in the next section. The main advantage of FWM over the other all-optical processing techniques is its transparency to the modulation format of the data signal. Moreover, compactness, low power consumption and the potential for monolithic integration are essential properties of SOAs. They are thus key components for high capacity all-optical communication networks.

10.2 Principle of four-wave mixing

Fig. 10.1 shows a typical FWM arrangement. Two optical waves, a pump

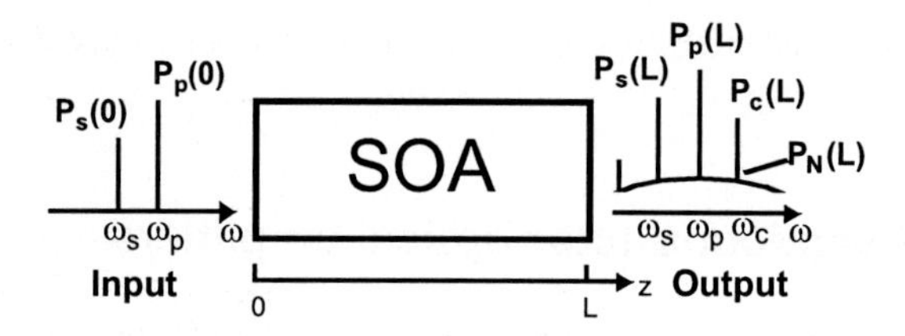

Fig. 10.1. FWM in SOAs. Here z denotes the spatial coordinate and L the amplifier length.

and a signal wave, are injected into the SOA with the same state of polarisation. The angular frequencies are ω_p and ω_s and the optical input powers are $P_p(0)$ and $P_s(0)$ for pump and signal wave, respectively. The beating of both waves influences the active medium, thereby generating index and gain modulation. The interaction of the injected waves with these modulations leads to new frequency components, the FWM signals that can be observed in the SOA output spectrum. Of main interest is the converted FWM signal $P_c(L)$ the optical angular frequency of which is $\omega_c = \omega_p + \Delta\omega$, with $\Delta\omega = \omega_p - \omega_s$

denoting the detuning of pump and signal wave angular frequencies. The non-linear gain dynamic in SOAs, responsible for the process of FWM, is based on inter- and intraband effects. Whereas the former refers to transitions between the conduction and the valence band, the latter are related to carrier-carrier and carrier-phonon scattering and modify the carrier distribution within one band. The interband effects change the carrier density due to the carrier depletion caused by the stimulated emission, which is referred to as carrier density pulsation (CDP). The characteristic time of CDP is the effective carrier lifetime, i.e. the lifetime due to all recombination processes, including stimulated emission. The intraband effects are associated with basically two phenomena. First, the stimulated emission burns a hole in the carrier distribution causing a deviation from the Fermi distribution. This is referred to as spectral hole burning (SHB). The time constant associated with this process is the time needed to restore the Fermi distribution (carrier-carrier scattering time), and is typically several tens of femtoseconds. Additionally, free carriers at low energy levels are removed by stimulated emission or transferred to higher levels due to free carrier absorption. This phenomenon leads to a heating of the electron and hole distributions and is called carrier heating (CH) [25]. The characteristic time needed to cool down to the lattice temperature, due to carrier-phonon scattering, is in the order of several hundred femtoseconds. It has been found that four-wave mixing is mainly mediated by CDP, CH and SHB [21, 26], for pump-signal frequency detuning $\Delta\nu$ up to a few THZ [5]. At higher detunings the two photon absorption and the Kerr effect can have a significant role [27].

10.3 Efficiency and signal-to-background ratio

Optical signal processing using the nonlinearities of a SOA, offers several advantages over the use of passive nonlinear devices, such as nonlinear crystals or optical fibres. One advantage is the high FWM conversion efficiency of the SOA which enables short interaction lengths of the injected waves. Phase matching is therefore not a crucial problem for co-propagating signals in SOAs. Moreover, only low optical input powers (in the range of 1 mW) are required for the pump and signal waves. The conversion efficiency

$$\eta_{FWM} = \frac{P_c(L)}{P_s(0)} \tag{10.1}$$

is defined as the ratio of the FWM signal output power to the signal input power. η_{FWM} therefore describes the efficiency by which an optical signal is converted from the angular frequency ω_s to the angular frequency ω_c. However, one drawback of using an active component rather than a passive one is the amplified spontaneous emission (ASE) noise, which is added by the SOA. In order to quantify the noise performance of an SOA, which is of crucial interest for practical applications, it is convenient to introduce the signal-to-background ratio SBR in addition to η_{FWM}. The signal-to-background ratio

$$SBR = \frac{P_c(L)}{P_N(L)} \tag{10.2}$$

is the ratio of the FWM signal power to the noise power $P_N(L)$ measured at the optical frequency ω_c within a certain optical bandwidth at the SOA output. Since the ASE noise is the main contribution to the noise in an SOA, $P_N(L) \approx P_{ASE}(L)$ is often used to determine the SBR. Note, that the SBR is not equivalent with the signal-to-noise ratio SNR, which is important for data transmission [28]. However, supposing that the ASE generated noise is dominant, a relation exists between these two quantities [29].

One of our main concerns has consisted to develop reliable criteria for an unambiguous characterisation of an SOA with respect to FWM. It has been decided to determine η_{FWM} and SBR at pump-signal detunings of $\Delta f = \Delta\omega/2\pi = 200$ GHz (about 1.6 nm) and 1000 GHz (about 8 nm) for both frequency up- and down conversion (see Table 10.1).

Detuning	Conversion efficiency (dB)		Signal-to-ASE (dB)	
	Freq. up	Freq. dn	Freq. up	Freq. dn
200 GHz				
1 THz				

Table 10.1. Table of comparison.

The measurements are performed under the following conditions:

- The SOA device is driven with the maximum current specified by the manufacturer.
- The pump wavelength is at the maximum of the unsaturated gain at maximum current.
- The temperature of the SOA is adjusted to +20 °C.
- Fibre input powers are +3 dBm for the pump and −7 dBm for the signal wave.
- The polarisation of the two input waves is adjusted to be the same, being in the plane of the SOAs maximum gain (if the device is polarisation dependent).
- The resolution bandwidth which was used to measure the ASE-noise and the exact pump wavelength must be indicated.

The idea of introducing such specific measurements conditions was to allow for an easy comparison between data from different research groups as well as to characterise different SOAs. The choice of these detunings has been driven by the fact that FWM is dominated by different physical effects at these two points. In order to perform comparable measurements, three SOAs have been made available by three partners in our group. We therefore have the unique opportunity to present experimental data from different devices that have been obtained by different research groups using different measurement techniques. The Table 10.1 was used as a comparison criterium between the measurements in the different labs during this exercise (referred to as round robin).

In order to give a comprehensive account at FWM, the next chapter is dedicated to a detailed theoretical analysis of the FWM process. Then, an extensive description and comparison between the set-ups as well as the general results of FWM measurements are given together with the round-robin results in Ch. 12. Ch. 13 presents some related topics, i.e., the parameter extraction, cross-gain modulation, nearly degenerate FWM, polarisation effects and FWM between optical pulses.

References

[1] M. Tatham, G. Sherlock, and L. Westbrook, "20 nm Optical Wavelength Conversion using Nondegenerate Four-Wave Mixing", *IEEE Photon. Technol. Lett.*, vol. 5, pp. 1303–1306, 1993.

[2] P. Cortes, M. Chbat, S. Artigaud, J. Beylat, and J. Chesnoy, "Below 0.3 dB Polarization Penalty in 10 Gbit/s Directly Modulated DFB Signal over 160 km using Mid-Span Spectral Inversion in a Semiconductor Laser Amplifier", in *ECOC '95*, pp. 271–274, 1995.

[3] R. Ludwig and G. Raybon, "All-Optical Demultiplexing Using Ultrafast Four-Wave Mixing in a Semiconductor Laser Amplifier at 20 Gbit/s", in *ECOC '93*, pp. 57–60, 1993.

[4] T. Morioka, H. Takara, S. Kawanishi, K. Uchiyama, and M. Saruwatari, "Polarisation-independent all-optical demultiplexing up to 200 Gbit/s using four-wave mixing in a semiconductor laser amplifier", *Electron. Lett.*, vol. 32, pp. 840–841, 1996.

[5] I. Koltchanov, S. Kindt, K. Petermann, S. Diez, R. Ludwig, and H. G. Weber, "Characterisation of Terahertz Four-Wave Mixing in a Semiconductor-Laser Amplifier", in *CLEO '96*, pp. 105–106, 1996.

[6] R. Schimpe, B. Bauer, C. Schanen, G. Franz, G. Kristen, and S. Pröhl, "1.5 μm InGaAsP tilted buried-facet optical amplifier", in *CLEO '91*, 1991.

[7] P. Doussière, F. Pommerau, D. Leclerc, R. Ngo, M. Goix, T. Fillion, P. Bousselet, and G. Laube, "Polarization independent 1550 nm semiconductor optical amplifier packaged module with 29 dB fiber to fiber gain", in *Optical Amplifiers and their Applications, OAA '95, Technical Digest*, pp. 119–122, 1995.

[8] F. Girardin, J. Eckner, G. Guekos, R. Dall'Ara, A. Mecozzi, A. D'Ottavi, F. Martelli, S. Scotti, and P. Spano, "Low noise and very high efficiency four-wave mixing in 1.5 mm long Semiconductor Optical Amplifiers", *IEEE Photon. Technol. Lett.*, vol. 9, pp. 746–748, 1997.

[9] A. E. Kelly, I. F. Lealman, L. J. Rivers, S. D. Perrin, and M. Silver, "Low noise figure (7.2 dB) and high gain (29 dB) semiconductor optical amplifier with a single layer AR coating", *Electron. Lett.*, vol. 33, pp. 536–538, 1997.

[10] T. Ducellier, M. Goix, J. P. Hebert, B. Mikkelsen, and M. Vaa, "Compact SOA-based preamplified receiver module for 20 Gbit/s applications", in *Optical Amplifiers and their Applications, OAA '97, Technical Digest*, pp. 120–123, 1997.

[11] L. F. Tiemeijer, P. J. A. Thijs, T. van Dongen, J. J. M. Binsma, E. J. Jansen, and S. Walczyk, "33 dB Fiber to Fiber Gain +13 dBm Fiber Saturation Power Polarization Independent 1310 nm MQW Laser Amplifier", in *Optical Amplifiers and their Applications, OAA '95, Technical Digest*, paper PD1, 1995.

[12] S. L. Danielsen, C. Joergensen, M. Vaa, K. E. Stubkjaer, P. Doussière, F. Pommerau, L. Goldstein, R. Ngo, and M. Goix, "Bit error rate assessment of a 40 Gbit/s all-optical polarisation independent Wavelength Converter", in *OFC'96*, paper PD12, 1996.

[13] M. Eiselt, W. Pieper, and H. G. Weber, "SLALOM: Semiconductor Laser Amplifier in a Loop Mirror", *J. Lightwave Technol.*, vol. 13, pp. 2099–2112, 1995.

[14] J. P. Sokoloff, P. R. Prucnal, I. Glesk, and M. Kane, "A Terahertz Optical Asymmetric Demultiplexer (TOAD)", *IEEE Photon. Technol. Lett.*, vol. 5, pp. 787–790, 1993.

[15] E. Jahn, N. Agrawal, W. Pieper, H. J. Ehrke, D. Franke, W. Furst, and C. M. Weinert, "Monolithically integrated nonlinear sagnac interferometer and its application as a 20 Gbit/s all-optical demultiplexer", *Electron. Lett.*, vol. 32, pp. 782–784, 1996.

[16] E. Jahn, N. Agrawal, H. J. Ehrke, R. Ludwig, W. Pieper, and H. G. Weber, "Monolithically integrated asymmetric Mach-Zehnder interferometer as a 20 Gbit/s all-optical add/drop multiplexer for OTDM systems", *Electron. Lett.*, vol. 32, pp. 216–217, 1996.

[17] C. Joergensen, S. L. Danielsen, T. Durhuus, B. Mikkelsen, K. E. Stubkjaer, N. Vodjdani, F. Ratovelomanana, E. Enard, G. Glastre, D. Rondi, and R. Blondeau, "Wavelength Conversion by Optimized Monolithic Integrated Mach-Zehnder Interferometer", *IEEE Photon. Technol. Lett.*, vol. 8, pp. 521–523, 1996.

[18] R. Hess, M. Dülk, W. Vogt, E. Gamper, H. Melchior, M. Vaa, B. Mikkelsen, K. S. Jepsen, K. E. Stubkjaer, and S. Bouchoule, "80 Gbit/s all optical demultiplexing using high-performance monolithically integrated Mach-Zehnder interferometer with semiconductor optical amplifiers", in *Optical Amplifiers and their Applications, OAA '97, Technical Digest*, pp. 198–201, 1997.

[19] M. Schilling, W. Idler, G. Laube, K. Daub, K. Dütting, E. Lach, D. Baums, and K. Wünstel, "10 Gbit/s monolithic MQW-based wavelength converter in Michelson-interferometer configuration", in *OFC '96*, pp. 122–124, 1996.

[20] G. P. Agrawal, "Population Pulsations and nondegenerate Four-Wave Mixing in Semiconductor Lasers and Amplifiers", *J. Opt. Soc. Am. B*, vol. 5, pp. 147–158, 1988.

[21] J. Zhou, N. Park, J. W. Dawson, K. J. Vahala, M. A. Newkirk, and B. I. Miller, "Terahertz four-wave mixing spectroscopy for study of ultrafast dynamics in a semiconductor optical amplifier", *Appl. Phys. Lett.*, vol. 63, pp. 1179–1181, 1993.

[22] A. Uskov, J. Mørk, and J. Mark, "Wave mixing in semiconductor laser amplifiers due to carrier heating and spectral-hole burning", *IEEE J. Quantum Electron.*, vol. 30, pp. 1769–1781, 1994.

[23] A. Mecozzi, S. Scotti, A. D'Ottavi, E. Iannone, and P. Spano, "Four-Wave Mixing in Traveling-Wave Semiconductor Amplifiers", *IEEE J. Quantum Electron.*, vol. 31, pp. 689–699, 1995.

[24] I. Koltchanov, S. Kindt, K. Petermann, S. Diez, R. Ludwig, R. Schnabel, and H. G. Weber, "Gain Dispersion and Saturation Effects in Four-Wave Mixing in Semiconductor Laser Amplifiers", *IEEE J. Quantum Electron.*, vol. 32, pp. 712–720, 1996.

[25] D. Bimberg and J. Mycielski, "The recombination-induced temperature change of non-equilibrium charge carriers", *J. Phys. C. Solid state phys.*, vol. 19, pp. 2363–2373, 1986.

[26] K. Kikuchi, M. Amano, C. E. Zah, and T. P. Lee, "Analysis of the Origin of Nonlinear Gain in 1.5 μm Semiconductor Active Layers by Highly Nondegenerate Four-Wave Mixing", *Appl. Phys. Lett.*, vol. 64, pp. 548–550, 1994.

[27] S. Scotti, L. Graziani, A. D'Ottavi, F. Martelli, A. Mecozzi, P. Spano, R. Dall'Ara, F. Girardin, and G. Guekos, "Effects of ultrafast processes on frequency converters based on four-wave mixing in semiconductor optical amplifiers", *IEEE J. of sel. topics in quantum electron.*, vol. 3, pp. 1156–1161, 1997.

[28] D. Marcuse, "Derivation of Analytical Expressions for the Bit-Error Probability in Lightwave Systems with Optical Amplifiers", *J. Lightwave Technol.*, vol. 8, pp. 1816–1823, 1990.

[29] R. Ludwig, W. Pieper, R. Schnabel, and H. G. Weber, "Ultrafast Wavelength Conversion and Switching by Four-Wave Mixing in Semiconductor Laser Amplifiers", in *Ultrafast Electronics and Optoelectronics '95*, pp. 39–41, 1995.

11 Theory of four-wave mixing

K. Obermann, A. Mecozzi, J. Mørk

The present chapter is devoted to presenting the theory of four-wave mixing in semiconductor optical amplifiers (SOAs). At first, a model describing the dynamics in terms of rate equations is developed. The rate equations are derived from density matrix equations, and after coupling with the propagation equation for the field envelope and its phase, a rather general description of the four-wave mixing is arrived at. In the limit of zero scattering loss, the model leads to a very simple description of the effects of four-wave mixing. In the succeeding sections we will concentrate on the advantages and draw-backs of some analytical models based on the coupled-mode theory and used to calculate the FWM performance of SOAs. In addition, the effects of amplified spontaneous emission (ASE) noise are considered. The theoretical models presented here should be of interest for all those performing experimental FWM investigations. Particular attention is given to the validity of the assumptions used in the models.

11.1 Rate equations

A rigorous approach of the interaction between a semiconductor material and a propagating electromagnetic wave would be based on the semiconductor Bloch Eqs. [1]. These equations incorporate the many-body aspects of the interacting electrons and holes in the semiconductor. Even so, the resulting set of equations is exceedingly complex and does not lend easily to physical interpretation; analytical or semi-analytical estimates can only be obtained in simple cases. Another problem with such a complex model is that it cannot easily be applied to simulations of actual devices since in this case there are often additional features that need to be accounted for, e.g., other loss mechanisms. It is therefore important to develop simplified models to help interpret measurement results and to develop simulation tools that can deal with complicated device geometries.

The semiconductor density matrix equations [2–4] are an approximation to the semiconductor Bloch equations that neglects excitonic effects (Coulomb interaction between electrons and holes) and treat carrier-carrier and carrier-phonon scattering (which lead to equilibration of the electron and hole gas towards Fermi-Dirac distribution functions) within a relaxation time approximation. The latter approximation is also often adopted when solving the full semiconductor Bloch equations and can be justified for not too large deviations from Fermi distributions [5]. It means, that one needs to supply phenomenological values for the relaxation times, which could be obtained from more basic calculations or by comparison with experimental results. Actually, the concept and the identification of characteristic relaxation times is a very useful way of summarising important features of the carrier equilibration, and their characteristic values often separate temporal regimes of qualitatively different behaviour. It has to be kept in mind also, that the semiconductor Bloch equations themselves are derived under various assumptions, an important one being the assumption of Markovian behaviour (no memory effects).

The density matrix equations have been found to be powerful enough to explain qualitatively, if not quantitatively, most of the experimental results observed in semiconductor samples at time scales larger than about 100 femtoseconds and at room temperature. An important point here is that most applications involving semiconductor waveguides rely on large injected carrier densities and room temperature operation which reduces the influence of more exotic effects due to Coulomb interaction and electron phase coherence.

Even after these simplifying approximations, the density matrix equations are still very complex and difficult to handle numerically. In the present section we shall describe a simplified set of rate equations that can be derived from the density matrix equations under certain approximations [6, 7].

The density matrix equations are formulated in terms of the probability of an electron occupying a state with wave vector k in the conduction band, $\rho_{c,k}$, the probability of a hole occupying the state k in the valence band, $\rho_{v,k}$ and the interband polarisation between those two states, $\rho_{cv,k}$. The approximate set of rate equations are derived by 1) adiabatical elimination of the off-diagonal elements $\rho_{cv,k}$, which requires the dephasing time τ_2 of the interband polarisation to be much shorter than the time scales of interest, and 2) the introduction of macroscopic (thermodynamic) quantities, which characterise the carrier distributions through a few of its moments, rather than following the individual carriers. We thus introduce the carrier density (zero'th order moment)

$$N(t) = \frac{1}{V} \sum_k \rho_{\beta,k}, \tag{11.1}$$

where V is the active region volume and the sum over the electron ($\beta = c$) and the hole ($\beta = v$) distribution gives identical results due to charge neutrality. The first order moment is the energy density

$$U_\beta = \frac{1}{V} \sum_k E_{\beta,k} \rho_{\beta,k}, \tag{11.2}$$

where $E_{\beta,k}$ is the energy corresponding to the wave vector k. The time evolution of these new variables can be found from the density matrix equations and if it is further assumed that the carrier-carrier scattering times, characterising the equilibration of the carrier distributions $\rho_{\beta,k}$ towards Fermi distribution functions, are small compared with the time scales of interest, a closed set of equations can be obtained. These equations include the dynamic changes of the carrier density and the carrier temperature (which can be calculated once the carrier and energy densities are known) due to interaction with the optical field in the waveguide. The approximation of neglecting the finite carrier-carrier scattering time means that the effect of spectral hole-burning is not included and this is often a poor approximation, in particular for the case of highly non-degenerate four-wave mixing *or* the case of strong optical fields.

The influence of spectral hole-burning can be accounted for by the introduction of "local carrier densities" n_β which effectively measure the number of carriers in the spectral region of the bandstructure that the optical field interacts with [6, 7]. The local densities are obtained by summation over the occupied electron and hole states within a bandwidth of the order of $\hbar/\tau_2$ around the mean carrier energies $E_{c,0}$ and $E_{v,0}$, where $\hbar\omega_0 = E_g + E_{c,0} + E_{v,0}$, with E_g being the bandgap energy (which is renormalised due to many-body effects). The obtained set of rate equations is the minimal set required to treat simultaneously the effects of carrier density depletion, carrier heating, and spectral hole-burning, as long as additional approximations are not made regarding the relative size and the temporal scale of the various processes.

The resulting set of rate equations reads:

$$\frac{\partial n_\beta}{\partial t} = -\frac{n_\beta - \overline{n}_\beta}{\tau_{1\beta}} - v_g g S\,, \tag{11.3}$$

$$\frac{\partial N}{\partial t} = \frac{I}{eV} - \frac{N}{\tau_s} - v_g g S, \tag{11.4}$$

$$\begin{aligned}\frac{\partial T_\beta}{\partial t} &= \left(\frac{\partial U_\beta}{\partial T_\beta}\right)_N^{-1} \left[\frac{\sigma_\beta N \hbar\omega_0}{g} + \left(\left(\frac{\partial U_\beta}{\partial N}\right)_{T_\beta} - E_{\beta,0}\right)\right] v_g g S \\ &\quad - \frac{T_\beta - T_L}{\tau_{h\beta}}\,. \end{aligned} \tag{11.5}$$

The gain g is a function of the local densities

$$g = \frac{1}{v_g} a_N (n_c + n_v - n_0)\,, \tag{11.6}$$

where n_0 is the density of available states in the optically coupled region and $a_N/v_g = \omega_0 |d_{cv}|^2 \tau_2/(\hbar\varepsilon_0 n c)$ is the gain cross section. Here, t is the time in a retarded time frame $t = t' - z/v_g$, t' is the time, z is the longitudinal coordinate, v_g is the group velocity, d_{cv} is the dipole moment (evaluated at the optical transition), ε_0 is the vacuum permittivity, n is the refractive index, T_β is the temperature of the corresponding carrier population and c is the velocity of light in vacuum. The subscripts N and T_β in the partial derivatives of the total energy density U indicate the variable to be held fixed. Henceforth, we will drop these subscripts, assuming the conditions for the derivative are clear from the context. The "Fermi densities" $\overline{n}_\beta$ appearing in Eq. (11.3) are defined through the Fermi function f evaluated at the instantaneous value of carrier density and carrier temperature

$$\overline{n}_\beta(N, T_\beta) = n_0 f(E_{\beta,0}; N, T_\beta) = n_0 \frac{1}{1 + \exp[(E_{\beta,0} - E_{f\beta})/(k_B T_\beta)]} . \quad (11.7)$$

Here, $E_{f\beta}$ is the Fermi level, which depends on the carrier density and the carrier temperature. Other parameters appearing in these equations are: carrier-carrier scattering times $\tau_{1\beta}$, temperature relaxation times $\tau_{h\beta}$, carrier lifetime τ_s, current I, active region volume V, free-carrier absorption coefficient σ_β, and lattice temperature T_L.

The corresponding propagation equation for the photon density $S = S(t, z)$ is [in units of m^{-3}, the photon number being normalised by the *modal* cross-sectional area of the waveguide]

$$\frac{\partial S}{\partial z} = (\Gamma g - \alpha_{\mathrm{i}}) S \quad (11.8)$$

where Γ is the confinement factor and α_{i} is the internal loss.

Let us briefly discuss some of the qualitative features of this rate equation model. First of all, the gain depends only on the values of the local carrier densities, n_c and n_v. The carrier density and carrier temperature influence the gain through the quasi-equilibrium variable $\overline{n}_\beta = \overline{n}_\beta(N, T_\beta)$ that the local carrier density equilibrates towards. From this description it is quite clear that the changes induced in the carrier density and the carrier temperatures by the optical field only affect the gain after a time delay of the order of the carrier-carrier scattering time [8, 9].

In the limit where both the carrier-carrier and the temperature relaxation (carrier-phonon) scattering times are much shorter than the timescale on which the photon density changes, the following expression for the gain can be derived by neglecting the time derivatives in Eqs. (11.3) and (11.5) [10]

$$g = \frac{a\,(N - N_0)}{1 + \epsilon S(t, z)} . \quad (11.9)$$

Here, a linear approximation for the carrier density dependence of the gain was adopted [11], i.e.,

$$\begin{aligned} g_l(N) &= \frac{a_N}{v_g} (\overline{n}_{c,l} + \overline{n}_{v,l} - n_0) \quad (11.10) \\ &= a(N - N_0), \quad (11.11) \end{aligned}$$

with a being the differential gain and N_0 the carrier density at transparency and we have introduced the well-known nonlinear gain suppression factor ϵ

$$\epsilon = \sum_{\beta} (\epsilon_{\mathrm{SHB},\beta} + \epsilon_{\mathrm{CH},\beta}) \, . \tag{11.12}$$

with contributions due to carrier heating (CH) and spectral hole-burning (SHB)

$$\epsilon_{\mathrm{CH},\beta} = -a_N K_\beta \tau_{h\beta} \frac{\partial n_\beta}{\partial T_\beta} = -v_g K_\beta \tau_{h\beta} \frac{\partial g}{\partial T_\beta} , \tag{11.13}$$

$$\epsilon_{\mathrm{SHB},\beta} = a_N \tau_{1\beta} . \tag{11.14}$$

We have neglected the influence of free carrier absorption, which is a good approximation as long as operation far from the transparency point is considered, and we introduced the parameter

$$K_\beta = \left(\frac{\partial U_\beta}{\partial T_\beta} \right)^{-1} \left[\frac{\partial U_\beta}{\partial N} - E_{\beta,0} \right] . \tag{11.15}$$

The dynamics of the carrier density N is still governed by Eq. (11.4), equivalent to Eq. (11.19). For pulses longer than the intraband scattering times, the set of three rate equations therefore reduce to the usual single rate equation for the carrier density, but with the gain given by the expression (11.9) in which the saturation induced by nonlinear gain is taken into account in the standard way.

By separating the total gain into contributions from carrier density induced changes, given by $g_l(N)$ above, carrier temperature induced changes

$$\Delta g_{\beta,h} = \frac{a_N}{v_g} \Delta n_{\beta,h} = \frac{a_N}{v_g} (\overline{n}_\beta - \overline{n}_{\beta,l}) = \frac{\partial g}{\partial T_\beta} \Delta T_\beta , \tag{11.16}$$

and changes due to non-equilibrium between the local carrier densities and the other carriers (spectral hole-burning)

$$\Delta g_\beta = \frac{a_N}{v_g} \Delta n_\beta = \frac{a_N}{v_g} (n_\beta - \overline{n}_\beta) , \tag{11.17}$$

i.e.,

$$g(t) = g_l(N) + \sum_\beta (\Delta g_{\beta,h} + \Delta g_\beta) , \tag{11.18}$$

we can derive the following set of equations for the different components [11]

$$\frac{\partial g_l}{\partial t} = -\frac{g_l}{\tau_s} - \frac{1}{S_s \, \tau_s} g \, S(t,z) + \frac{a \, (N_{\mathrm{st}} - N_0)}{\tau_s} , \tag{11.19}$$

$$\frac{\partial \Delta g_\beta}{\partial t} = -\frac{\Delta g_\beta}{\tau_{1\beta}} - \frac{\epsilon_{\mathrm{SHB},\beta}}{\tau_{1\beta}} g \, S(t,z) - \left(\frac{\partial \Delta g_{\beta,h}}{\partial t} + y_\beta \frac{\partial g_l}{\partial t} \right) , \tag{11.20}$$

$$\frac{\partial \Delta g_{\beta,h}}{\partial t} = -\frac{\Delta g_{\beta,h}}{\tau_{h\beta}} - \frac{\epsilon_{\mathrm{CH},\beta}}{\tau_{h\beta}} g \, S(t,z) . \tag{11.21}$$

Above, N_{st} is the unsaturated value of the carrier density, i.e. the value of the carrier density as set by current injection and spontaneous carrier recombination prior to the injection of any signal,

$$N_{\text{st}} = \frac{I\tau_s}{eV}. \tag{11.22}$$

The saturation photon density

$$S_s = \frac{1}{v_g a \tau_s}. \tag{11.23}$$

is related to the familiar saturation power by

$$P_s = \hbar\omega A_{\text{eff}} v_g S_s = \kappa S_s, \tag{11.24}$$

where $A_{\text{eff}} = dw/\Gamma$ is the effective area of the waveguide, d is the waveguide thickness, w the waveguide width, ω is the angular frequency, and $\kappa = \hbar\omega A_{\text{eff}} v_g$ is the conversion factor from photon density to power. The quantities y_β give the relative contribution of changes in the electron and hole density to the differential gain ($y_c + y_v = 1$)

$$y_\beta = \frac{a_N}{a\, v_g} \frac{\partial \overline{n}_{\beta,l}}{\partial N}. \tag{11.25}$$

Eqs. (11.19)-(11.21) are a set of differential equations once g is expressed in terms of g_l, Δg_β and $\Delta g_{\beta,h}$ using Eq. (11.18). Since, however, it is the input intensity that is given and not the intensity at a generic position z, the propagation problem requires the solution of a set of partial differential equations obtained by adding the propagation equation (11.8) to Eqs. (11.19)-(11.21). In the next section we shall see how this problem can be reduced to the solution of a set of ordinary differential equations, once the waveguide internal loss α_{i} is neglected.

The changes in the carrier distributions induced by the optical field also modify the refractive index of the waveguide and thereby the phase $\phi = \phi(t,z)$ of the propagating optical field

$$\frac{\partial \phi}{\partial z} = \Gamma \frac{\partial k}{\partial N} \Delta N + \sum_{\beta=c,v} \Gamma \frac{\partial k}{\partial T_\beta} \Delta T_\beta + \Gamma \Delta k_{\text{neq}}, \tag{11.26}$$

where k is the propagation constant of the field, $k = 2\pi\, n/\lambda$, n is the waveguide effective index, and λ is the vacuum wavelength. The term $\Gamma\Delta k_{\text{neq}}$ accounts for the non-equilibrium part of the phase dynamics. This term is necessary because our definitions of N and T refer to quasi-equilibrium. When the electromagnetic field is applied, those variables undergo instantaneous changes. One should, however, expect that these changes affect the gain and phase dynamics with a delay given by the time τ_1 taken by carrier-carrier scattering to establish quasi equilibrium within the bands. Before this time, only the non-equilibrium spectral holes burned in the carrier distributions may affect the gain and, possibly, the phase dynamics. The term $\Gamma \Delta k_{\mathrm{neq}}$ is therefore zero for $\tau_1 = 0$, and for finite τ_1 it comprises not only the SHB part, but also the delay that is associated with the establishment of the quasi-equilibrium corresponding to the values of N and T. For the gain, this delay is indeed predicted by the equations, as will become clear later.

In Eq. (11.26), we assumed that the amplifier is operating close to the gain peak, so that the spectral hole does not contribute to the nonlinear index change because its spectrum is symmetric.

Defining the phase-amplitude coupling coefficients as

$$\alpha_y = -2\frac{\partial \Delta k/\partial y}{\partial \Delta g/\partial y}, \tag{11.27}$$

where $y = N$, T_c, T_v, and inserting into Eq. (11.26) we get

$$\frac{\partial \phi}{\partial z} = -\frac{1}{2}\Gamma\left[\alpha_N a \Delta N + \sum_{\beta=c,v} \alpha_{T_\beta}\frac{\partial g}{\partial T_\beta}\Delta T_\beta\right] + \Gamma \Delta k_{\mathrm{neq}}. \tag{11.28}$$

The nonlinear carrier density change ΔN is the change of carrier density caused by stimulated emission. It can be expressed as $\Delta N = N - N_{\mathrm{st}}$, where N_{st} is given by Eq. (11.22). Writing $\Delta N = N - N_0 + N_0 - N_{\mathrm{st}}$ we get

$$\begin{aligned}\frac{\partial \phi}{\partial z} &= -\frac{1}{2}\Gamma\left[\alpha_N g_l(N) - \alpha_N g_l(N_{\mathrm{st}}) + \sum_\beta \alpha_{T_\beta}\Delta g_{\beta,h}\right] \\ &\quad + \Gamma \Delta k_{\mathrm{neq}}.\end{aligned} \tag{11.29}$$

The explicit expression for the phase including the non-equilibrium part will be given in the next section.

Our analysis neglects the optical nonlinearities arising from both two-photon absorption and the optical Stark effect, which produce instantaneous absorptive and refractive nonlinearities within a theory that adiabatically eliminates the medium polarisation. Also, we neglect the free-carrier absorption induced heating [8]. Finally, we neglect the spectral profile of the gain, and the group velocity dispersion of the waveguide.

Neglecting two photon absorption and the optical Stark effect is justified if the pulsewidth of the pump and probe pulses is larger than 1 ps, and if the pump-probe detuning does not exceed 4 THz. The phenomenological inclusion of these effects in the theory is not difficult, however.

If free-carrier absorption induced heating is included in the analysis, one gets larger FWM efficiencies for strong amplifier saturation [12, 13]. However, this approximation is not expected to introduce large errors if the case of very strongly saturated amplifiers is excluded. In ref. [11] it has been shown how the analysis can be extended to include free-carrier absorption induced heating.

The group velocity dispersion of the waveguide and the pulse walk-off is negligible for pulses longer than 1 ps and detunings lower than several THz [4]. For the pump at the maximum of the gain peak, gain dispersion has only a small effect. However, as will be shown in the next section, gain dispersion can be important if saturation of the carrier density induces a significant shift of the peak of the gain spectrum from its unsaturated value. An analysis which includes the effect of gain dispersion in pulsed FWM and analyses saturation effects within a perturbative approach can be found in [14].

11.2 Time-domain description

In this section we shall show how the solution of the coupled partial differential equations for the pulse propagation and the carrier density can be reduced to the solution of an ordinary differential equation. The general idea is well known from gas laser theory [15] and has been applied to semiconductor lasers by Agrawal and Olsson [16].

11.2.1 Derivation of the integral equation

The equations to be solved are Eqs. (11.19)-(11.21), along with Eq. (11.8) describing the propagation of the field along the amplifier waveguide. Let us neglect in Eq. (11.8) the waveguide internal loss α_{i}, which is a good approximation as long as the local material gain is significantly higher. The case in which the waveguide internal loss is non-negligible is treated in Ref. [11]. We obtain

$$\frac{\partial S(t,z)}{\partial z} = \Gamma g\, S(t,z). \tag{11.30}$$

The formal solution of Eq. (11.30) is

$$S(t,z) = S(t,0)G(t,z), \tag{11.31}$$

where we have defined

$$G(t,z) = \exp\left[g_m(t,z)\right], \tag{11.32}$$

$$g_m(t,z) = \Gamma \int_0^z dz' g(t,z'). \tag{11.33}$$

Let us introduce

$$h_N = \Gamma \int_0^z dz' g_l(N), \tag{11.34}$$

$$h_\beta = \Gamma \int_0^z dz' \Delta g_\beta, \tag{11.35}$$

$$h_{\beta,h} = \Gamma \int_0^z dz' \Delta g_{\beta,h}. \tag{11.36}$$

From Eq. (11.18) we have

$$g_m(t,z) = h_N + \sum_\beta (h_{\beta,h} + h_\beta). \tag{11.37}$$

Integrating over z both sides of Eqs. (11.19)-(11.21), using Eq. (11.34) and the result obtained by integration of both sides of Eq. (11.30),

$$\int_0^z dz' \Gamma g S = S(t,z) - S(t,0) = [G(t,z) - 1]\, S(t,0), \tag{11.38}$$

we obtain

$$\frac{dh_N}{dt} = -\frac{h_N}{\tau_s} - \frac{1}{S_s \tau_s}[G(t,z) - 1]\, S(t,0) + \frac{g_0(z)}{\tau_s}, \tag{11.39}$$

$$\frac{dh_\beta}{dt} = -\frac{h_\beta}{\tau_{1\beta}} - \frac{\epsilon_{\mathrm{SHB},\beta}}{\tau_{1\beta}}[G(t,z) - 1]\, S(t,0) - \frac{dh_{\beta,h}}{dt} - y_\beta \frac{dh_N}{dt}, \tag{11.40}$$

$$\frac{dh_{\beta,h}}{dt} = -\frac{h_{\beta,h}}{\tau_{h\beta}} - \frac{\epsilon_{\mathrm{CH},\beta}}{\tau_{h\beta}}[G(t,z) - 1]\, S(t,0), \tag{11.41}$$

where

$$g_0(z) = \Gamma g_l(N_{\mathrm{st}})\, z = \Gamma a(N_{\mathrm{st}} - N_0) z. \tag{11.42}$$

Eqs. (11.39)-(11.41) are a set of differential equations when they are combined with

$$G(t,z) = \exp\left[h_N + \sum_\beta (h_{\beta,h} + h_\beta)\right]. \tag{11.43}$$

Let us now give a formal solution of Eqs. (11.39)-(11.41). The first and the third equations can be formally solved as

$$h_N - g_0(z) = -\frac{1}{S_s}\int_{-\infty}^{\infty} dt' Q_N(t-t') \left[e^{g_m(t',z)} - 1\right] S(t',0) \tag{11.44}$$

$$h_{\beta,h} = -\epsilon_{\mathrm{CH},\beta}\int_{-\infty}^{\infty} dt' Q_{T_\beta}(t-t') \left[e^{g_m(t',z)} - 1\right] S(t',0), \tag{11.45}$$

where

$$Q_N(t) = \frac{u(t)}{\tau_s} \exp\left(-\frac{t}{\tau_s}\right), \tag{11.46}$$

$$Q_{T_\beta}(t) = \frac{u(t)}{\tau_{\beta,h}} \exp\left(-\frac{t}{\tau_{\beta,h}}\right), \tag{11.47}$$

with $u(t)$ being the Heaviside step function of unit amplitude. Then, the solution of the second equation in (11.39)-(11.41) can be found after some algebra

$$\begin{aligned} h_\beta &= -\int_{-\infty}^{\infty} dt' \left[\epsilon_{\mathrm{SHB},\beta} Q_{\mathrm{SHB},\beta}(t-t') + \epsilon_{\mathrm{CH},\beta} D_{T_\beta}(t-t') \right. \\ &\left. + \frac{1}{S_s} D_{N,\beta}(t-t')\right] \left[e^{g_m(t',z)} - 1\right] S(t',0) \end{aligned} \tag{11.48}$$

with

$$Q_{\mathrm{SHB},\beta}(t) = \frac{u(t)}{\tau_{1\beta}} \exp\left(-\frac{t}{\tau_{1\beta}}\right), \tag{11.49}$$

$$D_{T_\beta}(t) = -\frac{\tau_{1\beta}}{\tau_{\beta,h} - \tau_{1\beta}} \left[Q_{\mathrm{SHB},\beta}(t) - Q_{T_\beta}(t)\right], \tag{11.50}$$

$$D_{N,\beta}(t) = -y_\beta \frac{\tau_{1\beta}}{\tau_s - \tau_{1\beta}} \left[Q_{\mathrm{SHB},\beta}(t) - Q_N(t)\right]. \tag{11.51}$$

Let us define

$$R_N(t) = Q_N(t) + \sum_\beta D_{N,\beta}(t), \tag{11.52}$$

$$R_{\mathrm{SHB},\beta}(t) = Q_{\mathrm{SHB},\beta}(t), \tag{11.53}$$

$$R_{T_\beta}(t) = Q_{T_\beta}(t) + \sum_\beta D_{T_\beta}(t), \tag{11.54}$$

and

$$\epsilon_{\mathrm{tot}} R_{\mathrm{tot}}(t) = \frac{1}{S_s} R_N(t) + \sum_\beta \left[\epsilon_{\mathrm{CH},\beta} R_{T_\beta}(t) + \epsilon_{\mathrm{SHB},\beta} R_{\mathrm{SHB},\beta}(t)\right], \tag{11.55}$$

$$\epsilon_{\mathrm{tot}} = \frac{1}{S_s} + \epsilon, \tag{11.56}$$

with ϵ given by Eq. (11.12). It is immediate to verify that since $\int dt D_j(t) = 0$ and $\int dt Q_j(t) = 1$, definitions (11.12) and (11.56) imply that $\int R_{\mathrm{tot}}(t) dt = 1$. Using Eqs. (11.44) and the definition Eq. (11.37), we obtain

$$g_m(t,z) = g_0(z) - \epsilon_{\mathrm{tot}} \int_{-\infty}^{\infty} R_{\mathrm{tot}}(t-t') \left[e^{g_m(t',z)} - 1\right] S(t',0) dt'. \tag{11.57}$$

Eq. (11.57) is, in spite of its simplicity, very general. The integral Eq. (11.57) is exactly equivalent to the system Eq. (11.39)-(11.41), and its validity is therefore not restricted to small-signals. On one side, if one neglects the ultrafast saturation processes and considers only carrier density changes, this integral equation is exactly equivalent to the differential equation describing gain saturation given by Agrawal and Olsson in ref. [16], and it might therefore be considered as an extension of that approach to the cases in which the ultrafast dynamics become relevant, i.e. for pulses shorter than a few picoseconds [10] or for interacting pulses with frequency detunings in excess of a few THz. On the other hand, it is the extension into the nonlinear regime of approaches that include the nonlinear response of the amplifier gain to the input field, but linearise the response to first order in the field intensity. The linearisation approximation fails when short and intense pulses are injected into the amplifier. The simplicity of Eq. (11.57) makes it particularly useful for approximate analysis of semiconductor amplifier saturation and four-wave mixing between CW beams and pulses.

It is immediate to verify that terms $D_j(t)$, having zero area, contribute to the response function $R_{\mathrm{tot}}(t)$ adding just a delay of $\tau_{1\beta}$ to the carrier heating and carrier response functions $R_{T_\beta}(t)$ and $R_N(t)$. They are usually negligible, unless the total input field changes on a time-scale of the order of the intraband scattering time $\tau_{1\beta}$.

The description for the gain has to be supplemented with an expression for the temporal phase, which is next shown to be expressed in terms of the effective gain g_m. Let us define

$$h'_N - g_0(z) = \sum_\beta \frac{\tau_s}{\tau_s - \tau_{1\beta}} y_\beta \left[h_N - g_0(z)\right] - \sum_\beta \frac{\tau_{1\beta}}{\tau_s - \tau_{1\beta}} \frac{y_\beta}{S_s} \frac{A_\beta}{B_\beta}, \tag{11.58}$$

$$h'_{\beta,h} = \frac{\tau_{\beta,h}}{\tau_{\beta,h} - \tau_{1\beta}} h_{\beta,h} - \frac{\tau_{1\beta}}{\tau_{\beta,h} - \tau_{1\beta}} \epsilon_{\mathrm{CH},\beta} \frac{A_\beta}{B_\beta}, \tag{11.59}$$

$$h'_\beta = \epsilon_{\mathrm{SHB},\beta} \frac{A_\beta}{B_\beta} \tag{11.60}$$

where

$$A_\beta = h_\beta - \frac{\tau_{1\beta}}{\tau_{\beta,h} - \tau_{1\beta}} h_{\beta,h} - \frac{\tau_{1\beta}}{\tau_s - \tau_{1\beta}} y_\beta \left[h_N - g_0(z)\right], \tag{11.61}$$

$$B_\beta = \epsilon_{\mathrm{SHB},\beta} - \frac{\tau_{1\beta}}{\tau_{\beta,h} - \tau_{1\beta}} \epsilon_{\mathrm{CH},\beta} - \frac{\tau_{1\beta}}{\tau_s - \tau_{1\beta}} \frac{y_\beta}{S_s}. \tag{11.62}$$

We then have

$$h'_N - g_0(z) = -\frac{1}{S_s} \int_{-\infty}^{\infty} dt' R_N(t - t') \left[e^{g_m(t',z)} - 1\right] S(t',0), \tag{11.63}$$

$$h'_\beta = -\epsilon_{\mathrm{SHB},\beta} \int_{-\infty}^{\infty} dt' R_{\mathrm{SHB},\beta}(t - t') \left[e^{g_m(t',z)} - 1\right] S(t',0), \tag{11.64}$$

$$h'_{\beta,h} = -\epsilon_{\mathrm{CH},\beta} \int_{-\infty}^{\infty} dt' R_{T_\beta}(t-t') \left[e^{g_m(t',z)} - 1\right] S(t',0) \tag{11.65}$$

and

$$g_m = h_N + \sum_\beta [h_{\beta,h} + h_\beta] = h'_N + \sum_\beta [h'_{\beta,h} + h'_\beta], \tag{11.66}$$

The primed quantities are proportional to ϵ_j and $1/S_s$, and their sum gives the total amplifier gain. When one of the ϵ_j is zero (or S_s is infinity) the corresponding saturation process does not affect the gain, and therefore it is reasonable to assume that it does not affect the refractive index as well. These quantities may be considered proportional to the effect of changes in the carrier density, carrier temperature and spectral hole-burning, on the gain and index. A delay appears in the effect on gain and index of carrier density and carrier temperature changes. The primed quantities thereby naturally account for the delay in the establishment of the quasi equilibrium, which is the time required for carrier-carrier scattering to fill the spectral hole burned in the carrier distribution.

The phase Eq. (11.29) therefore becomes

$$\phi(t,z) = \phi(t,0) - \frac{1}{2}\alpha_N \left[h'_N - g_0(z)\right] - \frac{1}{2}\sum_\beta \alpha_{T_\beta} h'_{\beta,h}. \tag{11.67}$$

The primed quantities include the quasi equilibrium part, and so the term $\Gamma \Delta k_{\mathrm{neq}}$ is absorbed in them.

Using Eq. (11.67), the phase of the field is finally written in terms of the g_m and of the response functions $R_j(t)$

$$\begin{aligned} \phi(t,z) \; = \; & \phi(t,0) + \frac{1}{2}\left\{\frac{\alpha_N}{S_s}\int_{-\infty}^{\infty} R_N(t-t')\left[e^{g_m(t',z)} - 1\right] S(t',0)dt' \right. \\ & + \epsilon_{\mathrm{CH},c}\alpha_{T_c}\int_{-\infty}^{\infty} R_{T_c}(t-t')\left[e^{g_m(t',z)} - 1\right] S(t',0)dt' \\ & \left. + \epsilon_{\mathrm{CH},v}\alpha_{T_v}\int_{-\infty}^{\infty} R_{T_v}(t-t')\left[e^{g_m(t',z)} - 1\right] S(t',0)dt' \right\} \end{aligned} \tag{11.68}$$

Given the gain and phase temporal variation at location z, the electrical field at that location becomes (output field for $z = L$):

$$E(t,z) = E(t,0)\exp\left[\frac{1}{2}g_m(t,z) + i\phi(t,z)\right]. \tag{11.69}$$

Here and below, we use units for the electric field such that $|E|^2 = S$ is the photon density in m^{-3}. We have arrived at two equivalent formulations of the amplifier dynamics. Given an electric field at the waveguide input, at any particular location z along the amplifier, Eqs. (11.39)-(11.41) give the total integrated gain if solved for the unknown h_N, $h_{\beta,h}$ and h_β with initial conditions, at $t \to -\infty$, $h_N = g_0(z)$ and $h_\beta = h_{\beta,h} = 0$. Equation (11.69) along with the expression for the phase (11.68), and the algebraic relations Eqs. (11.58), (11.61), can thereby be used to calculate the time dependent complex electric field at that location. Alternatively, one can use the formulation in terms of the integral equation, Eq. (11.18), and use the expression (11.68) for the phase. While the formulation in terms of a set of ordinary differential equations allow an easy numerical integration, the integral equation approach is very useful to derive approximate analytical results.

11.2.2 Four-wave mixing between CW beams

In ref. [11] it was shown how the case of four-wave mixing between short optical pulses could be approximately treated using the theory presented above. Here we shall apply it to the case of four-wave mixing between CW beams, showing how the analytical and quite general results of references [12, 17] can be obtained, and that more general results that go beyond the conventional first order perturbation theory used in those papers can be consistently obtained as well.

The analysis of FWM between pulses using the integral equation (11.57) has many noteworthy advantages over descriptions based on the coupled-mode equations of refs. [12, 17]. This is especially true when the waveguide internal loss is negligible and equation (11.57) holds. The main advantage of the integral equation is that only the *input* field appears in equation (11.57). This permits an easier control of the validity of the truncation to the first order terms of a description of FWM based on perturbation theory. Also, non-perturbative results that, formally, arise from a summation of all orders of the perturbative expansion can be obtained as well.

By the coupled mode approach, the number of coupled equations required for a first order description of FWM are one for each field, both those injected and those generated by FWM. Injected and generated fields should be treated on equal footing, the only difference between them is that the first have a non-zero input. This easily gives a very large number of equations to deal with when more than two beams are injected. Using the approach based on the integral equation, instead, this case is analysed very simply. Only the input beams contribute to first order to the gain modulation, and, the effect of gain and index modulation on the output field is easily obtained.

Another advantage of the integral equation is that the results that we arrive at apply to both quasi degenerate FWM and to highly nondegenerate FWM. The interference between the fields generated by the FWM process and the injected fields, giving rise to the so-called Bogatov effect, are automatically taken into account.

Finally, a perturbation approach based on the integral equation is not limited to the small-signal case. If the detuning is low, even if the injected signal is much weaker than the pump and we may assume that the resulting gain modulation is small, the effect of the modulation may be enhanced by a large Henry factor α_N. In this case, we may have a significant depletion of the pump caused by FWM, and high order FWM sidebands may appear in the spectrum. If the injected field is much smaller than the pump, however, it is likely that the gain modulation is much smaller than the steady state gain and therefore a first order expansion of the integral equation is well founded (the validity of this assumption can be always self consistently checked by verifying that $\Delta g \ll 1$). Once the gain modulation is obtained, the amplifier output that corresponds to the given input is almost immediately found. Pump depletion and the appearance of high order sidebands, which are high order effects in the analysis of FWM based on coupled mode equations, are consistently described by a theory based on a first order expansion of the integral equation and the insertion of the first order gain and index modulation in the expression that gives the output field.

The input to the waveguide is composed of two CW beams at different optical frequencies,

$$E(t,0) = A_0(0) + A_1(0)\exp(-i\Omega t), \tag{11.70}$$

where $A_i(z)$ denote the time-independent amplitude of the pump ($i = 0$), probe ($i = 1$) and conjugate ($i = c$) fields. We will make use of Eq. (11.69) to get the four-wave mixing field at the amplifier output. The total intensity at the input is

$$\begin{aligned} S(t,0) &= S_{\text{in}} + \Delta S_{\text{in}}\exp(-i\Omega t) + \Delta S_{\text{in}}^*\exp(i\Omega t), && (11.71)\\ S_{\text{in}} &= |A_0(0)|^2 + |A_1(0)|^2, && (11.72)\\ \Delta S_{\text{in}} &= A_0^*(0)A_1(0). && (11.73) \end{aligned}$$

Using Eq. (11.70), we may write

$$\begin{aligned} g_m(t,z) &= \bar{g}(z) + \Delta g_1(t,z), && (11.74)\\ \phi(t,z) &= \bar{\phi}(z) + \Delta\phi_1(t,z), && (11.75) \end{aligned}$$

where the first order terms $\Delta g_1(t,z)$ and $\Delta\phi_1(t,z)$ are

$$\begin{aligned} \Delta g_1(t,z) &= \Delta g(z)\exp(-i\Omega t) + \Delta g^*(z)\exp(i\Omega t), && (11.76)\\ \Delta\phi_1(t,z) &= \Delta\phi(z)\exp(-i\Omega t) + \Delta\phi^*(z)\exp(i\Omega t). && (11.77) \end{aligned}$$

and $\Delta g(z)$ and $\Delta\phi(z)$ are the amplitudes of the beating terms at frequency Ω assuming that these contributions are small. This condition is met when the pump-probe detuning is much larger than the cut-off frequency of carrier density pulsations or otherwise when the probe is much weaker than the pump. Inserting the above equations into Eq. (11.69) we obtain

$$E(t,z) = [A_0(0) + A_1(0)\exp(-i\Omega t)]\exp\left[\frac{1}{2}\bar{g}(z) + i\bar{\phi}(z)\right]\Delta e(t,z), \tag{11.78}$$

with:

$$\Delta e(t,z) = \exp\left[\frac{1}{2}\Delta g_1(t,z) + i\Delta\phi_1(t,z)\right]. \tag{11.79}$$

Because the validity of the first order expansion of the integral equation requires that $\Delta g_1(t,z) \ll 1$, Eq. (11.78) can always be expanded to first order in $\Delta g_1(t,z)$. Expansion to first order in $\Delta\phi_1(t,z)$ may instead not always be well founded when the pump-probe detuning is lower than or of the order of the cutoff frequency of the carrier density pulsations. In this case, being $\Delta\phi_1(t,z) \simeq -\alpha_N \Delta g_1(t,z)$ with α_N that may easily assume values larger than 10, $\Delta g_1(t,z) \ll 1$ does not prevent $\Delta\phi_1(t,z)$ from being of the order of unity.

Inserting the expansion above for g_m in the integral Eq. (11.57), we get for the 0'th order term (time-independent)

$$g_0 - \bar{g} = \epsilon_{\mathrm{tot}}\left[\mathrm{e}^{\bar{g}} - 1 + \alpha_{\mathrm{i}}\int_0^z dz'\mathrm{e}^{\bar{g}}\right](|A_0(0)|^2 + |A_1(0)|^2), \tag{11.80}$$

and for the first-order term in Δg

$$\Delta g(z) = -\tilde{R}_{\mathrm{tot}}(\Omega)\frac{g_0(z) - \bar{g}(z)}{Z(\Omega)}\frac{A_0^*(0)A_1(0)}{|A_0(0)|^2 + |A_1(0)|^2}, \tag{11.81}$$

$$Z(\Omega) = 1 + \epsilon_{\mathrm{tot}}\tilde{R}_{\mathrm{tot}}(\Omega)\left[\mathrm{e}^{\bar{g}(z)} + \alpha_{\mathrm{i}}\int_0^z dz'\frac{\mathrm{e}^{\bar{g}(z')}\Delta g(z')}{\Delta g(z)}\right] \left[|A_0(0)|^2 + |A_1(0)|^2\right]. \tag{11.82}$$

In Eq. (11.81) the total input power $|A_0(0)|^2 + |A_1(0)|^2$ can be expressed in terms of the saturated gain using (11.80). To obtain Eqs. (11.81) and (11.82), we utilized the 0'th order result and $\tilde{R}(\Omega)$ is the Fourier transform of $R_{\mathrm{tot}}(t)$. In general, we define the Fourier transform pair as

$$f(t) = \int_{-\infty}^{\infty}\frac{d\Omega}{2\pi}\exp(-i\Omega t)\,\tilde{f}(\Omega), \quad \tilde{f}(\Omega) = \int_{-\infty}^{\infty} dt\exp(i\Omega t)\,f(t). \tag{11.83}$$

The terms proportional to ϵ_{tot} in $Z(\Omega)$ as given by Eq. (11.82) correspond effectively to a change of the characteristic times of the scattering processes, similar to the replacement of the spontaneous carrier lifetime with the stimulated lifetime when the rate of stimulated emission is high [11]. For the phase, from Eq. (11.68), we obtain

$$\Delta\phi(z) = \frac{1}{2}\left\{\frac{\alpha_N}{S_s}\tilde{R}_N(\Omega) + \alpha_{T_c}\epsilon_{CH,c}\tilde{R}_{T_c}(\Omega) + \alpha_{T_v}\epsilon_{CH,v}\tilde{R}_{T_v}(\Omega)\right\} \frac{1}{\epsilon_{\mathrm{tot}}}\frac{g_0(z) - \bar{g}(z)}{Z(\Omega)}\frac{A_0^*(0)A_1(0)}{|A_0(0)|^2 + |A_1(0)|^2}. \tag{11.84}$$

The expressions for the various components of $\tilde{R}_{\mathrm{tot}}(\Omega)$ are

$$\tilde{R}_N(\Omega) = \sum_\beta \frac{y_\beta}{(1 - i\Omega\tau_s)(1 - i\Omega\tau_{1\beta})}, \tag{11.85}$$

$$\tilde{R}_{T_\beta}(\Omega) = \frac{1}{(1 - i\Omega\tau_{\beta,h})(1 - i\Omega\tau_{1\beta})}, \tag{11.86}$$

$$\tilde{R}_{\mathrm{SHB},\beta}(\Omega) = \frac{1}{1 - i\Omega\tau_{1\beta}}. \tag{11.87}$$

If we now assume that, besides $\Delta g_1(t,z) \ll 1$, also the phase modulation is small, i.e. $\Delta\phi_1(t,z) \ll 1$, then the exponential at right hand side of Eq. (11.78) can be expanded to first order in its argument. Taking the term at frequency $-\Omega$ (the four-wave mixing term), we thereby get

$$A_c(z) = A_0(0)\left[\frac{1}{2}\Delta g^*(z) + i\Delta\phi^*(z)\right]\exp\left[\frac{1}{2}\bar{g}(z) + i\bar{\phi}(z)\right]. \tag{11.88}$$

For the conjugate power at the output we now obtain

$$|A_c(z)|^2 = \frac{1}{4}G\left(\ln\frac{G_0}{G}\right)^2\left[\frac{|A_1(0)|^2}{1 + |A_1(0)|^2/|A_0(0)|^2}\right]^2\frac{|\tilde{\mathcal{R}}(-\Omega)|^2}{\epsilon_{\mathrm{tot}}^2} \tag{11.89}$$

with the definition

$$\tilde{\mathcal{R}}(\Omega) = \frac{1}{Z(\Omega)}\left\{\sum_{\beta=c,v}\left[(1 - i\alpha_{T_\beta})\epsilon_{\mathrm{CH},\beta}\tilde{R}_{T_\beta}(\Omega) + \epsilon_{\mathrm{SHB},\beta}\tilde{R}_{\mathrm{SHB},\beta}(\Omega)\right] + \frac{(1 - i\alpha_N)}{S_s}\tilde{R}_N(\Omega)\right\}; \tag{11.90}$$

for high detunings where $Z(\Omega) \simeq 1$, this result is identical to the one derived in [12] using an approach based on the coupled mode equations.

Consider now the low detuning case. In this case, $Z(\Omega)$ is a closed function of the unsaturated gain and of the material parameters only if the waveguide internal loss is zero, $\alpha_{\mathrm{i}} = 0$. Assume therefore below that the waveguide internal loss is negligible. Using

$$\int_{-\pi}^{\pi}\frac{dt}{2\pi}\exp\left[ipt + a_1\exp(it) + a_2\exp(-it)\right] = \left(\frac{a_1}{a_2}\right)^{p/2} I_p\left(2\sqrt{a_1 a_2}\right) \tag{11.91}$$

where I_p is the Bessel function of imaginary argument and p designates an integer, we obtain

$$\Delta e(t,z) = \sum_{p=-\infty}^{\infty}\exp(-ip\Omega t)\,c_p, \tag{11.92}$$

$$c_p = \left(\frac{\Delta g^*/2 + i\Delta\phi^*}{\Delta g/2 + i\Delta\phi}\right)^{p/2} I_p\left[2(\Delta g^*/2 + i\Delta\phi^*)^{1/2}(\Delta g/2 + i\Delta\phi)^{1/2}\right]. \tag{11.93}$$

It is then easy to obtain the intensity of all the frequency components at the amplifier output using Eq. (11.78). The amplitude of the output field at frequency $p\Omega$, $A_p(z)$, is obtained as the sum of a term proportional to $A_0(0)c_p$ and a term proportional to $A_1(0)c_{p-1}$,

$$\begin{aligned} A_p(z)\exp(-ip\Omega t) &= [A_0(0)c_p + A_1(0)c_{p-1}] \\ &\quad \exp\left[\frac{1}{2}\bar{g}(z) + i\bar{\phi}(z)\right]\exp(-ip\Omega t). \end{aligned} \tag{11.94}$$

A result in terms of Bessel functions has been obtained by Diez in ref. [18], without including, however, the contribution of the stimulated emission to the lifetimes.

For $\alpha_i = 0$, $|A_p(z)|^2$ can be written in terms of the ratio between the signal intensity to the pump intensity, and of the saturated gain $\bar{g}(z)$ only. The final expression is not given here for brevity.

Let us address the issue of the validity of the first order expansion in Δg_1 of the integral equation, upon which Eqs. (11.81) and (11.84) were based. This expansion is well founded if $\Delta g_1 \ll 1$. Since, in general, $|\tilde{R}_{\rm tot}(\Omega)| \leq \tilde{R}_{\rm tot}(0) = \int R(t)dt = 1$, the gain modulation is deeper and the validity of the linear approximation more questionable when the modulation is below the cutoff frequency of carrier density pulsations. Assuming that the modulation of the input field is adiabatically slow, the response function $R(t) \simeq \delta(t)$ and hence the integral equation becomes the transcendental equation in the variable $g_m(t,z)$,

$$g_m(t,z) = g_0(z) - \epsilon_{\rm tot}\left[e^{g_m(t,z)} - 1\right] S(t,0), \tag{11.95}$$

where $\epsilon_{\rm tot} \simeq 1/S_s$. Let us compare the amplitude of the gain modulation at frequency Ω obtained by the exact solution of Eq. (11.95) assuming the input field given by Eq. (11.71), with that obtained by its first order perturbation expansion, equal to Eq. (11.81) evaluated at $\Omega = 0$. Assume that the unsaturated gain is $G_0 = 30$ dB, corresponding to $g_0 = 6.908$, and that $|A_1(0)|^2 = 0.1\,|A_0(0)|^2$. Consider four cases, $\epsilon_{\rm tot}S_{\rm in} = 0.001, 0.01, 0.1$ and 1, corresponding to saturation of the steady state gain of $g_0 - \bar{g} = 0.567, 1.742$, 3.363 and 5.1 respectively. Remember that the saturation corresponding at high frequency detuning to the maximum efficiency is $g_0 - \bar{g}_{\rm opt} = 2$ [12]. The amplitude of the modulation obtained with the first order approximation is $\Delta g = -0.104, -0.182, -0.217$ and -0.207, whereas the exact values are $\Delta g = -0.107, -0.196, -0.237$ and -0.229. The relative error is always less than 10 %. The amplitude of the harmonic at frequency 2Ω, Δg_2 to the harmonic at frequency Ω, Δg, both obtained by the exact equation, is for the four cases $\Delta g_2/\Delta g = -0.089, -0.136, -0.152$ and -0.172. The amplitude of the second harmonic of the gain modulation never exceeds 20 % of the fundamental. This numerical example shows that the first order expansion of the gain is a good approximation even at low detunings and high saturation, and that the low detuning case can be analysed within our theory, which is a first order theory only for what concerns the gain modulation. Eq. (11.94) may thereby consistently account for the appearance of high order harmonics in the output field and the significant pump depletion that may occur at low detunings, as well as for the strong effect that the gain modulation produces on the probe gain (the so-called Bogatov interaction). In the above example, for S_{in}, considering only carrier density pulsation and $\alpha_N = 12$, the intensity of the pump in the presence of a probe of intensity 10 dB lower, detuned by much less than the cutoff frequency of the carrier density pulsations ($\Omega \to 0$), is 0.0793 times its value without the probe, i.e. the weak probe depletes the pump by more than 92 %. The ratio of the second order FWM term, at frequency -2Ω, to the first order term at frequency $-\Omega$ is 0.82.

The above examples are not intended to cover all cases that can be profitably analysed by a perturbation expansion of the integral equation. For instance, the analysis of the experimental configuration in which more than one pump is injected into the amplifier is also easily analysed by a perturbation expansion of the integral Eq. (11.57), and the results will be presented elsewhere [19]. The examples given are only meant to show what we believe is the main advantage of this approach, namely the flexibility arising from the fact that the perturbation expansion is performed on an integral equation that is the exact solution of the equations governing the propagation of a generic field along the waveguide and that depends on time only. It is therefore easy to have a direct control on the validity of the truncation of the perturbation expansion to the linear terms. Another asset is that the theory is linear for the gain modulation only, and a nonlinear dependence of the output field on the gain and index modulations can be consistently retained. This permits the calculation, in a single step, of all FWM sidebands at once, going formally beyond the simple first order perturbation theory.

The price to be paid for its simplicity is that our approach has embedded some intrinsic simplifications. The waveguide internal loss and the free-carrier absorption induced heating can be considered, but the results are not in a closed form any more. Another effect that is difficult to include is gain dispersion, i.e. a frequency dependence of the SOA gain, which may become of some importance when FWM is used for spectroscopy and the pump-probe detuning exceeds some THz. The theory based on the integral equation thereby fits those cases in which waveguide internal loss, free-carrier absorption induced heating and gain dispersion are negligible. This approach has also a pedagogical aspect because of its simplicity and because it makes transparent the mechanisms behind the interaction of the fields injected and the generation of new fields in the SOA. The following sections will be devoted to a brief description of more general theories based on the coupled mode equations, which permit to relax some of the approximations behind the perturbation theory based on the integral equation (but inevitably adding others). The main purpose of the following section is not to explain the various models in details. Its aim is to review some of the techniques proposed to date and to give a brief account of the results obtained by them. The interested reader may refer to the original papers for details. Before going into a detailed discussion of the coupled-mode theory, let us highlight the relation between the coupled-mode formalism and the time-domain approach discussed above. Both approaches may be based on a description of the photon-carrier interaction on a microscopic level. Both assume that the nonlinearity is small and therefore use a first order expansion of the equations describing the dependence of the gain and refractive index on the *local* intensity. Both theories start from the same expression of the *local* gain coefficient and refractive index, namely a constant plus a term having the form of a convolution between a linear response function and the local intensity. At this point, the two methods follow different procedures. The time-domain approach integrates over z, exactly, the equations describing the propagation of the optical field in the presence of the nonlinearity, and the result is the integral Eq. (11.57) and the equation for the phase (11.68). Only the *input* field appears in these equations. Once the longitudinal coordinate z has been eliminated, a first-order perturbation theory gives the FWM response with a given input pump and signal. With the coupled mode approach, instead, one introduces in the *local* response of the gain and refractive index the Fourier expansion of the field *before* integration over z. After the expansion, one obtains a set of coupled differential equations in the variable z. Solution of these equations gives the output field of pump, signal and conjugate once the field of pump and signal is given at input ($z = 0$). To help the reader, we have summarised in Fig. 11.1 the relations between the analysis based on the time-domain description and the analysis based on the coupled mode equations.

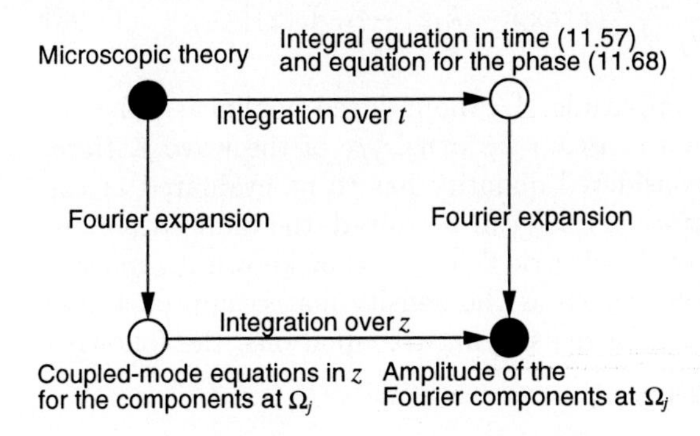

Fig. 11.1. Relation between the two analysis based on a time-domain approach and a coupled-mode approach

11.3 Coupled mode theory

The aim of this section is to provide an overview over the main analytical models of the FWM conversion efficiency based on the coupled mode approach (see Subsecs. 11.3.4-11.3.6). In addition to the FWM conversion efficiency, defined in 10.3 we introduce $\eta_N \equiv P_c(L)/[P_p^2(L)G_s]$, which is referred to as normalised FWM efficiency. η_N is a useful quantity for comparisons of analytical with numerical and experimental results [20, 21] as well as for parameter extraction [22]. It should be noted that in passive media, η_N is independent of the input powers. The consideration starts in Subsec. 11.3.1, where the fundamental equations of the coupled mode theory are presented in a rather general form. Thereafter, the assumptions made by the different analytical approaches are listed and discussed in Subsec. 11.3.2. Within all these approaches, the FWM conversion efficiency is calculated as a function of the saturated single-pass gain. Therefore, one first needs a model for the saturation behaviour of the single-pass gain, which is presented in Subsec. 11.3.3. Finally, the different analytical theories of the FWM efficiency are compared with a numerical model in Subsec. 11.3.7.

11.3.1 Fundamental equations

In Sec. 11.2, FWM has been described in the time domain. In contrast to that, the coupled mode approach is based on the assumption that the optical field is composed of a group of monochromatic input waves. Hence, this approach corresponds to a description of FWM in the frequency domain. In the following, the index j denotes the considered wave whereby $j = s, p, c$ corresponds to the signal, pump, and conjugate wave, respectively. Using the slowly varying envelope approximation, the following propagation equation for the wave j can be derived from Maxwell's equations [3, 23]

$$\frac{d\mathcal{A}_j}{dz} = \frac{1}{2}\left[\frac{i\omega_j \Gamma}{\varepsilon_0 c n_j}\mathcal{P}_j^{NL}(z)\exp(-ik_j z) - \alpha_i \mathcal{A}_j(z)\right] \tag{11.96}$$

with the slowly varying field amplitude $\mathcal{A}_j$, the induced nonlinear polarisation $\mathcal{P}_j^{NL}$, and the propagation constant $k_j = n_j\omega_j/c$ of the wave j. Here, the index j means that the considered quantity has to be evaluated at the angular frequency $\omega = \omega_j$. Before (11.96) can be solved, the induced nonlinear polarisation as a function of the electric field has to be known. In general, this requires a microscopic theory such as the density matrix approach (see Sec. 11.1). From the semiconductor density matrix equations, the following relationship can be derived [24]

$$\begin{aligned} \mathcal{P}_j^{NL}(z) &= \varepsilon_0 \left\{ \sum_{k,l,m} \chi_{jklm}^{(3)}(\omega_j = \omega_k - \omega_l + \omega_m)\mathcal{E}_k(z)\mathcal{E}_l^*(z)\mathcal{E}_m(z) \right. \\ &\qquad \left. + \chi_j^{(1)}\mathcal{E}_j(z) \right\} \end{aligned} \tag{11.97}$$

whereby every combination of k, l and m yielding the considered angular frequency ω_j contributes to the sum in (11.97). The electric field of wave j is related to the slowly varying amplitude by $\mathcal{E}_j(z) = \mathcal{A}_j(z)\exp(ik_j z)$. In contrast to Sec. 11.2, here the electric field (and therefore also the slowly varying field amplitude) is assumed to be normalised such that $|\mathcal{E}_j|^2 = P_j$ gives the optical power in Watts. Thus, the electric fields $\mathcal{E}$ and E as introduced in Sec. 11.2 are related by $\mathcal{E} = \sqrt{\hbar\omega v_g w d}E$. In (11.97), $\chi^{(1)}$ and $\chi^{(3)}$ denote the first- and the third-order nonlinear susceptibility. Thereby, $\chi^{(3)}$ consists of contributions due to the nonlinear mechanisms CDP, CH, SHB as well as the ultrafast processes of two photon absorption and Kerr effect [25]. As stated in Sec. 11.1, the ultrafast effects are negligible for detunings smaller than 4 THz. Therefore, these effects will not be considered here.

It should be mentioned that the first- and the third-order nonlinear susceptibility as introduced in (11.97) do not have the same meaning as in passive components. Usually, the introduction of a susceptibility χ implies that χ is independent of the electric field. In contrast to that, both $\chi^{(1)}$ and $\chi^{(3)}$ in (11.97) depend on the "saturated" carrier density N, which itself is a function of the field intensities due to the saturation (i.e. the depletion of the carrier density caused by stimulated emission). In the following it will be explained what is meant by saturated carrier density. Since in the case of FWM, the total intensity consists of the sum of the intensity of the participating waves plus an oscillatory part (the intensity beating, see Sec. 11.2.2), according to a small-signal approach also the total carrier density may be divided into a stationary and an oscillatory part. Thereby, the former corresponds to the saturated carrier density, and is determined by the sum of the intensity of the participating waves. The oscillatory part (which oscillates at the beat frequency Ω) is included in $\chi^{(3)}$ corresponding to the contribution of CDP. Obviously, this approach and therefore (11.97) yields a good approximation for CW input waves if the detuning is sufficiently high since in this case, the saturated carrier density is a stationary quantity and the oscillatory part is only a small perturbation of the total carrier density.

It is convenient to consider the different combinations of j, k and l in (11.97) yielding a contribution to the induced nonlinear polarisation at ω_j in order to provide an overview over the different nonlinearities in SOAs (see Fig. 11.2). Most important is the saturation of the carrier density because it is the strongest nonlinearity. Due to this effect the carrier density can be modulated by an intensity modulated input signal. This also yields a modulation of the gain (corresponding to the imaginary part of the first-order susceptibility) as well as the refractive index (described by the real part of the first-order susceptibility). The modulation of the gain and the refractive index by an intensity modulated input signal is exploited by all-optical wavelength converters based on XGM and XPM, respectively, whereas this effect is detrimental for FWM applications. Besides the saturation of the carrier density there are nonlinear effects which are described by the

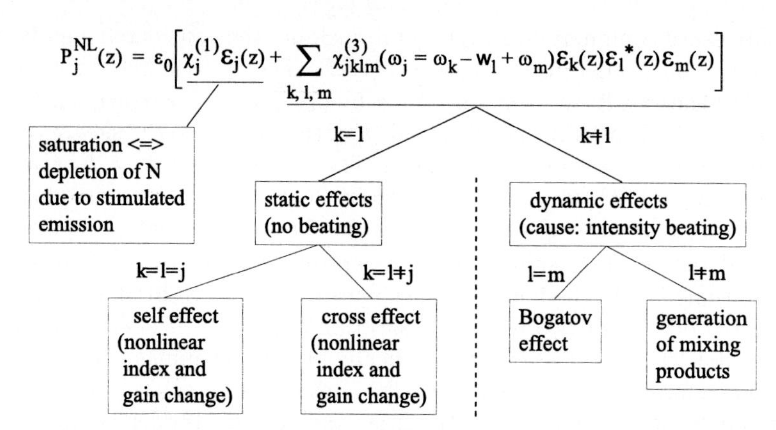

Fig. 11.2. Overview over nonlinear effects in SOAs: saturation (described by the first-order susceptibility), and higher order nonlinearities (described by the third-order susceptibility). The different higher order nonlinear effects are explained in the text.

third-order nonlinear susceptibility. It is convenient to distinguish the case $k = l$ yielding the static effects and $k \neq l$ yielding the dynamic effects. The static effects also exist if there is no intensity beating and therefore these effects are not dependent on the frequency detuning of the input waves (at least for all detunings of practical interest). Moreover, only SHB and CH contribute to these nonlinearities. The static effects can be further divided into the case $k = l = j$ which describes the nonlinear refractive index and gain change induced by the considered wave itself, and the case $k = l \neq j$ which accounts for the corresponding effect induced by all other waves. The nonlinear gain change induced by the considered wave itself is well known and is referred to as nonlinear gain suppression. This effect is often described by the nonlinear gain coefficient ϵ (see Sec. 11.1). Physically, the static effects may be explained by considering the effect of an optical wave on the carrier distribution. Due to SHB and CH the total carrier distribution within the energy band is changed. This causes a change of the gain and the refractive index at the energy level corresponding to the considered wave (self effect) as well as at all different energy levels (cross effect). In contrast to the static effects, the dynamic effects only exist if there is an intensity beating between the input waves. Consequently, these effects are dependent on the detuning of the input waves. Furthermore SHB, CH *and* CDP contribute to the dynamic effects. The case $k \neq l = m$ yields the Bogatov effect (see Sec. 11.2.2). Most important for FWM applications is the case $k \neq l \neq m$ since it describes the coupling of waves at different frequencies and consequently the generation of the FWM mixing products.

Inserting (11.97) into (11.96) yields a set of propagation equations for the slowly varying envelope fields $\mathcal{A}_j$ which are coupled by the third-order nonlinear susceptibility according to

$$\frac{dA_j}{dz} = i\frac{\Gamma\omega_j}{2cn_j}\sum_{k,l,m}\chi^{(3)}_{jklm}(\omega_j = \omega_k - \omega_l + \omega_m)A_k(z)A_l^*(z)A_m(z)e^{i\Delta k_{jklm}z} + \frac{1}{2}[\Gamma g_{lj}(N)(1 - i\alpha_N) - \alpha_i]A_j(z) \tag{11.98}$$

with $\Delta k_{jklm} = k_k - k_l + k_m - k_j$. g_{lj} corresponds to the linear gain coefficient introduced in (11.11) evaluated at the angular frequency ω_j. Within the description according to (11.98), the nonlinear reduction of g_{lj} due to CH and SHB (yielding the gain coefficient g as introduced in (11.9)) is included in the third-order nonlinear susceptibility. A similar expression could be obtained from (11.9) assuming that $(1+\epsilon S)^{-1} \approx (1-\epsilon S)$. For the derivation of (11.98) it has been used that the first-order susceptibility and the linear gain coefficient are related by [16]

$$\frac{i\omega_j}{cn_j}\chi_j^{(1)}(N) = g_{lj}(N)(1 - i\alpha_N) \tag{11.99}$$

Additionally, the propagation equations for the fields (11.98) are coupled by the rate equation for the carrier density. Since here we are focussing on continuous waves, the temporal derivative in (11.4) vanishes yielding

$$0 = \frac{I}{eV} - \frac{N}{\tau_s} - \sum_j \frac{g_{lj}(N)}{\hbar\omega_j A_{\text{eff}}}P_j(z) \tag{11.100}$$

where the total photon density S has been expressed as a sum of the power of the participating waves. The first term on the right-hand side of (11.100) describes the carrier injection, whereas the second term accounts for the carrier consumption due to the spontaneous radiative and nonradiative recombination. Finally, the last term describes the carrier depletion due to the stimulated emission caused by the optical field and hence accounts for the saturation. Since the optical intensity changes along the amplifier, the carrier density will generally be dependent on z.

At this point, the formulation of the fundamental equations is complete. Eqs. (11.98) and (11.100) represent the basis for the modelling of FWM in the frequency domain according to the coupled- mode approach. Before this set of equations can be solved, one needs an expression for the third-order nonlinear susceptibility as a function of the carrier density and the frequencies of the participating waves. This will be the subject of the next section.

11.3.2 General assumptions

A calculation of the FWM conversion efficiency which takes all nonlinear effects mentioned in Sec. 11.3.1 into account requires a numerical solution.

There are several numerical models dealing with FWM in SOAs in the frequency domain [4, 20] as well as in the time domain [26]. However, with some reasonable assumptions also analytical solutions can be derived. Analytical theories are desirable because they generally provide more physical insight and can be evaluated much more efficiently than numerical solutions. Furthermore, they allow extraction of the nonlinear SOA parameters by comparison with measurements (see Sec. 13.1). In the following, the assumptions that are made by the different analytical approaches based on the coupled-mode theory are listed and discussed.

1. The static effects of CH and SHB are negligible. Since the efficiency of these effects is proportional to the intensity of the optical waves, this assumption is valid in particular when the power associated with the considered waves is low.
2. The Bogatov effect can be neglected. This effect can be important for small frequency detunings due to the large contribution associated with CDP in this case [17, 27, 28]. Consequently, the neglect of the Bogatov effect yields a good approximation when the detuning of the input waves is sufficiently large. In the following, this case will also be referred to as high detuning limit.
3. The effect of FWM on the pump wave is negligible, which is the case for $P_p(z) \gg P_s(z), P_c(z)$. This is a reasonable assumption because in practice the pump input power has to be significantly larger than the signal power in order to avoid performance degradation caused by bit-pattern effects [29, 30].
4. The back conversion of the conjugate onto the signal wave can be neglected. This assumption is good as long as $P_s(z) \gg P_c(z)$, or equivalently $P_s(L) \gg P_c(L)$ because the ratio $P_s(z)/P_c(z)$ decreases with z. Under these conditions, also the saturation caused by the FWM mixing products is small.
5. Suppression of the scattering losses. The effect of the scattering losses on the FWM efficiency was investigated in [17]. It has been found that the influence of α_i on the FWM efficiency directly is negligible as long as the scattering losses are taken into account when calculating the single-pass gain.
6. Suppression of the gain dispersion. In [20] it has been shown that the spectral dependence of the gain becomes important for detunings $|\Delta\nu| > 1$ THz due to saturation effects. However, for moderate detunings this effect can be neglected and consequently the index j can be omitted. In this case both CDP and SHB vanish when the SOA is completely saturated and the only contribution to the third-order nonlinear susceptibility stems from CH due to FCA.

Due to the assumptions 1-4, $\chi^{(3)}_{jklm} = 0$ in Eq. (11.98) for $j = s, p$ (i.e all higher order nonlinearities are neglected for the signal and the pump wave).

Consequently, the only higher order nonlinear effect which is accounted for is the term describing the generation of the FWM mixing product at ω_c. The corresponding part of $\chi^{(3)}_{jklm}$ is obtained for the combination $\chi^{FWM} = \frac{i\omega_c}{cn_c}\chi^{(3)}(\omega_c = \omega_p - \omega_s + \omega_p)$. The susceptibility χ^{FWM} can be expressed as a sum of contributions from the different nonlinear processes $y = N, T_c$ and SHB according to [31, 32]

$$i\chi^{FWM}(N,\Omega) = -\sum_y \rho_y(N)(1 - i\alpha_y)\tilde{R}_y(\Omega). \tag{11.101}$$

with the α-parameter α_y as defined in (11.27). In contrast to Sec. 11.1, here only the contribution of carrier heating in the conduction band has been accounted for whereas carrier heating in the valence band has been neglected. The weight factors ρ_y describe the strength of the nonlinear process y, which are related to the more familiar gain compression coefficient by $\epsilon_y(N) = \rho_y(N)/g_l(N)$. The frequency responses $\tilde{R}_y(\Omega)$ are given by (11.85)-(11.87) with $\tau_{1c} = \tau_{1v} = \tau_1$ and $\tau_{c,h} \gg \tau_{v,h}$ corresponding to the neglect of CH in the valence band

$$\tilde{R}_N^{-1}(\Omega) = (1 - i\Omega\tau_s)(1 - i\Omega\tau_1) \tag{11.102}$$

$$\tilde{R}_{T_c}^{-1}(\Omega) = (1 - i\Omega\tau_{c,h})(1 - i\Omega\tau_1) \tag{11.103}$$

$$\tilde{R}_{\mathrm{SHB}}^{-1}(\Omega) = (1 - i\Omega\tau_1) \tag{11.104}$$

Usually, when (11.101) is substituted in (11.98) it is assumed that $\Delta k_{cpsp} = 2k_p - k_s - k_c = 0$. This corresponds to the assumption that the phase matching condition is fulfilled. For a pump wavelength of 1.5μm and the value of the group index dispersion reported in [33], this assumption is valid for detunings $\Delta\nu = \Omega/2\pi$ with $(\Delta\nu/\mathrm{THz}) \ll \sqrt{63/(L/\mathrm{mm})}$ [4].

In [20] it has been shown, that both the differential gain a and the carrier density at transparency N_0 defined in (11.11) are different for different waves j when the spectral dependence of the gain coefficient should be taken into account. In this case, the gain coefficient at ω_j can be expressed as $g_{lj}(N) = a_j(N - N_{0j})$ with a_j denoting the differential gain and N_{0j} the transparency carrier density of the wave j. Finally, one needs to know the dependence of the weight factors ρ_y as introduced in (11.101) on the carrier density. In order to obtain analytical solutions, it is advantageous to assume a linear dependence according to [20]

$$\rho_y(N) = \rho_{Ny}(N - N_y) \tag{11.105}$$

with $y = N, T_c$ and SHB. When ρ_y and g_l have the same transparency points, or equivalently $\rho_y(N) \propto g_l(N)$, the gain compression coefficient ϵ_y is independent of the carrier density. Hence $P_y^{sat} = \epsilon_y^{-1}$ yields a characteristic power of the nonlinear process y [32, 34]. In these conditions, the characteristic power associated with CDP corresponds to the saturation power as introduced in

(11.24). Otherwise, the nonlinear process has to be described by two characteristic powers P_{y1}^{sat} and P_{y2}^{sat} [17, 27]. It should be mentioned that there is one analytical study where ρ_y is allowed to have a quadratic dependence on the carrier density [13].

This concludes the consideration of the general assumptions. The specific assumptions made by the different analytical models of the FWM conversion efficiency are explicitly listed in the corresponding subsections.

11.3.3 Saturation of the single-pass gain

The FWM conversion efficiency can be calculated as a function of the single-pass gain $G_j = P_j(L)/P_j(0)$, which itself can be determined by evaluating an implicit relationship. It is noteworthy that the single-pass gain as defined above corresponds to $G(t, L)$ as defined in (11.32) at the considered frequency ω_j for CW signals. When the spectral dependence of the gain is neglected, the gain of all waves is equal. With (11.98) as well as (11.100) and the assumptions made in Sec. 11.3.2 the following transcendental equation can be derived [35, 12]

$$\frac{P_{in}}{P_s} = (\frac{\Gamma g_0}{\alpha_i} - 1)\frac{1 - (G/G_0)^{\frac{\Gamma g_0}{\alpha_i}}}{G - (G/G_0)^{\frac{\Gamma g_0}{\alpha_i}}}. \tag{11.106}$$

with g_0 and $G_0 = e^{(\Gamma g_0 - \alpha_i)L}$ denoting the unsaturated linear gain coefficient and single-pass gain, respectively, and $P_{in} = P_s(0) + P_p(0)$ the total input power. Eq. (11.106) describes the saturation of the single-pass gain, i.e. the reduction of G due to the saturation caused by intense input fields. For the following treatment it is useful to consider a specific case of (11.106). For vanishing scattering losses (11.106) reduces to [17, 20]

$$\frac{P_{in}}{P_s} = \frac{1}{G-1}\ln\left(\frac{G_0}{G}\right) \tag{11.107}$$

whereby in this case, the unsaturated single-pass gain is given by $G_0 = e^{\Gamma g_0 L}$.

11.3.4 Gain-cube theory

This theory predicts a cubic dependence of the FWM efficiency on the single-pass gain and is therefore called G^3-theory. Within this approach all static effects, the Bogatov effect, the effect of FWM on the pump and the signal wave as well as the gain dispersion and the scattering losses are neglected. Moreover, it is assumed that $\chi^{FWM}(N) \propto g(N)$ yielding $N_y = N_0$ for all nonlinear processes $y = N, T_c$ and SHB. Under these conditions, the following analytical expression for $\eta_{FWM} = |\mathcal{A}_c(L)|^2/|\mathcal{A}_s(0)|^2$ can be derived from (11.98) and (11.100) [12]

$$\eta_{FWM} = \frac{1}{4} G_s (G_p - 1)^2 P_p^2(0) \left| \sum_y \frac{1 - i\alpha_y}{P_y^{sat}} \tilde{R}_y(\Omega) \right|^2 \tag{11.108}$$

with $P_N^{sat} = P_s$ and $P_y^{sat} = a/\rho_{Ny}$ for $y = T_c$ and SHB. Since the spectral dependence of the gain coefficient has been neglected it follows that $G_s = G_p \equiv G$, which is determined by (11.106). Furthermore, in [34, 32] it is assumed that $G \gg 1$ yielding a cubic dependence of η_{FWM} on the single-pass gain. In this limit, (11.108) is called G^3-theory. The G^3-theory has the advantage of being simple but, on the other hand, is restricted to the case of an unsaturated (or slightly saturated) SOA due to the assumptions made. However, at high input powers the SOA is strongly saturated. One problem arising from this is that according to the G^3-theory, the normalised FWM efficiency should be independent of the input pump power, which is in contradiction with experimental results. Measurements published in [20] reveal that the normalised FWM efficiency versus the pump power increases for small detunings, whereby the slope decreases with increasing detuning. For $\Delta\nu = 375$ GHz, the normalised FWM efficiency is increased by approx. 8 dB when the input pump power is increased from -20 dBm to +5 dBm. The increase of η_N with the pump power is attributed to the generally different transparency points $N_y \neq N_0$.

11.3.5 Inclusion of saturation effects

More sophisticated theories of the FWM conversion efficiency which include saturation effects have been published in [12, 13, 17]. In all of these works, the static effects, the effect of FWM on the pump and on the signal wave as well as the gain dispersion are neglected. However, the scattering losses and the different transparency points $N_y \neq N_0$ are taken into account in [12] and [13], respectively. The model published in [17] considers all previously mentioned effects and additionally includes the Bogatov effect. Therefore this approach is valid also for small frequency detunings. For simplicity and in order to compare the result with the expressions presented in Subsec. 11.3.4 and 11.3.6, the solution is given in the high detuning limit $\Omega\tau_s \gg 1$ and for $\alpha_i = 0$ yielding [17]

$$\begin{aligned} \eta_{FWM} &= \frac{G}{4} P_p(0)^2 \left| (G-1) \sum_y \frac{1 - i\alpha_y}{P_{y1}^{sat}} \tilde{R}_y(\Omega) \right. \\ &\quad \left. + \ln\left(\frac{G_0}{G}\right) \frac{(G+1)}{2} \sum_y \frac{1 - i\alpha_y}{P_{y2}^{sat}} \tilde{R}_y(\Omega) \right|^2 \end{aligned} \tag{11.109}$$

where $P_{N1}^{sat} = P_s$, $P_{T_c1}^{sat} = g_0/\rho_{T_c}(N_{st})$, and $P_{SHB1}^{sat} = a/\rho_{N,SHB}$. Note that the only contribution in the second sum of (11.109) stems from carrier heating with $P_{T_c2}^{sat} = g_0/\rho_{T_c}(N_0)$ due to the fact that $\rho_{T_c}(N_0) > 0$. For $y =$

CDP and SHB, P_{y2}^{sat} tends to infinity since $N_N = N_{SHB} = N_0$ as discussed in Sec. 11.3.2. Moreover, the refractive index change associated with SHB is assumed to be zero (see Sec. 11.1). In order to include also the ultrafast nonlinear effects two photon absorption and Kerr effect one has to add an additional term $y = ul$ in both sums in (11.109) [13]. Thereby, these effects are assumed to be instantaneous, i.e. $\tilde{R}_{uf}(\Omega) = 1$. When the ultrafast nonlinear effects are independent of the carrier density, the two characteristic powers are equal with $P_{uf1}^{sat} = P_{uf2}^{sat} = \epsilon_{uf}^{-1}$. The single-pass gain G in (11.109) is determined by (11.106). Obviously, (11.109) differs from (11.108) by the second sum which takes into account that CH does not vanish when the SOA is completely saturated. For an unsaturated or a slightly saturated SOA the FWM conversion efficiency is determined by the first term in (11.109) since $\ln(G_0/G) \to 0$ for $G \to G_0$. With increasing pump power the first term is reduced due to the gain saturation, and vanishes when the SOA is completely saturated (i.e. $G = 1$). However, in this case $\ln(G_0/G)(G+1)/2 \to \ln(G_0)$ yielding a constant contribution to η_{FWM} of the second term. Consequently, the FWM efficiency according to (11.108) shows a maximum as a function of $P_p(0)$, whereas (11.109) predicts a local maximum followed by a further increase of η_{FWM} (see e.g. [36]).

11.3.6 Inclusion of gain dispersion effects

In [20, 27] an analytical model of the FWM conversion efficiency has been developed which takes the gain dispersion and also the different transparency points $N_y \neq N_0$ into account. However, all static effects, the Bogatov effect, the effect of FWM on the pump and the signal wave as well as the scattering losses are neglected. In these conditions, an analytical solution can only be derived assuming that one wave (i.e. the pump) predominantly saturates the SOA. The solution writes [20]

$$\begin{aligned} \eta_{FWM} &= \frac{1}{4} G_s G_p^2 P_p(0)^2 \frac{\exp(2\sigma x_L)}{(\sigma x_L)^{2\beta}} \left| \Gamma(\beta, \sigma x_0, \sigma x_L) \sum_y \frac{1 - i\alpha_y}{P_{y1}^{sat}} \tilde{R}_y(\Omega) \right. \\ &\quad \left. + \frac{\Gamma(\beta+1, \sigma x_0, \sigma x_L)}{\sigma} \sum_y \frac{1 - i\alpha_y}{P_{y2}^{sat}} \tilde{R}_y(\Omega) \right|^2 \end{aligned} \tag{11.110}$$

where $x_0 = P_p(0)/P_{sp}$ and $x_L = G_p x_0$ denote the normalised pump input and output power, respectively. P_{sp} represents the saturation power according to (11.24) at the pump frequency ω_p. The pump gain G_p can be calculated with (11.106) by replacing $G \to G_p, G_0 \to G_{0p}, g_0 \to g_{0p}, P_s \to P_{sp}$ and $P_{in} = P_p(0)$. The gain dispersion is described by the factors

$$\beta \equiv 1 + \frac{g_{0s} - g_{0c}}{2 g_{0p}} \qquad \sigma \equiv 1 - \beta + \frac{a_s - a_c}{2 a_p}. \tag{11.111}$$

Evidently, one obtains $\beta \to 1$ and $\sigma \to 0$ for small detunings. In (11.110), $\Gamma(a, x_1, x_2)$ represents the incomplete Gamma function, which is defined as

$$\Gamma(a, x_1, x_2) \equiv \int_{x_1}^{x_2} x^{a-1} \exp(-x) dx. \tag{11.112}$$

Since the transparency points N_{0j} for $j = s, c$ differ from N_{0p}, even CDP and SHB do not vanish when the SOA is completely saturated by the pump ($N \to N_{0p}$). Consequently, *all* nonlinear processes have to be described by two characteristic powers with $P_{y1}^{sat} = g_{0p}/\rho_y(N_{st})$ and $P_{y2}^{sat} = g_{0p}/\rho_y(N_{0p})$ because $N_y \neq N_{0p}$ for $y =$ CDP, CH, and SHB in this case. Furthermore, for large detunings $\alpha_{SHB} \neq 0$ which follows from a generalisation [24] of the microscopic theory given in [4]. It is noteworthy that for $\beta \to 1$ and $\sigma \to 0$, i.e. when the gain dispersion is neglected, (11.110) and (11.109) are identical. In agreement with the results of Subsec. 11.3.5, (11.110) also predicts a local maximum of η_{FWM} followed by a further increase at large pump powers which is due to the nonvanishing nonlinear susceptibility at transparency.

11.3.7 Comparison with a numerical model

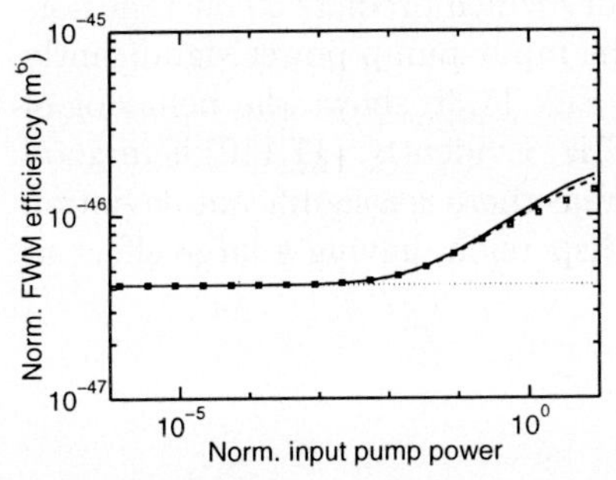

Fig. 11.3a. Comparison of the normalised FWM efficiency (expressed in photon densities) versus input pump power normalised to the saturation power according to a numerical solution (▪) with the analytical models presented in Subsec. 11.3.4 (·····), Subsec. 11.3.5 (——), and Subsec. 11.3.6 (– –). Pump-signal detuning: 0.4 THz

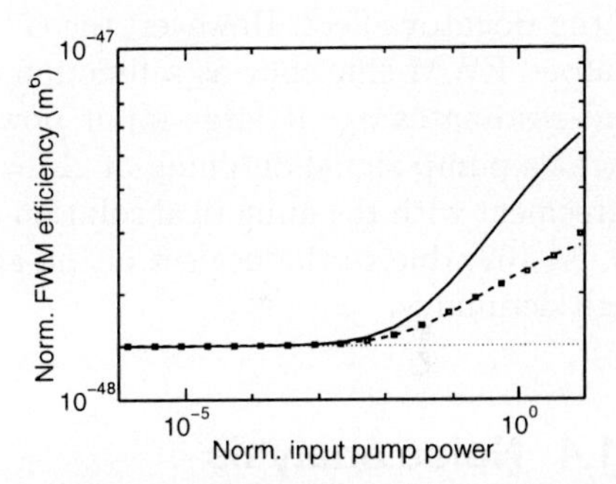

Fig. 11.3b. Same as (a) with pump-signal detuning: 4 THz

A comparison of the FWM conversion efficiency according to the different analytical models presented with experimental data is not trivial since the nonlinear SOA parameters are generally not known. Therefore it is preferable to compare the analytical models with a numerical solution which allows to calculate the values of these parameters. In this subsection the analytical results will be compared with numerical calculations based on the density-matrix formalism. The numerical model used for the evaluation was published in [20] and is similar to the models of [4, 23]. The numerical solution includes:

- all static effects and the Bogatov effect;
- the back-conversion of the conjugate onto the signal wave due to the FWM interaction and the effect of FWM on the pump wave;
- the exact dependence of the gain coefficient and the third-order nonlinear susceptibility on the carrier density as well as the wavelength;
- the contribution of all waves (i.e. pump, signal and conjugate wave) to the saturation of the SOA.

Fig. 11.3a and 11.3b show the normalised FWM efficiency (expressed in photon densities) versus the normalised input pump power for two different detunings. The density matrix equations calculations were performed assuming the same material and SOA parameters as in [4] and $P_P(0) = 10P_S(0)$. It can be seen from Fig. 11.3a that the analytical theories according to Subsec. 11.3.5 and 11.3.6 yield nearly the same results for $\Delta\nu = 0.4$ THz, and are in good agreement with the numerical solution of η_N. This justifies the assumptions made for the analytical models (see Sec. 11.3.2). It has been found that the small deviation at large pump powers $x_0 > 1$ is due to the neglect of the Bogatov effect. However, the G^3-theory which predicts a constant normalised FWM efficiency as a function of the input pump power significantly underestimates η_N at large input powers. Fig. 11.3b shows the behavior of η_N at a pump signal detuning of $\Delta\nu = 4$ THz. Evidently, (11.110) is in good agreement with the numerical solution whereas there is a significant deviation to (11.109) due to the neglect of the gain dispersion, having a large effect at high detunings.

11.4 Noise analysis

Since SOAs are active components they add noise in terms of amplified spontaneous emission (ASE) to the signals. In many cases the ASE is the dominant noise source, and therefore it is desirable to have accurate models. The ASE can be described by the noise power spectral density ρ_{ASE}. In the case of SOAs used for signal amplification the ASE spectral density is well known [37–39]. There, the analysis is based on the assumption that the inversion parameter is spatially homogeneous which is justified for an unsaturated (or slightly saturated) SOA only. However, SOAs used for signal processing generally operate under conditions of strong excitation, where the theory [37–39] cannot be applied due to saturation [36, 40] and gain dispersion effects [28].

In this section it will be shown how the ASE spectral density at the end facet of a SOA can be calculated (Subsec. 11.4.1), and the results according to the different analytical approaches will be presented (Subsecs. 11.4.2–11.4.4). The different analytical models of the ASE spectral density are compared with experimental results in Subsec. 11.4.5. Besides, the ASE at the frequency of the conjugate wave there are several other noise contributions. finally, these contributions are discussed and an expression for the signal-to-background ratio (11.4.6) as well as for the noise Fig. (11.4.7) are given.

11.4.1 Calculation of the ASE spectral density

In order to calculate the ASE spectral density of a SOA one has to add a noise term in the propagation equations for the slowly varying envelope fields. In this case (11.96) writes [36, 40]

$$\frac{d\mathcal{A}_j}{dz} = \frac{1}{2}\left[\frac{i\omega_j\Gamma}{\varepsilon_0 c n_j}\mathcal{P}_j^{NL}(z)\exp(-ik_j z) - \alpha_i \mathcal{A}_j(z)\right] + \mu_j(z,t) \tag{11.113}$$

where $\mu_j(z,t)$ denotes the spontaneous emission noise which is coupled into the fundamental waveguide mode and added to the signal wave at ω_j. Thereby, the spontaneous emission is modeled as a white, Gaussian distributed noise process with an autocorrelation function given by

$$\langle\langle \mu_j(z,t)\mu_j^*(z-z',t-t')\rangle\rangle = \rho_{spon,j}(N(z))\delta(z')\delta(t'), \tag{11.114}$$

and the noise power spectral density of the spontaneous emission per unit length $\rho_{spon,j}(N) = \hbar\omega_j\Gamma g_{lj}(N)n_{sp,j}(N)$ [38]. Here, δ denotes the Dirac Delta-function. The inversion parameter $n_{sp,j}$ at ω_j can be expressed as [41]

$$n_{sp,j}(N) = \{1 - \exp([\hbar\omega_j - \Delta\varepsilon_f(N)]/k_B T)\}^{-1}. \tag{11.115}$$

where $\Delta\varepsilon_f$ denotes the difference of the quasi-Fermi levels and k_B and T the Boltzmann constant and the absolute temperature, respectively. Neglecting the nonlinearities described by the third-order nonlinear susceptibility, the following expression for the electrical field at the SOA end facet can be derived [40]

$$\begin{aligned}\mathcal{A}_j(L) &= \mathcal{A}_j(0)\exp\left[\int_0^L h_j(z)dz\right]\\ &\quad + \underbrace{\int_0^L \mu_j(z,t)\exp\left[\int_z^L h_j(z')dz'\right]dz}_{\equiv \mathcal{A}_{ASE,j}(t)}\end{aligned} \tag{11.116}$$

with $h_j(z) \equiv 0.5[\Gamma g_{lj}(N(z))(1 - i\alpha_N) - \alpha_i]$. The first term in (11.116) describes the amplification of the input signal wave, whereas the second term denotes the slowly varying envelope field due to the ASE. Using

$\langle \mathcal{A}_{ASE,j}(t)\mathcal{A}_{ASE,j}(t-t')\rangle = \rho_{ASE,j}\delta(t')$ one obtains for the ASE spectral density

$$\rho_{ASE,j} = \int_0^L \rho_{spon,j}(N(z)) \exp\left[\int_z^L (\Gamma g_{lj}(N(z') - \alpha_i)dz')\right] dz \quad (11.117)$$

which is consistent with the result of [28]. Eq. (11.117) describes the sum of the locally generated spontaneous emission at the angular frequency ω_j which coupled into the fundamental waveguide mode and is amplified while propagating through the SOA. The spatial distribution of the carrier density $N(z)$ is determined by the input field intensities at frequencies which generally differ from ω_j in the case of nonlinear applications.

For the calculation of ρ_{ASE} it is essential to know the dependence of the spontaneous emission spectral density on the carrier density. According to the theory [37–39], the inversion parameter is assumed to be spatially homogeneous and consequently ρ_{spon} shows the same dependence on the carrier density as the gain. This *clarifies* the importance to account for the correct dependence of the inversion parameter on the carrier density especially under saturated conditions and frequencies which are larger than the wave predominantly saturating the SOA. Therefore, let us consider a wave at the angular frequency ω_j. At the carrier density of transparency N_{0j} the gain coefficient g_{lj} is zero and the inversion parameter tends to infinity yielding a finite spontaneous emission spectral density at $N = N_{0j}$. Moreover, for $N < N_{0j}$ the gain coefficient becomes negative whereas the spontaneous emission is always positive for $N > 0$ and only vanishes for $N = 0$. Note that $N < N_{0j}$ only occurs at frequencies ω_j which are larger than the wave predominantly saturating the SOA. Concerning the dependence of the spontaneous emission spectral density on the carrier density, numerical calculations based on the density-matrix theory reveal that ρ_{spon} varies nearly linearly with N [28]. Hence the approach

$$\rho_{spon,j} = \rho_{sp,Nj}(N - N_{sp,j}) \quad (11.118)$$

yields a good approximation. Note that a "transparency carrier density" $N_{sp,j}$ different from zero can occur when $\rho_{spon,j}$ is expanded around a given bias point.

It is interesting to note that the derivation of an analytical expression for the ASE spectral density can be performed analogous to the calculation of the FWM efficiency. Hence, the expressions obtained for η_{FWM} and ρ_{ASE} show a similar structure. Moreover, also the ASE spectral density is calculated as a function of the single-pass gain, which itself is determined by (11.106).

11.4.2 Uniform inversion parameter

As mentioned above, the conventional theory has been developed for SOAs used for linear applications where the amplifier operates under moderate saturation conditions. This allows the assumption of a spatially homogeneous inversion parameter as well as the neglection of the scattering losses. Furthermore, one is interested in the amount of ASE at the frequency ω of the injected signal. It should be noted that these are the only assumptions made by the conventional theory. No further assumptions, e.g. concerning the dependence of the gain coefficient or spontaneous recombination rate on the carrier density are needed. Under these conditions the integral (11.117) can be solved yielding [37–39]

$$\rho_{ASE} = \hbar\omega \langle n_{sp} \rangle (G-1) \tag{11.119}$$

with $\langle n_{sp} \rangle$ denoting an appropriate value of the inversion parameter which can be either the unsaturated [28] or an average value [40]. When $\langle n_{sp} \rangle$ corresponds to the unsaturated inversion parameter $n_{sp}(N_{st})$, (11.119) significantly underestimates the ASE spectral density for large input powers. According to calculations based on the density-matrix theory the deviation of (11.119) and (11.121) is larger than 4 dB for an input power $P(0) = P_s$, whereby the deviation rapidly increases for $P(0) > P_s$ [28].

11.4.3 Nonuniform inversion parameter

In order to model the ASE spectral density for nonlinear applications, a theory has been developed which accounts for the spatial dependence of both the gain coefficient and the inversion parameter [36]. Within this approach the spectral dependence of the gain is neglected whereas the scattering losses are taken into account. Using the assumptions listed in Sec. 11.3.2 and assuming a linear dependence of ρ_{spon} on the carrier density, the integral (11.117) can be solved analytically. For simplicity the result is given for $\alpha_i = 0$ yielding [36]

$$\rho_{ASE} = \hbar\omega n_{sp}(N_{st})(G-1) + b_{sp}\frac{P(0)}{P_s}G\ln(G) \tag{11.120}$$

with the parameter $b_{sp} \equiv n_{sp}(N_{st}) - \rho_{sp,N}/(\Gamma\hbar\omega a) = n_{sp}(N_{st})[(n_{sp}(N_{st}) - 1)(\Delta\varepsilon_f(N_{st}) - \hbar\omega)/k_B T]$. Obviously (11.120) differs from (11.119) with $\langle n_{sp} \rangle = n_{sp}(N_{st})$ by the second term describing the enhancement of ρ_{ASE} due to the spatially increasing inversion parameter under highly saturated conditions. For small input powers $P(0)$, the ASE spectral density is determined by the first term in (11.120). With increasing $P(0)$ this term decreases since $G \to 1$, and ρ_{ASE} is mainly determined by the second term in (11.120). In contrast to the first term, this term first increases with $P(0)$ and then decreases again for extremely large input powers on the order of $P(0) \approx 10P_s$.

11.4.4 Effect of gain dispersion

For sufficiently large detunings, the gain dispersion may play an important role. According to experimental results presented in [28], the ASE spectral density within the gain bandwidth of the SOA varies by more than 7 dB due to saturation effects. Therefore, a theory has been developed which accounts for saturation effects as well as the spectral dependence of ρ_{ASE} [28]. With the same assumptions as in Sec. 11.3.2 and the linear approximation of the spontaneous emission spectral density on the carrier density according to (11.118), a closed form solution for ρ_{ASE} can be derived when the scattering losses are neglected. In this conditions one finally obtains [28]

$$\begin{aligned}\rho_{ASE,j} &= \hbar\omega_j(\hat{b}_j x_L)^{\hat{a}_j}\exp(\hat{b}_j x_L)\Big[a_{spj}\Gamma(-\hat{a}_j,\hat{b}_j x_0,\hat{b}_j x_L) \\ &\quad + \frac{b_{spj}}{\hat{b}_j}\Gamma(1-\hat{a}_j,\hat{b}_j x_0,\hat{b}_j x_L)\Big] \end{aligned} \tag{11.121}$$

with $x_0 = P_p(0)/P_{sp}$ and $x_L = G_p x_0$ denoting the normalised pump input and output power, respectively, as defined in Sec. 11.3.6. The spectral dependence of the gain is described by the factors

$$\hat{a}_j \equiv g_{0j}/g_{0p}\ , \qquad \hat{b}_j \equiv \hat{a}_j - a_j/a_p \tag{11.122}$$

whereby $\hat{a}_j$ describes the spectral dependence of the unsaturated gain coefficient and $\hat{b}_j$ the nonuniform gain saturation [20]. The parameters a_{spj} and b_{spj} account for the corresponding effects for the spontaneous emission spectral density and are related to the gain parameters $\hat{a}_j$ and $\hat{b}_j$ by [28]

$$a_{spj} = \hat{a}_j n_{sp,j}(N_{st}) \tag{11.123}$$

$$b_{spj} = a_{spj}\left\{\frac{\hat{b}_j}{\hat{a}_j} + \left[1 - \frac{a_{spj}}{\hat{a}_j}\right]\frac{\hbar\omega_p - \Delta\varepsilon_f(N_{st})}{k_B T}\right\}. \tag{11.124}$$

Hence, a_{spj} and b_{spj} can be calculated from $\hat{a}_j$ and $\hat{b}_j$ once the difference of the quasi-Fermi levels for the unsaturated carrier density is known. Eq. (11.121) describes the ASE spectral density at an angular frequency ω_j which is different from the pump frequency ω_p. Indeed, for $\omega_j \to \omega_p$ the gain parameter $\hat{a}_j$ tends to one whereas $\hat{b}_j$ tends to zero. In this limit, (11.121) corresponds to (11.120), i.e. to the theory [36] for vanishing scattering losses. In order to evaluate (11.121) one needs to know the pump gain G_p, which is determined by (11.107) by replacing $G \to G_p, G_0 \to G_{0p}, g_0 \to g_{0p}, P_s \to P_{sp}$ and $P_{in} = P_p(0)$. Thus, only the saturation caused by the pump is taken into account whereas the theory (11.120) additionally considers the saturation due to the signal wave.

11.4.5 Comparison with measurements

In contrast to the FWM conversion efficiency the different analytical theories describing the ASE spectral density can directly be compared with

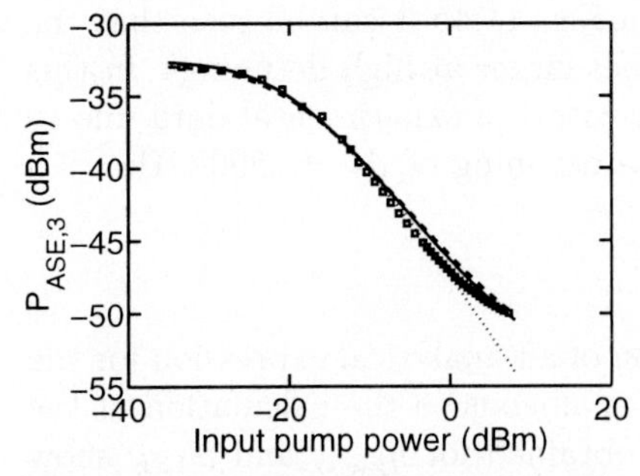

Fig. 11.4a. Comparison of the ASE power at the frequency of the conjugate wave versus the input pump power according to experimental data (▪) with the analytical theories according to Subsec. 11.4.2 (·), Subsec. 11.4.3 (—) and Subsec. 11.4.4 (– –). Pump-signal detuning: 300 GHz.

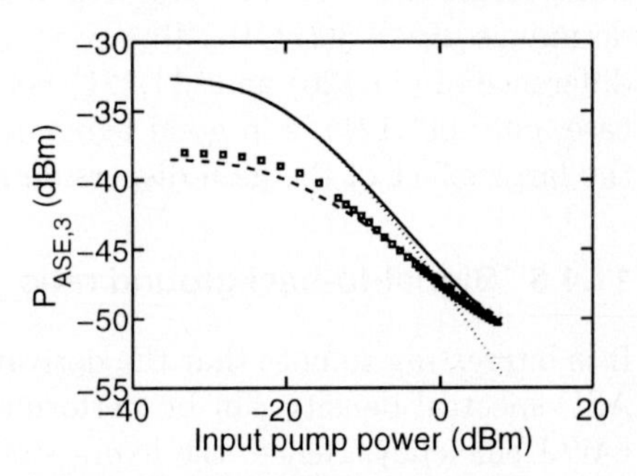

Fig. 11.4b. Same as in (a) with pump-signal detuning: 2.9 THz.

experimental data since all parameters required for the calculation can be determined easily [28]. The experimental data were taken from the FWM experiment published in [22]. The SOA investigated was a bulk InGaAsP amplifier [42] with length $L = 500\mu$m and polarisation independent fibre-to-fibre gain of $\approx$ 20 dB for an injection current of 80 mA. The ASE power was measured by an optical spectrum analyser into a 0.08 nm resolution bandwidth, and the ratio of the pump and the signal input power was kept constant with $P_p(0) = 10P_s(0)$.

In order to calculate the ASE spectral density with (11.119) and (11.120), the effective noise bandwidth B_o as well as the difference of the quasi-Fermi levels for the unsaturated carrier density $\Delta\varepsilon_f(N_{st})$ have to be known. In [28] it was shown how these parameters can be determined yielding $B_o = 0.092$ nm and $\Delta\varepsilon_f(N_{st}) = 0.823$ eV. Note that compared to (11.119), no further parameter is required when the ASE spectral density is calculated with (11.120). However, in order to apply (11.121) additionally the gain parameters $\hat{a}_j$ and $\hat{b}_j$ have to be known. The values of these parameters can be found as described in [20].

A comparison of the measured dependence of the ASE power $P_{ASE,c} = \rho_{ASE,c}B_o$ at the frequency of the conjugate wave on the input pump power $P_p(0)$ with the different analytical theories is given in Fig. 11.4a and 11.4b. According to Fig. 11.4a, both the theory presented in Subsec. 11.4.3 and 11.4.4 are in good agreement with the experimental results, whereas the theory described in Subsec. 11.4.2 significantly underestimates the ASE for input pump powers $P_p(0) > 0$ dBm due to the assumption of a spatial homogeneous inversion parameter. Moreover, (11.120) and (11.121) yield nearly the same result since the spectral dependence of the gain is negligibly small at detunings $\Delta\nu = 300$ GHz. However, from Fig. 11.4b it can be seen that the difference of (11.120) and (11.121) becomes larger at high detunings. In this case, only (11.121) is in good agreement with the experimental data due to the large effect of the gain dispersion at a detuning of $\Delta\nu = 2890$ GHz.

11.4.6 Signal-to-background ratio

It is interesting to note that the derivation of an analytical expression for the ASE spectral density can be performed analogous to the calculation of the FWM efficiency. Hence, the expressions obtained for η_{FWM} and ρ_{ASE} show a similar structure. Moreover, also the ASE spectral density is calculated as a function of the single-pass gain, which itself is determined by (11.106).

Since the FWM process converts the noise in the same way as the signal power, the total noise in the frequency converted channel consists of the ASE at ω_c plus several contributions which are converted via FWM. Consequently, the total noise power spectral density at the frequency ω_c can be written as [28]

$$\rho_{N,c} = \underbrace{\eta_{FWM}\rho_{N,s} + \eta_{c,p}\rho_{N,p}}_{\text{external noise sources}} + \underbrace{\sum_{j=s,p,c} \widetilde{\eta}_{c,j}\rho_{ASE,j}}_{\text{internal noise sources}} \tag{11.125}$$

with $\rho_{N,s}$ and $\rho_{N,p}$ denoting the input noise power spectral density of the signal and the pump, respectively. $\eta_{c,p}$ describes the transmission function of the pump onto the conjugate wave. Hence, the first two terms in (11.125) account for the input intensity noise of the signal and the pump wave. The sum describes the total ASE added by the SOA corresponding to the ASE at ω_c plus the ASE at ω_c and ω_p, respectively, which is converted to ω_c due to the FWM interaction. Thereby, the internal conversion efficiencies $\widetilde{\eta}_{c,j}$ are given by [43]

$$\widetilde{\eta}_{c,s} = \frac{\eta_{FWM}}{G_s}, \quad \widetilde{\eta}_{c,p} = 4\frac{\eta_{FWM}}{G_p}\frac{P_s(0)}{P_p(0)} \quad \text{and} \quad \widetilde{\eta}_{c,c} = 1. \tag{11.126}$$

The SBR as defined in 10.3 is the FWM signal power over the total noise power $P_{N,c}(L)$ at the SOA output and the angular frequency ω_c within the optical bandwidth B_o, i.e.

$$\text{SBR} = \frac{P_c(L)}{P_{N,c}(L)} = \frac{\eta_{FWM}P_s(0)}{\varrho_{N,c}B_o}. \tag{11.127}$$

In many cases of practical interest, the conversion efficiencies in (11.125) are much smaller than unity and therefore SBR $\approx \eta_{FWM}P_s(0)/\varrho_{ASE,c}B_o$ is a good approximation. As mentioned in Subsec. the SBR is not equivalent to the signal-to-noise ratio (SNR). However, if the signal-spontaneous beat noise is assumed to be the dominant noise source, SBR/2 is a good estimation of the SNR for a signal of bandwidth equal to B_o [44]. For FWM applications in optical transmission systems it may be more important to obtain maximum signal-to-noise ratio than maximum conversion efficiency [45].

11.4.7 Noise figure

A more system-related quantity which describes the cascadability of several components is the noise figure. The noise figure is defined as the ratio of the input to the output SNR. Considering either a direct or a coherent detection system where the dominant noise contribution comes from the beating of the ASE with the signal and the local oscillator, respectively, the following expression for the noise figure can be derived [45]

$$\mathrm{NF} = \frac{2c_{out}\rho_{N,c} + \hbar\omega_c}{c_{in}c_{out}\eta_{FWM}\hbar\omega_s}. \tag{11.128}$$

In (11.128) the power coupling efficiencies at the input (c_{in}) and output (c_{out}) of the SOA have been taken into account since they will be unavoidable in practice yielding an increase of the noise figure. In many investigations dealing with FWM the noise figure is used as a figure of merit [45, 44, 46]. Thus, the optimisation criterion for components based on FWM is to obtain minimum noise figure.

It should be noted that in real communications systems it is most likely that the signal to be frequency converted will first be amplified [47]. Consequently, the quantity of interest is the effective noise figure of the cascade amplifier-frequency converter which is given by [48]

$$\mathrm{NF}_{eff} = \mathrm{NF}_A + \frac{\mathrm{NF} - 1}{G_A} \tag{11.129}$$

with NF_A and G_A denoting the noise figure and the gain of the amplifier preceding the frequency converter. Hence, the overall noise figure is mainly determined by the amplifier as long as the noise figure of the frequency converter is lower than the gain of the amplifier. For instance, assuming an amplifier with $\mathrm{NF}_A = 6$ dB and $G_A = \mathrm{NF} - 1$ yields an effective noise figure of 7 dB, which is mainly due to the amplifier.

References

[1] W. W. Chow, S. W. Koch, and M. Sargent, *Semiconductor-laser physics*, Springer-Verlag, Berlin, 1994.

[2] M. Asada and Y. Suematsu, "Density matrix theory of semiconductor lasers with relaxation broadening model - gain and gain-suppression in semiconductor lasers", *IEEE J. Quantum Electron.*, vol. 21, pp. 434–441, 1985.

[3] G. P. Agrawal, "Population Pulsations and nondegenerate Four-Wave Mixing in Semiconductor Lasers and Amplifiers", *J. Opt. Soc. Am. B*, vol. 5, pp. 147–158, 1988.
[4] A. Uskov, J. Mørk, and J. Mark, "Wave mixing in semiconductor laser amplifiers due to carrier heating and spectral-hole burning", *IEEE J. Quantum Electron.*, vol. 30, pp. 1769–1781, 1994.
[5] K. Henneberger, F. Herzel, S. W. Koch, R. Binder, A. E. Paul, and D. Scott, "Spectral holeburning and gain saturation in short-cavity semiconductor lasers", *Phys. Rev. A*, vol. 45, pp. 1853–1859, 1992.
[6] J. Mark and J. Mørk, "Subpicosecond gain dynamics in InGaAsP optical amplifiers: Experiment and theory", *Appl. Phys. Lett.*, vol. 61, pp. 2281–2283, 1992.
[7] J. Mørk and J. Mark, "Time-resolved spectroscopy of semiconductor laser devices: Experiments and modelling", in *Proc. SPIE, vol. 2399*, pp. 146–159, 1995.
[8] K. Hall, G. Lenz, A. M. Darwish, and E. P. Ippen, "Subpicosecond gain and index nonlinearities in InGaAsP diode lasers", *Opt. Commun.*, vol. 111, pp. 589–612, 1994.
[9] J. Mørk and A. Mecozzi, "Theory of the ultrafast optical response of active semiconductor waveguides", *J. Opt. Soc. Am. B.*, vol. 13, pp. 1803–1816, 1996.
[10] A. Mecozzi and J. Mørk, "Saturation induced by picosecond pulses in semiconductor optical amplifiers", *J. Opt. Soc. Am. B*, vol. 14, pp. 761–770, 1997.
[11] A. Mecozzi and J. Mørk, "Saturation effects in non-degenerate four-wave mixing between short optical pulses in semiconductor laser amplifiers", *J. Sel. Topics Quantum Electron.*, vol. 3, pp. 1190–1207, 1997.
[12] A. Mecozzi, "Analytical Theory of Four-Wave Mixing in Semiconductor Amplifiers", *Optics Lett.*, vol. 19, pp. 892–894, 1994.
[13] A. Mecozzi, A. D'Ottavi, F. Cara Romeo, P. Spano, R. Dall'Ara, G. Guekos, and J. Eckner, "High Saturation Behaviour of the Four-Wave Mixing Signal in Semiconductor Amplifiers", *Appl. Phys. Lett.*, vol. 66, pp. 1184–1186, 1995.
[14] J. Mørk and A. Mecozzi, "Theory of nondegenerate four-wave mixing between pulses in a semiconductor waveguide", *IEEE J. Quantum Electron.*, vol. 33, pp. 545–555, 1997.
[15] A. E. Siegman, *Lasers*, University Science Books, Mill Valley, California, 1986.
[16] G. P. Agrawal and N.A. Olsson, "Self-Phase Modulation and Spectral Broadening of Optical Pulses in Semiconductor Laser Amplifiers", *IEEE J. Quantum Electron.*, vol. 25, pp. 2297–2306, 1989.
[17] A. Mecozzi, S. Scotti, A. D'Ottavi, E. Iannone, and P. Spano, "Four-Wave Mixing in Traveling-Wave Semiconductor Amplifiers", *IEEE J. Quantum Electron.*, vol. 31, pp. 689–699, 1995.
[18] S. Diez, *Vierwellenmischung in InGaAsP-Halbleiterlaserverstärken*, Diplomarbeit, Fachbereich Physik, TU Berlin, Germany, 1996.
[19] A. Mecozzi and J. Mørk, *unpublished*, 1998.
[20] I. Koltchanov, S. Kindt, K. Petermann, S. Diez, R. Ludwig, R. Schnabel, and H. G. Weber, "Gain Dispersion and Saturation Effects in Four-Wave Mixing in Semiconductor Laser Amplifiers", *IEEE J. Quantum Electron.*, vol. 32, pp. 712–720, 1996.
[21] S. Diez, C. Schmidt, R. Ludwig, H. G. Weber, S. Kindt, I. Koltchanov, and K. Petermann, "Four-Wave Mixing in Semiconductor Optical Amplifers for Frequency Conversion and Fast Optical Switching", *IEEE J. of sel. topics in quantum electron.*, vol. 3, pp. 1131–1145, 1997.

[22] I. Koltchanov, S. Kindt, K. Petermann, S. Diez, R. Ludwig, and H. G. Weber, "Characterisation of Terahertz Four-Wave Mixing in a Semiconductor-Laser Amplifier", in *CLEO '96*, pp. 105–106, 1996.

[23] A. Uskov, J. Mørk, J. Mark, M. C. Tatham, and G. Sherlock, "Terahertz four-wave mixing in semiconductor optical amplifiers: Experiment and theory", *Appl. Phys. Lett.*, vol. 65, pp. 944–946, 1994.

[24] I. Koltchanov, *private communication.*

[25] S. Scotti, L. Graziani, A. D'Ottavi, F. Martelli, A. Mecozzi, P. Spano, R. Dall'Ara, F. Girardin, and G. Guekos, "Effects of ultrafast processes on frequency converters based on four-wave mixing in semiconductor optical amplifiers", *IEEE J. of sel. topics in quantum electron.*, vol. 3, pp. 1156–1161, 1997.

[26] S. Kindt, I. Koltchanov, K. Obermann, and K. Petermann, "New Time-Domain Model of a Semiconductor Laser Amplifier Suitable for System Simulations", in *OAA '96*, vol. 11, pp. 170–173, 1996.

[27] I. Koltchanov, S. Kindt, K. Petermann, S. Diez, R. Ludwig, R. Schnabel, and H. G. Weber, "Analytical theory of terahertz four-wave mixing in semiconductor-laser amplifiers", *Appl. Phys. Lett.*, vol. 68, pp. 2787–2789, 1996.

[28] K. Obermann, I. Koltchanov, K. Petermann, S. Diez, R. Ludwig, and H. G. Weber, "Noise Analysis of Frequency Converters Utilizing Semiconductor-Laser Amplifiers", *IEEE J. Quantum Electron.*, vol. 33, pp. 81–88, 1997.

[29] M. A. Summerfield and R. S. Tucker, "Optimization of Pump and Signal Powers for Wavelength Converters Based on FWM in Semiconductor Optical Amplifiers", *IEEE Photon. Technol. Lett.*, vol. 8, pp. 1316–1318, 1996.

[30] D. Nesset, D. D. Marcenac, and A. E. Kelly, "Improved system performance of wavelength conversion via four-wave mixing in a tandem semiconductor optical amplifier configuration", *Electron. Lett.*, vol. 33, pp. 148–149, 1997.

[31] K. Kikuchi, M. Amano, C. E. Zah, and T. P. Lee, "Analysis of Origin of nonlinear Gain in 1.5 μm Semiconductor Active Layers by highly nondegenerate Four-Wave Mixing", *Appl. Phys. Lett.*, vol. 64, pp. 548–550, 1994.

[32] J. Zhou, N. Park, J. W. Dawson, K. J. Vahala, M. A. Newkirk, and B. I. Miller, "Efficiency of Broadband Four-Wave Mixing Wavelength Conversion Using Semiconductor Traveling-Wave Amplifiers", *IEEE Photon. Technol. Lett.*, vol. 6, pp. 50–52, 1994.

[33] K. L. Hall, G. Lenz, and E. P. Ippen, "Femtosecond time domain measurements of group velocity dispersion in diode lasers at 1.5 μm", *J. Lightwave Technol.*, vol. 10, pp. 616–619, 1992.

[34] K. Kikuchi, M. Kakui, C. E. Zah, and T. P. Lee, "Observation of Highly Nondengenerate Four-Wave Mixing in 1.5 μm Traveling-Wave Semiconductor Optical Amplifier and Estimation of Nonlinear Gain Coefficient", *IEEE J. Quantum Electron.*, vol. 28, pp. 151–156, 1992.

[35] G. Eisenstein, N. Tessler, U. Koren, J. M. Wiesenfeld, G Raybon, and C. A. Burrus, "Length Dependence of the Saturation Characteristics in 1.5-μm Multiple Quantum Well Optical Amplifiers", *IEEE Photon. Technol. Lett.*, vol. 2, pp. 790–792, 1990.

[36] D'Ottavi, E. Iannone, A. Mecozzi, S. Scotti, P. Spano, R. Dall'Ara, J. Eckner, and G. Guekos, "Efficiency and Noise Performance of Wavelength Converters Based on FWM in Semiconductor Optical Amplifiers", *IEEE Photon. Technol. Lett.*, vol. 7, pp. 357–359, 1995.

[37] T. Mukai, Y. Yamamoto, and T. Kimura, "S/N and Error Rate Performance in AlGaAs Semiconductor Laser Preamplifier and Linear Repeater Systems", *IEEE J. Quantum Electron.*, vol. 18, pp. 1580–1568, 1982.

[38] C. H. Henry, "Theory of Spontaneous Emission Noise in Open Resonators and its Application to Lasers and optical Amplifiers", *J. Lightwave Technol.*, vol. 3, pp. 228–297, 1986.

[39] N. A. Olson, "Lightwave Systems With Optical Amplifiers", *J. Lightwave Technol.*, vol. 7, pp. 1071–1082, 1989.

[40] M. Shtaif and G. Eisenstein, "Noise Characteristics of Nonlinear Semiconductor Optical Amplifiers in the Gaussian Limit", *IEEE J. Quantum Electron.*, vol. 32, pp. 1801–1809, 1996.

[41] K. Petermann, *Laser Diode Modulation and Noise*, Kluwer Academic Publishers, Dordrecht, 1988.

[42] R. Schimpe, B. Bauer, C. Schanen, G. Franz, G. Kristen, and S. Pröhl, "1.5 μm InGaAsP tilted buried-facet optical amplifier", in *CLEO '91*, 1991.

[43] A. D'Ottavi, A. Mecozzi, S. Scotti, F. Cara Romeo, F. Martelli, P. Spano, R. Dall'Ara, J. Eckner, and G. Guekos, "Four-Wave Mixing Efficiency in Traveling Wave Semiconductor Optical Amplifiers at high Saturation", *Appl. Phys. Lett.*, vol. 67, pp. 2753–2755, 1995.

[44] F. Martelli, A. Mecozzi, A. D'Ottavi, S. Scotti, R. Dall'Ara, J. Eckner, and G. Guekos, "Noise of Wavelength Converters using Four-Wave Mixing in Semiconductor Optical Amplifiers", *Appl. Phys. Lett.*, vol. 70, 1997.

[45] M. A. Summerfield and R. S. Tucker, "Noise figure and conversion efficiency of four-wave mixing in semiconductor optical amplifiers", *Electron. Lett.*, vol. 31, pp. 1159–1160, 1995.

[46] F. Girardin, J. Eckner, G. Guekos, R. Dall' Ara, A. Mecozzi, A. D'Ottavi, F. Martelli, S. Scotti, and P. Spano, "Low noise and very high efficiency four-wave mixing in 1.5 mm long Semiconductor Optical Amplifiers", *IEEE Photon. Technol. Lett.*, vol. 9, pp. 746–748, 1997.

[47] E. Iannone and R. Sabella, "Performance Evaluation of an Optical Multi-Carrier Network Using Wavelength Converters Based on FWM in Semiconductor Optical Amplifier", *J. Lightwave Technol.*, vol. 13, pp. 312–324, 1995.

[48] A. Mecozzi, "Frequency converters based on four-wave mixing in semiconductor optical amplifiers", in *Optical Amplifiers and their Applications, OAA '96, Technical Digest*, vol. 11, pp. 118–120, 1996.

12 Measurement techniques and results

F. Girardin, T. Ducellier, S. Diez

Before measuring one has to answer a few important questions such as: Which set-up to use? What are the possible methods? One of the goals of this chapter is to help to answer these questions by showing and comparing measurements. FWM performance of SOAs has been measured with different set-ups and techniques in different laboratories.

In the Sec. 12.1, various experimental set-ups will be described together with the critical points impacting the FWM measurements. Experimental FWM results will be presented in a rather general matter in Sec. 12.2. Detailed measurements that were performed by different research laboratories on a number of SOA devices will be reported and compared in Sec. 12.3.

12.1 Set-up

The basic blocks of an experimental set-up used to characterise FWM are the sources, the SOA and the detector (see Fig. 12.1).

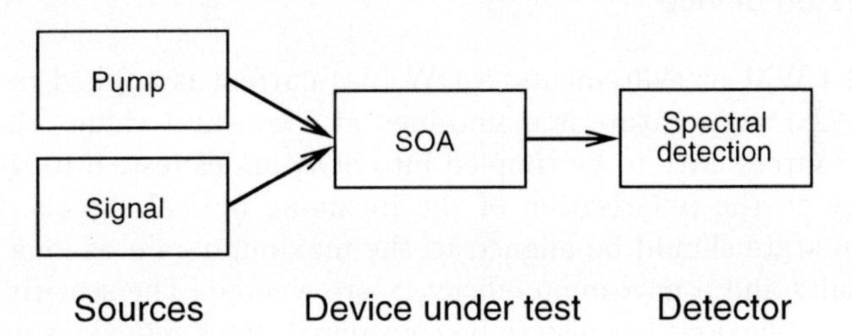

Fig. 12.1. Block scheme of the FWM characterisation set-up.

12.1.1 Sources

There are three main concerns about the optical sources used in FWM experiments: their power, their spectral quality (single-mode behaviour), and their state of polarisation.

Concerning the input powers, it will be shown in the following that powerful sources with controllable power levels are required. To avoid pattern effects that are due to gain saturation caused by the modulated signal wave in FWM applications, the signal power is generally kept 10-12 dB below the CW pump power (see Ch. 11). This insures a gain saturation mainly due to the pump, and reduces the potential intersymbol interference effects. If realistic operating conditions are to be applied in basic FWM experiments, it is therefore often necessary to amplify the pump wave optically. This can generally be done either by erbium doped fibre amplifiers (EDFAs) or SOAs. Mostly EDFAs are used due to their larger output power. However, in any case, optical bandpass filters are needed to supress the ASE of the amplifiers.

Another problem that arises from a strong pump power is that it must be obtained together with a very large side-mode suppression. Indeed, since the efficiency of the FWM process can be relatively weak, the optical signal-to-background ratio (SBR) of the conjugate is reduced if the input pump SBR itself is not sufficient. From our experience, we recommend to have an input side-mode suppression of the pump of at least 55 dB in 0.2 nm to avoid these problems. The same suppression ratio is required for the noise that is added by the optical amplifiers used to amplify the sources. All this can be achieved for example by using a single-mode semiconductor laser with a high side-mode suppression ratio and by strongly filtering the optical pump if it gets amplified.

To adjust the states of polarisation (SOP) for pump and signal waves, polarisation controllers are to be implemented at the source output. As controllers half- and quarter-wave plates are usually used in either fibre technique [1] or crystal optics.

12.1.2 Tested device

For typical FWM measurements, a CW bias current is applied to the SOA and the device temperature is maintained at a constant value. The outputs of the two sources have to be coupled into SOA under test. If its gain shows a sensitivity to the polarisation of the incoming optical signals, the input polarisation state should be aligned to the maximum gain as long as reproducible results and a maximum efficiency are wanted. The sensitivity of the SOA to back-reflexions has also to be considered. Back reflexions into it have to be avoided. The simplest way to do this is to connect its output directly to an optical isolator by means of a low-reflexion connection such as an angled connector or a fibre splice. When the SOA is strongly saturated, as it is often the case for FWM measurements, the sensitivity to back-reflexions is reduced.

In the measurements presented below, the devices were pigtailed and only fibre-to-fibre performance is considered.

12.1.3 Detection and filtering

For the detection and measurement of FWM outputs, two main issues are of interest: the power dynamic range of the detector and the spectral resolution of the detection scheme. The power level range in the FWM measurement is very wide: typically +10 dBm for the output pump, and −40 dBm for the noise level. This means that the power dynamics of the detector should be at least 50 dB to measure weak FWM signals. In order to meet these requirements a monochromator or an optical spectrum analyser (OSA) can be used. As explained in the next section, monochromators have a higher resolution and allow the use of a lock-in amplifier for synchronised detection. On the other hand, the resolution bandwidth of the detection scheme gives a minimum detuning at which a measurement set-up can be used. To perform measurements with lower detunings one has to use another detection scheme. This point is treated in Sec. 13.3.

In principle, by using an OSA it is possible to determine the powers of the optical waves directly from the displayed peak values. However, there is one obvious mistake that must be avoided in FWM measurements. It arises from the fact that the displayed peak values P_{peak} always represent the sum of any signal power P_{signal} and of the noise power P_{ASE} within the selected optical bandwidth of the OSA. To obtain the real signal power P_{signal}, it is therefore necessary to substract P_{ASE} from P_{peak}. Using the logarithmic dBm values the correction can be performed using the following equation:

$$P_{signal} = 10\log\left(10^{\frac{P_{peak}}{10}} - 10^{\frac{P_{ASE}}{10}}\right). \tag{12.1}$$

Note that using (12.1) the correction for a signal that reaches 10 dB out of the noise level is still 0.5 dB.

12.1.4 Set-up examples

Following the general remarks on the basic building blocks of a FWM measurement, two detailed set-up examples will be described in this subsection. The first one is shown in Fig. 12.2.

The set-up is based on external cavity semiconductor laser sources for pump and signal. Both waves are amplified by EDFAs and subsequently bandpass filtered (bandwidth about 1nm) in order to suppress the ASE of the EDFAs. Their optical powers are adjusted by tuneable attenuators. The polarisation controllers serve to align the SOPs of the input waves, which are then combined with a 3 dB coupler and injected into the SOA. Most simply, the detection unit is an OSA. To improve the repeatability, an additional pathway containing an optical switch was optionally implemented to measure the optical input powers without disconnecting any fibre connectors. In that case, the switch S2 was closed, while S1 was open. To measure the SOA

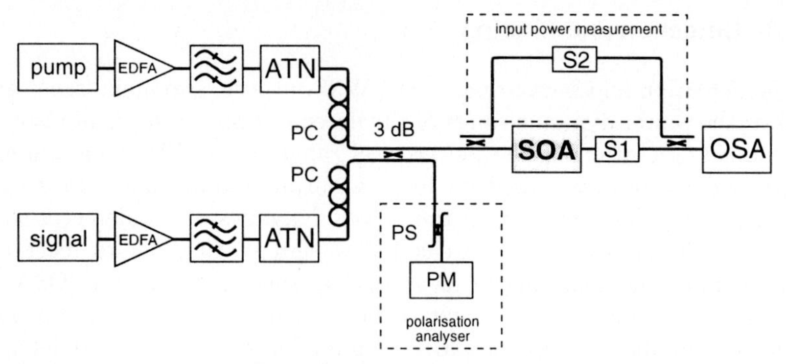

Fig. 12.2. FWM measurement set-up with two semiconductor laser sources for pump and signal. The optical input and output powers are measured with an OSA. PC-Polarisation controller, ATN-Attenuator, PS-Polarisation splitting coupler, PM-Power meter, S1 and S2-optical switches.

output spectrum, S1 was closed and S2 was open. To guarantee the same SOP for all measurements, a polarisation analyser (consisting of a polarisation splitting coupler and a power-meter at one output arm) can be used. The way to achieve a reproducible setting of the SOP is to first switch off the strong pump source and to minimise the optical power at the power meter after the polarisation splitting coupler. This setting corresponds to a well defined SOP of the signal wave at the 3 dB coupler in the set-up. Then the pump is switched on and the procedure is repeated, guaranteeing the same SOP for the pump at the 3 dB coupler. The SOP of pump and signal are thereby aligned in parallel to each other at the 3 dB coupler and thus also at the SOA input.

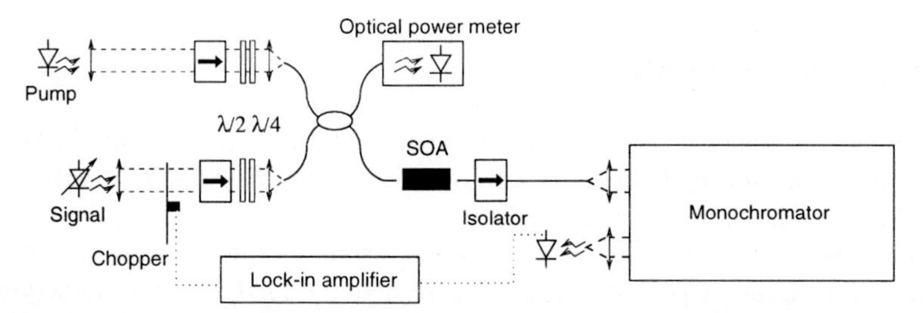

Fig. 12.3. FWM measurement set-up with two semiconductor laser sources for pump and signal. The optical output powers are measured with a monochromator by using a lock-in amplifier.

In the set-up shown in Fig. 12.3, the two sources are not amplified, yielding a limited input source power but also the highest possible input SBR. The polarisation control is done with a $\lambda/2$ and a $\lambda/4$ plates in the air. The detection is performed by using a lock-in amplifier and the spectral filtering is done with a monochromator. The SOP of the pump is tuned to the gain maximum and after that the SOP of the signal is tuned in order to reach the maximum FWM efficiency. Thereby, the use of the lock-in amplifier allows to detect with a high sensitivity. The ASE background is measured by suppressing the signal and placing the chopper just after the monochromator. When using the lock-in technique, one has to take care that a saturation due to the signal may lead to a perturbation of the measurements. This is important in particular when measuring the ASE background.

12.2 General results

As pointed out earlier, the FWM performance of a SOA strongly depends on a number of parameters. In order to characterise and compare the devices under test, the set-up of Fig. 12.2 was used. We decided to perform experiments where the injection current of the SOAs, the optical input powers of the injected waves as well as their detuning were varied. In this section, some typical measurement results are provided that show the interdependencies of those parameters that are important for the applications. Although the general shape of the curves measured is similar for all devices investigated, the absolute values of the parameters can vary significantly from device to device, as it becomes evident in Sec. 12.3. A typical FWM output spectrum is shown in Fig. 12.4. One can see the pump, the signal and the generated side-bands on both sides.

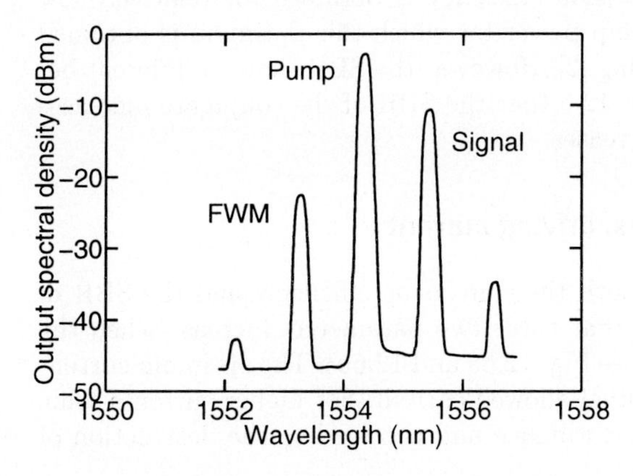

Fig. 12.4. Typical FWM spectrum at the output of the SOA, the resolution bandwidth of the OSA is set to 0.2 nm.

12.2.1 FWM performance vs. optical input power

Fig. 12.5 shows a typical dependence of the pump and signal output powers $P_{p,s}(L)$ of the generated FWM signal $P_c(L)$ and of the ASE $P_{ASE}(L)$ (measured with a resolution bandwidth of 0.08 nm) vs. the input power of the pump wave $P_p(0) = 10P_s(0)$. The injection current and the detuning were fixed. The dotted line represents the signal input power. This line can be used to determine the gain of pump G_p and signal G_s as well as η_{FWM}. The SBR is the ratio of $P_c(L)$ over $P_{ASE}(L)$. In this case, as the detuning is large, $P_c(L)$ only becomes detectable at high input powers. When the ratio

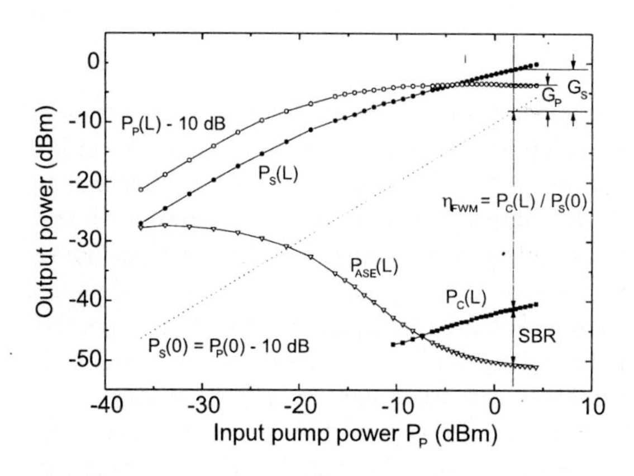

Fig. 12.5. The output powers $P_{p,s,c}(L)$, $P_{ASE}(L)$ vs. the pump input power $P_p(0)$; $\lambda_s = 1580$ nm, $\lambda_p = 1555$ nm, $\Delta\nu = 3.05$ THz.

between the pump and the signal is kept constant at 10 dB, one can show that an optimum in the conversion efficiency is obtained for relatively low input powers and that the pump power for which this optimum is obtained does not depend on the detuning [2]. However, the SBR shows a different behaviour. It can be seen on Fig. 12.5 that the SBR of the conjugate increases as long as the input power increases.

12.2.2 FWM performance vs. driving current

The gain strongly influences both the conversion efficiency and the SBR of the conjugate. It can be seen that these two parameters increase when the injection current is increased (see Fig. 12.6a and 12.6b). The optimum current is therefore generally the maximal allowed current. For higher currents some thermal effects degrade the performance and can lead to the destruction of the device.

In the case of a SOA with a significant residual gain ripple at high injection currents, a high input pump power is needed in order to saturate the gain significantly and then reduce the ripple. But, as pointed out in the previous section, a strong pump is also required to have a high SBR. Thus, a residual ripple does not bring significant perturbation of the FWM performance measurement.

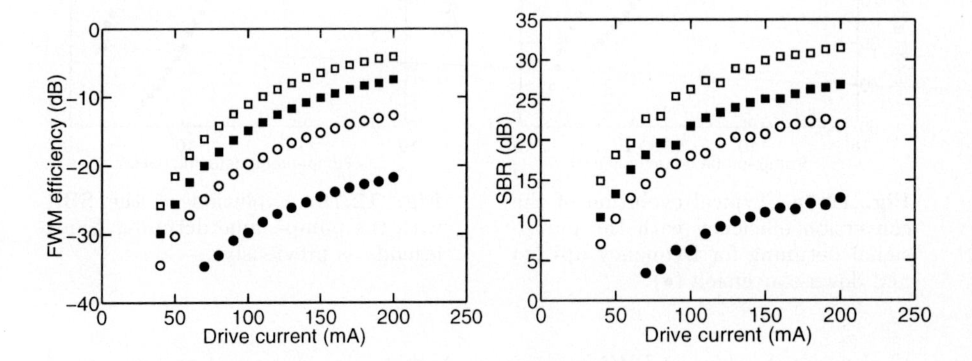

Fig. 12.6a. Evolution of the conversion efficiency with the driving current of the SOA for four different pump-signal detunings: +200 GHz (□), −200 GHz (■), +1 THz (○), −1 THz (●).

Fig. 12.6b. Evolution of the signal-to-ASE background ratio with the driving current of the SOA for four different pump-signal detunings. Same legends as in (a).

12.2.3 FWM performance vs. detuning

Due to the different time constants of the dynamic physical processes in SOAs, both η_{FWM} and SBR change upon a variation of the detuning between the injected light waves. This is also observed experimentally as shown in Figs. 12.7a and 12.7b. The input powers of the injected waves were kept constant in this experiment. The difference between frequency up- and down-conversion stems from the different phases of the different physical effects involved in the process of FWM (see Sec. 11.3.1).

12.3 Round robin results

A key activity in the framework of the COST 240 Action has been the comparison of different types of measurements in different laboratories. This exercise, referred to as round robin, allows a precise comparison of the different measurement techniques and points out the influence of some measurement parameters that may be otherwise forgotten. The different experimental setups have been presented in the previous section, in the following the results

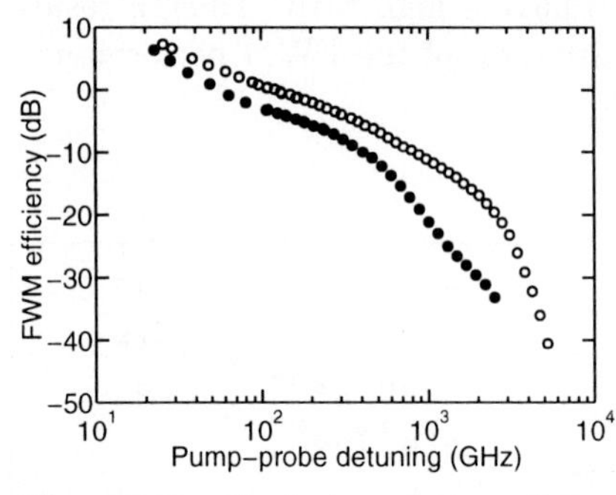

Fig. 12.7a. Typical evolution of the conversion efficiency with the pump-signal detuning for frequency up- (∘) and down-conversion (•).

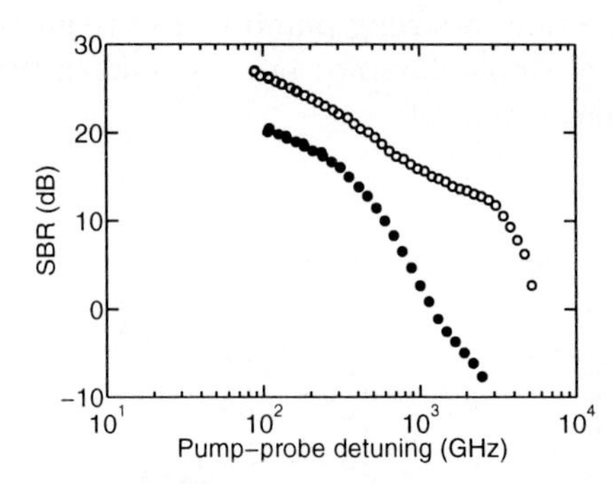

Fig. 12.7b. Evolution of the SBR with the pump-signal detuning. Same legends as previously.

of the round robin on FWM are given. A short description of the measured devices is provided first.

12.3.1 Device description

Three different devices where measured, one from Alcatel Alsthom Recherche, one from British Telecom (BT) and one from the Swiss Federal Institute of Technology Zurich in collaboration with Opto Speed SA.

Device A. This SOA is an amplifier module with Peltier cooler and double fibre pigtail. The active region of the SOA consisted of square bulk material providing polarisation insensitive amplification in the 1.55 μm region [3]. Its length was 400 μm. The facets were antireflection coated and tilted with respect to the optical axis. The fibre-to-fibre gain versus output power at 100 mA is given in Fig. 12.8. The output saturation power was around +7 dBm. The peak gain wavelength was 1530 nm, with around 1 dB max. ripple, together with a polarisation sensitivity below 0.7 dB (all values are indicated for 100 mA driving current). This sample device has circulated many times through European laboratories without any noticeable variation in its performance.

Device B. This amplifier is made of bulk 1.55 μm quaternary material, and is 500 μm long. Fig. 12.9 shows the buried heterostructure used. The active region dimensions, measured from a scanning electron microscopy photograph, are 1.17 μm x 0.21 μm. The confinement factor was calculated to be $\Gamma = 40$ %. Approximate characteristics are as follows:

– saturated output power 10 dBm
– maximum small-signal gain 25 dB
– facets are not angled or buried, but AR-coated.

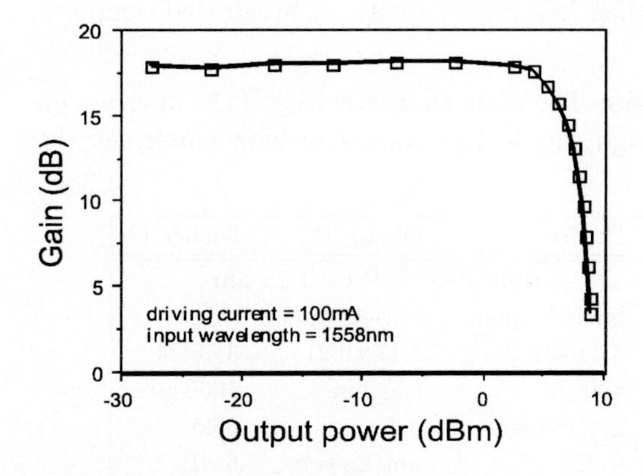

Fig. 12.8. Gain and output saturation power of the device A.

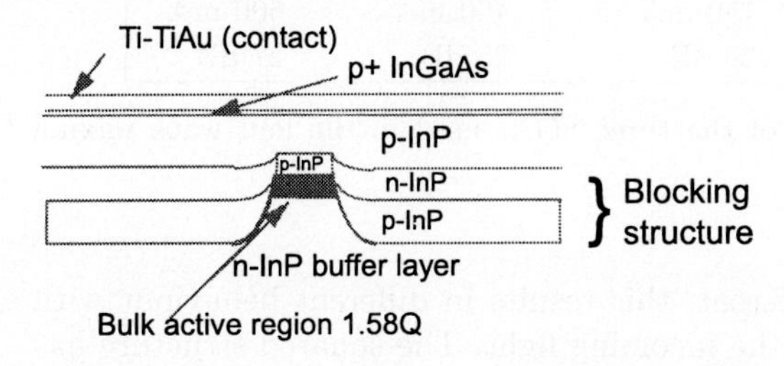

Fig. 12.9. Buried heterostructure of the device B.

This amplifier has been used for many experiments using nonlinear optical properties of semiconductor amplifiers (e.g. [4–6]). It is mounted on a heat sink, with connections for a Peltier temperature controller.

Device C. The basic structure is a bulk ridge waveguide SOA designed for operating in the 1.55 μm wavelength window [7]. The Fig. 12.10 shows a view of the 1 mm long device. The ridge width is 3 μm and the active layer thickness is 0.24 μm. The measured characteristics are the following:

– saturated output power ≈10 dBm
– maximum small-signal fibre-to-fibre gain 27 dB at 1550 nm
– residual gain ripple 2 dB

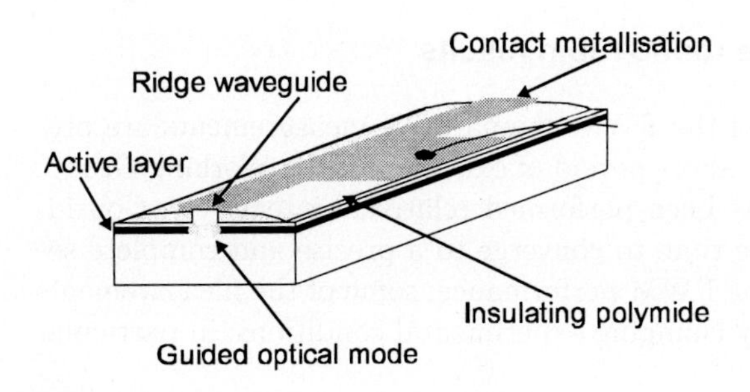

Fig. 12.10. Schematic description of the device C used for the round robin experiments.

The SOA is pigtailed and supplied in a housing with an integrated temperature controller.

Comparison of the three SOAs. The main characterstics of the devices can be found in Table 12.1. Although the devices consist of bulk materials, the

Characteristic	Device A	Device B	Device C
material	Bulk InGaAsP for 1.55 μm		
waveguide	buried square	buried ridge	ridge
cross-section (μm$\times\mu$m)	0.4$\times$0.4	1.17$\times$0.21	3$\times$0.24
length	400 μm	500 μm	1000 μm
confinement factor	not known	0.4	0.45
total coupling losses	6 dB	not known	6 dB
max. injection current	150 mA	400 mA	500 mA
max. gain	20 dB	25 dB	27 dB

Table 12.1. Comparison table of the three SOAs used in the four-wave mixing round robin.

form of the waveguides is different; this results in different behaviour with respect to the polarisation of the incoming light. The squared structure has the advantage of an intrinsic polarisation insensitivity, whereas the two other structures require an adjustment of the ratio width-thickness in order to obtain a reduction of the polarisation sensitivity [7]. Regarding the fabrication, the non-buried waveguide requires a simpler processing and presents a good suitability for monolithic integration. One advantage of the buried structure, however, is a good control of the carrier injection resulting in smaller leakage currents.

The reader should keep in mind, however, that the goal of the interlaboratory comparative measurements was to examine the experimental techniques employed through the results obtained with the same device. The devices provided by the companies were test samples for the specific purpose of the round robin and are different from the devices offered commercially by these companies.

12.3.2 Comparison of the round robin results

In this section, the results of the FWM comparative measurements are presented. Due to the relatively short period of existence of the Working Group, the round robin exercise has been performed relatively intensively. Considering that it has taken some time to converge to a precise and complete set of parameters to measure the FWM performance, some of the measurements were performed with slightly changing experimental conditions; in particular for the BT SOA, which has been measured at the beginning of the exercise, the comparison is limited to qualitative descriptions.

It should also be mentionned that since the devices are all pigtailed, all values reported in the following correspond to fibre-to-fibre performance.

Performed FWM characterisations. The laboratories who participated in this measuring exercise are the following:

- Alcatel Alsthom Recherche (AAR), Marcoussis, F
- British Telecom Laboratories (BT), Ipswich, UK
- Ecole Polytechnique Fédérale de Lausanne (EPFL), Lausanne, CH
- Swiss Federal Institute of Technology (ETHZ), Zurich, CH
- Fondazione Ugo Bordoni (FUB), Rome, I
- Heinrich-Hertz-Institut (HHI), Berlin, D

All the measurements of the same laboratory are identified with the same symbol in the following.

Table 12.2 lists the laboratories where the measurements have been performed for the three devices whereas Table 12.3 lists the type of measurement set-up used in the different laboratories. The results presented in the following paragraphs are sorted by SOA. This means that we are not interested in comparing the performance of the SOAs, but more the measurements techniques by themselves. We will often make reference to the *normalised conditions*, they correspond to the agreement and are listed in the Sec. 10.3.

Device	A	B	C
Laboratory	AAR, BT, ETH FUB, HHI	AAR, EPFL, ETH FUB, HHI	AAR, BT, ETH FUB

Table 12.2. List of the measurements made for the four-wave mixing round robin exercise.

Laboratory	▽	◇	△	+	□	○
Pump	EEO	EEO	DFB	DFB+OF	ECL	EEO
Signal	EEO	EEO	ECL	ECL	ECL	EEO
Polarisation	PC	PC	PC	PC	$\lambda/4$+$\lambda/2$	PC+PA
Detection	OSA	OSA	OSA	OSA	MC+LIA	OSA

Table 12.3. Type of set-up used in the different laboratories. EEO corresponds to ECL+EDFA+OF with: external cavity laser (ECL), optical filter (OF), fibre polarisation controller (PC), polarisation analyser (PA), λ/n plates (λ/n), monochromator (MC) and lock-in amplifier (LIA). Cf. Subsec. 12.1.4 for more explanations.

Measurements on the device A. Measurements corresponding almost to the normalised conditions have been performed in all laboratories. The results are given in Table 12.4 for a driving current of 150 mA and in Table 12.5 for 100 mA. The agreement for the measurement at 150 mA is good. For

Detuning	Conversion efficiency (dB)				Signal-to-ASE (dB)			
	Freq. up		Freq. dn		Freq. up		Freq. dn	
200 GHz	-6.5	(▽)	-10.1	(▽)	29.9	(▽)	25.1	(▽)
	-4.5	(+)	-8	(+)	29	(+)	24	(+)
	-6.2	(○)	-10.1	(○)	29.3	(○)	24	(○)
1 THz	-15.2	(▽)	-24.5	(▽)	20.7	(▽)	11.0	(▽)
	-15	(+)	-24	(+)	20	(+)	11	(+)
	-15.7	(○)	-24.0	(○)	20.5	(○)	10.1	(○)

Table 12.4. Table of comparison for the device A for a 150 mA biasing current.

the FWM efficiency, a maximum difference of 2 dB is measured at 200 GHz but this difference is reduced below 1 dB at 1 THz. Considering the SBR, the agreement is very good. All three measurements were performed with an optical spectrum analyser used for the detection. Even if absolute calibration of the device is made (or if all the powers are measured with this device), the polarisation sensitivity and the repeatability of the measurement can lead to a variation of more than 1 dB. This can explain the small difference, but we will see in the following that such a good agreement in not easily reached! For

Detuning	Conversion efficiency (dB)				Signal-to-ASE (dB)			
	Freq. up		Freq. dn		Freq. up		Freq. dn	
200 GHz	-11.1	(▽)	-14.9	(▽)	26.3	(▽)	21.7	(▽)
	-8	(◇)	-12.0	(◇)	25.5	(◇)	19.5	(◇)
	-9.5	(+)	-13.6	(+)	27.2	(+)	22.0	(+)
	-8.1	(□)	-11.2	(□)	NM		NM	
	-10.8	(○)	-14.4	(○)	24	(○)	19	(○)
1 THz	-19.8	(▽)	-29.8	(▽)	18.0	(▽)	6.2	(▽)
	-18.6	(◇)	-23.6	(◇)	19.3	(◇)	10	(◇)
	-19.3	(+)	-28.2	(+)	17.8	(+)	7.8	(+)
	-17.4	(□)	-25.2	(□)	NM		NM	
	-20.9	(○)	-29.1	(○)	17	(○)	4	(○)

Table 12.5. Table of comparison for the device A with a biasing current of 100 mA, the difference to the normalised conditions are: +2 dBm pump power, pump-to-signal power ratio 8 dB and a 5.4 GHz optical resolution bandwidth (□) and pump walvelength 1530 nm (○). NM: not measured.

the comparison of the results at 100 mA (cf. Table 12.5), the experimental conditions of the measurements were slightly different from the normalised conditions:

- ○ measurements were performed with a 1530 nm (peak gain wavelength) pump instead of ≈ 1555 nm for the others
- □ used a +2 dBm (instead of +3 dBm) pump power and a 8 dB (instead of 10 dB) pump-to-signal ratio at the input

Here, the agreement is not so good as previously. A difference of about 4 dB is found between the values obtained for the efficiency as well as for the SBR in frequency up-conversion and in frequency down-conversion at 200 GHz detuning. A larger difference is observed for frequency down-conversion with 1 THz detuning. We can observe that the discrepancy is higher when the efficiency and the SBR are smaller.

The saturated gain peak in the measurement conditions lies approximately at 1555 nm. As measured in [8], a 25 nm difference between the saturated gain peak and the pump wavelength as is the case for the measurement ○, can bring a reduction of the efficiency and of the SBR of a few dB.

The FWM efficiency versus the pump-signal detuning is given in Fig. 12.11a for the frequency up-conversion and in Fig. 12.11b for the frequency down-conversion. The conditions for the measurements are the same as in the Table 12.5. A similar shape of the spectral dependence of the FWM efficiency

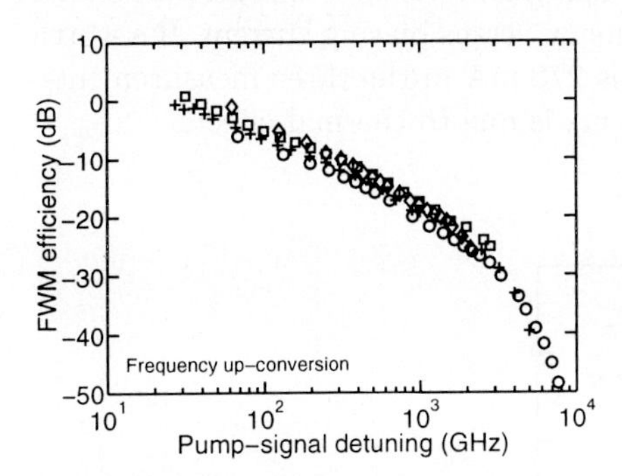

Fig. 12.11a. Comparison of the measured frequency up-conversion efficiency on the device A by four groups at 100 mA.

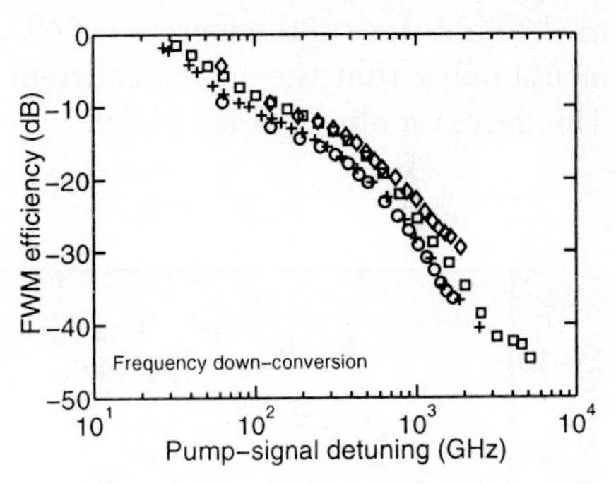

Fig. 12.11b. Comparison of the measured frequency down-conversion efficiency on the device A by four groups at 100 mA.

is obtained in all measurements. The pretty large discrepancy observed in the Table 12.5 can be found also in Fig. 12.11b. The higher discrepancy in frequency down-conversion may be explained by the fact that the efficiency is lower, especially at high detuning. In this case the SBR is rather low, and then, the value of the detected signal depends strongly on the performance of the receiver. The polarisation dependency of the receiver can influence the SBR and therefore the calculated value of the efficiency. A second reason for the discrepancy observed at high detunings may be the influence of the birefringence on the FWM effect as explained in Sec. 13.4.

Measurements on the device B. As explained before, only a few of the measurements performed on this device can be compared quantitatively; at the time where most of the measurements were made the agreement on the table of comparison (cf. 10.3) was not yet definitive. The results obtained at HHI are given in Table 12.6 for a 150 mA driving current and a 0.08 nm noise bandwidth. In Fig. 12.12 the results of three measurements of the FWM

Detuning	Conversion efficiency (dB)		Signal-to-ASE (dB)	
	Freq. up	Freq. dn	Freq. up	Freq. dn
200 GHz	-15.5	-19	20.6	15.1
1 THz	-26	-35	10.6	0.4

Table 12.6. Table of comparison for the device B.

efficiency versus the bias current of the SOA are shown. The pump-probe detunings are 80, 140 and 200 GHz in frequency up-conversion, the pump and signal powers are also different in each case. Therefore, it is not possible to compare the obtained values of the efficiency. Nevertheless this graphic shows us the limitation of the increase of efficiency versus biasing current. It's worth mentionning that the optimal current is 270 mA in the three measurements. The decrease observed for higher currents is due to thermal effects.

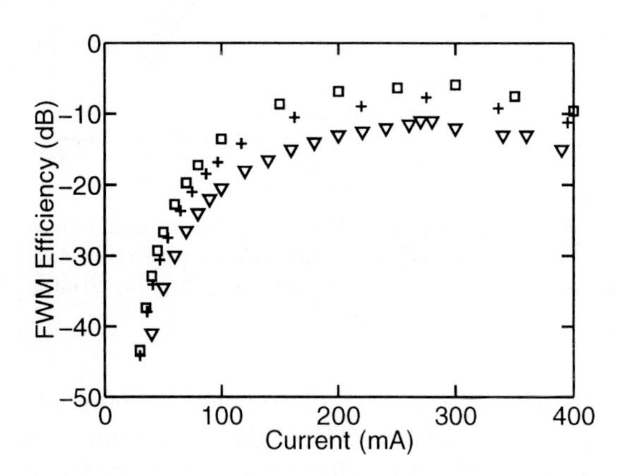

Fig. 12.12. FWM efficiency versus biasing current for the device B. The three curves have been measured with three different pump-probe detunings: 140 (▽), 80 (□) and 200 GHz (+).

Measurement on the device C. Table 12.7 shows the results of the comparison for the different measurements. The agreement on the efficiency is in the range of 3 dB. Concerning the SBR, the agreement seems to be very good when comparing the values. In fact the opical resolution bandwidth used for □ is 2 times smaller than for ▽ and +, this should lead to a 3 dB difference in the noise level and in the SBR. In order to have a more precise idea

Detuning	Conversion efficiency (dB)				Signal-to-ASE (dB)			
	Freq. up		Freq. dn		Freq. up		Freq. dn	
200 GHz	-4.3	(▽)	-8.2	(▽)	28	(▽)	22.8	(▽)
	-2	(+)	-5.5	(+)	27	(+)	21	(+)
	-4.2	(□)	-7.8	(□)	26.5	(□)	20.8	(□)
1 THz	-13.6	(▽)	-24.5	(▽)	19.5	(▽)	7.0	(▽)
	-11.5	(+)	-21	(+)	19	(+)	5.6	(+)
	-13.4	(□)	-23.4	(□)	18.3	(□)	6.3	(□)

Table 12.7. Table of comparison for the device C for a 490 mA biasing current, with a 5.4 GHz optical resolution bandwidth for (□).

of the origin of the discrepancy, it can be worth plotting the curves giving the efficiency and the SBR versus the pump-probe detuning. That is done in Figs. 12.13 and 12.14 respectively. A shift between the efficiency curves is clear from Fig. 12.13. This means that a calibration problem may be the origin of this discrepancy. Apart from this, there is a very good agreement for the shape of the curves. The curves obtained for the SBR in frequency

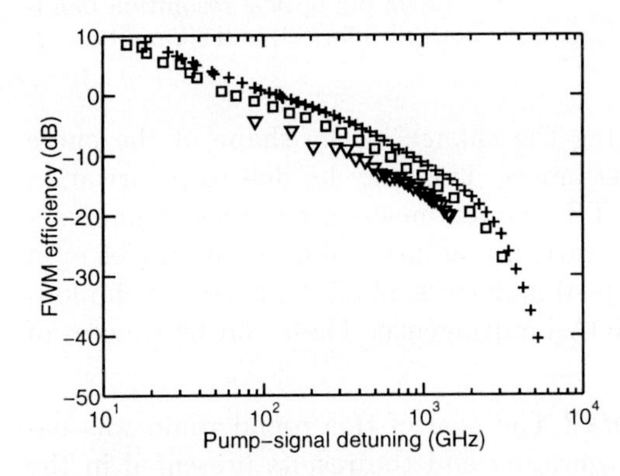

Fig. 12.13. Comparison of the measured frequency up-conversion efficiency on the device C by three groups at 490 mA.

up-conversion are plotted in Fig. 12.14. Due to the nominal resolution bandwidth difference, the values measured at + should be approximately 3 dB below the values measured at □ which is not the case according to the Fig.

12.14. This points out the very critical role of a precise determination of the optical resolution bandwidth of the receiver. The observed decrease of the measured SBR curves at low detunings is due to the fact that the pump and the signal are too close to each other to be properly separated. In this case, the detected noise level is due to the pump wave and not to the ASE of the SOA. The fact that the maximum of SBR is observed approximately at the same detuning for the two set-up means that the optical resolution bandwidth around −30 dB is almost the same. In fact, the resolution bandwidth is always specified at −3 dB, but in this case it is the width of the detected pump around −30 dB which is important. They are in this case equal for both set-up. This shows that the filtering functions of the two detection systems must be different. This results indicates that the SBR measurement can only be performed down to a certain minimum detuning which is determined not only by the optical resolution bandwidth, but also by the shape of the filtering function of the detectors.

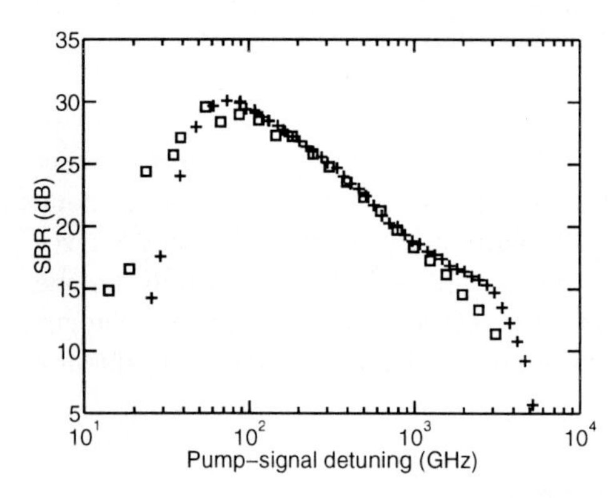

Fig. 12.14. Comparison of the measured frequency up-conversion SBR on the ETH-OS SOA by two groups at 490 mA in a 43 pm optical resolution bandwidth for (□) and 0.1 nm for (+).

Finally, it is worth mentionning the change in the shape of the curve observed in Fig. 12.14 at high detunings. This may be due to polarisation problems. At detunings of a few THz, which means a few tens of nanometers, the polarisation needs to be carefully adjusted due to the wavelength dependency of the polarisation adjusting devices, of all the fibres which modify the polarisation and, to the SOA birefringence. These can be sources of discrepancy at high detunings.

Can we trust our measurements? The goal of this round robin was basically to respond to this crucial question and the results presented in the previous paragraphs can help us to give an answer. The results show in general an agreement inside a margin of approximately 3 dB. The critical points are listed hereunder:

- The precise calibration of the instruments used for measuring the powers is a must. The measurements made in this round robin and in the ones listed in the second part of this book show how important this point is.
- The polarisation of the detected signal influences the detected level. Almost all the spectrally selective devices show a polarisation sensitivity, which is typically of the order of magnitude of 1 dB for standard optical spectrum analysers. The systematic use of a polarisation controller in front of the detector in order to find a maximum in the detection may be a solution. However when measuring non-polarised light such as ASE, this does not solve this problem, in particular when measuring the SBR.
- All the fibre-based set-ups need a very careful adjustement of the polarisation. The components of the set-up such as single-mode fibre and polarisation controller have a wavelength sensitive change of the polarisation. Due to the large wavelength range usually investigated in this kind of measurements, this can be a critical point.
- The problem of the optical bandwidth which has a large influence on the measurement of the noise characteristics has to be pointed out. Even if all the set-ups used here were grating-based monochromators, the filter function of all of them has been shown to be different. This can be caused by different adjustments of the spatial distribution of the light coming onto the gratings. Prior to the characterisation, a measurement of the monochromator filter function may be useful.
- The optical resolution bandwidth of the detector has to be precisely calibrated for each instrument. The influence of the optical bandwidth on the SBR is larger than on the conversion efficiency.

References

[1] H.C. Lefevre, "Single-mode fibre fractional wave devices and polarisation controllers", *Electron. Lett.*, vol. 16, pp. 778–780, 1980.

[2] A. Mecozzi, "Analytical theory of four-wave mixing in semiconductor amplifiers", *Opt. Lett.*, vol. 19, pp. 892–894, 1994.

[3] P. Doussière, P. Garabedian, C. Graver, D. Bonnevie, T. Fillion, E. Derouin, M. Monnot, J.G. Provost, D. Leclerc, and M. Klenk, "1.55 μm Polarization Independent Semiconductor-Optical Amplifier with 25 dB Fiber to Fiber Gain", *IEEE Photon. Technol. Lett.*, vol. 6, pp. 170–172, 1994.

[4] M.C. Tatham, X. Gu, L.D. Westbrook, G. Sherlock, and D.M. Spirit, "200 km Transmission of 10 Gb/s Directly Modulated DFB Signals using Mid-Span Spectral Inversion in a Semiconductor Optical Amplifier", *Electron. Lett.*, vol. 30, pp. 1335–1336, 1994.

[5] A.D. Ellis, M.C. Tatham, D.A.O. Davies, D. Nesset, D.G. Moodie, and G. Sherlock, "40 Gbit/s transmission over 202 km of standard fibre using mid-span spectral inversion", *Electron. Lett.*, vol. 31, pp. 299–301, 1995.

[6] D. Nesset, M.C. Tatham, and D. Cotter, "High Bit-Rate Operation of an All-Optical AND Gate Using Four-wave Mixing in a Semiconductor Laser Amplifier with Degenerate input signals", in *OFC'95*, San Diego, p. TuD2, 1995.

[7] C. Holtmann, P.A. Besse, T. Brenner, R. Dall'Ara, and H. Melchior, "Polarization insensitive bulk ridge-type semiconductor optical amplifiers for 1.3 μm wavelength", in *OAA '93*, pp. 8–11, 1993.
[8] F. Martelli, A. D'Ottavi, L. Graziani, A. Mecozzi, P. Spano, R. Dall'Ara, J. Eckner, and G. Guekos, "Pump wavelength dependence of FWM performance in semiconductor optical amplifiers", *IEEE Photon. Technol. Lett.*, vol. 9, pp. 743–745, 1997.

13 Related topics

S. Diez, K. Obermann, S. Scotti, D. Marcenac, F. Girardin

As already pointed out in Ch. 10, studying Four-Wave Mixing (FWM) in Semiconductor Optical Amplifiers (SOAs) can be of interest for numerous reasons. In Ch. 11 and 12, FWM among continuous waves was treated both theoretically and experimentally. These investigations helped to develop a better understanding of the underlying physical processes involved in FWM. Moreover, it became possible to compare different techniques to measure the conversion efficiency as well as the signal-to-background ratio. From an application point of view, the presented investigations are of interest for frequency conversion and optical phase conjugation arrangements. However, there are many more aspects of FWM in SOAs that have not been covered in the previous chapters. Ch. 13 aims to serve as an introduction to some related topics involving theoretical and experimental FWM observations. The results that are presented in this chapter are not meant to cover all related aspects. Still, they were also obtained within the activities of our Working Group and we regard them to be a valuable addition to the results presented before.

It will be demonstrated how nonlinear SOA parameters can be extracted by FWM (Sec. 13.1) and cross-gain modulation measurements (Sec. 13.2). The Sec. 13.3 is dedicated to the study of FWM for low detunings. Polarisation effects due to birefringence in SOAs will be investigated in Sec. 13.4. Experimental results of FWM among picosecond optical pulses, which are of particular interest if FWM in SOAs is to be applied for fast all-optical switching (e.g. demultiplexing and sampling), will be presented in Sec. 13.5.

13.1 Parameter extraction

In the previous chapters frequency conversion by FWM in SOAs has been described. Experimental and theoretical investigations lead to the conclusion that the converter performance is characterised by two main quantities: the conversion efficiency η_{FWM} and the signal-to-background ratio (SBR) of the converted field. As shown in Ch. 8, these quantities depend on a set of parameters related to the material and the operating conditions. Controlling these parameters is very useful for device optimisation. Parameter evaluation can be obtained by fitting the theoretical expression of η_{FWM} to experimental data. An activity on parameter extraction was carried out in the context of

Working Group 3 of the COST 240 project. In the framework of this activity the results of two different parameter evaluation procedures performed at the Fondazione Ugo Bordoni (FUB) and the Technical University of Berlin (TUB) are comparatively studied. The FUB and TUB procedures for parameter extraction are described in Subsec. 13.1.1 and 13.1.2, respectively. In each Subsection the used model of FWM in SOAs is first shortly reviewed. Then a list of parameters along with a schematic description of the fit procedure is reported.

13.1.1 Parameter extraction: first approach

The parameter extraction procedure performed at the Fondazione Ugo Bordoni is based on the analytical model of FWM described in Subsec. 11.3.5. To the reader's convenience, the main phases of the theory and the analytical solution [1] are herein shortly reviewed. The equations needed for the extraction procedure are specified together with a list of parameters that can be extracted. The nonlinear response of a SOA is theoretically described by the nonlinear susceptibility $\chi^{(3)}$, whose expression is given in Subsec. 11.3.2. Such a relationship can include the contributions of various processes: carrier density pulsation (CDP), carrier heating (CH), spectral hole burning (SHB) and ultrafast (UF) processes, i.e. two photon absorption (TPA) and Kerr effect. Each effect is characterised by time constants and nonlinear parameters, such as amplitude and phase of the effect. The expression of the nonlinear susceptibility is inserted in the equations describing the propagation of the fields along the waveguide (see Subsec. 11.3.1). One obtains a system of coupled equations [(11.98), (11.100) in Subsec. 11.3.1], whose solution yields to the theoretical prediction of the field power at the output of the device.

The equation system can be solved by standard numerical methods, but it has been demonstrated that various analytical approaches are also possible. It is worth reminding that the assumptions required for the analytical solution presented in Subsec. 11.3.5 are: constant saturation power along the propagation direction, and conjugate power smaller than the signal power at each point z along the propagation direction, i.e. $P_c(z) << P_s(z)$. Moreover variations of the single-pass gain G with frequency detuning (i.e. the gain dispersion) have been neglected. Under these conditions a closed form expression for the FWM efficiency and the FWM yield can be obtained (see Subsec. 11.3.5). The latter is defined as the ratio between the conjugate and the signal power at the output of the device, i.e. $\rho_{FWM} = P_c(L)/P_s(L)$, which is related to the FWM efficiency by $\rho_{FWM} = \eta_{FWM}/G$. The obtained expressions allow to calculate η_{FWM} and ρ_{FWM} with a good approximation as shown by a comparison with the results of a numerical solution [1]. As discussed in Ch. 11, it is convenient to consider the case of negligible scattering losses ($\alpha_i = 0$). In this case the expression obtained for η_{FWM} is (11.109) in Subsec. 11.3.5. Usually within the FUB approach ε_{CH2} is assumed to be equal to zero. In fact, the contribution of free carrier absorption (FCA) to the FWM yield is negligible at pump power used in these experiments [2]. The adjustment of the different parameters is realised by a fit computer program which is the core of the extraction procedure.

Within the FUB procedure it is chosen to fit experimental values of ρ_{FWM} that allow to take into account the effects of gain saturation induced by the pump field. Specifically a set of ρ_{FWM} values at various pump-probe detunings is considered. Given that each nonlinear process is dominant in a different range of detuning [3], the whole set of parameters can be determined if the behavior of ρ_{FWM} is considered over a wide pump-probe detuning range. In the fit procedure a standard least squares method is exploited. The required experimental input data are as follows: a set of measured values of the FWM yield at various frequency detunings, the length of the waveguide L, the device unsaturated gain G_0, the saturated gain G, the pump and the signal-power at the device input. All these measured quantities need particular care before their insertion in the computer code:

- The fit procedure must be run considering both up- and down-conversion because their asymmetrical behavior allows to determine the relative weight of the nonlinear processes with respect to each other [3]. It is important to verify that the values of ρ_{FWM} in the matrix satisfy the assumptions for the validity of the analytical solution. The condition $P_c(z) < P_s(z)$ is usually fulfilled at frequency detuning higher than a few tens of GHz. This threshold value $\Delta\nu_{min}$ depends on the working conditions and on the length of the amplifier. Experimental results obtained have shown that $\Delta\nu_{min}$ ranges from 10 to 100 GHz for an amplifier length ranging from 0.5 to 2 mm [4]. Thus it is necessary to evaluate $\Delta\nu_{min}$ and to insert in the input matrix only the data at a detuning $\Delta\nu > \Delta\nu_{min}$. It is also important to determine which of the nonlinear parameters can be extracted from the experimental data. To extract nonlinear parameters related to the nonlinear effect y, experimental data in the detuning range where y is the main process contributing to the generation of the conjugate wave are required. For example if the set of experimental data does not extend beyond 1 THz of detuning the parameters related to the ultrafast effects cannot be extracted.
- The value of G_0 is experimentally obtained by measuring the small-signal gain (the small-signal gain is the gain measured when the SOA is not saturated by an input field). However, the measured value of the small-signal gain corresponds to the unsaturated gain only if the amplified spontaneous emission (ASE) does not affect the device behavior in low saturation conditions at which the small-signal gain is measured [5]. Therefore it is important, before estimating G_0, to verify the operating conditions of the amplifier: if the ASE does not affect the device behavior, the small-signal gain is a good estimate of G_0. Otherwise G_0 must be deduced from other kinds of measurements. From experimental investigations small-signal gain

values of 12 dB, 24 dB, 35 dB , and 35 dB were obtained corresponding to device lengths of 250 μm, 500 μm, 1 mm, and 2 mm, respectively [4, 6]. The effect of ASE saturation in long devices is evident. The value of G_0 used to fit experimental data for the 1 mm and 2 mm devices were 48 dB and 96 dB, respectively.

- The value of the saturated gain used as input for the fit procedure is the value of G corresponding to a pump-signal detuning of zero. The value of G is extrapolated from measurements of the signal gain at various detunings in the presence of pump saturation. For the sake of simplicity, the analytical solution of the propagation equations has been carried out considering a constant signal gain. However, G depends on the detuning. There is an exponential dependence of the net gain experienced by the conjugate and the signal fields on the amplifier length at a given frequency. Experimental results have shown that the variation of G affects ρ_{FWM} at high detuning ($\Delta\nu > 1$ THz for a 500 μm long device). This effect is more evident in devices with long waveguides due to the narrowing of the gain lineshape [4, 7]. Thus the dependence of the gain on the frequency can be neglected when extracting the nonlinear parameters related to CDP, CH, and SHB, since they act at detunings smaller than 1 THz. The gain dependence on frequency could indeed be important in order to extract parameters related to ultrafast processes.

Scheme of the fit procedure. The scheme of the fit procedure implemented at the FUB is as follows:

- The first step is the evaluation of the saturation power. The total input power and the saturated gain G are related by (11.107) in Ch. 11. Once G and P_{in} are measured, the value of P_s is obtained by solving (11.107) with respect to P_s. The assumption of a constant P_s is a good approximation in devices with a length $L < 500$ μm. Actually P_s depends on the carrier density and varies along the propagation direction. The variation is stronger in longer devices operating in very high carrier injection conditions, where the carrier density at the device input can differ significantly from that at the device output. It is still possible to solve the propagation equations analytically considering the variation of P_s and to obtain a good quantitative prediction of ρ_{FWM} [8]. On the other hand, the assumption of a constant P_s leads to a simpler model and gives an adequate estimate of the FWM yield (and conversion efficiency) also in long devices. Thus in the fit procedure only the model with a constant P_s was used.
- The second step consists of the calculation of the theoretical FWM yield followed by the application of a standard least-squares method. The minimisation is done in two phases: first the nonlinear parameters are set to standard values taken from the literature and only the response times are evaluated by fitting. In the second phase, the response times are set to the values obtained from the first phase and the fit algorithm is run again to

estimate the other nonlinear parameters. The minimisation in two phases has the advantage of reducing the number of parameters to be estimated.

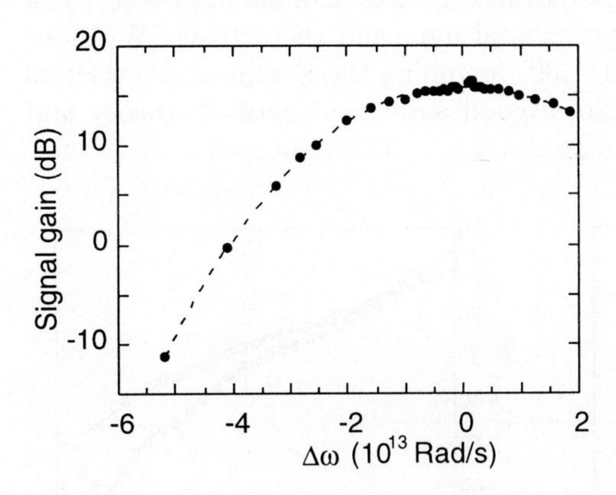

Fig. 13.1. Experimental signal gain (dB) vs. pump-probe angular frequency detuning for an input pump power of -2.2 dBm. $\Delta\omega = 0$ corresponds to the pump wavelength λ_p. Dots refer to experimental data and the dashed line refers to the third order polynomial which fits the experimental data.

Example of parameter extraction. Some measurement results obtained at the FUB labs [7] are reported and the parameter extraction procedure is described in practical implementation. The experimental data refer to a device with a ridge waveguide structure. The thickness of the InGaAsP active layer is 250 nm, the width of the ridge is 3 μm and the device is 500 μm long. The waveguide has been designed to obtain a device gain independent of the mode polarisation [9]. During the measurements the bias current was set to 480 mA/mm. The pump wavelength was $\lambda_p = 1.55\mu$m, the pump power at the input was -2.2 dBm and the ratio between pump and signal power was kept constant and set to 12 dB as suggested in [10]. The measured small-signal gain was 24 dB, and the unsaturated gain used as input was set equal to the small-signal gain ($G_0 = 24$ dB). Hence the effect of saturation caused by the ASE has been neglected as discussed above. The saturated gain G is set to 16 dB. This value is extrapolated from the experimental data of the saturated gain vs. the frequency detuning as reported in Fig. 13.1.

The experimental data of ρ_{FWM} vs. detuning are shown in Fig. 13.2b. The saturation power obtained by the first step of the fit procedure is $P_s = 10.8$

y	CDP	CH	SHB	UF
τ_y (fs)	$70 \cdot 10^3$	480	100	0
α_y	1.55	0.94	0	-9.17
P_{1y}^{sat} (mW)	10.8	512	854	10^5
P_{2y}^{sat} (mW)	$\to \infty$	$\to \infty$	$\to \infty$	10^5

Table 13.1. Extracted nonlinear parameters of the SOA investigated in [7].

mW. The set of parameters obtained by the second step of the extraction procedure is reported in Table 13.1. These values are used in the expression of ρ_{FWM} in order to calculate its behavior vs. the frequency detuning. The obtained theoretical curves are reported in Fig. 13.2b as solid lines. Fig. 13.2a shows a comparison between experimental data and theoretical FWM efficiency, calculated according to (11.109) assuming the nonlinear parameters listed in Table 13.1. In Fig. 13.2b, a good agreement between theory and

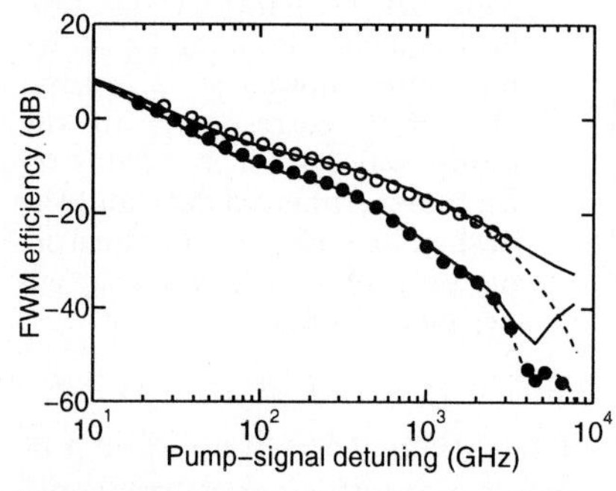

Fig. 13.2a. FWM efficiency as a function of pump-signal detuning. Experimental data for frequency up- (∘) and down-conversion (•) and corresponding fitted curves including (−−) and not including (—) the gain dispersion.

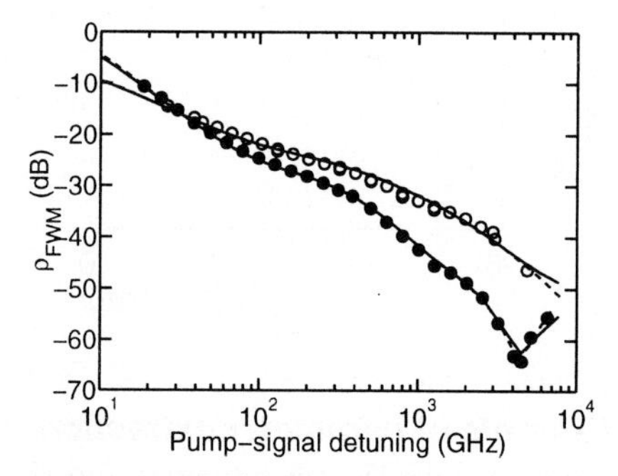

Fig. 13.2b. FWM yield as a function of pump-signal detuning. Same legends as in (a).

experiment is evident. In Fig. 13.2a the theoretical curves generally fit the experimental data, although a discrepancy occurs in the down-sideband at detuning higher than 2 THz. Further considerations of the results reported in Fig. 13.2a have shown that the poor fitting at high detuning has not to be ascribed to a wrong estimate of the nonlinear parameters but, rather, to the fact that gain dispersion has been neglected in the analytical formula. To prove this, theoretical curves of the FWM yield and efficiency were calculated taking into account the gain variation. Specifically, the experimental gain variation as shown in Fig. 13.1 was introduced in the propagation equations. Thereafter, the theoretical behavior of ρ_{FWM} and η_{FWM} was evaluated numerically assuming the parameter listed in Table 13.1. The obtained theoretical curves are reported as dashed lines in Fig. 13.2b for ρ_{FWM} and in Fig. 13.2a for η_{FWM}. In contrast to solid lines, both dashed lines for the conversion efficiency show a very good agreement with the experimental data. In Fig. 13.2b too dashed lines follow more closely the experimental data at high detunings. This result indicates that parameter extraction is not significantly affected if the gain dispersion is neglected in the theoretical analysis.

13.1.2 Parameter extraction: second approach

The extraction of the nonlinear SOA parameters performed at the Technical University of Berlin (TUB) is based on the analytical model of the FWM efficiency described in Subsec. 11.3.6. As mentioned previously, the theory used at the TUB includes the gain dispersion. This has several consequences: firstly, the α-factor associated with SHB differs from zero and secondly, all nonlinear processes have to be described by two characteristic powers due to the different transparency points for the pump, the signal and the conjugate wave (see Subsec. 11.3.6). Moreover, the TUB approach considers CDP, CH and SHB as the dominant nonlinear mechanisms and does not include the ultrafast nonlinear effects TPA and Kerr effect. Thus, the total number of nonlinear parameters to be extracted is 12 (time constants, α-factors and two characteristic powers for the nonlinear mechanisms y = CDP, CH, and SHB). At the TUB, the normalised FWM efficiency as defined in Subsec. 11.3 is used for comparison with experimental data mainly due to the following reasons. The normalised FWM efficiency can easily be accessed experimentally, and the analytical expression for η_N is somewhat simpler compared to the expressions for η_{FWM} and ρ_{FWM}, respectively. Furthermore, the deviation of the experimental results to the gain cube theory can easily be seen when the normalised FWM efficiency is plotted against the input pump power (see Subsec. 11.3.7). The experimental data is firstly fitted as a function of the input pump power. This procedure is repeated for every detuning, and the nonlinear parameters are then determined in a second step.

Example of parameter extraction. Before the nonlinear parameters can be extracted several other SOA parameters have to be known. The internal saturation power P_s can be determined by fitting the measured dependence $G(P_{in})$ according to Eq. (11.107), compare with Sec. 13.1.1. The gain dispersion parameters β and σ are related to the parameters $\hat{a}_j$ and $\hat{b}_j$ as introduced in Subsec. 11.4.4 by

$$\begin{aligned} \beta &= 1 + (\hat{a}_s - \hat{a}_c)/2 \\ \sigma &= (\hat{b}_c - \hat{b}_s)/2 \quad . \end{aligned}$$

Since the gain $G_{j,dB}$ of the wave j in dB depends linearly on the gain of the pump wave according to [11] $G_{j,dB} = (\hat{a}_j - \hat{b}_j)G_{p,dB} + b_j G_{0p,dB}$, the parameters $\hat{a}_j, \hat{b}_j$ can be found by making a linear fit of $G_{j,dB}(G_{p,dB})$. Subsequently, the dependence of $\hat{a}_j$ and $\hat{b}_j$ on the detuning can be approximated by a second-order polynom [12].

Now the nonlinear SOA parameters can be determined in two steps. According to Eq. (11.110), the normalised FWM efficiency η_N consists of a sum of two detuning characteristics $H_k(\Delta\omega) \equiv \sum_y (1 - i\alpha_y)\tilde{R}_y(\Delta\omega)/P_{yk}^{sat}$ for $k = 1, 2$ which are both independent of the input power. In contrast to that, the coefficients of $H_{1,2}(\Delta\omega)$ represented by the incomplete Gamma functions in (11.110) are a function of both the detuning and the input power. At first, $H_1(\Delta\omega)$ and $H_2(\Delta\omega)$ for a given detuning are obtained by a comparison of (11.110) with the measured dependences $\eta_N(P_p(0), \Delta\omega)$. In [13] this was done for about 50 different detunings. At the second step $H_{1,2}(\Delta\omega)$ are fitted using (11.102)-(11.104) in order to determine the nonlinear SOA parameters. Thereby, the characteristic powers of $H_1(\Delta\omega)$ and $H_2(\Delta\omega)$ can be different but the time constants and α-parameters should be equal. This can serve as a self-consistency test. Table 13.2 shows the resulting values of the nonlinear SOA parameters of the amplifier investigated in [13]. Figs. 13.3a and

y	CDP	CH	SHB
τ_y (fs)	$200 \cdot 10^3$	510	50
α_y	5.0	1.7	-1.5
P_{1y}^{sat} (mW)	2.94	181.8	476.2
P_{2y}^{sat} (mW)	2.70	263.3	909.1

Table 13.2. Extracted Nonlinear Parameters of the SOA investigated in [13].

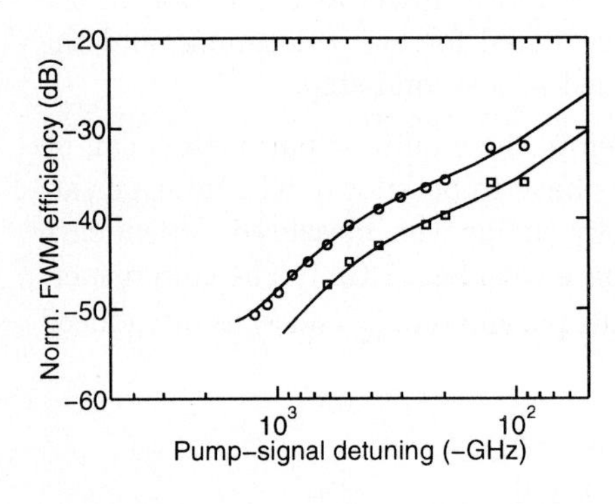

Fig. 13.3a. Comparison of the measured (○and □) and calculated (—) normalised FWM efficiency versus detuning for two different pump input powers: $P_p(0) = -7.5$ dBm (□) and $P_p(0) = 4.3$ dBm (○). For the calculations, the nonlinear parameters listed in Table 13.2 have been assumed.

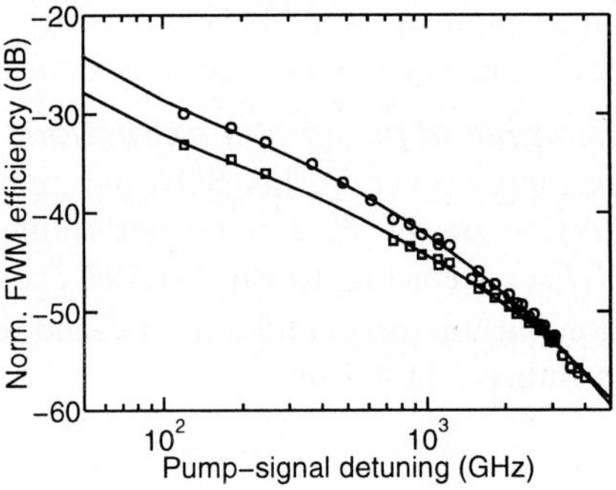

Fig. 13.3b. Same legend as (a) for frequency up-conversion.

13.3b show the normalised FWM efficiency versus detuning calculated with (11.110) and the parameters according to Table 13.2 (lines) as well as the experimental data for two different pump input powers $P_p(0)$. Evidently, a good agreement between theory and experiment is obtained for both input powers $P_p(0) = -7.5$ dBm and $P_p(0) = 4.3$ dBm. However, this is not possible when the experimental data is fitted with the gain cube theory (predicting that η_N is independent of the input power), though the measured dependence of the FWM efficiency can well be fitted for a fixed input pump power.

13.1.3 Interpretation of the results

Evidently, the theoretical predictions according to Subsecs. 11.3.5 and 11.3.6 are in good agreement with the experimental data. Hence, the same hints for the device optimisation can be drawn from the theoretical results. In conclusion the theoretical approaches are equivalent as far as the design of an optimised device is concerned. Such conclusion does not imply that the extraction procedures show similar reliability for other purposes as well, e.g. for spectroscopical investigation. The main problem is the great number of unknown parameters. An improvement of the extraction procedures could possibly be obtained by reducing the number of parameters to be optimised during the fitting procedure. For example, some parameters could be determined by other kind of measurements on the same device.

13.2 Cross gain modulation measurements

Small-signal measurements of the cross-gain modulation (XGM) response of the device A were performed using the experimental set-up shown in Fig. 13.4. A CW beam at wavelength λ_2 is modulated at a frequency which is slowly swept between 0.1 and 20 GHz. The resulting intensity modulation transferred through XGM onto wavelength λ_2 is detected with a high speed photodiode, and the resulting frequency response is plotted by the network analyser.

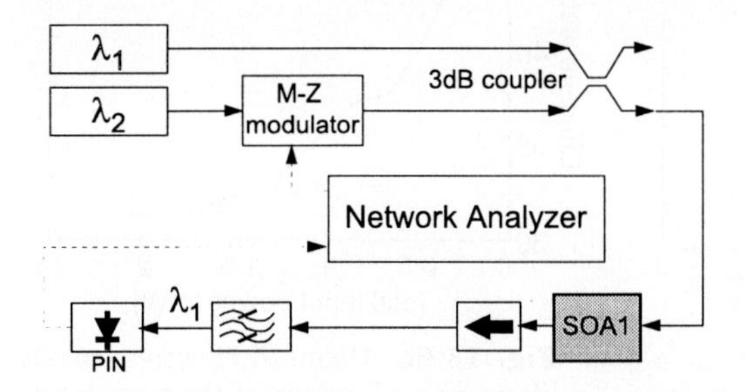

Fig. 13.4. Experimental set-up for XGM measurement.

A typical measured response is shown in Fig. 13.5. No resonance is visible, and the response resembles that of a low-pass filter. The 3dB electrical

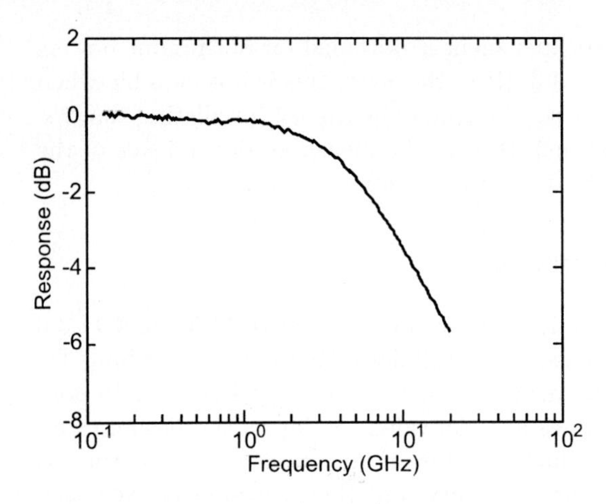

Fig. 13.5. XGM response of the device A.

bandwidth for this response is related to the effective carrier lifetime in the SOA (including spontaneous and stimulated emission, enhancements due to travelling wave effects, and even contributions from ultra-fast processes such as spectral hole burning). Assuming that carrier density fluctuation is the most important mechanism for XGM in this device, the measured responses can be fitted to a simple low-pass filter function, and a value for the carrier lifetime can be deduced. A weak dependence of XGM bandwidth on optical input power was measured. The bandwidth increased from 5.5 GHz to 9 GHz when the optical power was increase over an order of magnitude from −6 to +4 dBm. The effective carrier lifetime decreased under the same conditions from 50 to 30 ps. For a higher drive current of 150 mA, bandwidths

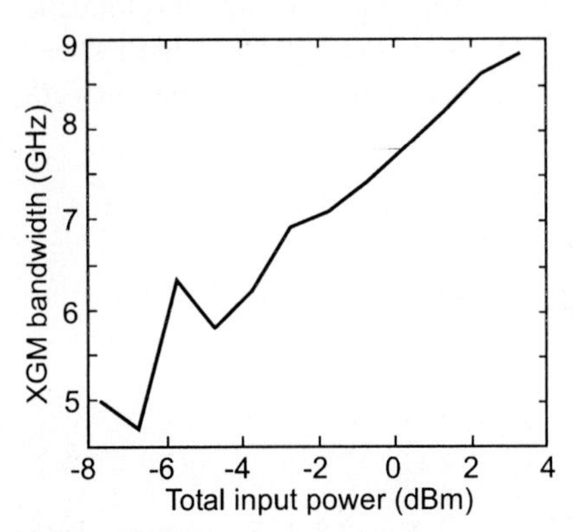

Fig. 13.6a. XGM bandwidth as a function of the total input power for a drive current of 100 mA.

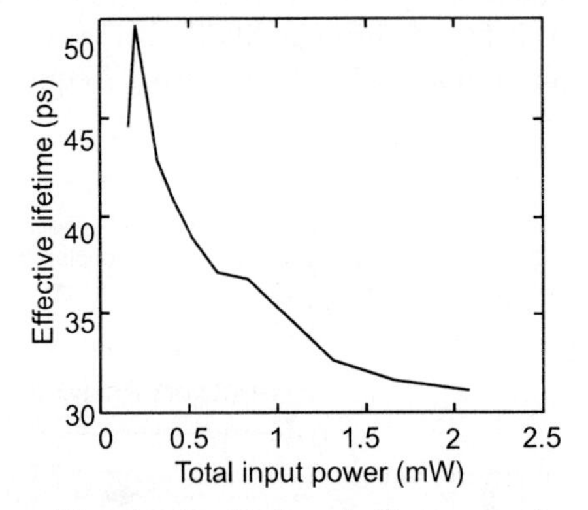

Fig. 13.6b. Deduced effective carrier lifetime as a function of the total input power for a drive current of 100 mA.

were larger, and carrier lifetimes were measured down to 21 ps, as shown in Fig. 13.7b.

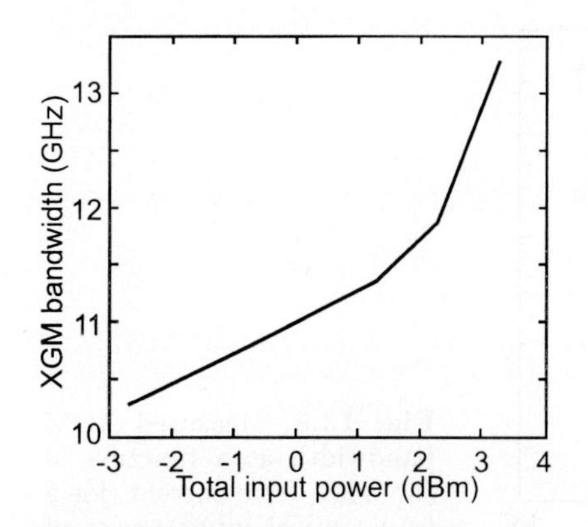

Fig. 13.7a. XGM bandwidth as a function of the total input power for a drive current of 150 mA.

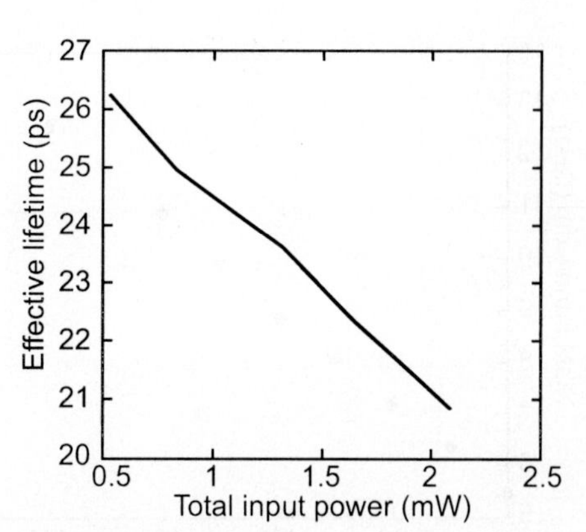

Fig. 13.7b. Deduced effective carrier lifetime as a function of the total input power for a drive current of 150 mA.

Fig. 13.8 shows the measured XGM bandwidth as a function of SOA drive current. A much stronger dependence is seen here. The bandwidth is roughly proportional to the drive current. This can be understood by noting that the output power of the SOA will vary approximately linearly with the drive current (before complete band-filling or thermal considerations take over), and therefore the stimulated emission rate – and the XGM bandwidth – will also increase linearly with drive current. The measured XGM bandwidths, up to nearly 14 GHz, indicate that this device would be suitable for wavelength conversion using XGM at bit rates of 10 Gbit/s with very little penalty or distortion.

The conclusions from this study are that the XGM bandwidth does not vary enormously with optical input power: this is beneficial since it means that no excessive deformation of the data would occur for cross gain modulation due to widely different carrier recovery rates on data ones and zeros. A far more important parameter is the drive current. For fast nonlinear effects, as high a drive current as possible should be used. The increase of XGM bandwidth with drive current, however, is only expected to hold true for a given device up to a certain point. Overheating of the device at high currents will diminish the gain. At a given wavelength, the differential gain can only increase up to a certain value, beyond which the semiconductor medium completely inverted: this will also stop any increase in the stimulated emission rate. Finally, the SOA blocking structure may break down at high injection currents due to the excessive voltage required. The use of longer SOAs is one simple way to increase the XGM bandwidth, by allowing greater injection currents to be used, as well as by utilising travelling wave, pulse shaping effects to sharpen edges for wavelength conversion.

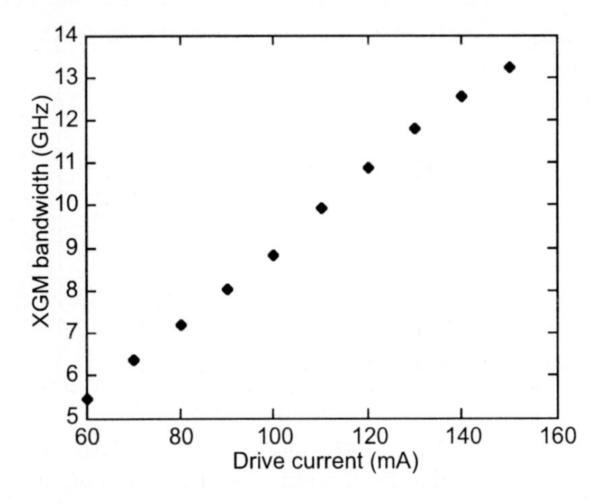

Fig. 13.8. Measured XGM bandwidth as a function of the SOA drive current, for a total optical input power of +3 dBm.

13.3 Nearly degenerated FWM measurements

At first it may be worth defining what *nearly degenerated* means. In the set-up described in the Ch. 12, the wavelength resolution of the detection is determined by the resolution bandwidth of the optical spectrum analyser, which is typically in the range of 0.1 nm. That means that two peaks may be well distinguished only if they are a few tens of GHz away one from the other. Nearly degenerated FWM corresponds to detunings lower than a few tens of GHZ. In order to allow an analysis with smaller pump-probe detunings one has to use another detection scheme.

13.3.1 Measurement set-up

The most obvious alternative to the optical spectrum analyser is the scanning Fabry-Perot filter as it gives the optical spectrum of the received light. The resolution of this measurement is determined by the finesse and the free spectral range of the Fabry-Perot. The obtained resolution is typically of a few tens of MHz. A possible configuration for using a Fabry-Perot filter is described on the Fig. 13.9. In the measurements presented afterwards, the used Fabry-Perot filter had a finesse of 50 and a free spectral range of 2 GHz, which results in a resolution of 40 MHz. The highest reachable detuning is determined mainly by the efficiency of the four-wave mixing process and the detection. Fabry-Perot analysers are generally used with linear detectors, the

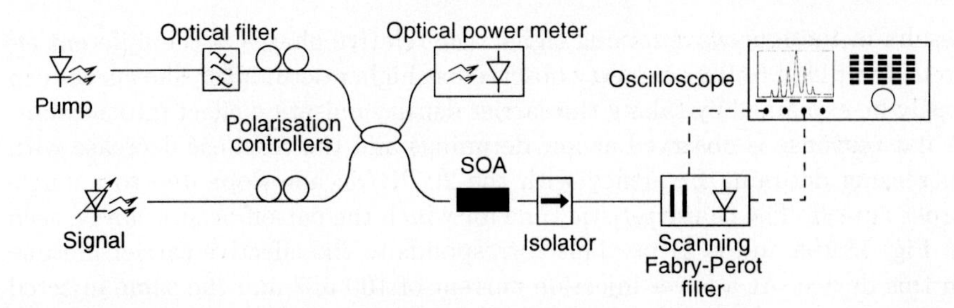

Fig. 13.9. Typical experimental set-up used for the characterisation of SOAs by four-wave mixing at low detunings.

dynamic range of the detection is then small, around 15 dB. It means that the conjugated signal must not be much smaller than the other signals at the output of the SOA in order to be properly detected. That is the main limitation that determines the highest measurable detuning.

The optical sources have to be very well stabilised in frequency in order to be able to perform these measurements. In order to vary the optical frequency of the probe signal, one can use a tunable laser or a temperature adjustment. The use of a tunable external cavity laser has the advantage that the tuning range is much wider and the same laser can be used for low and high detunings measurements, but even for finely tunable lasers, the precision of the tuning is sometimes not high enough to perform detuning measurements below 1 GHz – corresponds to a 10 pm wavelength change! Changing the temperature allows the tuning of the wavelength of single-mode DFB lasers. As long as the temperature controller permits a precise choice of the laser temperature, this method allows a very exact choice of the freqency of the laser. A typical value for the variation of the emitting frequency is 10 GHz/^{o}C, a temperature stabilisation with a 0.01 ^{o}C resolution, which is not a strict requirement, allows a 0.1 GHz controlled variation of the optical frequency.

13.3.2 Results

The three SOAs used in the framework of this Working Group have been characterised by using the set-up described in 13.9. A precise description of the SOAs can be found in Sec. 12.3.1. The investigated detuning is in the range of a few hundred of MHz up to approximately 100 GHz, depending on the device. In all the experiments presented in this chapter, the injected pump and probe powers are +3 dBm and −7 dBm respectively.

Experimental results. The Fig. 13.10a shows the dependence of the conjugated power versus detuning for the device A. As one can see in this figure, the curves for frequency up- and down-conversion are almost identical. This is due to the fact that the effect that dominates the four-wave mixing process is, in this frequency range, the carrier density pulsation. As this only effect results in the four-wave mixing signal, the relative phases of the different effects don't lead to the asymetry observed at higher detunings. The curves can easily be explained by taking the carrier density pulsation effect into account. A flat response is observed at low detunings and the response decrease with increasing detuning frequency with the 20 dB/decade slope due to a single order cut-off. The characteristic time for which the cut-off occurs can be seen in Fig. 13.10a and is 21 ps. This corresponds to the effective carrier lifetime in this device. At a lower injection current of 100 mA and the same injected powers, the measured lifetime is 26 ps on the same device.

The results of the same measurements on the device B are given in Fig. 13.10b. The curves are no more identical for up- and down-conversion and the decrease is slower than 20 dB/decade for the frequency up-conversion curve. This can be due to the fact that the carrier heating effect begins to play a significant role in the generation of the four-wave mixing signal. The phase difference resulting of the different phase-amplitude coupling factors for these two effects leads to this asymetry (cf. Ch. 11).

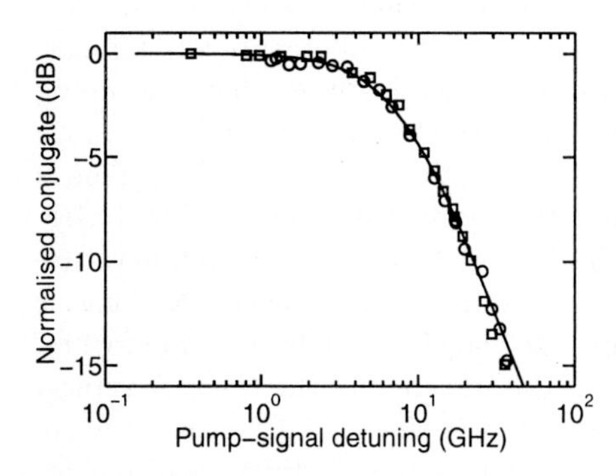

Fig. 13.10a. Dependence of the conjugated signal power versus pump-probe detuning for the device A at 150 mA drive current. □ are for frequency down-conversion and ○ are for frequency up-conversion, the line corresponds to the theoretical calculation with an effective carrier lifetime of 21 ps.

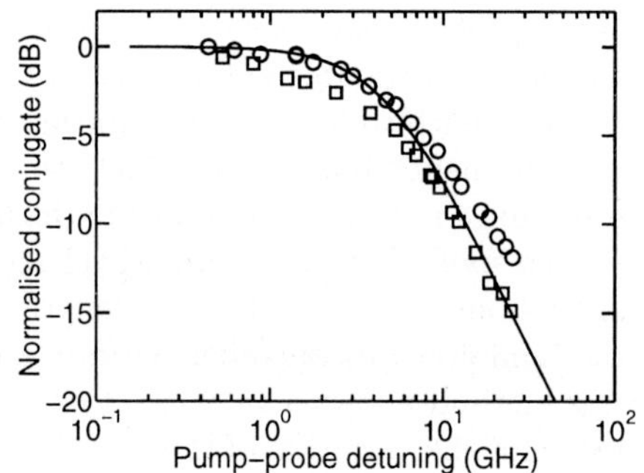

Fig. 13.10b. Dependence of the conjugated signal power versus pump-probe detuning for the device B at 395 mA drive current. The line corresponds to the theoretical calculation with an effective carrier lifetime of 35 ps.

The third SOA, showed a different behavior as the two previous ones. The measured normalised conjugated power as a function of the detuning is given in Fig. 13.11a. As in the case of the device B, the two curves for the up- and down-conversion are not identical, but in this case the basic shape of the curve is also not the same as expected. A maximum of efficiency is observed for a detuning of 4 GHz for the down-conversion and of 5 GHz for the up-conversion. This maximum can not be explained by the usual theory taking into account only the first order generated signal. The Fig. 13.11b

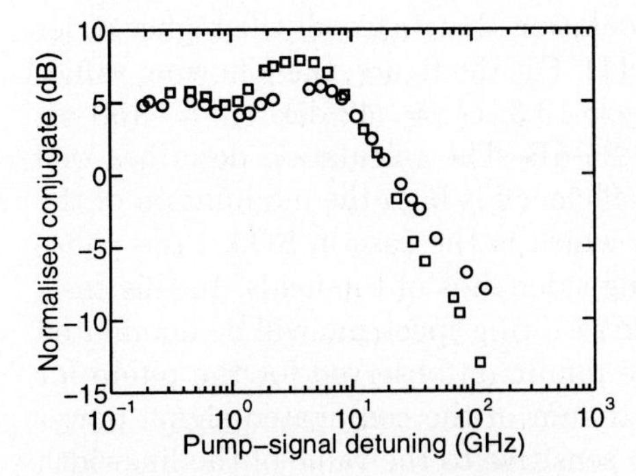

Fig. 13.11a. Dependence of the conjugated signal power versus pump-probe detuning for the device C at 490 mA drive current. Same legends as previously.

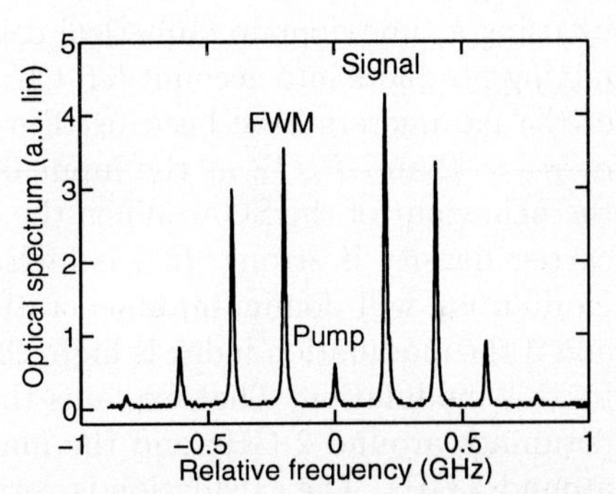

Fig. 13.11b. Measured optical spectrum with a strong four-wave mixing generating higher order modulation side-bands. The detuning is 200 MHz.

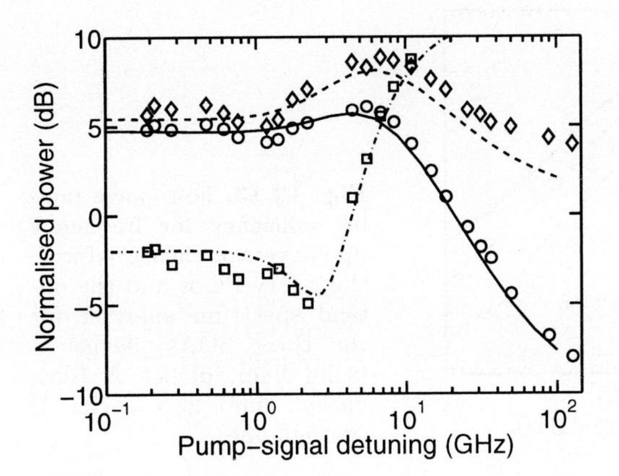

Fig. 13.12. Dependence of the pump (□), the signal (◇) and the conjugate powers (○) at the output for frequency up-conversion. The lines show the theoretical behaviour.

shows the measured spectrum at the output of the device C for a detuning of 200 MHz. One can easily observe in this figure that the pump signal is strongly suppressed and that the second and third order mixing products are of the same order of magnitude as the first and the probe. This means that a complete understanding of this low detuning behaviour requires to take the higher order products into account in the analysis. The inclusion in the modelling of the higher order signals, at the frequencies $\omega_0 + p\Delta\omega$

where p is an integer and ω_0 the frequency of the pump, gives rise to such a behaviour. One can see in Fig. 13.12 the evolution of the pump, the signal and the conjugated powers at the output of the SOA as a function of the detuning for frequency up-conversion. The lines in the figure are obtained by using a time domain numerical calculation that takes all the higher order mixing products into account (cf. Ch. 11). For the figure, the following values of the parameters have been used: $\alpha = 10.3$, $\bar{G} = 42$ dB, $\tau_s = 150$ ps, $\alpha_{CH} = 2$ and P_s/P_0 at the input is 25 dB. The calculation describes well the behaviour of the SOA. When the efficiency is high the modulation of the carrier density is strong. If α is high, which is the case in SOAs, the phase modulation will dominate in generating sidebands of the fields. In this case, and if the modulation index is high, the resulting spectrum will be dominated by this modulation. That explains the minimum observed for the pump for detunings around 2 GHz and the maximum of the conjugated signal power around 5 GHz. The calculation is very sensitive to the value of the linewidth enhancement factor. This is then a very good way to estimate the value of this parameter. The value used for the calculations of Fig. 13.12 is chosen by trying to adjust the calculated curve to the theoretical ones. The value then obtained is in a good agreement with previous measurement made on similar devices with different lengths [14].

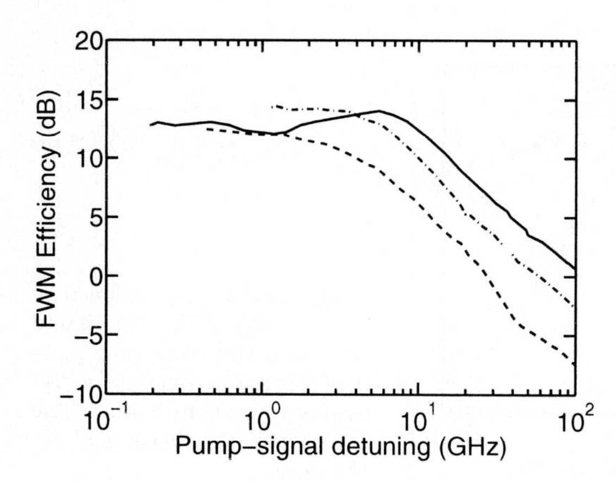

Fig. 13.13. Four-wave mixing efficiency for frequency up-conversion measured with the Fabry-Perot and the optical spectrum analyser for the three SOAs: device C (solid line), device A (dot-dashed line) and device B (dashed line).

In order to compare the three SOAs, the graphs giving the efficiency measured on the spectral range from under 1 GHz up to 100 GHz for the frequency up-conversion are plotted in Fig. 13.13. The absolute value of the efficiency for the measurement using the Fabry-Perot is found by superposing the values in the detuning range where both Fabry-Perot and spectrum analyser measurements are available. If the maximum efficiency for the three SOAs is within a difference of 3 dB, the highly different cut-off frequencies for the different devices appear clearly in the figure. This difference gives rise to a larger efficiency difference after the cut-off.

13.4 Effect of birefringence on four-wave mixing

One problem of FWM in SOAs that has not been discussed in the previous sections is the polarisation sensitivity of the FWM process. To obtain maximum conversion efficiency it is mandatory that both input waves possess the same state of polarisation while they interact in the SOA[1]. If that is the case, no dependence of the conversion efficiency on the common state of polarisation is expected, if a bulk SOA with a polarisation insensitive gain is used. However, investigating the device A for frequency detunings larger than 2 THz a strong variation in the conversion efficiency was observed if the polarisation of both copolarised waves (i.e. both waves have the same state of polarisation at the SOA input) is changed with respect to the SOA [20]. Since the polarisation sensitivity of the SOA gain is less than 0.5 dB in the case of our experimental conditions this cannot be held responsible for the observed effect. We, therefore, attribute our results to different refractive indices for two principal axes, i.e. to birefringence in the SOA. Birefringence causes a polarisation walkoff of the two injected waves within the SOA due to their different wavelengths, if the light is injected off-axis (i.e. off a principal axis). The degradation of the FWM efficiency with respect to on-axis light injection can be as high as 10 dB for detunings of about 6 THz. These results are in coherence with polarisation resolved measurements of the amplified spontaneous emission (ASE) noise power, where we also observed different refractive indices for TE and TM light propagation.

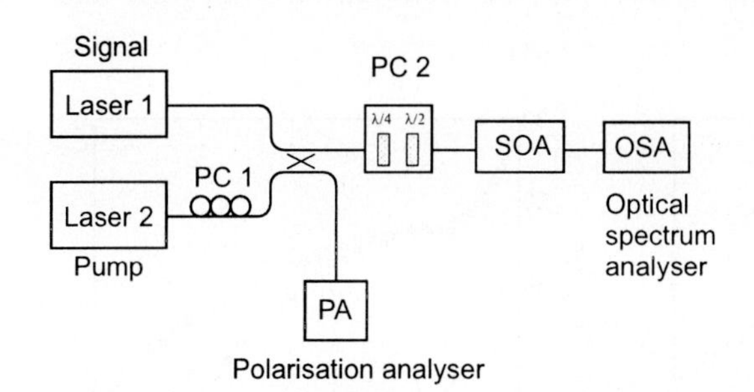

Fig. 13.14. Experimental set-up for FWM measurements of the effect of birefringence in SOAs (PC 1 and PC 2 - polarisation controller).

[1] However, since the optical polarisation of a data signal at a certain position in a real transmission system cannot be predicted, polarisation insensitive FWM schemes have to be applied. Besides several approaches utilising polarisation diversity arrangements [15], the use of two pump waves with either parallel [16, 17] or orthogonal [18] polarisation promise potential applicability. For pulsed FWM applications a further method is described in [19].

13.4.1 Experimental set-up

The experimental set-up for the FWM experiments is shown in Fig. 13.14. Two continuous wave external cavity lasers are used as sources for the pump and signal waves. The pump laser is adjusted to have the same state of polarisation as the signal by means of the polarisation controller PC 1 and the polarisation analyser PA. The polarisation controller PC 2 consists of a $\lambda/4$ and a $\lambda/2$ plate and enables a reproducible setting of the common state of polarisation of both input waves with respect to the SOA. The SOA is an InGaAsP-bulk SOA [21] with a length of 400 μm. The unsaturated fibre-to-fibre gain is 23 dB (at an injection current of 100 mA and a wavelength of 1530 nm) with a polarisation sensitivity of less than 1.5 dB. The pump and signal waves were injected into the SOA with input powers of $P_p(0) = +3$ dBm and $P_s(0)$ = -7 dBm, respectively. The wavelength of the pump was fixed at λ_p=1530 nm whereas the wavelength of the signal λ_s was varied between 1530 nm and 1580 nm. In this configuration the polarisation sensitivity of the gain reduces to less than 0.5 dB due to gain saturation effects.

13.4.2 Impact of birefringence on the conversion efficiency and the signal-to-background ratio

To investigate the FWM behavior of the SOA, η_{FWM} and SBR were experimentally determined as functions of the frequency detuning $\Delta\nu$. To obtain the data shown in Fig. 13.15a and 13.15b for frequency up-conversion (i.e. the FWM signal under investigation possesses a larger optical frequency than the two input waves), only the setting of the polarisation controller PC 2 was varied, while pump and signal waves remained copolarised with respect to each other. Fig. 13.15a and 13.15b show, that the variation of both, SBR and

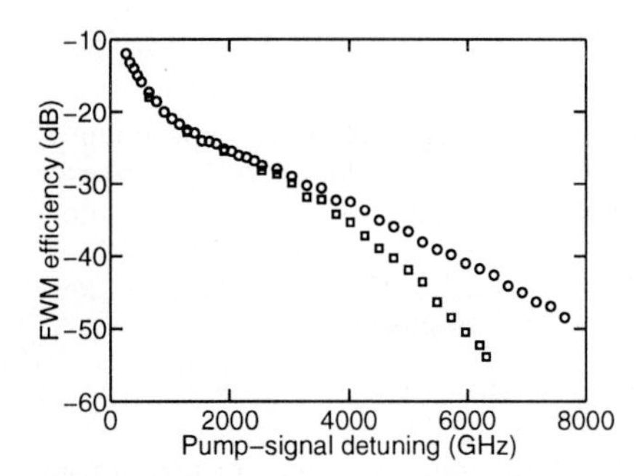

Fig. 13.15a. η_{FWM} vs. pump-signal detuning for different input polarisations of copolarised pump and signal waves. Maximum (O) and minimum (□) FWM signal.

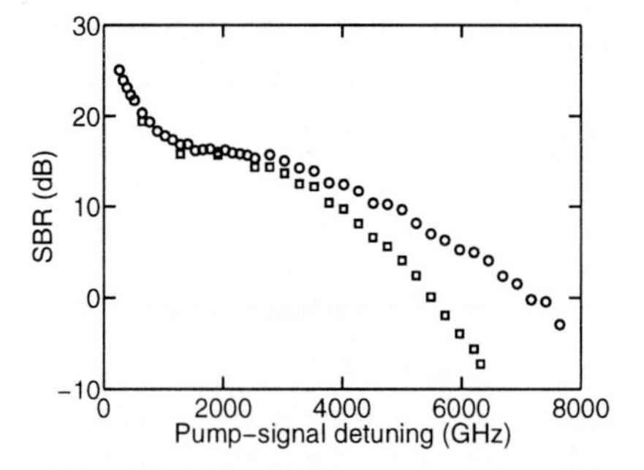

Fig. 13.15b. SBR vs. pump-signal detuning for different input polarisations of copolarised pump and signal waves. Same legends as (a).

η_{FWM} can be as large as 10 dB for detunings of about 6 THz. An additional experiment was performed at a fixed detuning of 6 THz. We adjusted the quarterwave- and halfwave-plate of PC 2 for maximum FWM signal. Then the rotation angle of the halfwave-plate was changed. Fig. 13.16 shows the output power of the FWM signal as a function of the rotation angle of the halfwave-plate. Notice the strong variation of the FWM signal power with a 45 degree periodicity (i.e. a 90 degree periodicity of the optical input polarisation). The variation of the FWM signal can therefore not be attributed to a polarisation sensitivity of the gain, since in that case one would observe a 90 degree periodicity (corresponding to 180 degrees polarisation periodicity). However, the slight variation of about 1.5 dB that can be seen in the max-

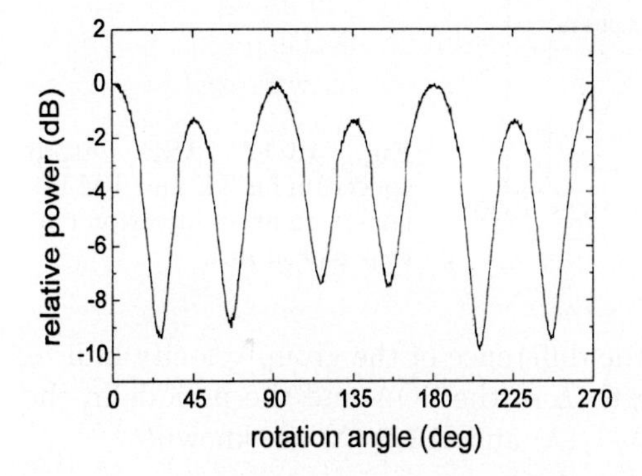

Fig. 13.16. Experimental data of the FWM-signal power as a function of the rotation angle of the halfwave-plate.

imum values is likely to result from the residual polarisation sensitivity of the gain. Since the FWM signal output power is approximately proportional to the cube of the gain, this corresponds well with the residual polarisation sensitivity of the SOA gain of 0.5 dB. The variation in the minima is not yet fully understood but will be a subject of future work. To exclude any misreading of the observed effects, we replaced the SOA by a polarisation analyser and thoroughly checked, that the observed effect is not caused by the short piece of fibre between PC 2 and the SOA. Both experiments strongly suggest the presence of birefringence in the polarisation insensitive SOA under investigation. Maximum FWM efficiency occurs, when both copolarised input waves are polarised linear and parallel to one of the principal axes, whereas a minimum FWM efficiency refers to a 45 degree off axis injection.

13.4.3 Polarisation resolved ASE measurements

An additional proof is the birefringence seen in polarisation resolved measurements of the ASE noise power of the same device. For that, a polariser was used to measure the unsaturated ASE output spectrum of the SOA for two orthogonal states of polarisation. Here, the weak polarisation sensitivity of the gain can be used to identify the two orthogonal axes as the TE and TM modes of the amplifier structure. The two spectra are shown in Fig. 13.17. Note, that the small ripples on the curves, which are due to the residual reflections on the SOA facets, show different periodicities for TE and TM modes. While both curves are in phase at wavelength $\lambda_1 = 1501$ nm, they are out of phase at $\lambda_2 = 1524$ nm. Using the general formulas for a Fabry-

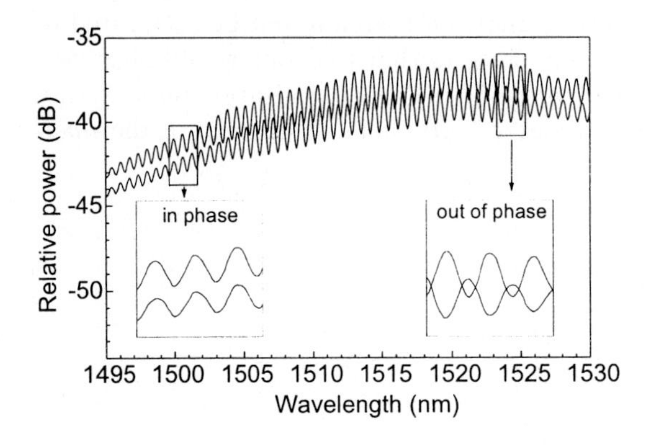

Fig. 13.17. ASE output spectrum for TE and TM Polarisation at an injection current of 100 mA.

Perot resonator an expression for the difference of the group velocity indices $\Delta N(\lambda)$ can be obtained if the length L of the SOA and the periods in the ripple (i.e. the modal spacings) $\Delta\lambda_{TM}(\lambda)$ and $\Delta\lambda_{TE}(\lambda)$ are known:

$$\begin{aligned} \Delta N(\lambda) &= |N_{TM}(\lambda) - N_{TE}(\lambda)| \\ &= \frac{\lambda^2}{2L}\left|\frac{1}{\Delta\lambda_{TE}(\lambda)} - \frac{1}{\Delta\lambda_{TM}(\lambda)}\right|. \end{aligned} \tag{13.1}$$

Here $N_{TM,TE}(\lambda) = (n_{TM,TE}(\lambda) - \lambda\frac{\partial n_{TM,TE}(\lambda)}{\partial\lambda})$ denote the group velocity indices for TM and TE light with $n_{TM,TE}(\lambda)$ being the corresponding refractive indices. Since the curves in Fig. 13.17 indicate that $\Delta\lambda_{TM}(\lambda) \neq \Delta\lambda_{TE}(\lambda)$, it can be concluded that $\Delta N(\lambda) \neq 0$, i.e. the group velocity indices for TE and TM propagation are different. However, it is not possible to determine definite values for $\Delta N(\lambda)$, since $\Delta\lambda_{TM,TE}(\lambda)$ cannot be obtained precisely enough for one particular wavelength and no precise data about the dispersion relations $\frac{\partial N_{TM}(\lambda)}{\partial\lambda}$ and $\frac{\partial n_{TE}(\lambda)}{\partial\lambda}$ are known. Experiments were also performed with two other polarisation insensitive SOAs but the birefringence effects were most pronounced in the described device. One reason for that might be the particular waveguide structure of the SOA under investigation.

In conclusion, birefringence effects in SOAs have to be taken care of, when FWM is used for the various optical signal processing applications. The described effects should also be considered if the nonlinear SOA parameters are to be extracted from high detuning FWM experiments (as performed in Sec. 13.1).

13.5 Four-wave mixing experiments with picosecond optical pulses

All experimental and theoretical considerations presented here assumed continuous waves as SOA input signals. However, the various applications of FWM show, that it is useful to distinguish between FWM under continuous wave (CW) and under pulse conditions (see Fig.13.18).

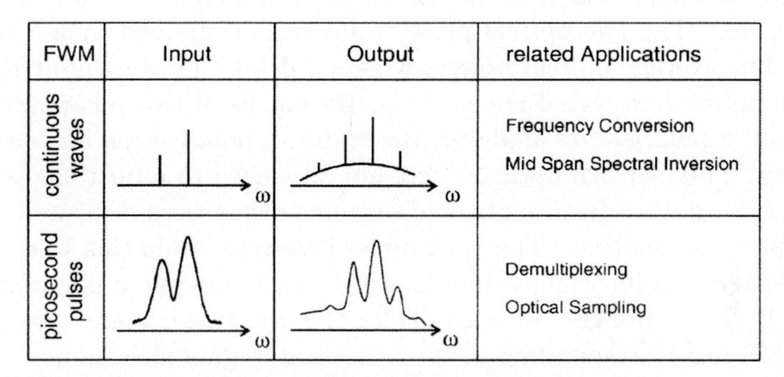

Fig. 13.18. Comparison of FWM with continuous waves (CW) and with picosecond pulses (pulsed).

Under CW FWM we understand all arrangements, where the pump is a continuous wave. The signal can either be a continuous wave (CW FWM - as used in basic FWM experiments and for parameter extraction) or it can be a modulated data stream (quasi-CW FWM - as in actual signal processing). Since in most applications the pump power is significantly larger than the signal power, the gain of the amplifier is mainly saturated by the (continuous) pump wave. In this case, the influence of the fast gain dynamics of the SOA on the temporal and spectral shape of the data signals is negligible. The results of FWM experiments with continuous waves can therefore be used to describe the $quasi - CW$ FWM behaviour of an SOA. The main FWM applications in this configuration are frequency conversion [22, 23] and mid-span spectral inversion (MSSI) [24–26].

The term *pulsed FWM* is used, if pump and signal are both short optical pulses in the range of some picoseconds. In this case the fast gain dynamic of the SOA substantially influences the temporal and spectral shape of pump and signal pulses. As will be discussed in this section, the performance of an SOA used in pulsed FWM applications can no longer be described by CW experiments. Different experimental and theoretical approaches have to be chosen in order to investigate pulsed FWM in SOAs. Pulsed FWM finds applications in optical time division demultiplexing [19, 27] and optical sampling [28].

13.5.1 Experimental set-up

A schematic diagram of the experimental set-up used for the pulsed FWM measurements is shown in Fig. 13.19. Two external cavity modelocked semiconductor lasers [29] were used as sources for pump and signal. Both lasers were hybridly modelocked at a repetition rate of 2 GHz. The wavelengths of the pump and signal waves were 1545 nm and 1552 nm, respectively. The temporal width of the optical pulses after amplification by an EDFA and subsequent filtering was 3 ps (FWHM) as determined by autocorrelation measurements. The two optical pulse trains were combined by a 3 dB fibre coupler. The average optical powers were +3 dBm and -4 dBm at the SOA input, for pump and signal respectively. By means of two polarisation controllers and a polarisation analyser, the states of polarisation for both input waves were matched. An optical delay line was used to adjust the temporal overlap of the pulses. Both input and output powers were measured with an optical spectrum analyser. The SOA under investigation in this section was a bulk InGaAsP amplifier [30] with a length $L = 980$ μm and a polarisation insensitive fibre-to-fibre gain of ≈ 28 dB for the injection current $I_c = 200$ mA at $\lambda = 1545$ nm corresponding to the unsaturated gain maximum.

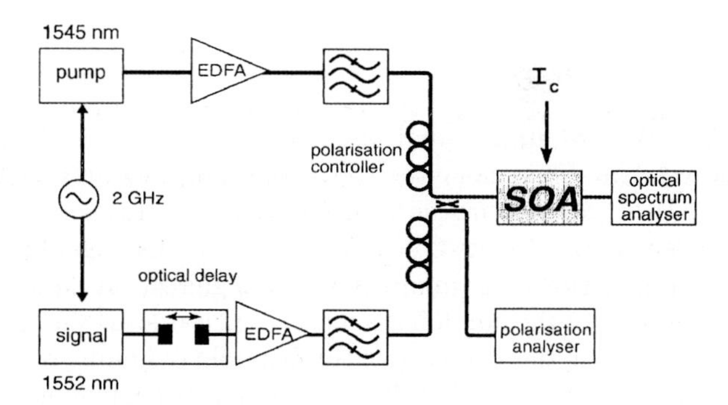

Fig. 13.19. Experimental set-up for FWM measurements with optical picosecond pulses. All experiments were performed with $P_p(0) = +3$ dBm and $P_s(0) = -4$ dBm. The temporal width (FWHM) of the optical pulses at the SOA input was 3 ps.

13.5.2 Short pulse amplification in SOAs

There are only few studies on FWM using optical pulses. The main reason is the high complexity of both experimental and theoretical investigations if short optical pulses are used to perform FWM in SOAs. In contrast to CW FWM, additional parameters, such as temporal width, temporal overlap, peak power and chirp of the injected pulses become important quantities that influence both η_{FWM} and SBR. In particular, if the temporal pulse width is in the order of the SOA time constants, the amplifier parameters change on the same timescale and non-static conditions apply. The complex dynamic of the carrier density and carrier distribution, which influences both gain [31] and refractive index [32], can be studied by pump-probe measurements with ultrashort optical pulses. Detailed descriptions of the gain dynamics (including a comprehensive list of references) can be found in [33–36]. Another difference in comparison to CW FWM is the generally larger conversion efficiency, which, however, depends on the pulse width and the temporal overlap. Surprisingly, the maximum η_{FWM} does not occur for a perfect overlap of pump and signal pulses but for a slight advance of the generally weaker signal pulse before the pump pulse [37, 38]. Analytical results, including ultrafast effects are presented in [39, 40]. It is also shown theoretically there, that some chirp is added to the signal. While in the mentioned references mainly η_{FWM} has been of interest, other optimisation criteria are added if the SBR is investigated [41]. Recent experimental results indicate an optimum injection current I_c that is not identical with the maximum current [42] (see also paragraph 13.5.3). A sophisticated theoretical approach that is in qualitative agreement with the presented experimental findings was recently developed by Mecozzi and Mørk [43].

13.5.3 Comparison of CW and pulsed FWM measurements

In this paragraph, we show that the conversion efficiency and the signal-to-background ratio behave differently if picosecond-optical pulses are used instead of CW-light. By use of the optical delay line the temporal overlap of the injected pulses (pulse width 3 ps) was adjusted to give maximum FWM signal. Fig. 13.20 depicts the SOA output spectra for different injection currents I_c. As can be seen in Fig. 13.20, almost no FWM signal is observed

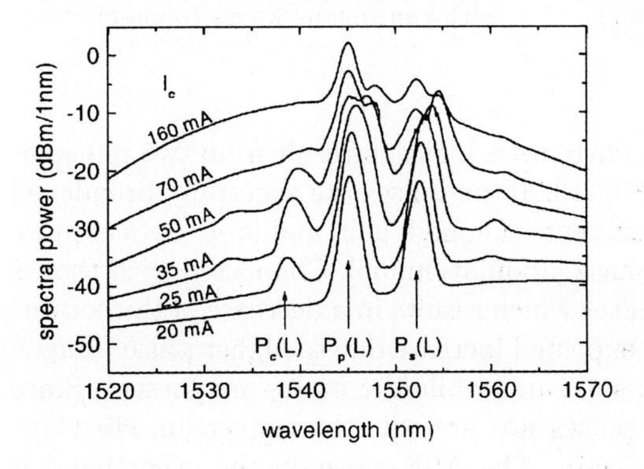

Fig. 13.20. FWM output spectra for different injection currents I_c of the SOA. The optical input powers were $P_p(0) = +3\ dBm$ and $P_s(0) = -4\ dBm$. The resolution bandwidth of the optical spectrum analyser was adjusted to 1 nm.

in the SOA output spectrum for low injection currents (20 mA). Increasing the injection current leads to an enhancement of the ASE noise power but also to a visible FWM signal due to an increased SBR. However, if the driving current exceeds 35 mA the SBR degrades. For currents larger than 70 mA almost no FWM signal can be seen. Note the distortion and the broadening of the pump and signal pulse spectra at the SOA output.

Both, η_{FWM} and SBR are plotted as functions of the injection current in Fig. 13.21a and 13.21b. In our experiment all optical powers are time averaged peak powers measured by an optical spectrum analyser (resolution bandwidth of 1 nm for pulsed and 0.1 nm for CW measurements). In the CW measurements of η_{FWM} and SBR the pulse sources in Fig. 13.19 were replaced by CW sources. The continuous wave measurements reveal an increase of η_{FWM} and SBR for increasing injection current as seen in 12.2. The curves that represent the pulsed measurements, however, show a clear maximum of the SBR for about 35 mA, while the conversion efficiency η_{FWM} seems to remain constant for injection currents larger than 45 mA. Note also that the conversion efficiency is significantly larger in the case of pulsed FWM as compared to CW measurements for injection currents $I_c < 70$ mA. The

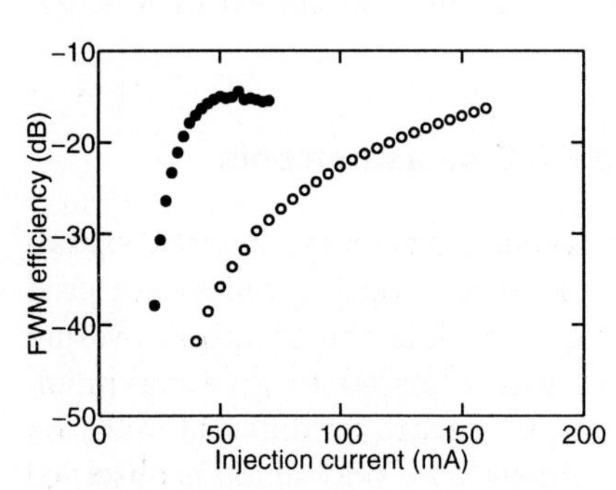

Fig. 13.21a. FWM efficiency, η_{FWM}, for FWM with short optical pulses (solid circles) and with continuous waves (open circles).

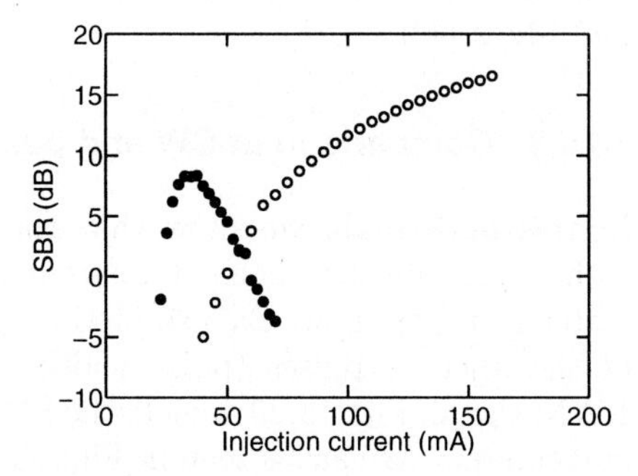

Fig. 13.21b. SBR for FWM with short optical pulses (solid circles) and with continuous waves (open circles).

degradation of the SBR at high currents is likely to result from two phenomena. Firstly, as can be seen in Fig. 13.20 the pulses are spectrally broadened and distorted at high injection currents (i.e. high gain and large optical pulse powers) due to self- and cross-phase modulation [35]. Connected with that is a temporal broadening of the pulses which results in a decrease of the conversion efficiency counteracting an expected increase due to higher pulse powers. Secondly, the amplifier strongly saturates while the pulses are present. Since FWM occurs only while both pulses are present, the conversion efficiency mainly depends on a saturated gain. The ASE noise on the other hand is time averaged (i.e. it experiences a higher average gain than the conversion efficiency due to gain recovery). For an increasing current, this difference between the pulse and the ASE gain increases and the ASE noise therefore grows faster than the conversion efficiency leading to the maximum in the SBR curve [43]. We repeated our experiments at 10 GHz which is currently a repetition rate of large interest for demultiplexing applications. Although the optimum injection current for maximum SBR shifted to 45 mA, the SBR again strongly degraded by up to 10 dB for larger injection currents. Theoretical results that well describe both, the maximum in the SBR curve as well as the shift of the maximum towards higher gain for higher repetition rates, can be found in [43].

Although our presented results are presumably due to the low duty cycle operation of the SOA (3 ps pulses and 500 or 100 ps in between subsequent pulses) our experimental conditions are realistic if FWM is to be applied for fast optical demultiplexing. For fast optical switching applications based on pulsed FWM in SOAs, it can therefore be more advantageous to use a modest injection current, i.e. a modest gain, rather than the maximum current.

References

[1] A. Mecozzi, S. Scotti, A. D'Ottavi, E. Iannone, and P. Spano, "Four-Wave Mixing in Traveling-Wave Semiconductor Amplifiers", *IEEE J. Quantum Electron.*, vol. 31, pp. 689–699, 1995.

[2] S. Scotti and A. Mecozzi, "Frequency Converters Based on FWM in Traveling-Wave Optical Amplifiers: Theoretical Aspects", *Fiber and Integrated Optics*, vol. 15, pp. 243–256, 1996.

[3] A. Mecozzi, A. D'Ottavi, F. Cara Romeo, P. Spano, R. Dall'Ara, G. Guekos, and J. Eckner, "High Saturation Behaviour of the Four-Wave Mixing Signal in Semiconductor Amplifiers", *Appl. Phys. Lett.*, vol. 66, pp. 1184–1186, 1995.

[4] A. D'Ottavi, F. Girardin, L. Graziani, F. Martelli, P. Spano, A. Mecozzi, S. Scotti, R. Dall'Ara, J. Eckner, and G. Guekos, "Four-wave mixing in semiconductor optical amplifiers: a practical tool for wavelength conversion", *IEEE J. of sel. topics in quantum electron.*, vol. 3, pp. 522–528, 1997.

[5] A. D'Ottavi, F. Martelli, P. Spano, A. Mecozzi, S. Scotti, R. Dall'Ara, J. Eckner, and G. Guekos, "Very high efficiency four-wave mixing in a single semiconductor traveling-wave amplifier", *Appl. Phys. Lett.*, vol. 68, pp. 2186–2188, 1996.

[6] T. Liu, K. Obermann, K. Petermann, F. Girardin, and G. Guekos, "Effect of saturation caused by amplified spontaneous emission on semiconductor optical amplifier performance", *Electron. Lett.*, vol. 33, pp. 2042–2043, 1997.

[7] S. Scotti, L. Graziani, A. D'Ottavi, F. Martelli, A. Mecozzi, P. Spano, R. Dall'Ara, F. Girardin, and G. Guekos, "Effects of ultrafast processes on frequency converters based on four-wave mixing in semiconductor optical amplifiers", *IEEE J. of sel. topics in quantum electron.*, vol. 3, pp. 1156–1161, 1997.

[8] A. Mecozzi, A. D'Ottavi, F. Martelli, S. Scotti, P. Spano, R. Dall'Ara, J. Eckner, and G. Guekos, "Highly non-degenerate four-wave mixing in semiconduc-

tor laser amplifier", in *Physics and Simulation of Optoelectronic Devices IV*, pp. 288–302, 1996.

[9] C. Holtman, P. A. Besse, T. Brenner, and H. Melchior, "Polarization independent bulk active region semiconductor optical amplifiers for 1.3 μm wavelength", *IEEE Photon. Technol. Lett.*, vol. 8, pp. 343–345, 1996.

[10] R. S. Tucker, M. A. Summerfield, and J. P. R. Lacey, "Advanced optical frequency converters for lightwave communication networks", *Digests of International Quantum Electronics Conference*, p. 223, 1996.

[11] I. Koltchanov, S. Kindt, K. Petermann, S. Diez, R. Ludwig, R. Schnabel, and H. G. Weber, "Gain Dispersion and Saturation Effects in Four-Wave Mixing in Semiconductor Laser Amplifiers", *IEEE J. Quantum Electron.*, vol. 32, pp. 712–720, 1996.

[12] K. Obermann, I. Koltchanov, K. Petermann, S. Diez, R. Ludwig, and H. G. Weber, "Noise Analysis of Frequency Converters Utilizing Semiconductor-Laser Amplifiers", *IEEE J. Quantum Electron.*, vol. 33, pp. 81–88, 1997.

[13] I. Koltchanov, S. Kindt, K. Petermann, S. Diez, R. Ludwig, and H. G. Weber, "Characterisation of Terahertz Four-Wave Mixing in a Semiconductor-Laser Amplifier", in *CLEO '96*, pp. 105–106, 1996.

[14] F. Girardin, S. Pajarola, and G. Guekos, "Nonlinear parameters of bulk InGaAsP photonic devices", in *Materials for nonlinear optics, EOS topical meeting*, 1997.

[15] J.P.R. Lacey, S.J. Madden, and M.A. Summerfield, "Four-channel polarization-insensitive optically transparent wavelength converter", *IEEE Photon. Technol. Lett.*, vol. 9, pp. 1355–1357, 1997.

[16] R. Schnabel, U. Hilbk, Th. Hermes, P. Meißner, C. von Helmolt, K. Magari, F. Raub, W. Pieper, F. J. Westphal, R. Ludwig, L. Küller, and H. G. Weber, "Polarization Insensitive Frequency Conversion of a 10-Channel OFDM Signal Using Four-Wave-Mixing in a Semiconductor Laser Amplifier", *IEEE Photon. Technol. Lett.*, vol. 6, pp. 56–58, 1994.

[17] S. Diez, R. Ludwig, E. Patzak, H. G. Weber, G. Eisenstein, and R. Schimpe, "Four-wave mixing in semiconductor-laser amplifiers: phase matching in configurations with three input waves", in *CLEO '96*, pp. 505–506, 1996.

[18] P. Cortes, M. Chbat, S. Artigaud, J. Beylat, and J. Chesnoy, "Below 0.3 dB Polarization Penalty in 10 Gbit/s Directly Modulated DFB Signal over 160 km using Mid-Span Spectral Inversion in a Semiconductor Laser Amplifier", in *ECOC '95*, pp. 271–274, 1995.

[19] T. Morioka, H. Takara, S. Kawanishi, K. Uchiyama, and M. Saruwatari, "Polarisation-independent all-optical demultiplexing up to 200 Gbit/s using four-wave mixing in a semiconductor laser amplifier", *Electron. Lett.*, vol. 32, pp. 840–841, 1996.

[20] S. Diez, C. Schmidt, R. Ludwig, P. Doussière, and T. Ducellier, "Effect of birefringence in a bulk semiconductor optical amplifier on four-wave mixing", *IEEE Photon. Technol. Lett.*, vol. 10, pp. 212–214, 1998.

[21] P. Doussière, P. Garabedian, V. Colson, O. Legouezigou, F. Leblond, J. L. Lafragette, M. Monnot, and B. Fernier, "Polarisation insensitive semiconductor optical amplifier with buried lateraly tapered active waveguide", in *Optical Amplifiers and their Applications, OAA '93, Technical Digest*, pp. 140–143, 1993.

[22] M. Tatham, G. Sherlock, and L. Westbrook, "20 nm Optical Wavelength Conversion using Nondegenerate Four-Wave Mixing", *IEEE Photon. Technol. Lett.*, vol. 5, pp. 1303–1306, 1993.

[23] R. Schnabel, U. Hilbk, T. Hermes, P. Meissner, C. von Helmholt, K. Magari, F. Raub, W. Pieper, F. J. Westphal, R. Ludwig, L. Küller, and H. G. Weber,

"Polarization Insensitive Frequency Conversion of a 10-Channel OFDM Signal using Four-Wave Mixing in a Semiconductor-Laser Amplifier", *IEEE Photon. Technol. Lett.*, vol. 6, pp. 56–58, 1994.

[24] W. Pieper, C. Kurtzke, R. Schnabel, D. Breuer, R. Ludwig, K. Petermann, and H. G. Weber, "Nonlinearity-insensitive standard-fibre transmission based on optical- phase conjugation in a semiconductor-laser amplifier", *Electron. Lett.*, vol. 30, pp. 724–726, 1994.

[25] A. Ellis, M. Tatham, D. Davies, D. Nesset, D. Moodie, and G. Sherlock, "40 Gbit/s transmission over 202 km of standard fibre using midspan spectral inversion", *Electron. Lett.*, vol. 31, pp. 299–301, 1995.

[26] D. D. Marcenac, D. Nesset, A. E. Kelly, M. Brierley, A. D. Ellis, D. G. Moodie, and C. W. Ford, "40 Gbit/s transmission over 406 km of NDSF using mid-span spectral inversion by four-wave-mixing in a 2 mm long semiconductor optical amplifier", *Electron. Lett.*, vol. 33, pp. 879–880, 1997.

[27] R. Ludwig and G. Raybon, "All-Optical Demultiplexing Using Ultrafast Four-Wave Mixing in a Semiconductor Laser Amplifier at 20 Gbit/s", in *ECOC '93*, pp. 57–60, 1993.

[28] M. Jinno, J. B. Schlager, and D. L. Franzen, "Optical sampling using nondegenerate four-wave mixing in a semiconductor laser amplifier", *Electron. Lett.*, vol. 30, pp. 1489–1491, 1994.

[29] R. Ludwig and A. Ehrhardt, "Turn-key-ready wavelength-, repetition rate- and pulsewidth- tunable femtosecond hybrid modelocked semiconductor laser", *Electron. Lett.*, vol. 31, pp. 1165–1167, 1995.

[30] P. Doussière, P. Garabedian, C. Graver, D. Bonnevie, T. Fillion, E. Derouin, M. Monnot, J. G. Provost, D. Leclerc, and M. Klenk, "1.55 μm Polarization Independent Semiconductor- Optical Amplifier with 25 dB Fiber to Fiber Gain", *IEEE Photon. Technol. Lett.*, vol. 6, pp. 170–172, 1994.

[31] J. Mark and J. Mørk, "Subpicosecond gain dynamics in InGaAsP optical amplifiers: Experiment and theory", *Appl. Phys. Lett.*, vol. 61, pp. 2281–2283, 1992.

[32] K. L. Hall, A. M. Darwish, E. P. Ippen, U. Koren, and G. Raybon, "Femtosecond index nonlinearities in InGaAsP optical amplifiers", *Appl. Phys. Lett.*, vol. 62, pp. 1320–1322, 1993.

[33] A. Uskov, J. Mørk, and J. Mark, "Theory of Short-Pulse Gain Saturation in Semiconductor Laser Amplifiers", *IEEE Photon. Technol. Lett.*, vol. 4, pp. 442–446, 1992.

[34] P. J. Delfyett, A. Dienes, J. P. Heritage, M. Y. Hong, and Y. H. Chang, "Femtosecond Hybrid Mode-Locked Semiconductor Laser and Amplifier Dynamics", *Appl. Phys. B*, vol. 58, pp. 183–195, 1994.

[35] G. P. Agrawal and N.A. Olsson, "Self-Phase Modulation and Spectral Broadening of Optical Pulses in Semiconductor Laser Amplifiers", *IEEE J. Quantum Electron.*, vol. 25, pp. 2297–2306, 1989.

[36] M. Y. Hong, Y. H. Chang, A. Dienes, J. P. Heritage, and P. J. Delfyett, "Subpicosecond Pulse Amplification in Semiconductor Laser Amplifiers: Theory and Experiment", *IEEE J. Quantum Electron.*, vol. 30, pp. 1122–1131, 1994.

[37] M. Shtaif and G. Eisenstein, "Analytical solution of wave mixing between short optical pulses in a semiconductor optical amplifier", *Appl. Phys. Lett.*, vol. 66, pp. 1458–1460, 1995.

[38] M. Shtaif, R. Nagar, and G. Eisenstein, "Four-Wave Mixing Among Short Optical Pulses in Semiconductor Optical Amplifiers", *IEEE Photon. Technol. Lett.*, vol. 7, pp. 1001–1003, 1995.

[39] A. Mecozzi and J. Mørk, "Saturation induced by picosecond pulses in semiconductor optical amplifiers", *J. Opt. Soc. Am. B*, vol. 14, pp. 761–770, 1997.

[40] J. Mørk and A Mecozzi, "Theory of Nondegenerate Four-Wave Mixing Between Pulses in a Semiconductor Waveguide", *IEEE J. Quantum Electron.*, vol. 33, pp. 83–195, 1997.

[41] M. Shtaif and G. Eisenstein, "Calculation of Bit Error Rates in All-Optical Signal Processing Applications Exploiting Nondegenerate Four-Wave Mixing in Semiconductor Optical Amplifier", *J. Lightwave Technol.*, vol. 14, pp. 2069–2077, 1996.

[42] S. Diez, C. Schmidt, R. Ludwig, H. G. Weber, S. Kindt, I. Koltchanov, and K. Petermann, "Four-Wave Mixing in Semiconductor Optical Amplifers for Frequency Conversion and Fast Optical Switching", *IEEE J. of sel. topics in quantum electron.*, vol. 3, pp. 1131–1145, 1997.

[43] A. Mecozzi and J. Mørk, "Saturation effects in non-degenerate four-wave mixing between short optical pulses in semiconductor laser amplifiers", *IEEE J. of sel. topics in quantum electron.*, vol. 3, pp. 1190–1207, 1997.

In the chapters of Part III, the main topics that may be encountered in studying nonlinear effects in semiconductor optical amplifiers on a theoretical as well as on an experimental point of view have been discussed. The different models presented here for the treatment of four-wave mixing in semiconductor optical amplifiers have shown in which case which effect has to be taken into account. For example, the gain dispersion at high pump-signal detuning. The time-domain approach compared to the coupled-mode approach has been shown to give a more accurate explanation of the nearly degenerate four-wave mixing behaviour. The comparison of different experimental works in different groups has been presented. A systematic metrological comparison has allowed us to evaluate the importance of the choice of the measurement methods and the range of precision in which the results can be compared.

As the goal of this Part was to present the different facets of the non-linearity based effects in semiconductor optical amplifiers in general and the four-wave mixing in particular, we did try to give the basis for further work in this direction. Beyond the results and examples presented here, there are other topics of interest that should be studied in order to bring these techniques to the level of industrial applications. Among them some work on the influence of the state of polarisation of the signal, on the detailed effects of pulsed signals and pumps on the behaviour of the device, and on the spectral properties of output signals for example should be done.

Appendix. An Optical Data Interchange Format

P. Verhoeve

A.1 Introduction

The problem of defining a data format appears each time when a model is implemented or when a measurement is automated. Both the input data and the results need to be stored on a computer disk and all persons or applications involved need to understand the content of the file.

Most applications have their own specific format, the numbers in a certain order, in the best case accompanied with some comments but mostly it is just a set of numbers without any reference to their meaning nor their units. When another application needs the results but wants them in a different order or unit, a conversion is needed.

When the data is to be exchanged with other parties, conversion difficulties cause the exchange format to be a paper copy, even when all parties involved have full access to electronic media. This situation is far from optimal, since each conversion (either automatically or manual retyping) has a big error potential. Furthermore there is the loss of manpower due to the constant need for conversion (programs). In order to prevent this proliferation of conversion applications, the error potential, and the manpower spent on deciphering large sets of numbers, a file format is needed which can cope with the complexity of the optical problems yet is not a limiting factor for the actual data storage. For this purpose ODIF, an Optical Data Interchange Format was developed.

A.2 Terms and conditions of a common interchange format

The problem of data interchange has actually two aspects: the first aspect is what you put in a file, the second aspect is how you put it in the file.

Deciding what is in a file is specifying the content of that file, e.g. "L" stands for the laser cavity length. Since content is a highly problem-specific issue, it is almost impossible to cast content specification into a set of rigid yet non limiting rules.

Deciding on how to put the information in a file however can be treated in a generic way and thus it can be cast into strict rules, which are commonly known as a "syntax description". All programming languages have a strict syntax on how to write a program (i.e. which characters and in what order they can be in the file), yet it doesn't limit what the actual program achieves (i.e. the content or the performance of the algorithm the program describes). In the same way, a data file syntax description can be made such that it is a strict definition of how to write the file yet doesn't limit what can be stored in the file. In order to successfully design this syntax, the boundary conditions and restrictions for the files must be outlined on a general level.

The demands can be summarised as follows:

- *Flexibility* : the syntax must be designed such that all types of results (from modelling, measurements and parameter extraction) can be incorporated into the same file. This without increasing the overall complexity of the file or syntax.
- *Uniformity* : Since all types of results can be incorporated into one file, they should be treated in a similar way, in order to simplify the usage of the syntax and reduce the number of rules.
- *Platform compatibility* : The syntax must be designed such that every partner can generate and read files on the computer platform he normally uses (UNIX-Workstation, Mac, PC, ...).
- *E-mail compatibility* : Possibility to use modern media for transmission of data (e-mail) is essential to the long term success of such a syntax.
- *Readable by "normal" applications* : The user of the syntax must be able to read a file and manipulate its data without specialised applications or translators. Most of the "normal" programs on the different platforms (e.g. commercially available text editors, word processors, spreadsheets, ...) should be able to read the files and plot the data inside.
- *Not only results* : The origin of results in a file is often as relevant as the results themselves (e.g. the used model and its parameters, the measurement method and settings of the instruments, extraction tools and reference to original data, ...). In order to keep this information during data exchange, the best way is to enclose the information in the "result" file in a structured way.
- *Comments are useful* : As with programming languages, data files need comments added to the results (e.g. used methods, definitions, etc.). The syntax should also incorporate a standard way to include such comments, without damaging the integrity or readability of the data.
- *Simplicity first* : Although the syntax must be able to cope with highly complex data (e.g. a combination of several measurement, modelling and extraction results), it should still be able to store simple results (e.g. a one value measurement) with a minimal overhead.
- *Dynamic to the future* : The syntax must be designed such that it has no hard constraints to future needs, nor should it have hard constraints towards the complexity of the file contents.

- *No strict order* : When storing and recalling results, most errors occur when the numbers need to be in a strict order, referencing the results by a name is more natural and reduces the errors.

Each of these demands has implications on the definition of the syntax. Fig. A.1 gives a summary of how the demands are linked to the most important restrictions of the syntax definition.

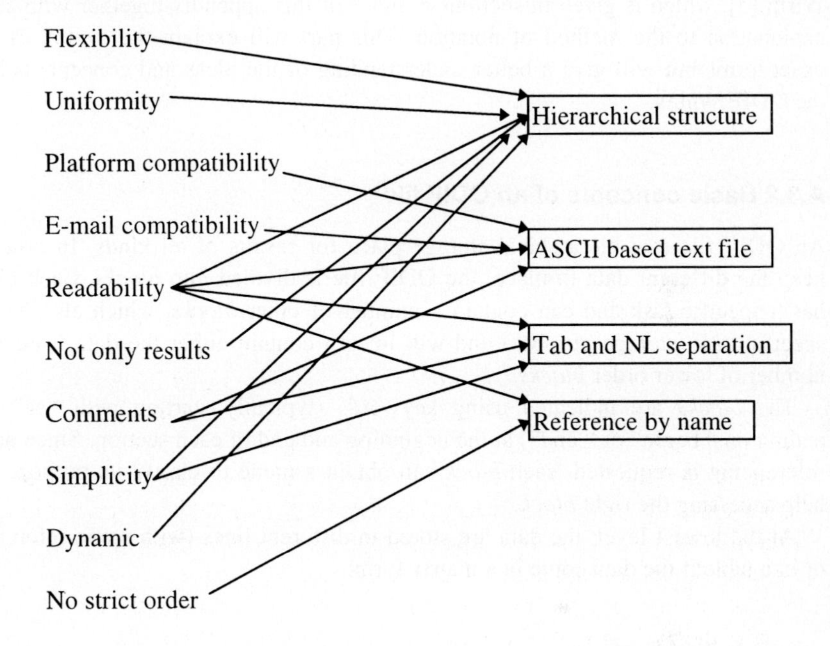

Fig. A.1. Summary of the demands and restrictions of the syntax defintion.

The ODIF syntax like most syntax definitions, will be a hierarchically structured file format stored in ASCII. It will allow to store and retrieve information based on a name rather than a position in the file. But contrary to most syntax definitions, it will have a strict limitation with respect to white spaces : since most spreadsheet applications use tab and new line characters as separators, ODIF will have to impose restrictions to enable the spreadsheet compatibility.

If the demands for readability (both software and human) would be given up, the strict limitations towards white spaces as well as the ASCII limit can be removed and a more compact, yet binary version of ODIF could be specified.

A.3 The syntax description

A.3.1 Introduction

Describing a syntax is not a trivial task and errors are very likely to occur due to misinterpretation of the standard. For this reason, a formal specification of the ODIF syntax has been written using the classical syntax notation according to N. Wirth [5], which is given in sections 6 and 7 of this appendix together with some explanation to the method of notation. This part will explain the syntax in less exact terms but will give a better understanding of the ideas and concepts behind the ODIF syntax.

A.3.2 Basic concepts of an ODIF file

An ODIF file is a hierarchical storage place for results of all kinds. In order to keep the different data in place, the ODIF file is divided into *blocks*. Each block has a specific task and can contain a number of other *blocks*, which also have a specific task (storage purpose) and will in turn contain either the data itself or a number of lower order *blocks*.

The *blocks* are indicated using keywords (typically starting with "@" and ending on ".begin" or ".end") at the beginning and end of each section. Since name referencing is requested, each *block* can obtain a name to clarify its purpose and help accessing the right *block*.

At the lowest level, the data are stored in different lines (with name reference) or in a table if the data come in a matrix form.

A.3.3 The highest level

On the highest level is the *file block*, which can consist of a number of *global blocks*. Each *global block* is a container of results and all results stored in such a container are considered to be related. E.g. several measurements on a device can be stored within one *global block*, measurements on several devices would typically be stored in several *global blocks*.

Since a *global block* is describing a device and measurement/modelling results it will contain a section (*block*) describing the parameters of that device : the *parameter block*. Following that section, multiple sections for results of measurements, modelling or extraction experiments are possible.

In Fig. A.2, a schematic of the ODIF file on the highest level is shown. It should be noted that the figure shows a more complex example and that most ODIF files will consist of only one *global block* which contains a *parameter block* and one *measurement/model/extraction block.*

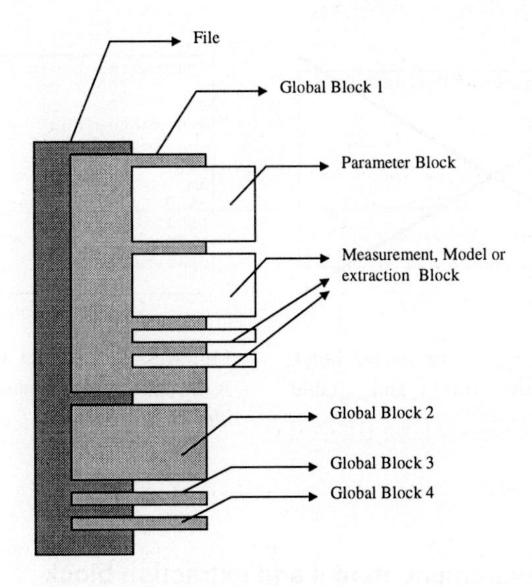

Fig. A.2. The high level structure of an ODIF file

A.3.4 The parameter block

The *parameter block* is meant for storing information about the device parameters as the name indicates. Device parameters are typically "data" and will be stored directly in the *parameter block* which cannot contain other blocks.

Data can take two forms : simple numbers (with indication of name and unit for each one) or more complex tables of numbers (with indication of name and unit for each column). Both can be stored in a parameter block, but only one at a time. When the need exists to combine both, two consecutive parameter blocks should be used. Some examples of valid and invalid *parameter blocks* are shown in Fig. A.3 to A.5.

```
@parameter.begin <waveguide>
L       300    um
W       10     um
@parameter.end
```

```
@parameter.begin <loss>
wav        int_loss
nm         cm^-1
1550       2
1551       2.1
1552       2
@parameter.end
```

Fig. A.3. Two correct *parameter blocks*.

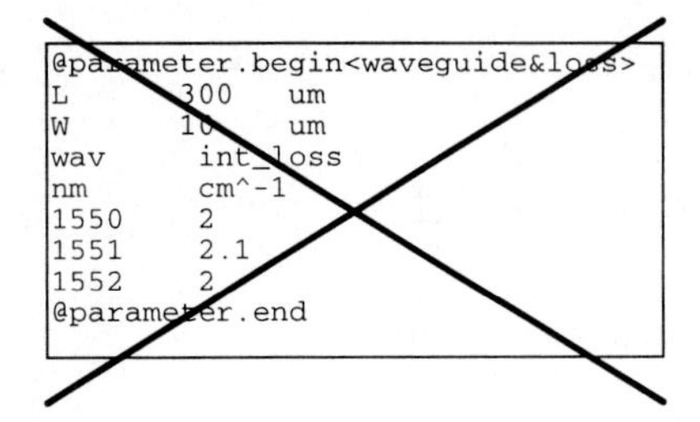

Fig. A.4. An incorrect *parameter block* containing both line and table information.

```
@parameter.begin <waveguide>
L        300    um
W        10     um
@parameter.end
@parameter.begin <loss>
wav      int_loss
nm       cm^-1
1550     2
1551     2.1
1552     2
@parameter.end
```

Fig. A.5. The correct version of Fig. A.4, using two separate *parameter blocks*.

A.3.5 The measurement, model and extraction block

In an abstract approach, data originating from a measurement, data calculated by a model or data obtained by parameter extraction methods can be treated in a similar fashion. For this reason, the three blocks associated with these data have a similar internal structure and differ only by the used keywords which indicate the kind of the data inside.

As all blocks, the measurement, model and extraction blocks have keywords which indicate the beginning and the end of the block. (respectively : "@measurement.begin", "@model.begin", "@extraction.begin" and "@measurement.end", "@model.end", "@extraction.end"). The blocks can again consist of a number of *setting blocks* and *data blocks.*

The purpose of the *setting block* is to store the information about how the data were obtained e.g. the settings of the measurement equipment (current sources, power meters, ...), the calculation parameters (step size, number of iterations, ...) or the method used for extraction (fitted using downhill simplex algorithms with minimax cost function, ...).

Following the *setting blocks*, the data itself can be stored in *data blocks*, which either can contain a number of single lines or a table with data. As it was the case with the *parameter block*, it is one or the other and no combinations (lines and data) are possible; if there is a need to store both, two *data blocks* should be used.

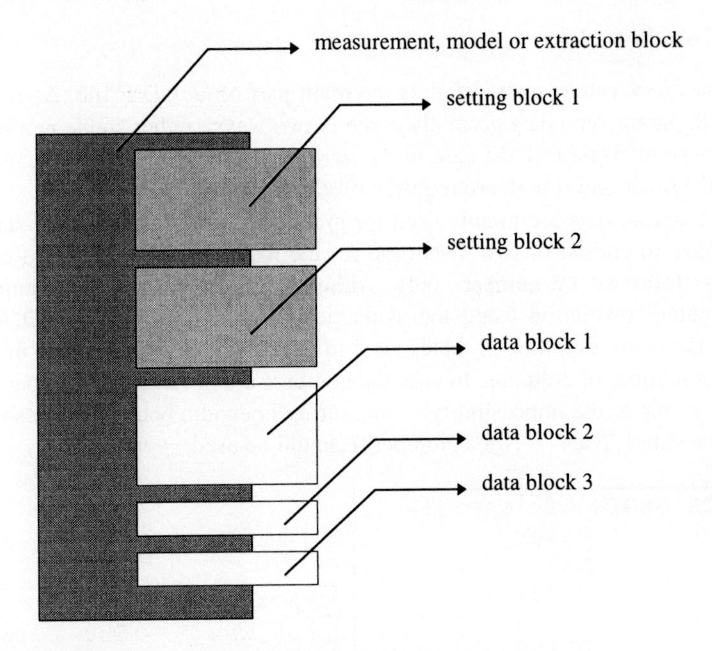

Fig. A.6. Structure of a measurement, model or extraction block.

A.3.6 The setting block

The *setting block*, designed for storing the information which explains how the results are obtained, stores that information using lines. In general, tables are never used to store "setting" information and therefore tables are not allowed in a *setting block*.

Information about instrument or software settings, is probably the least general information stored in a file, e.g. similar instruments (like optical spectrum analysers) still have a largely different set of control parameters. For this reason, the name reference of the *setting block* has been extended with an obligatory "labname". This ensures that the interpretation of the numbers inside the *setting block* can always be related to the person, software or standard that defined the content of the block. The "labname" is written between square brackets ("[" and "]") to distinguish it from regular reference names (denoted between "<" and ">").

```
@setting.begin [OLIMPEX] <HP90751>
RBW        5     nm
SPAN       100   nm
@setting.end
```

Fig. A.7. Example of a valid *setting block.*

A.3.7 The data block

The *data block* can be considered as the main part of an ODIF file. As it was the case with parameters, data normally come in two ways : either single numbers or a tabular format. Similar to the case of the *parameter block*, a *data block* can contain one or the other and if both are required two *data blocks* need to be used.

Since tabular data are mainly used for graphing purposes, the syntax states that tables have to consist of two lines (one for the name of the column and one with the unit) followed by numbers only. Although it is not possible to impose the "rectangular" restriction using the syntactical notation by N. Wirth, ODIF does impose the restriction that all tables need to be rectangular i.e. all rows must have the same number of columns. In case the numbers are not available for parts of the table (e.g. due to the impossibility of measuring linewidth below the threshold of a laser) the value "NaN" ("Not a Number") should be used.

```
@DATA.BEGIN LI<laser_1>
I             Power
mA            mW
0.0           0.001
1.0           0.013
2.0           0.159
3.0           0.957
4.0           1.568
@DATA.END
```

```
@DATA.BEGIN LI<laser_1>
peak_wav     1550    nm
@DATA.END
```

Fig. A.8. Two correct *data blocks*.

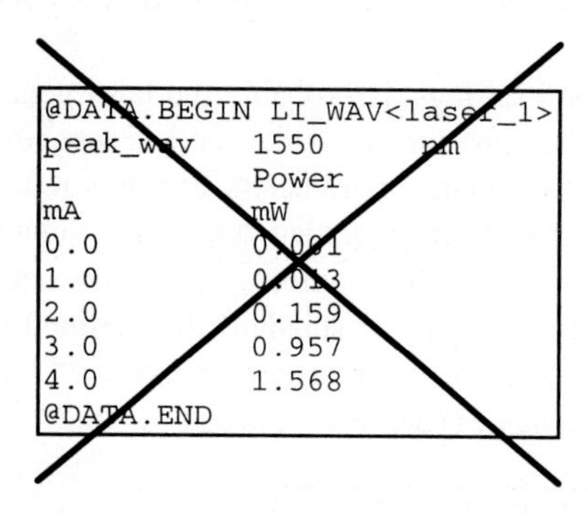

```
@DATA.BEGIN LI_WAV<laser_1>
peak_wav     1550      nm
I            Power
mA           mW
0.0          0.001
1.0          0.013
2.0          0.159
3.0          0.957
4.0          1.568
@DATA.END
```

Fig. A.9. An incorrect *data block* containing both line and table information.

```
@DATA.BEGIN LI<laser_1>
I             Power
mA            mW
0.0           0.001
1.0           0.013
2.0           0.159
3.0           0.957
4.0           1.568
@DATA.END
@DATA.BEGIN WAV<laser_1>
peak_wav     1550       nm
@DATA.END
```

Fig. A.10. The correct version of Fig. A.9, using two separate *data blocks*.

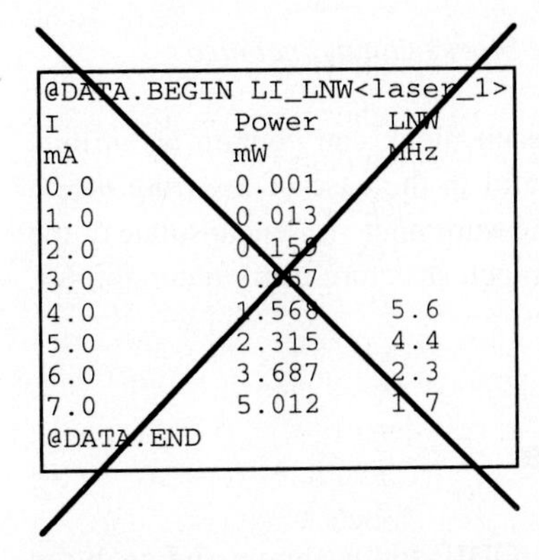

```
@DATA.BEGIN LI_LNW<laser_1>
I       Power   LNW
mA      mW      MHz
0.0     0.001
1.0     0.013
2.0     0.159
3.0     0.957
4.0     1.568   5.6
5.0     2.315   4.4
6.0     3.687   2.3
7.0     5.012   1.7
@DATA.END
```

Fig. A.11. An incorrect *data block*, the missing numbers where linewidth could not be measured create a non rectangular table.

```
@DATA.BEGIN LI_LNW<laser_1>
I       Power   LNW
mA      mW      MHz
0.0     0.001   NaN
1.0     0.013   NaN
2.0     0.159   NaN
3.0     0.957   NaN
4.0     1.568   5.6
5.0     2.315   4.4
6.0     3.687   2.3
7.0     5.012   1.7
@DATA.END
```

Fig. A.12. The correct version of Fig. A.10, the non existing measurement results are indicated with the value "NaN" which stands for "Not a Number".

A.3.8 Rules for comments

Comments in ODIF files are quite similar to C++ style comments, both the "block style" comment (between "/*" and "*/") as well as the "line style" comment (all text after "//") exist. However, they are not identical in usage : contrary to the C++ way where comments are disregarded and therefore can appear almost everywhere in the text, the comments in an ODIF file can only appear on very specific locations.

The formal syntax notation is very clear on these locations, but as a rule of thumb it can be stated that:

"comments are allowed after every line of ODIF syntactical text, ***except*** *in tables"*

The latter to ensure compatibility with spreadsheet applications.

A.3.9 Advanced aspects of ODIF syntax

As is the case with programming languages, there are extra features in ODIF besides the general structure. By providing features like inclusion and recursion, the ODIF syntax is more powerful and provides sufficient flexibility to the future.

Inclusion allows to use and reference files, without having to copy the contents each time. This example can be very useful for default settings or extractions based on many measurements. The rule for inclusion is the following:

"Each block on its own and only entire blocks can be included"

Recursion is provided on the block level : each block can contain an infinite number of the same kind. This is especially useful in the case of a *setting block*. When a program is using another to do the measurement, it "encapsulates" the latter. This property is duplicated in the setting block structure using recursion (cf. the example in section 4).

A.4 Example of a realistic ODIF file

In this section a complete example of a realistic ODIF file is shown and analysed into its components. It should be noted that this example illustrates most facets of ODIF files, yet it cannot be seen as a template showing how all ODIF files should look like. Information on how to adapt a specific file format problem to ODIF syntactical rules, is given in the next section (section 5).

In order to make the example more comprehensive, Fig. A.13 gives a schematic overview of the structure of the file and the positions of all the blocks used in the example.

The first part of the data, which describes the device that has been measured, is stored in the *parameter block*, inside the *global block*. Following the device specification, there is a *measurement block* with the actual measurement data, in turn followed by an *extraction block* which contains the results of the performed parameter extraction (loss calculation in this case). Inside the *measurement block*, there are a number of *setting blocks* followed by a *data block*. These are describing the measurement conditions, the equipment and the actual results, respectively.

The *extraction block* contains a *setting block* describing the extra inputs needed to make the extraction possible, followed by a number of *data blocks* containing the results of the extraction.

In order to keep the example compact in this text, only a few data points are shown (the original file counts over 500 points).

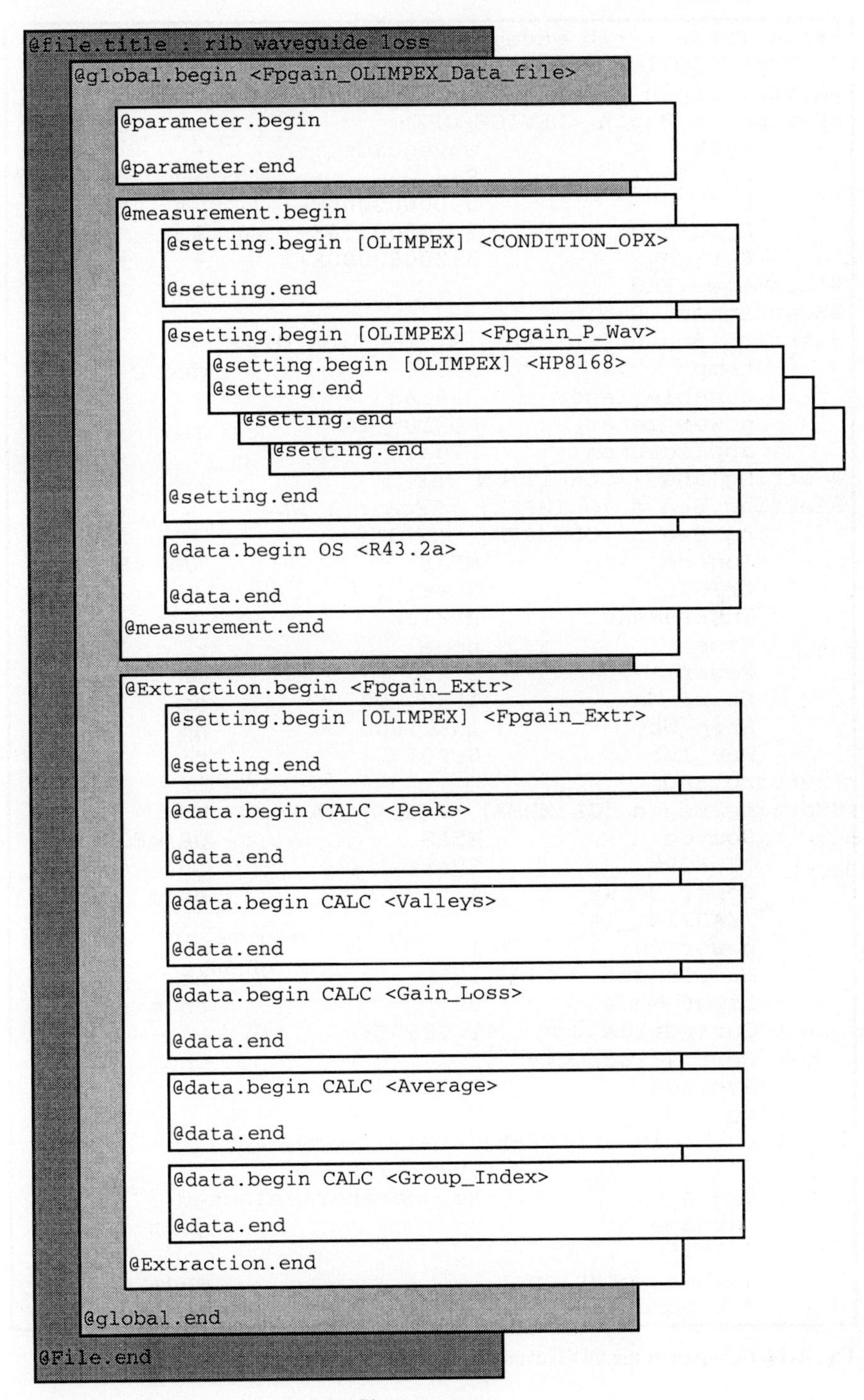

Fig. A.13. Schematic overview of the file structure.

```
@File.Title : rib waveguide loss
// FPgain Olimpex Data File $Ver: 1.0
@Global.Begin <Fpgain_Olimpex_Data_File>
@Parameter.Begin <DEVICE_OPX>
      type              waveguide          -
      ID                R34.2a             -
      L                 5.00000000E-1      mm
      R_left            3.20000000E+1      %
      R_right           3.20000000E+1      %
@Parameter.End
@Measurement.Begin
@Setting.Begin [OLIMPEX] <CONDITION_OPX>
      temp              25.00              deg C
      tunable_laser     HP8168             -
      power_meter       AQ2105_1A          -
      application       FPgain_P_Wav       -
@Setting.End //CONDITION_OPX
@Setting.Begin [OLIMPEX] <FPgain_P_Wav>
@Setting.Begin [OLIMPEX] <HP8168>
      Source            MEAS               WAV <FTLS>
      // FTLS           $Ver: 1.0
      Instrument        HP8168             -
      Name              none               -
      Power             2.000              mW
      Start_Wav         1550.000           nm
      Stop_Wav          1551.000           nm
      Wav_Inc           0.001              nm
@Setting.End
@Setting.Begin [OLIMPEX] <AQ2105_1A>
      Source            MEAS               OP <FOPM>
      // FOPM           $Ver: 1.0
      Power_Meter       1                  enum
      //AQ2105_1A
      power_units       1                  enum   //mW
      Wavelength        1550               nm
      Light_Mode        0                  enum   //CW
      Corr_Value        1.0000E+0          -
      Corr_Value_unit   1                  enum   //---
      Average           1                  -
@Setting.End
@Setting.Begin [OLIMPEX] <Data_Specs>
      //                $Ver: 1.0
      Operator          No_operator_selected    -
      Labname           0                  enum
      //INTEC
      Number_Notation   1                  enum
      //Exponential
```

Fig. A.14. First part of the ODIF example.

```
    Precision            8                  -
    Temp_Units           1                  enum   //degC
@Setting.End //Data_Specs
@Setting.End //FPgain_P_Wav
@Data.Begin OS <R34.2a>
    Wav                  P_out
    nm                   dBm
    1.54273900E+3        -7.41000000E+1
    1.54278000E+3        -7.33200000E+1
@Data.End //R34.2a
@Measurement.End
@Extraction.Begin <FPgain_Extr>
@Setting.Begin [OLIMPEX] <FPgain_Extr>
    L                    5.00000000E-1      mm
    R_left               3.20000000E+1      %
    R_right              3.20000000E+1      %
@Setting.End
@Data.Begin CALC <Peaks>
    Wav                  P_peak
    nm                   dBm
    1.54293852E+3        -6.88180255E+1
    1.54348402E+3        -6.91819083E+1
@Data.End
@Data.Begin CALC <Valleys>
    Wav                  P_valley
    nm                   dBm
    1.54325013E+3        -7.47994704E+1
    1.54379930E+3        -7.49567065E+1
@Data.End
@Data.Begin CALC <Gain_Loss>
    Wav                  Gain
    nm                   1/cm
    1.54309433E+3        6.95776054E-1
    1.54336707E+3        -4.70907270E-1
@Data.End
@Data.Begin CALC <Average>
    Avg_Gain             1.60992509E+0      1/cm
@Data.End
@Data.Begin CALC <Group_Index>
    Wav                  Group_Index
    nm                   -
    1.54321127E+3        4.36572404E+0
    1.54352471E+3        4.33829148E+0
@Data.End
@Extraction.End
@Global.End
@File.End
```

Fig. A.15. Second part of the ODIF example.

A.5 How to build ODIF files

A.5.1 Introduction

When applying the ODIF syntax to specific data storage problems, one should always keep in mind that ODIF is designed as a file interchange format with enormous flexibility and possibilities for implementing very complex structures. However, although it is possible to realise these complex structures, ODIF can still implement simple things without adding a big overhead. E.g. omitting the information in the *setting blocks* and leaving out all reference to the origin of the numbers can provide a far less complex looking ODIF file which consists almost purely of data itself.

A.5.2 The minimal ODIF file for storing one number

In order to save one measurement result only, a measurement block is needed, containing a data block with one entry line. Putting this a measurement block into an ODIF file requires, a *global block* and a *file block*. So the resulting file should look like this

```
File.Title : one measurement
global.begin
measurement.begin
data.begin POWER<single>
power 10    mW
data.end
measurement.end
global.end
file.end
```

Fig. A.16. Minimal ODIF file example.

A.5.3 The minimal ODIF file for storing one table of numbers

This is an identical problem as the one in the previous point, with as only difference the fact that the data block will now contain a table instead of an entry line. The file should look like this :

```
File.Title : PI measurement
global.begin
meas.begin
data.begin PI<sweep>
I           P
mA          mW
0           0
1           0.001
2           0.002
3           0.003
data.end
meas.end
global.end
file.end
```

Fig. A.17. Minimal ODIF file example for a table.

A.5.4 Multiple results

Although ODIF can put multiple results of different nature (or even on different devices) into one file, it is not an obligation. The user can still create a new file for each measurement, modelling or extraction result that needs to be saved. It is up to the user of ODIF to decide how big the file may get.

A.5.5 The creative process of making an ODIF file

A.5.5.1 The problem

Suppose one wants to put the following into the ODIF format :

1. One file for each measurement (simplest case).
2. Each measurement has defined conditions
3. Each measurement consists of a number of results (e.g. power vs. wavelength).

A.5.5.2 Basic form of the ODIF file

Since the data come in a tabular form (power vs. wavelength) and it will be one measurement result only, a file containing a *measurement block* is needed. As explained in section 5.3, this implies that we need to add a *global block* and a *file title*.

The base of the file will be :

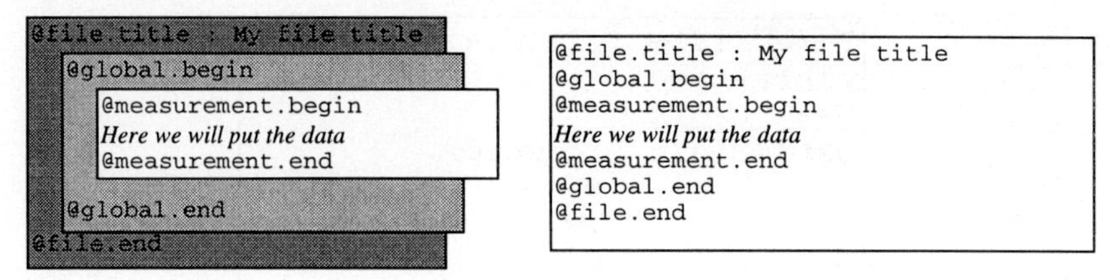

Fig. A.18. Base structure and file contents of the ODIF file.

A.5.5.3 Dividing the information

When the base is laid out, one can start thinking about the nature of the actual information that needs to be stored in the ODIF file. In this case (as in most cases) this information can be split into two categories :

1. Settings or stuff you need to specify when you do the measurement
2. The actual measurement results, the table of power versus wavelength

This translates directly into ODIF syntax elements :

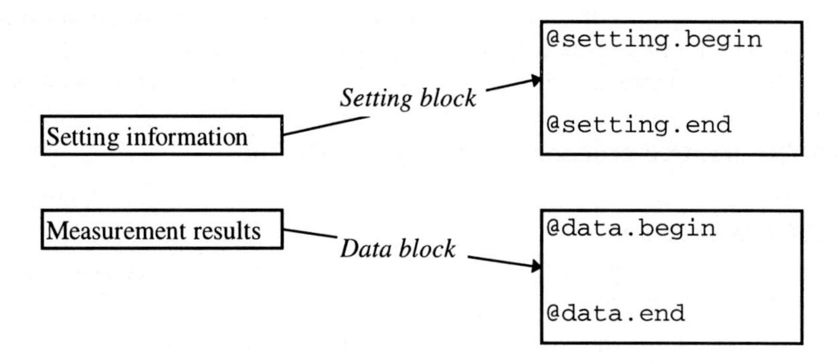

Fig. A.19. Mapping information to syntax elements.

So the ODIF file looks now like this :

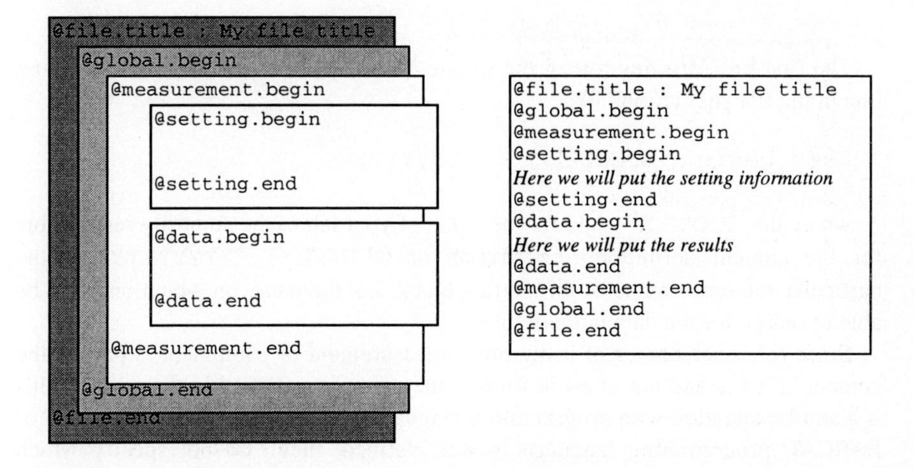

Fig. A.20. Schematic and textual overview of a high level ODIF file.

A.5.5.4 The syntactical details of each block

Once the detailed structure of the ODIF file is available, one has to check the formal syntax in order to be able to write the correct strings into the file. In this example only the more complicated rules for the setting block and the data block will be treated in detail. The other blocks (measurement, global and file block) can be treated in the same fashion.

The setting block. The syntactical definition of a setting block is:

Setting_Block = Setting_Begin LabName Reference_name NewLine
{Setting_Line | Source_Line | Comment}
Setting_End (Comment | NewLine)

These refer to the following syntax elements :
Setting_Begin = "@" **Setting_key** "." **Begin_key**
Reference_name = "<" **Name** ">"
LabName = "[" **Name** "]"
Setting_End = "@" **Setting_key** "." **End_key**

Which in turn refer to the following terminal symbols
Setting_key = "Set" | "Setting"
Begin_key = "B" | "Begin"
End_key = "E" | "End"

The first line (opening line of the *setting block*) specifies that the corresponding line in the file should look like :

```
@set.begin [ XXXXXX ] < YYYYYYYY >
```

where the "XXXXXX" stand for the name of your lab or the authority responsible for the content definition (e.g MyLab or OLIMPEX). "YYYYYYYY" is the particular reference name of this setting block, i.e. the name on which one will be able to search for the data in the future.

Since reference names of institutions, measurement or equipment is part of the content of a file and not of a file format, they are not specified in the syntax. This is a similar situation with programming languages : although the syntax of the C or PASCAL programming language is well defined, they do not specify which variable names should be used.

The data block. The syntactical definition of a data block is:

Data_Block = **Data_Begin Data_ID NewLine**
{**Comment**}
(**Table** | { **Entry_Line** })
Data_End (**Comment** | **NewLine**)

These refer to the following syntax elements [1] :
Data_ID = **Data_enum Reference_name**
Reference_name = "<" **Name** ">"
Data_enum = "PI" | "RIN" | "OS" | "LNW" | "PARAM" | "CALC".....
Data_Begin = "@" **Data_key** "." **Begin_key**
Data_End ="@" **Data_key** "." **End_key**

Which in turn refer to the following terminal symbols:
Data_key = "Dat" | "Data"
Begin_key = "B" | "Begin"
End_key = "E" | "End"

The first line (opening line of the *data block*) specifies that the corresponding line in the file should look like :

```
@data.begin XXXXXX < YYYYYYYY >
```

where the "YYYYYYYY" stands for the referencing name (E.g. "device1") as in the setting block case and the "XXXXXX" stands for the data enum. The latter is designed to be an identifier common to all similar data, a special name. Although

ODIF does not provide a complete list of possible identifiers, it is recommended to use a limited list

The ODIF file with syntactical details.

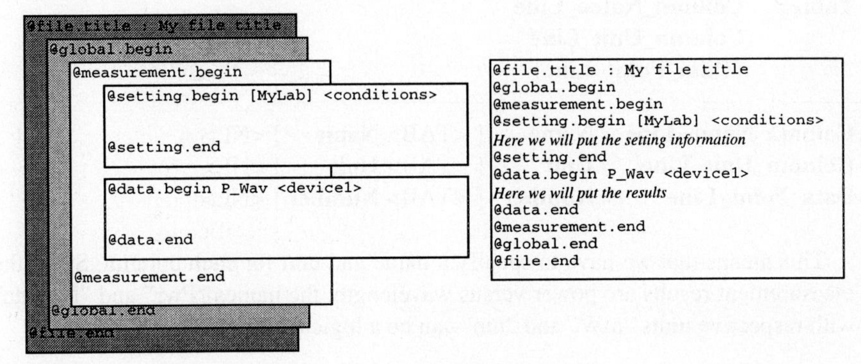

Fig. A.21. Schematic and textual overview of a high level ODIF file with correct syntax.

A.5.5.5 Putting in the data : settings and results

According to the syntactical definition (cf. 5.5.4.2) settings can be written using multiple *setting lines* or *source lines*. The first to be used to store information, the latter to be used to store a reference to the origin of the data (in case this is wanted).

The syntax specifies [2]:
Setting_Line = Entry_Line | Hex_Block

Entry_Line =
Name <TAB> (**Number** | **Plain_Text**)<TAB> **Unit** ([<TAB> **CommentLine**] | **NewLine**)

Source_Line = **Source_key** <TAB> **Source_type** <TAB> **Source_Names NewLine**
Source_Names = **Data_ID** { <TAB> **Data_ID** }

Since the *Hex block* is meant for storing binary data and since our example is numbers only, we can focus on the *entry line* and disregard the *hex block* option as well as the *source line* (no reference needed). As previously indicated, the name itself is not specified by the syntax and should be chosen wisely. In our case we want to store the temperature and the resolution bandwidth of the measurement, so the two resulting lines could look like :

```
temp    20   degC
RBW     0.1  nm
```

Each item separated by tab-characters

As for the data which come in a tabular form the syntactical rule states [3], [4]:

Table = **Column_Name_Line**
Column_Unit_Line
{Data_Point_Line}

Column_Name_Line = Name { <TAB> **Name** } <NL>
Column_Unit_Line = Unit { <TAB> **Unit** } <NL>
Data_Point_Line = Number { <TAB> **Number** } <NL>

This means that we have to specify a name and unit for each column. Since the measurement results are power versus wavelength, the names "Pwr" and "lambda" with respective units "mW" and "nm" can be a logical choice.

A.5.5.6 The ODIF file : end result

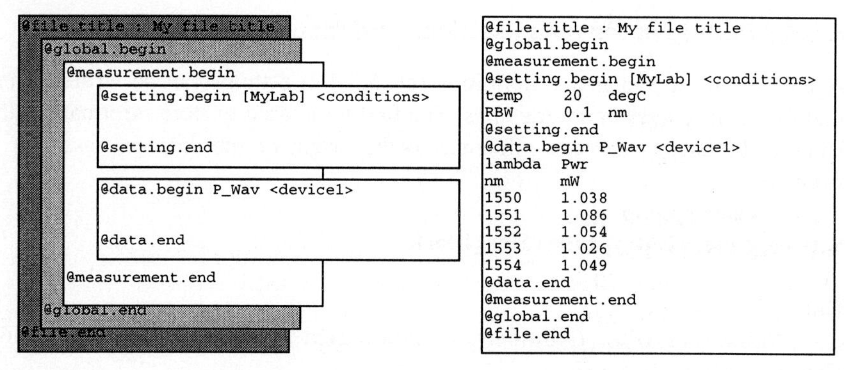

Fig. A.22. Structure and contents of the complete ODIF file.

A.6 Introduction to syntax notation

The syntax is written using a meta-language, which describes the total format from a top-down hierarchical approach. This notation method was first proposed by N. Wirth in 1977 [5] and treats syntax using a top down description. It first expresses the syntax in an abstract way using "non-terminals", i.e. words or symbols that will not appear as such in the file. These are then expanded in several levels, until the terminal level is reached, where the exact characters that appear in the file are specified.

In order to make a formal notation possible the following rules are used :

- Words written in bold are "non-terminal"-symbols (this means an abstract symbol).
- Tokens written between double quotes (e.g. “@Parameter”) are "terminal symbols" : tokens that will appear literally (without adding any extra spaces) in the resulting file.
- If the syntax rule lets the user choose *exclusively* between two or more things, it will use the symbol : “|”. E.g. the rule specifying : “a” | “b” | “c”, indicates that a file containing "a" is correct according to the syntax, a file containing "b" also, but a file containing "ab" is not allowed.
- The square brackets “[“ and “]” are used to denote the optional presence of something. e.g. [“a”] means that “” (empty string) and “a” are valid, but a repetition “aa” is not.
- The curly brackets “{“ and “}” are used to denote infinite repetition (i.e. zero, one, two or more times). E.g. the rule specifying : {a}, means that files containing : “”, “a”, “aa”, “aaa”, “aaaa”, etc. are correct.

A.7 The complete syntax description

File = **File_Begin** ":" **Text** **NewLine**
{**Global_Set**}
File_End

Global_Set = **Global_Block** | **Global_Include** | **Comment**
Parameter_Set = **Parameter_Block** | **Parameter_Include** | **Comment**
Model_Set = **Model_Block** | **Model_Include** | **Comment**
Measurement_Set = **Measurement_Block** | **Measurement_Include** | **Comment**
Extraction_Set = **Extraction_Block** | **Extraction_Include** | **Comment**
Setting_Set = **Setting_Block** | **Setting_Include** | **Comment**
Data_Set = **Data_Block** | **Data_Include** | **Comment**

Global_Block = **Global_Begin** [**Reference_name**] **NewLine**
{**Parameter_Set**}
{**Model_Set** | **Measurement_Set** | **Extraction_Set**}
{**Global_Set**}
Global_End (**Comment** | **NewLine**)

Parameter_Block = **Parameter_Begin** [**Reference_name**] **NewLine**
{**Setting_Line** | **Source_Line** | **Data_Set**}
{**Parameter_Set**}
Parameter_End (**Comment** | **NewLine**)

Model_Block = **Model_Begin** [**Reference_name**] **NewLine**

{Setting_Set}
Data_Set}
{Model_Set}
Model_End (Comment | NewLine)

Measurement_Block= Measurement_Begin [Reference_name] NewLine
{ Setting_Set}
{Data_Set}
{Measurement_Set}
Measurement_End (Comment | NewLine)

Extraction _Block = Extraction_Begin [Reference_name] NewLine
{Setting_Set}
{Data_Set}
{Extraction_Set}
Extraction_End (Comment | NewLine)

Data_Block = Data_Begin Data_ID NewLine
{Comment}
(Table | { Entry_Line })
Data_End (Comment | NewLine)

Setting_Block = Setting_Begin LabName Reference_name NewLine
{Setting_Line | Source_Line | Comment}
Setting_End (Comment | NewLine)

Data_ID = Data_enum Reference_name

Reference_name = "<" Name ">"

LabName = "[" Name "]"

Source_Line = Source_key <TAB> Source_type <TAB> Source_Names NewLine

Source_Names = Data_ID { <TAB> Data_ID }

Setting_Line = Entry_Line | Hex_Block
Hex_Block = ":" Hex_digit {Hex_digit} ";" <NL> ([6])
Entry_Line =
Name <TAB> (Number | Plain_Text)<TAB> Unit ([<TAB> CommentLine] | NewLine)

Data_enum = "PI" | "RIN" | "OS" | "LNW" | "PARAM" | "CALC".....
([7])
Source_type = "MODL" | "MEAS" | "EXTR" | "PARAM"

Table = **Column_Name_Line**
Column_Unit_Line
{Data_Point_Line}

Column_Name_Line = **Name** { <TAB> **Name** } <NL>
Column_Unit_Line = **Unit** { <TAB> **Unit** } <NL>
Data_Point_Line = **Number** { <TAB> **Number** } <NL>

Comment = **CommentBlock** | **CommentLine** ([8])
CommentBlock = "/*"{**Text** | **NewLine** } "*/" **NewLine**
CommentLine = "//" **TextLine**

File_Begin = "@"**File_key**"."**Title_key**
File_End = "@"**File_key**"."**End_key**

Global_Begin = "@"**Global_key**"."**Begin_key**
Global_End = "@"**Global_key**"."**End_key**
Global_Include = "@"**Global_key**"."**Include_key** **FileName** **NewLine**

Parameter_Begin = "@"**Parameter_key**"."**Begin_key**
Parameter_End = "@"**Parameter_key**"."**End_key**
Parameter_Include = "@"**Parameter_key**"."**Include_key** **FileName** **NewLine**

Model_Begin = "@"**Model_key**"."**Begin_key**
Model_End = "@"**Model_key**"."**End_key**
Model_Include = "@"**Model_key**"."**Include_key** **FileName** **NewLine**

Measurement_Begin = "@"**Measurement_key**"."**Begin_key**
Measurement_End = "@"**Measurement_key**"."**End_key**
Measurement_Include = "@"**Measurement_key**"."**Include_key** **FileName** **NewLine**

Extraction_Begin = "@"**Extraction_key**"."**Begin_key**
Extraction_End = "@"**Extraction_key**"."**End_key**
Extraction_Include = "@"**Extraction_key**"."**Include_key** **FileName** **NewLine**

Setting_Begin = "@"**Setting_key**"."**Begin_key**
Setting_End = "@"**Setting_key** "."**End_key**
Setting_Include = "@"**Setting_key** "."**Include_key** **FileName NewLine**

Data_Begin = "@"**Data_key**"."**Begin_key**
Data_End = "@"**Data_key**"."**End_key**
Data_Include = "@"**Data_key**"."**Include_key FileName NewLine**

File_key ="Fil" | "File" ([9])
Global_key ="Glo" | "Global"
Parameter_key ="Par" | "Param" | "Parameter"
Model_key ="Mod" | "Model"
Measurement_key ="Mea" | "Meas" | "Measurement"
Extraction_key ="Ext" | "Extr" | "Extraction"
Setting_key ="Set" | "Setting"
Data_key ="Dat" | "Data"

Title_key = "T" | "Title"
End_key = "E" | "End"
Begin_key = "B" | "Begin"
Include_key = "I" | "Include"

Source_key = "Source"

NewLine = <NL> { <NL> }

TextLine = [**Text**] **NewLine**
Text = {**Plain_Text** | <TAB>}
Plain_Text = {**letter** | **digit** | **non_control_char**}

Name = **letter** {**letter** | **digit** | "_" | "."} ([10])

letter = **uppercase** | **lowercase**
lowercase = "a" | "b" | "c" | "d" | "e" | "f" | "g" | "h" | "i" | "j" | "k" | "l" | "m" | "n" | "o" | "p" | "q" | "r" | "s" | "t" | "u" | "v" | "w" | "x" | "y" | "z"
uppercase = "A" | "B" | "C" | "D" | "E" | "F" | "G" | "H" | "I" | "J" | "K" | "L" | "M" | "N" | "O" | "P" | "Q" | "R" | "S" | "T" | "U" | "V" | "W" | "X" | "Y" | "Z"
non_control_char = "_" | "!" | "@" | "#" | "$" | "%" | "^" | "&" | "*" | "(" | ")" | "=" | """ | "'" | "[" | "]" | " <" | ">" | "," | "." | "/" | "?" | "‘" | "~" | ";" | ":" | " " | sign ([11])

Unit = **Number Unit_combination**

Unit_combination = Simple_Unit {("/" **Simple_Unit**) | **Simple_Unit**}
Simple_Unit = (**prefix Plain_Unit Real**) | "(" **Unit** ")"
prefix = "f" | "n" | "u" | "m" | "c" | "d" | "da" | "h" | "k" | "M" | "G" | "T"
Plain_Unit = "-" | "g" | "m" | "s" | "A" | "Hz" | "W" | "dB" | "deg" | "rad" |"C" | "%"
| "YYMMDD" | .... ([12])

Number = ([**sign**] (**positive_real** | **positive_integer** | "inf")) | "NaN"
([13])
positive_real= positive_integer [**fraction**] [**exponent**]
fraction = "." **positive_integer**
exponent = ("e"|"E") **integer**
integer = [**sign**] **positive_integer**
positive_integer = **digit** {**digit**}
sign = "+" | "-"
Hex_digit = **digit** | "A" | "B" | "C" | "D" | "E" | "F" | "a" | "b" | "c" | "d" | "e" | "f"
digit = "0" | "1" | "2" | "3" | "4" | "5" | "6" | "7" | "8" | "9"

note on terminal symbols :
<NL> stands for one and only one NewLine ('\n' from C)
<TAB> stands for one and only one tab ('\t' from C).

References

[1] **Data_enum** describes the type of the data, a complete list is available and can be extended when needed.

[2] Remark: the Hex block is mainly introduced as an alternative way to write down (compressed) binary information. The actual interpretation of this binary information is up to the user.

[3] Although the syntax doesn't describe it, a table has to have a rectangular form, which means that all rows must have equal number of columns (divided by the tabs).

[4] Remark: <NL> means one and only one linefeed, this means that empty lines are not allowed in tables.

[5] N. Wirth, "What can we do about the unnecessary diversity of notation for syntactic definitions ?", Communications of the ACM, vol. 20, 11 (1977).

[6] Format of the binary data according to the INTEL-Hex standard.

[7] **Source_type** describes how the original data has been obtained, a complete list is available and can be extended when needed.

[8] Although they resemble to common C++ comment identifiers, the usage is only allowed at very strict locations through the file.

[9] Not explicitly mentioned in the syntax, but keywords are case Insensitive.

[10] Standard programming language convention.

[11] Probably some chars will have to be eliminated.

[12] Remark: space is NOT allowed for no unit, use "-" instead.

[13] Classical numbers, extended with two special numbers: NaN (Not a Number) and infinity.

Unit_combination = Simple_Unit [illegible] Simple_Unit [illegible]
Simple_Unit = [prefix] Plain_Unit [illegible] Unit [illegible]
prefix = [illegible]
Plain_Unit = [illegible]
[illegible]

[illegible] positive_real [illegible] positive_integer [illegible]

positive_real = positive_integer [illegible] [exponent]
fraction = [illegible] positive_integer
exponent = [illegible]
integer = [sign] positive_integer
positive_integer = digit {digit}
[illegible]
hex_digit = [illegible]
digit = [illegible]

[illegible] notational symbols
[illegible]
[illegible]

References

[1] Plain_Unit [illegible] a complete list is [illegible] available and [illegible] extended when needed.
[2] [illegible] introduced as an alternative [illegible]
[3] Although [illegible]
[4] [illegible] means one and only one [illegible]
[5] [illegible] the unnecessary diversity of notation for syntactic definitions. Communications of the ACM vol. 20 [illegible] 1977.
[6] [illegible]
[7] [illegible]
[8] [illegible]
[9] [illegible]
[10] [illegible]
[11] [illegible] eliminated
[12] [illegible]
[13] [illegible]

Symbols

α	material linewidth enhancement factor	
	or absorption coefficient (α=$-$g)	m^{-1}
α_i	internal loss	m^{-1}
α_y	α associated with the nonlinear process labelled y	
β	propagation constant	m^{-1}
β_{sp}	spontaneous emission coupling factor	
$\Gamma = \Gamma_v.\Gamma_h$	confinement factor	
γ	resonance damping factor	s^{-1}
γ_{sc}	scattering losses	m^{-1}
γ_N	inverse effective differential carrier lifetime	s^{-1}
Δ	vectorial Laplace operator, $\nabla\nabla\cdot-\nabla\times\nabla\times$	
	or pump - signal angular detuning	rad/s
$\Delta\varepsilon_f$	difference of the quasi-Fermi levels	eV
$\Delta\omega$	pump-signal pulsation detuning	rad/s
$\delta(x)$	Dirac's Delta-function	
δ	complex detuning from Bragg wave number	m^{-1}
ϵ	gain suppression factor	m^3
ε	nonlinear gain coefficient	m^3
ε_y	gain suppression factor associated with the nonlinear process labelled y	m^3
ε_0	permittivity of vacuum	F/m
$\varepsilon_r = n^2$	relative permittivity, square of the refractive index	
$\eta(x,y)$	spatial noise	
$\eta_0 = \sqrt{\mu_0/\epsilon_0}$	free-space wave impedance	Ω
η_{inj}	injection efficiency	
η_{int}	internal efficiency	
η_{FWM}	FWM conversion efficiency	dB
η_N	normalized FWM conversion efficiency	mW^{-2}

Θ_i	grating facet phase	
θ	damping of resonance	s^{-1}
κ_g	gain/loss grating coupling coefficient	m^{-1}
κ_i	index grating coupling coefficient	m^{-1}
Λ	grating period	nm
λ	wavelength	m
μ_0	permeability of vacuum	H/m
ν	optical frequency	Hz
ρ_{ASE}	ASE noise power spectral density	W/Hz
ρ_N	total noise power spectral density	mW/Hz
τ_{cap}	local carrier capture time into quantum well	s
$\tau_{cap,e}$	effective carrier capture time into quantum well	s
τ_d	differential carrier lifetime	s
τ_{diff}	carrier diffusion time across SCH layers to the well	s
$\tau_{diff(well)}$	carrier diffusion time across SCH layers from the well	s
τ_{esc}	local carrier escape time from quantum well	s
τ_n	carrier lifetime	s
τ_p	photon lifetime	s
τ_y	characteristic time of the nonlinear process labelled y	s
$\underline{\phi(x)}$	mode field distribution	
χ	linear susceptibility	
	or carrier transport correction factor	
χ^2	total chi square error	
$\chi^{(3)}$	third order nonlinear susceptibility	
χ^{FWM}	part of $\chi^{(3)}$ describing the FWM interaction	
ψ	slowly varying envelope function	
Ω	angular frequency of the modulation	rad/s
ω	optical angular frequency	rad/s
$\mathcal{A}$	slowly varying envelope field	V/m
A	linear recombination coefficient	s^{-1}
	or active region cross section	μm^2
a or g_N	differential gain	m^2
a	linear material gain coefficient	
B	bimolecular recombination coefficient	m^3s^{-1}

b	normalized propagation constant	
C	Auger recombination coefficient	m^6s^{-1}
c	velocity of light in vacuum	m/s
c_{in}	input power coupling efficiency fibre-chip	dB
c_{out}	output power coupling efficiency chip-fibre	dB
D	chromatic dispersion	
$\boldsymbol{D}$	difference operator matrix	
d	thickness active region or waveguide	μm
E	electric field	V/m
e	elementary charge	C
F_S	Langevin photon density noise term	$m^{-3}s^{-1}$
F_N	Langevin carrier density noise term	$m^{-3}s^{-1}$
f	modulation frequency	Hz
f_r	resonance frequency	Hz
$G(\nu)$	spectral density of the intensity	
G	single-pass saturated chip gain	dB
G_0	single-pass unsaturated chip gain	dB
g	material gain	m^{-1}
g_m	modal gain	m^{-1}
$g(x,y)$	spatial impulse response	
$\overline{g}$	unsaturated material gain	m^{-1}
$\boldsymbol{H}$	magnetic field intensity vector	A/m
$H(\nu)$	function of dispersion	
h	Fourier transform of the function of dispertion	
h_p	Planck's constant	J.s
$\hbar$	reduced Planck's constant: $h_p/2\pi$	J.s
I	electric current	A
	or optical intensity	W/m^2
I_l	leakage current	A
i	imaginary unit	
J	injection current density	A/cm^2
j	imaginary unit	
$\mathbf{k}$	propagation vector	
K	Fabry-Perot cavity contrast $(= T_{max}/T_{min})$	
	or laser modulation K-factor	s
$k_0 = 2\pi/\lambda$	free-space angular wave number	m^{-1}
k_B	Boltzmann constant	$J.K^{-1}$

k_x	transversal propagation constant	m^{-1}
L	laser or SOA length	μm
L_D	carrier diffusion length	m
L_g	Bragg grating length	m
L_i	length of i-th section in case of multi-section lasers	μm
l	length of an optical path	m
$\boldsymbol{M}$	transfer matrix of a longitudinal section of a waveguide	
M	grating order	
N	carrier density	m^{-3}
$\overline{N}$	unsaturated carrier density	m^{-3}
N_0	transparency carrier density	m^{-3}
NF	noise figure for FWM	dB
n	refractive index	
n_{eff}	waveguide effective index	
n_g	waveguide group index	
n_{sp}	population inversion parameter	
$\boldsymbol{O}$	overlap integral matrix	
$\mathcal{P}$	induced macroscopic polarisation	
P	optical power	mW, dBm
	or optical power flux (chap 1-5)	W/m^2
P_N	total noise power	mW
P_r	reference optical power flux	W/m^2
P_s	internal saturation power	mW, dBm
R_i	facet power reflectivities	
R_L	photodetector load resistance	Ω
R_N	non-stimulated recombination rate	m^{-3}.s^{-1}
R_{sp}	spontaneous emission rate into the mode	m^{-3}.s^{-1}
R_y	frequency response of the nonlinear process labelled y	
r	photodetector responsivity	A/W
r_i	facet field reflectivities	
S	photon density	m^{-3}
	or phase of an optical field (chap 1-5)	
S_{ASE}	recombination rate due to the amplification of the spontaneous emission	m^{-3}.s^{-1}
$S_{sd}(\omega)$	spectral density of photon density noise	m^{-3}.s$^{-1/2}$

T	transmittance	
	or absolute temperature	K
U	optical field amplitude	
u	mode amplitude	
V	active layer volume	
v_g	group velocity	$m.s^{-1}$
W	width of the waveguide (chap 1-5)	W/m^2
w	width active region or waveguide	μm
x, y, z	Cartesian coordinates	
$\boldsymbol{Y}$	wave admittance matrix	
$\boldsymbol{y}$	normalized wave admittance matrix	
$\boldsymbol{Z}$	wave impedance matrix	
$\boldsymbol{z}$	normalized wave impedance matrix	
∇	operator nabla	
∇^2	scalar Laplace operator, $\nabla \cdot \nabla$	
$\otimes$	convolution operator	

Index

Printing: Mercedesdruck, Berlin
Binding: Buchbinderei Lüderitz & Bauer, Berlin